Theory and Decision Library C

Game Theory, Social Choice, Decision Theory, and Optimization

Volume 50

The series covers formal developments in game theory, social choice, and decision theory, paying particular attention to the role of optimization in these disciplines. Applications to economics, politics, and other social sciences are also welcome.

All titles in this series are peer-reviewed.

For further information on the series and to submit a proposal for consideration, please contact Johannes Glaeser (Senior Editor Economics and Political Science) Johannes.glaeser@springer.com.

More information about this series at https://link.springer.com/bookseries/6618

David W. K. Yeung · Leon A. Petrosyan

Durable-Strategies Dynamic Games

Theory, Solution Techniques and Applications

Springer

David W. K. Yeung
Center of Game Theory
St. Petersburg State University
St. Petersburg, Russia

SRS Consortium for Advanced Study
in Cooperative Dynamic Games
Hong Kong Shue Yan University
Hong Kong, Hong Kong

Leon A. Petrosyan
Faculty of Applied Mathematics—Control
Processes
St. Petersburg University
St. Petersburg, Russia

ISSN 0924-6126 ISSN 2194-3044 (electronic)
Theory and Decision Library C
ISBN 978-3-030-92744-8 ISBN 978-3-030-92742-4 (eBook)
https://doi.org/10.1007/978-3-030-92742-4

This Springer imprint is published by the registered company Springer Nature Switzerland AG
The registered company address is: Gewerbestrasse 11, 6330 Cham, Switzerland

Foreword

Since the appearance of differential games and discrete-time dynamic games in the middle of last century, dynamic games have been adopted as an efficient and effective instrument in analyzing dynamic interactive problems. Given the prevalence of the dynamic nature in economic, social and political interactive behaviours, dynamic game applications have been growing continually in the past four decades. Durable strategies that have effects lasting over time are very common in real-life situations. Revenue generating investments, durable goods and services, advertising and promotion, payment by instalments, contracts and government regulations, toxic waste disposable and deforestation are commonly seen examples of durable strategies in day to day life. The introduction of durable strategies leads to fundamental changes in the structure of conventional dynamic games. Significantly new dynamic optimization techniques, non-cooperative equilibria, cooperative solutions and subgame consistent payoff distribution procedures are required to accommodate the phenomenon of durable strategies.

The paradigm of *Durable-Strategies Dynamic Games* developed by Professors Leon Petrosyan and David Yeung provides a new perspective approach with a new class of strategies in dynamic game theory. The addition of durable strategies in the strategies set of dynamic games does not only make game theory applications possible for analysis of real-life problems with durable strategies but also establish a new theoretical framework with novel optimization theory, solution concepts and mathematical game techniques. This book indeed fills a significant gap in conventional dynamic game theory. In addition, the text also presents theories, solution techniques and application examples of durable-strategies dynamic games in advanced topics involving random game horizon, cross-generations of players and stochastic analysis. Application outcomes that could not be obtained by conventional dynamic games are derived in the study of environmental degradation and business investments.

The book *Durable-strategies Dynamic Games* is truly a world-leading addition to the field of dynamic games. It is a timely and much-needed publication to tackle the increasingly crucial problems of dynamic interactions amid the reality of durable strategies.

Vladimir Mazalov
Director of Karelian Institute of
Applied Mathematical Research of
Russian Academy of Sciences
President of International Society of
Dynamic Games
Petrozavodsk, Russia

Preface

With rapid advancements in communication and information technology, real-life strategic interactions become intertemporal and dynamic. One particularly complex and fruitful branch of game theory is dynamic games which investigates interactive decision-making over time. Since human beings live in time and decisions generally lead to effects over time, it is perhaps only a slight exaggeration to maintain that "life is a dynamic game". The origin of dynamic games traces back to the late 1940s from the work of Rufus Isaacs (publicly published in 1965) in continuous-time dynamic games, which is known as differential games. Discrete-time dynamic games (usually known as dynamic games) were developed so that dynamic games can be applied to continuous time and discrete time problems. The applications and development of differential games and dynamic games in various fields have been accumulating at an increasing rate over the past six decades.

Durable strategies that have effects lasting over an extended period of time are prevalent in real-life situations. Practically, almost all investments have effects lasting over a certain number of stages. Durable goods and services occupy a considerably large proportion of all goods and services. Diffusion of technology, build-up of physical capital stocks, and reputation building through promotion are known to require a certain period of time. In reality, the existence of durable strategies in the decision-maker's set of controls is likely to be a rule rather than an exception. Given the prevalence of durable strategies in the decision-maker's strategy space, dynamic games form an innate host framework for durable strategies to be built in. In analyzing interactive behaviour over time, the development of dynamic games with durable strategies is a much needed natural progress of the field. Yeung and Petrsoyan (2020) and Petrosyan and Yeung (2020) established the new paradigm of durable-strategies dynamic games.

This book is the first text on the new paradigm of durable-strategies dynamic games. It expounds in greater detail the authors' origination of durable-strategies dynamic games. In the presence of durable strategies, significant modification of the dynamic optimization techniques is required to accommodate the lagged effects of the strategies. Novel dynamic optimization theorems for solving problems with durable controls are provided for game-theoretic analysis. New theorems to characterize the feedback Nash equilibria for durable-strategies dynamic games are derived. Cooperation solutions under strategies with durable effects are developed. Additional features in durable-strategies dynamic games, including random game horizon, asynchronous players' horizons and stochastic setting, are also covered in the text. The analyses in this book open up a wide range of potential applications in real-life interactive problems.

We are extremely grateful to Prof. Vladimir Mazalov, President of the International Society of Dynamic Games, for authoring the Foreword of this book. Our gratitude to Aloysius Lui and Yingxuan Zhang for their excellent assistance in compiling this book. We thank Nina and Ovanes (LP) and Stella and Patricia (DY) for their love and patience during this book project which on occasion might have diverted our attention from them.

St. Petersburg, Russia Leon A. Petrosyan
David W. K. Yeung

Reference

Isaacs R (1965) Differential games: a mathematical theory with applications to warfare and pursuit, control and optimization (SIAM studies in applied mathematics). Wiley, New York

Contents

Chapter 1
Introduction

Durable strategies that have effects lasting over an extended period of time are prevalent in real-life situations. Revenue generating investments, toxic waste disposal, durable goods, emission of pollutants, regulatory measures, coalition agreements, diffusion of knowledge, advertisement and investments to build up physical capital are vivid examples of durable controls. In reality, the existence of durable strategies in the decision-maker's controls is likely to be a rule rather than an exception. In many practical situations, both the decision-maker's payoff and the evolution of the state dynamics can be subjected to effects arising from durable strategies. For instance, durable goods, revenue generating investments, toxic waste disposal and long term effects of business contracts and regulations affect the payoff of the decision-maker. Tariffs have delayed impacts on the economy's growth dynamics, knowledge diffusion takes time in building up knowledge-based capital, conversion of investments into physical capital stock involves a certain number of stages and advertisement enhances the firm's reputation gradually over time. Moreover, if the costs of durable strategies affecting the state dynamics are paid by instalments, they would also affect both the payoffs and dynamics of the decision-maker. Dynamic and differential games without durable strategies have been the majority of game analyses since the classic work of Rufus Isaacs in 1949 which appeared systematically in his classic book (1965). A large number of dynamic games (both continuous time and discrete time) had appeared in the past seven decades with standard non-durable strategies. Examples can be found in Basar and Olsder (1999), Dockner et al. (2000), Jørgensen and Zaccour (2004), Yeung and Petrosyan (2012, 2016a, 2016b).

Petrosyan and Yeung (2021) and Yeung and Petrosyan (2019, 2021a) introduced control lags in dynamic games and developed a new paradigm of durable-strategies dynamic games. This game paradigm allows strategies with lagged effects to affect both the player's payoffs and the state dynamics. In the presence of durable strategies, significant modification of the dynamic optimization techniques is required to accommodate the lagged effects of the strategies. A dynamic optimization theorem for solving problems with durable controls affecting both the payoff and state dynamics

D. W. K. Yeung and L. A. Petrosyan, *Durable-Strategies Dynamic Games*,
Theory and Decision Library C 50, https://doi.org/10.1007/978-3-030-92742-4_1

is provided for the game-theoretic analysis. A novel theorem characterizing the feedback Nash equilibrium for durable-strategies dynamic games is derived. This opens up a wide range of potential applications in real-life interactive problems involving durable strategies.

It is well known that non-cooperative behaviours among players would, in general, lead to an outcome that is not Pareto optimal. In a dynamic world, non-cooperative behaviours guided by short-sighted individual rationality could be a source for series of disastrous consequences in the future, especially when the effects of their strategies are durable. Cooperation suggests the possibility of obtaining socially optimal and group efficient solutions to decision problems involving durable strategic actions. To guarantee dynamic stability, a cooperative solution has to be subgame consistent so that the solution for a subgame that starts at a later time with the state brought about by prior optimal behaviours would remain optimal (see Yeung and Petrosyan (2004, 2010, 2016a). In the presence of durable strategies, a subgame consistent cooperative solution has to guarantee that the solution policy in a subgame starting at a later time with the state and previously executed strategies brought about by prior optimal behaviours would remain optimal. A Payoff Distribution Procedure (PDP) leading to subgame consistent solutions under durable strategies is presented. Cooperative dynamic games with durable strategies are presented and their subgame consistent solutions are characterized.

This book is the world's first text focusing on durable-strategies dynamic games developed by the authors. It presents a comprehensive treatment of this game paradigm. Mathematical theorems and solution concepts pertaining to this class of games are provided. The organization of the book is given as follows.

Chapter 2 presents the theory and solution techniques of a general class of cooperative dynamic games with multiple durable strategies of different lags affecting both the players' payoffs and the state dynamics. A dynamic optimization theorem involving durable strategies affecting both the players' payoffs and the state dynamics is developed. A computational illustration of the derivation of the corresponding optimal strategies is provided. The game equilibrium solution and the corresponding Hamilton–Jacobi-Bellman (HJB) equations are demonstrated. An illustrative example of the computational details is presented.

Chapter 3 presents a generic class of durable-strategies dynamic games of firms with durable interactive investments. Four major types of investments with durable effects are considered. These include investments in private and public capitals, revenue generating investments and debt-financed project investments. The equilibrium solution of a generic class of dynamic games of durable interactive investments is derived and a novel dynamic theory of the firm with durable investment strategies is provided.

Chapter 4 considers dynamic environmental games under anthropogenic eco-degradation in the presence of strategies lags. A generic class of durable-strategies environmental games of eco-degradation and green technology development is developed. Overviews of pollution and ecosystem degradation are discussed. Anthropogenic eco-degradation and green technology developments under strategy lags are analyzed in a game-theoretic framework. An illustrative game model under durable strategies with explicit functional forms is also presented.

Non-cooperative behaviours among participants would, in general, lead to an outcome that is not Pareto optimal. In a dynamic world, non-cooperative behaviours with durable effects guided by short-sighted individual rationality could be a source for a series of disastrous consequences in the future. Chapter 5 presents a novel game paradigm of cooperative dynamic games with multiple durable strategies of different lags affecting both the players' payoffs and the state dynamics. The Pareto optimal solution and conditions that guarantee individual rationality are derived. The fundamentals of dynamically consistent solutions and the corresponding payment mechanism are presented.

Chapter 6 deals with eco-degradation management under durable strategies. Anthropogenic eco-degradation has become one of the world's most complex and urgent problems to be dealt with. Self-seeking maximization by individual agents leads to environmental degradation that would destroy the environmental conditions for the existence of human beings. Unilateral response of a single nation or region deems to be ineffective in providing a viable solution. A general class of cooperative dynamic environmental games involving anthropogenic eco-degradation and green technology development under durable strategies is presented. An overview of collaboration in ecosystem degradation management is provided. A dynamically stable payoff distribution procedure is developed to fulfil the realization of the agreed-upon payoff imputations so that no region will have inventive to deviate from the cooperation scheme within the cooperation duration. Efficiency gain and sustainable imputation under the cooperation scheme are examined.

Chapter 7 studies durable-strategies dynamic games under uncertain game horizons. It presents a new class of dynamic games that incorporates two frequently observed real-life phenomena—durable strategies and uncertain horizons. A novel dynamic optimization theorem is provided and a new set of equations characterizing a non-cooperative game equilibrium is derived. A subgame consistent solution for the cooperative game under random horizon and durable strategies is obtained with a new theorem for the derivation of a payoff distribution procedure. A number of new application results in dynamic games are derived to reflect practical considerations in making decisions.

In many game situations, the players' time horizons differ. This may arise from different life spans, different entry and exit times in different markets and the different duration for leases and contracts. Asynchronous horizons game situations occur frequently in economic and social activities. Chapter 8 considers durable-strategies dynamic games in which the players' game horizons are asynchronous. Durable strategies, combined with asynchronous horizons, make the analysis of strategic behaviour over time for players entering the game in different stages extremely complicated. A group optimal, individually ration and dynamically stable solution is examined. A generalized game paradigm with $n-$player and complex patterns of players' game horizons is presented.

Chapter 9 considers stochastic durable-strategies dynamic games. An essential characteristic of time—and hence decision-making over time—is that although the individual may invest in gathering past and current information, random future events are unavoidable. An empirically meaningful theory must therefore incorporate these uncertainties. A new class of durable strategies dynamic games of the Petrosyan and Yeung (2020) paradigm with stochastic state dynamics and randomly furcating payoffs is formulated. A dynamic optimization theorem involving durable strategies problems with stochastic elements appearing in the state dynamics and payoff structures is presented. A general class of durable strategies dynamic games with stochastic state dynamics and random payoffs is provided. Non-cooperative equilibria and cooperative solutions are developed. Computational illustrations are given.

This book provides techniques for solving durable strategies dynamic games and for deriving relevant cooperative solution mechanisms. Non-cooperative equilibrium analysis provides rigorous study on the dynamic externalities of durable strategies. Cooperative solutions suggest efficient policy initiatives to resolve the problem. While only a limited number of applications are presented within the limited space of this book, a list of potentially applicable real-life dynamic interactive problems with durable strategies are readily observable in the fields of economics, business, politics, technology, health studies and environment:

The use of the environmental commons.
Green technology development
Paris Agreement on climate change.
European Union.
NATO.
ASEAN.
BREXIT.
Nuclear disarmament.
Elected governments.
Trade wars.
Territorial disputes.
China's Belt-road initiatives.
Military confrontation.
Peaceful use of outer space.
Cyber security.
Cryptocurrency.
Global financial reform.
Disease control.
Business operation.
International trade and investment.
Research and development.
Technology spillover
Advertising and promotion.

This book, as the first comprehensive disquisition of the field, not only widens the scope of application of dynamic game theory but also provides a solid foundation for further theoretical and technical development.

References

Başar T, Olsder GJ (1999) Dynamic noncooperative game theory, 2nd. Philadelphia, Pa., SIAM

Dockner EJ, Jorgensen S, Long NV, Sorger G (2000) Differential games in economics and management science. Cambridge University Press, Cambridge

Jørgensen S, Zaccour G (2004) Differential games in marketing. Kluwer Academic Publishers

Petrosyan LA, Yeung DWK (2020) Cooperative dynamic games with durable controls: theory and application. Dyn Games Appl 10:872–896. https://doi.org/10.1007/s13235-019-00336-w

Petrosyan LA, Yeung DWK (2021) Shapley value for differential network games: theory and application. J Dyn Games 8(2):151–166. https://doi.org/10.3934/jdg.2020021

Yeung DWK, Petrosyan LA (2012) Subgame consistent economic optimization: an advanced cooperative dynamic game analysis. Birkhäuser, Boston, p 395. ISBN 978-0-8176-8261-3

Yeung DWK, Petrosyan LA (2004) Subgame consistent cooperative solution in stochastic differential games. J Optim Theory Appl 120(3):651–666

Yeung DWK, Petrosyan LA (2010) Subgame consistent solutions for cooperative stochastic dynamic games. J Optim Theory Appl 145:579–596

Yeung DWK, Petrosyan LA (2016) A cooperative dynamic environmental game of subgame consistent clean technology development. Int Game Theor Rev 18(2):164008.01-164008.23

Yeung DWK, Petrosyan LA (2016) Subgame consistent cooperation: a comprehensive treatise. Springer

Yeung DWK, Petrosyan LA (2019) Cooperative dynamic games with control lags. Dyn Games Appl 9(2):550–567. https://doi.org/10.1007/s13235-018-0266-6

Yeung DWK, Petrosyan LA (2021) Generalized dynamic games with durable strategies under uncertain planning horizon. J Comput Appl Math 395:113595

Chapter 2
Durable-Strategies Dynamic Games: Theory and Solution Techniques

Durable strategies that have effects lasting over a certain period of time are prevalent in real-life situations. Revenue generating investments, toxic waste disposal, durable goods, emission of pollutants, regulatory measures, coalition agreements, diffusion of knowledge, advertisement and investments to build up physical capital are vivid examples of the many durable strategies. Durable strategies may affect both the decision-makers' payoffs and the evolution of the state dynamics. This Chapter presents a new class of dynamic games developed by Petrosyan and Yeung (2020) with multiple durable strategies of different lags affecting both the players' payoffs and the state dynamics. Section 1 presents a dynamic optimization theorem involving durable strategies affecting both the players' payoffs and the state dynamics. Section 2 gives a computational illustration of the derivation of the corresponding optimal strategies. Section 3 provides a general class of dynamic games with multiple durable strategies of different lags affecting both the players' payoffs and the state dynamics. The game equilibrium solution and the corresponding Hamilton–Jacobi-Bellman (HJB) equations are demonstrated. Section 4 gives an illustrative example of the computation details. Mathematical appendices are relegated to Sect. 5 and chapter notes are given in Sect. 6. Problem sets are supplied in Sect. 7.

1 Optimization with Durable Strategies

Durable strategies that have influences lasting over a certain period of time are frequently observed in real-life situations. Revenue generating investments, toxic waste disposal, durable goods, emission of pollutants, business contracts, taxes, regulations, coalition agreements, diffusion of knowledge, advertising and investments to build up physical capital are vivid examples of durable strategies. In many practical cases, both the decision-maker's payoff and the evolution of the state dynamics can be subjected to effects arising from durable strategies. For instance, durable goods,

D. W. K. Yeung and L. A. Petrosyan, *Durable-Strategies Dynamic Games*,
Theory and Decision Library C 50, https://doi.org/10.1007/978-3-030-92742-4_2

revenue generating investments, toxic waste dispersal, business contracts and regulations affect the decision-maker's payoffs for more than one stage. At the same time, strategies with durable effects can impact the state dynamics. For instance, tariffs have delayed impacts on the economy's growth dynamics, knowledge diffusion takes time in building up knowledge-based capital, conversion of investments into physical capital stock involves a certain number of stages and advertisement enhances the firm's reputation gradually over time. Moreover, if the costs of durable strategies affecting the state dynamics are paid by instalments, they would also affect both the payoffs and dynamics of the decision-maker. In the presence of durable strategies, significant modification of the dynamic optimization techniques is required to accommodate the lagged effects of the strategies. In the sequel, we present the dynamic optimization technique for durable strategies developed in Petrosyan and Yeung (2020a).

1.1 Problem Formulation

Consider a general $T-$ stage dynamic optimization problem in which there exist non-durable strategies and durable strategies of different lag durations. They may affect the payoff, the state dynamics or both. We use $u_k \in U \subset R^m$ to denote the set of non-durable control strategies. We use $\overline{u}_k = (\overline{u}_k^{(2)}, \overline{u}_k^{(3)}, \cdots, \overline{u}_k^{(\omega)})$ to denote the set of durable control strategies, where $\overline{u}_k^{(\zeta)} \in \overline{U}^{\zeta} \subset R^{m_\zeta}$ for $\zeta \in \{2, 3, \cdots, \omega\}$. In particular, the strategies $\overline{u}_k^{(2)}$ are durable strategies that have effects in stages k and $k+1$. The strategies $\overline{u}_k^{(3)}$ are durable strategies that have effects within stages k, $k+1$ and $k+2$. The strategies $u_k^{(\omega)}$ are durable strategies that have effects within the duration from stage k to stage $k+\omega-1$. The single-stage payoff received in stage k can then be expressed as $g_k(x_k, u_k, \overline{u}_k; \overline{u}_{k-})$, where $x_k \in X \subset R^m$ is the state at stage k and $\overline{u}_{k-}$ is the set of durable controls which are executed before stage k but still in effect in stage k.

The state dynamics is

$$x_{k+1} = f_k(x_k, u_k, \overline{u}_k; \overline{u}_{k-}), \quad x_1 = x_1^0. \tag{1.1}$$

The payoff to be maximized becomes

$$\sum_{k=1}^{T} g_k(x_k, u_k, \overline{u}_k; \overline{u}_{k-})\delta_1^k + q_{T+1}(x_{T+1}; \overline{u}_{(T+1)-})\delta_1^{T+1}, \tag{1.2}$$

where $q_{T+1}(x_{T+1}; \overline{u}_{(T+1)-})$ is the terminal payoff at stage $T+1$ and δ_1^k is the discount factor from stage 1 to stage k.

The controls executed before the start of the operation in stage 1, that is u_{1-}, are known and some or all of them can be zeros. The functions $g_k(x_k, u_k, \overline{u}_k; \overline{u}_{k-})$, $f_k(x_k, u_k, \overline{u}_k; \overline{u}_{k-})$ and $q_{T+1}(x_{T+1}; \overline{u}_{(T+1)-})$ are differentiable functions.

1.2 Optimization Methodology

A solution theorem for obtaining the optimal control strategies in the dynamic optimization problem (1.1)–(1.2) can be characterized as follows.

Theorem 1.1. Durable-Strategies Dynamic Optimization
Let $V(k, x; \overline{u}_{k-})$ be the maximal value of the payoff

$$\sum_{t=k}^{T} g_t(x_t, u_t, \overline{u}_k; \overline{u}_{t-})\, \delta_1^t + q_{T+1}(x_{T+1}; \overline{u}_{(T+1)-})\, \delta_1^{T+1}$$

for problem (1.1)–(1.2) *starting at stage k with state $x_k = x$ and previously executed controls $\overline{u}_{k-}$, then the function $V(k, x; \overline{u}_{k-})$ satisfies the following system of recursive equations*:

$$V(T+1, x; \overline{u}_{(T+1)-}) = q_{T+1}(x_{T+1}; \overline{u}_{(T+1)-})\delta_1^{T+1}, \tag{1.3}$$

$$\begin{aligned} V(k, x; \overline{u}_{k-}) &= \max_{u_k, \overline{u}_k} \left\{ g_k(x, u_k, \overline{u}_k; \overline{u}_{k-})\delta_1^k + V[k+1, f_k(x, u_k, \overline{u}_k; \overline{u}_{k-}); \overline{u}_{(k+1)-}] \right\} \\ &= \max_{u_k, \overline{u}_k} \left\{ g_k(x, u_k, \overline{u}_k; \overline{u}_{k-})\, \delta_1^k + V[k+1, f_k(x, u_k, \overline{u}_k; \overline{u}_{k-}); \overline{u}_k, \overline{u}_{(k+1)-} \cap \overline{u}_{k-}] \right\}, \\ &\text{for } k \in \{1, 2, \cdots, T\}. \end{aligned} \tag{1.4}$$

Proof To prove Theorem 1.1, we adopt the technique of backward induction. Consider first the last operational stage T, invoking Theorem 1.1 we have

$$\begin{aligned} V(T, x; \overline{u}_{T-}) &= \max_{u_T, \overline{u}_T} \left\{ g_T(x, u_T, \overline{u}_T; \overline{u}_{T-})\delta_1^T + q_{T+1}[f_T(x, u_T, \overline{u}_T; \overline{u}_{T-}); \overline{u}_{(T+1)-}] \right\} \\ &= \max_{u_T, \overline{u}_T} \left\{ g_T(x, u_T, \overline{u}_T; \overline{u}_{T-})\, \delta_1^T + q_{T+1}[f_T(x, u_T, \overline{u}_T; \overline{u}_{T-}); \overline{u}_T, \overline{u}_{(T+1)-} \cap \overline{u}_{T-}] \right\}. \end{aligned} \tag{1.5}$$

The maximization operator in stage T involves u_T and $\overline{u}_T$ only and $\overline{u}_{(T+1)-} \cap \overline{u}_{T-}$ is a subset of $\overline{u}_{T-}$. The current state x and the previously executed controls $\overline{u}_{T-}$ appear in the stage T maximization problem as given parameters. If the first order conditions of the maximization problem in (1.5) satisfy the implicit function theorem, one can obtain the optimal controls u_T and $\overline{u}_T$ as functions of x and $\overline{u}_{T-}$. Substituting these optimal controls into the function on the RHS of (1.5) yields the function $V(T, x; \overline{u}_{T-})$, which satisfies the optimal conditions of a maximum for given x and $\overline{u}_{T-}$.

Consider the second last operational stage $T-1$, using $V(T, x; u_{T-})$ derived from (1.5) and invoking Theorem 1.1 we have

$$V(T-1, x; \overline{u}_{(T-1)-}) = \max_{u_{T-1}, \overline{u}_{T-1}} \{g_{T-1}(x, u_{T-1}, \overline{u}_{T-1}; \overline{u}_{(T-1)-})\delta_1^{T-1} + V[T, f_{T-1}(x, u_{T-1}, \overline{u}_{T-1}; \overline{u}_{(T-1)-}); \overline{u}_{T-}]\}$$
$$= \max_{u_{T-1}, \overline{u}_{T-1}} \{g_{T-1}(x, u_{T-1}, \overline{u}_{T-1}; \overline{u}_{(T-1)-})\delta_1^{T-1} + V[T, f_{T-1}(x, u_{T-1}, \overline{u}_{T-1}; \overline{u}_{(T-1)-}); \overline{u}_{T-1}, \overline{u}_{T-} \cap \overline{u}_{(T-1)-}]\}. \tag{1.6}$$

The maximization operator in stage $T-1$ involves u_{T-1} and $\overline{u}_{T-1}$. The current state x and the previously executed controls $\overline{u}_{(T-1)-}$ appear in the stage $T-1$ maximization problem as given parameters. If the first order conditions of the maximization problem in (1.6) satisfy the implicit function theorem, one can obtain the optimal controls u_{T-1} and $\overline{u}_{T-1}$ as functions of x and previously determined controls $\overline{u}_{(T-1)-}$. Substituting these optimal controls into the function on the RHS of (1.6) yields the function $V(T-1, x; \overline{u}_{(T-1)-})$.

Now consider stage $k \in \{T-2, T-3, \cdots, 2, 1\}$, invoking Theorem 1.1 we have

$$V(k, x; \overline{u}_{k-}) = \max_{u_k, \overline{u}_k} \left\{g_k(x, u_k, \overline{u}_k; \overline{u}_{k-})\delta_1^k + V[k+1, f_k(x, u_k, \overline{u}_k; \overline{u}_{k-}); \overline{u}_{(k+1)-}]\right\}$$
$$= \max_{u_k, \overline{u}_k} \left\{g_k(x, u_k, \overline{u}_k; \overline{u}_{k-})\delta_1^k + V[k+1, f_k(x, u_k, \overline{u}_k; \overline{u}_{k-}); \overline{u}_k, \overline{u}_{(k+1)-} \cap \overline{u}_{k-}]\right\}. \tag{1.7}$$

The maximization operator involves u_k and $\overline{u}_k$. Again, the current state x and the previously executed controls $\overline{u}_{k-}$ appear in the stage k optimization problem. If the first order conditions of the maximization problem in (1.7) satisfy the implicit function theorem, one can obtain the optimal controls u_k and $\overline{u}_k$ as functions of x and $\overline{u}_{k-}$. Substituting these optimal controls into the function on the RHS of (1.7) yields the function $V(k, x; \overline{u}_{k-})$. ■

Theorem 1.1 is established by Petrosyan and Yeung (2020) and it yields a new optimization technique that can be used to solve durable control problems with lagged effects in both the payoffs and state dynamics of the decision-maker. It is worth noting that both the current state x_k and previously executed controls $\overline{u}_{k-}$ appear as given in the stage k optimization problem. While the state variables x_k have transition equations governing their transition from one stage to another, the previously executed controls $\overline{u}_{k-}$ have no such equations of motion.

In Bellman's (1957) standard dynamic programming technique, the non-durable strategies executed in stage k will affect the state x_{k+1} in stage $k+1$ through the dynamic equation and the payoff in stage k. In Theorem 1.1, the durable strategies $\overline{u}_k$ executed in stage k could affect the state x_{k+1} in stage $k+1$, the state x_{k+2} in stage $k+2$, up to $x_{k+\omega-1}$ in stage $k+\omega-1$. Similarly, the durable strategies $\overline{u}_k$ executed in stage k could affect the single-stage payoff function in stage k up to stage $k+\omega-1$. In particular, the previously executed durable strategies controls $\overline{u}_{k-}$ in stage k act as a vector of idiosyncratic state variables that will not change but will last for some finite stages.

2 A Computational Illustration

Consider a monopoly in which planning horizon involves T stages. It uses knowledge-based capital $x_k \in X \subset R^+$ to produce its output. Capital investment takes time to be built up the capital stock. The accumulation process of the capital stock is governed by the dynamics

$$x_{k+1} = x_k + a_k^{|k|} \overline{u}_k^{(\omega)} + \sum_{\tau=k-\omega+1}^{k-1} a_k^{|\tau|} \overline{u}_\tau^{(\omega)} - \lambda_k x_k, \quad x_1 = x_1^0, \tag{2.1}$$

where $u_k^{(\omega)}$ is the investment in the knowledge-based capital in stage k which is effective within ω stages, $a_k^{|k|} \overline{u}_k^{(\omega)}$ is the addition to the capital in stage k by capital building investment $\overline{u}_k^{(\omega)}$ in stage k, $a_k^{|\tau|} \overline{u}_\tau^{(\omega)}$ is the addition to the capital in stage k by capital building investment $\overline{u}_\tau^{(\omega)}$ in stage $\tau \in \{k-1, k-2, \cdots, k-\omega+1\}$ and λ_k is the rate of obsolescence of the capital.

The firm uses knowledge-based capital together with non-durable productive inputs u_k to produce outputs. The revenue from the output produced is $R_k(u_k)^{1/2}(x_k)^{1/2}$. The non-durable input cost is $c_k u_k$ and the cost of capital investment is $\phi_k(\overline{u}_k^{(\omega)})^2$.

The payoff of the firm is

$$\sum_{k=1}^{T} \left([R_k(u_k)^{1/2}(x_k)^{1/2} - c_k u_k - \phi_k(\overline{u}_k^{(\omega)})^2] \right) \delta^{k-1}$$
$$+ \left(Q_{T+1} x_{T+1} + \sum_{\tau=T+1-\omega+1}^{T} v_{T+1}^{|\tau|} \overline{u}_\tau^{(\omega)} + \varpi_{T+1} \right) \delta^T, \tag{2.2}$$

where $\left(Q_{T+1} x_{T+1} + \sum_{\tau=T+1-\omega+1}^{T} v_{T+1}^{|\tau|} \overline{u}_\tau^{(\omega)} + \varpi_{T+1} \right)$ is the salvage value of the firm in stage $T+1$ and $\delta = (1+r)^{-1}$ is the discount factor.

In particular, the salvage value of the firm in stage $T+1$ depends on the capital stock of the firm and Q_{T+1} is positive. It also depends on the previous capital investment $\sum_{\tau=T+1-\omega+1}^{T} v_{T+1}^{|\tau|} \overline{u}_\tau^{(\omega)}$. In particular, some or all $v_{T+1}^{|\tau|}$, for $\tau \in \{T, T-1, \cdots, T-\omega+2\}$, can be zero.

Invoking Theorem 1.1, we obtain.

Corollary 2.1. *Let $V(k, x; \overline{u}_{k-}^{(\omega)})$ be the maximal value of the payoff*

$$\sum_{k=1}^{T}\left([R_k(u_k)^{1/2}(x_k)^{1/2} - c_k u_k - \phi_k(\overline{u}_k^{(\omega)})^2]\right)\delta^{k-1}$$
$$+\left(Q_{T+1}x_{T+1} + \sum_{\tau=T+1-\omega+1}^{T} v_{T+1}^{|\tau|}\overline{u}_\tau^{(\omega)} + \varpi_{T+1}\right)\delta^T,$$

for the dynamic optimization problem (2.1)–(2.2) *starting at stage k with state* $x_k = x$ *and previously executed controls* $\overline{u}_{k-}^{(\omega)} = (\overline{u}_{k-1}^{(\omega)}, \overline{u}_{k-2}^{(\omega)}, \cdots, \overline{u}_{k-\omega+1}^{(\omega)})$, *then the function* $V(k, x; \overline{u}_{k-}^{(\omega)})$ *satisfies the following system of recursive equations:*

$$V(T+1, x; \overline{u}_{(T+1)-}^{(\omega)}) = \left(Q_{T+1}x_{T+1} + \sum_{\tau=T+1-\omega+1}^{T} v_{T+1}^{|\tau|}\overline{u}_\tau^{(\omega)} + \varpi_{T+1}\right)\delta^T, \quad (2.3)$$

$$V(k, x; \overline{u}_{k-}^{(\omega)}) = \max_{u_k, \overline{u}_k^{(\omega)}}\Bigg\{[R_k(u_k)^{1/2}(x)^{1/2} - c_k u_k - \phi_k(\overline{u}_k^{(\omega)})^2]\delta^{k-1}$$
$$+ V\left[k+1, x + a_k^{|k|}\overline{u}_k^{(\omega)} + \sum_{\tau=k-\omega+1}^{k-1} a_k^{|\tau|}\overline{u}_\tau^{(\omega)} - \lambda_k x; \overline{u}_k^{(\omega)}, \overline{u}_{(k+1)-}^{(\omega)} \cap \overline{u}_{k-}^{(\omega)}\right]\Bigg\},$$
$$\text{for } k \in \{1, 2, \cdots, T\}. \quad (2.4)$$

■

Performing the indicated maximization operator in Corollary 2.1, the value function $V(k, x; \overline{u}_{k-}^{(\omega)})$ which reflects the value of the firm covering stage $k \in \{1, 2, \cdots, T\}$ to stage $T+1$ can be obtained as follows.

Proposition 2.1. *System* (2.3)–(2.4) *admits a solution with the optimal payoff of the firm being*

$$V(k, x; \overline{u}_{k-}^{(\omega)}) = (A_k x + C_k)\delta^{k-1}, \quad (2.5)$$

where $A_{T+1} = Q_{T+1}$ *and* $C_{T+1} = \sum_{\tau=T+1-\omega+1}^{T} v_{T+1}^{|\tau|}\overline{u}_\tau^{(\omega)} + \varpi_{T+1}$;

$$A_k = \frac{(R_k)^2}{4c_k} + \delta A_{k+1}(1-\lambda_k), \quad \text{for } k \in \{1, 2, \cdots, T\}, \quad (2.6)$$

and C_k *is an expression that contains the previously executed controls* $\overline{u}_{k-}^{(\omega)}$.

In particular, the term involving $\overline{u}_{k-}^{(\omega)}$ *in the* C_k *is given in Appendix* I.

Proof See Appendix I. ■

Using Proposition 2.1 and Corollary 2.1 one can obtain the optimal strategies of the firm in stage k as (see derivation details in Appendix I):

$$u_k = \frac{(R_k)^2 x}{4(c_k)^2}, \quad \text{for } k \in \{1, 2, \cdots, T\}; \text{ and}$$

$$\overline{u}_k^{(\omega)} = \frac{\sum_{\tau=k}^{k+\omega-1} \delta^{\tau-k+1} A_{\tau+1} a_\tau^{|k|} + \delta^{(T+1-k)} v_{T+1}^{|k|}}{2\phi_k},$$

for $k \in \{T-\omega+2, T-\omega+3, \cdots, T\}$ and $A_\tau = 0$ for $\tau > T+1$;

$$\overline{u}_k^{(\omega)} = \frac{\sum_{\tau=k}^{k+\omega-1} \delta^{\tau-k+1} A_{\tau+1} a_\tau^{|k|}}{2\phi_k}, \quad \text{for } k \in \{1, 2, \cdots, T-\omega+1\}. \tag{2.7}$$

The optimal capital investment strategies $\overline{u}_k^{(\omega)}$ in (2.7) are positively related to the sum of marginal benefits of investment from stage k to stage $k+\omega-1$. Substituting the optimal level of capital investment $\overline{u}_k^{(\omega)}$ in (2.7) into (2.1), the optimal capital accumulation path can be obtained as:

$$x_{k+1} = x_k + a_k^{|k|} \overline{u}_k^{(\omega)} + \sum_{t=k-\omega+1}^{k-1} a_k^{|t|} \overline{u}_t^{(\omega)} - \lambda_k x_k, \quad x_1 = x_1^0, \tag{2.8}$$

where $\overline{u}_k^{(\omega)}$, for $k \in \{1, 2, \cdots, T\}$, are given in (2.7).

Equation (2.8) is a first order linear difference equation which can be solved by standard techniques. Using the solution $\{x_k\}_{t=1}^{T+1}$ from (2.8), the optimal non-durable input can be obtained from $u_k = \frac{(R_k)^2 x_k}{4(c_k)^2}$, for $k \in \{1, 2, \cdots, T\}$.

3 Durable-Strategies Dynamic Game: Formulation and Equilibria

In this Section, we first develop a general class of non-cooperative dynamic game with durable strategies. Then, we present a theorem characterizing the equilibrium game solution.

3.1 Game Formulation

Consider a $T-$ stage $n-$ player nonzero-sum discrete-time non-cooperative dynamic game with durable and non-durable strategies affecting the players' payoffs and the state dynamics. We use $u_k^i \in U^i \subset R^{m^i}$ to denote the set of non-durable control strategies of player i. We use $\overline{u}_k^i = (\overline{u}_k^{(2)i}, \overline{u}_k^{(3)i}, \cdots, \overline{u}_k^{(\omega_i)i})$ to denote the set of durable strategies of player i, where $\overline{u}_k^{(\zeta)i} \in \overline{U}^{(\zeta)i} \subset R^{m_{(\zeta)i}}$ for $\zeta \in \{2, 3, \cdots, \omega_i\}$. In particular, $\overline{u}_k^{(2)i}$ are non-durable strategies that have effects in stages k and $k+1$. The strategies $\overline{u}_k^{(3)i}$ are durable strategies that have effects within stage k to stage $k+2$. The strategies $\overline{u}_k^{(\omega)i}$ are durable strategies that have effects within stages k, $k+1$,

$\cdots, k+\omega-1$. The state at stage k is $x_k \in X \subset R^m$ and the state space is common for all players. The single-stage payoff of player i in stage k is

$$g_k^i(x_k, \underline{u}_k, \overline{u}_k; \overline{u}_{k-}), \text{ for } k \in \{1, 2, \cdots, T\} \text{ and } i \in \{1, 2, \cdots, n\} \equiv N,$$

where $\underline{u}_k = (u_k^1, u_k^2, \cdots, u_k^n)$ is the set of durable strategies of all the n players, $\overline{u}_k = (\overline{u}_k^1, \overline{u}_k^2, \cdots, \overline{u}_k^n)$ is the set of durable strategies of all the n players and $\overline{u}_{k-} = (\overline{u}_{k-}^1, \overline{u}_{k-}^2, \cdots, \overline{u}_{k-}^n)$ is the set of strategies which are executed before stage k by all players but still in effect in stage k.

The payoff of player i is:

$$\sum_{k=1}^{T} g_k^i(x_k, \underline{u}_k, \overline{u}_k; \overline{u}_{k-})\delta_1^k + q_{T+1}^i(x_{T+1}; \overline{u}_{(T+1)-})\delta_1^{T+1}, \tag{3.1}$$

where $q_{T+1}^i(x_{T+1}; \overline{u}_{(T+1)-})$ is the terminal payoff of player i.

The state dynamics is characterized by a vector of difference equations:

$$x_{k+1} = f_k(x_k, \underline{u}_k, \overline{u}_k; \overline{u}_{k-}), \quad x_1 = x_1^0, \tag{3.2}$$

for $k \in \{1, 2, \cdots, T\}$.

The controls executed before the start of the operation in stage 1, that is $\overline{u}_{1-}$, are known and some or all of them can be zeros. The $g_k^i(x_k, \underline{u}_k, \overline{u}_k; \overline{u}_{k-})$, $f_k^i(x_k, \underline{u}_k, \overline{u}_k; \overline{u}_{k-})$ and $q_{T+1}^i(x_{T+1}; \overline{u}_{(T+1)-})$ are continuously differentiable functions.

The information set of every player includes the knowledge in.

(i) all the possible moves by himself and other players, that is u_k^i and $\overline{u}_k^i$, for $k \in \{1, 2, \cdots, T\}$ and $i \in N$;
(ii) the set of controls which are executed before stage k by all players but still in effect in stage k, that is $\overline{u}_{k-} = (\overline{u}_{k-}^1, \overline{u}_{k-}^2, \cdots, \overline{u}_{k-}^n)$, for $k \in \{1, 2, \cdots, T\}$;
(iii) the payoff functions of all players, that is $\sum_{k=1}^{T} g_k^i(x_k, \underline{u}_k, \overline{u}_k; \overline{u}_{k-})\delta_1^k + q_{T+1}^i(x_{T+1}; \overline{u}_{(T+1)-})\delta_1^{T+1}$, for $i \in N$; and
(iv) the state dynamics $x_{k+1} = f_k(x_k, \underline{u}_k, \overline{u}_k; \overline{u}_{k-})$ and the values of present and past states $(x_k, x_{k-1}, \cdots, x_1)$.

3.2 Game Equilibria

The non-cooperative payoffs of the players in a Nash equilibrium of the dynamic game (3.1)–(3.2) can be characterized by the following theorem.

Theorem 3.1. *Let $(\underline{u}_k^{**}, \overline{u}_k^{**})$ be the set of feedback Nash equilibrium strategies and $V^i(k, x; \overline{u}_{k-}^{**})$ be the feedback Nash equilibrium payoff of player i at stage k in the* *non-cooperative dynamic game* (3.1)–(3.2), *then the function* $V^i(k, x; \underline{\overline{u}}_{k-}^{**})$ *satisfies the following recursive equations*:

$$V^i(T+1, x; \underline{\overline{u}}_{(T+1)-}^{**}) = q_{T+1}^i(x; \underline{\overline{u}}_{(T+1)-}^{**})\delta_1^{T+1}; \tag{3.3}$$

$$V^i(k, x; \underline{u}_{k-}^{**}) = \max_{u_k^i, \overline{u}_k^i} \Big\{ g_k^i(x, u_k^i, \overline{u}_k^i, \underline{u}_k^{**(\neq i)}, \underline{\overline{u}}_k^{**(\neq i)}; \underline{\overline{u}}_{k-}^{**})\, \delta_1^k$$
$$+ V^i[k+1, f_k(x, u_k^i, \overline{u}_k^i, \underline{u}_k^{**(\neq i)}, \underline{\overline{u}}_k^{**(\neq i)}; \underline{\overline{u}}_{k-}^{**}); \overline{u}_k^i, \underline{\overline{u}}_k^{**(\neq i)}, \underline{\overline{u}}_{(k+1)-}^{**} \cap \underline{\overline{u}}_{k-}^{**}] \Big\}, \tag{3.4}$$

for $k \in \{1, 2, \cdots, T\}$ *and* $i \in N$,
where $\underline{u}_k^{(\neq i)**} = (u_k^{1**}, u_k^{2**}, \cdots, u_k^{i-1**}, u_k^{i+1**}, \cdots, u_k^{n**})$ *and.*
$\underline{\overline{u}}_k^{(\neq i)**} = (\overline{u}_k^{1**}, \overline{u}_k^{2**}, \cdots, \overline{u}_k^{i-1**}, \overline{u}_k^{i+1**}, \cdots, \overline{u}_k^{n**})$.

Proof Conditions (3.3)–(3.4) show that $V^i(k, x; \underline{\overline{u}}_{k-}^{**})$ is the maximized payoff of player $i \in N$ according to Theorem 1.1 given the game equilibrium strategies of the other $n-1$ players. Hence a Nash equilibrium results. ■

System (3.3)–(3.4) can be regarded as the Hamilton–Jacobi-Bellman equations for durable-strategies dynamic games. Worth noting is that this class of games cannot be handled by the standard approach of dynamic programming. The definitive contributions of this game theory paradigm include:

First, durable-strategies dynamic games provide a novel and realistic dynamic game paradigm game framework which allows the presence of durable strategies. Given that durable strategies would be involved in most of the real-life situations, the paradigm yields a better framework in characterizing interactive decision making over time.

Second, the game paradigm establishes novel mathematical results. A dynamic optimization theorem involving durable strategies is supplied as the essential foundation of corresponding game theory analysis. The derived dynamic optimization technique predominates the classic dynamic programming by providing a more comprehensive solution mechanism. In addition, a novel system of Hamilton–Jacobi-Bellman equations is presented to characterize the feedback Nash equilibrium solution.

Third, given the prevalence of durable strategies, the game paradigm yields a wide scope of applications in many practical scenarios. For instance, technology innovation under knowledge diffusion over time, long-lasting pollution-generating effects in global environmental management, advertising campaigns with lagged effects, oligopoly competition with investments requiring several stages to be converted into productive physical capital, business transactions involving payment by instalments. Applications can also be readily made by constructing relevant dynamic game counterparts with durable strategies in marketing games from Jørgensen and Zaccour (2004), in economics games from Long (2010), in various dynamic games from Başar and Zaccour (2018) and in economic optimizations from Yeung and Petrosyan (2012).

Finally, novel results, which cannot be generated by any of the existing dynamic games, can be obtained. For an example application given in Petrosyan and Yeung (2020), the emission of environmentally damaging industrial waste (like toxic and chemical wastes and deforestation) could continue for many years, hence it is possible that pollution would persist or exacerbate even after the operation stops completely. A series of new results are given in chapters 3–9 in this book.

4 A Computational Illustration in Public Capital Investment

Consider an illustrative example of an oligopoly with n firms. The planning horizon of these firms involves T stages. The firms use a knowledge-based public capital $x_k \in X \subset R^+$ to produce its output. Capital investment takes ω stages to complete its built up into the capital stock. The accumulation process of the capital stock is governed by the dynamics

$$x_{k+1} = x_k + \sum_{j=1}^{n} a_k^{|k|j} \overline{u}_k^{(\omega)j} + \sum_{j=1}^{n} \sum_{\tau=k-\omega+1}^{k-1} a_k^{|\tau|j} \overline{u}_\tau^{(\omega)j} - \lambda_k x_k, \qquad x_1 = x_1^0, \tag{4.1}$$

where $\overline{u}_k^{(\omega)j}$ is the investment in the knowledge-based capital by firm j in stage k, $a_k^{|k|j} \overline{u}_k^{(\omega)j}$ is the addition to the capital in stage k by firm j's capital building investment $\overline{u}_k^{(\omega)j}$ in stage k, $a_k^{|\tau|j} \overline{u}_\tau^{(\omega)j}$ is the addition to the capital in stage k by capital building investment $\overline{u}_\tau^{(\omega)j}$ in stage $\tau \in \{k-1, k-2, \cdots, k-\omega+1\}$ and λ_k is the rate of obsolescence of the capital.

Firm $i \in N$ uses the knowledge-based capital together with a non-durable productive inputs u_k^i to produce outputs. The revenue from the output produced is $R_k^i (u_k^i)^{1/2} (x_k)^{1/2}$. The non-durable input cost is $c_k^i u_k^i$ and the cost of capital investment is $\phi_k^i (\overline{u}_k^{(\omega)i})^2$.

The payoff of the firm i is

$$\sum_{k=1}^{T} \left([R_k^i (u_k^i)^{1/2} (x_k)^{1/2} - c_k^i u_k^i - \phi_k^i (\overline{u}_k^{(\omega)i})^2] \right) \delta^{k-1}$$
$$+ \left(Q_{T+1}^i x_{T+1} + \sum_{\tau=T+1-\omega+1}^{T} v_{T+1}^{|\tau|i} \overline{u}_\tau^{(\omega)i} + \varpi_{T+1}^i \right) \delta^T, \tag{4.2}$$

where $\left(Q_{T+1}^i x_{T+1} + \sum_{\tau=T+1-\omega+1}^{T} v_{T+1}^{|\tau|i} \overline{u}_\tau^{(\omega)i} + \varpi_{T+1}^i \right)$ is the salvage value of firm i in stage $T+1$ and $\delta = (1+r)^{-1}$ is the discount factor.

The salvage value of the firm i in stage $T+1$ depends on the public capital stock, with Q_{T+1}^i being positive. It also depends on the previous capital investment

$\sum_{\tau=T+1-\omega_3+1}^{T} v_{T+1}^{|\tau|i} \overline{u}_{\tau}^{(\omega)i}$. In particular, some or all $v_{T+1}^{|\tau|}$, for $\tau \in \{T, T-1, \cdots, T-\omega+2\}$, can be zero.

Invoking Theorem 3.1, we obtain.

Corollary 4.1. *Let* $(\underline{u}_k^{**}, \underline{u}_k^{(\omega)**}) = [(u_k^{1**}, u_k^{2**}, \cdots, u_k^{n**}), (u_k^{(\omega)1**}, u_k^{(\omega)2**}, \cdots, u_k^{(\omega)n**})]$ *be the set of feedback Nash equilibrium strategies and* $V^i(k, x; \underline{\overline{u}}_{k-}^{(\omega)**})$ *be the feedback Nash equilibrium payoff of player i in the non-cooperative game* (4.1)–(4.2), *then the function* $V^i(k, x; \underline{\overline{u}}_{k-}^{(\omega)**})$ *satisfies the following recursive equations:*

$$V^i(T+1, x; \underline{\overline{u}}_{(T+1)-}^{(\omega)**}) = \left(Q_{T+1}^i x + \sum_{\tau=T+1-\omega+1}^{T} v_{T+1}^{|\tau|i} \overline{u}_{\tau}^{(\omega)i} + \varpi_{T+1}^i \right) \delta^T; \quad (4.3)$$

$$V^i(k, x; \underline{\overline{u}}_{k-}^{(\omega)**}) = \max_{u_k^i, \overline{u}_k^{(\omega)i}} \Bigg\{ [R_k^i (u_k^i)^{1/2} (x)^{1/2} - c_k^i u_k^i - \phi_k^i (\overline{u}_k^{(\omega)i})^2] \delta^{k-1}$$

$$+ V^i[k+1, x + a_k^{|k|i} \overline{u}_k^i + \sum_{j \in N, j \neq i} a_k^{|k|j} \overline{u}_k^{(\omega)j**} + \sum_{j=1}^{n} \sum_{\tau=k-\omega+1}^{k-1} a_k^{|\tau|j} \overline{u}_{\tau}^{(\omega)j**}$$

$$- \lambda_k x; , \overline{u}_k^{(\omega)i}, \underline{\overline{u}}_k^{(\omega)(\neq i)**}, \underline{\overline{u}}_{(k+1)-}^{(\omega)**} \cap \underline{\overline{u}}_{k-}^{(\omega)**}] \Bigg\},$$

for $k \in \{1, 2, \cdots, T\}$ and $i \in N$. (4.4)

■

Performing the indicated maximization operator in Corollary 4.1, the value function $V^i(k, x; \underline{\overline{u}}_{k-}^{(\omega)**})$ which reflects the value of the firm covering stage $k \in \{1, 2, \cdots, T\}$ to stage $T+1$ can be obtained as follows.

Proposition 4.1. *System* (4.3)–(4.4) *admits a solution with the game equilibrium payoff of firm i being*

$$V^i(k, x; \underline{\overline{u}}_{k-}^{(\omega)**}) = (A_k^i x + C_k^i) \delta^{k-1}, \quad (4.5)$$

where $A_{T+1}^i = Q_{T+1}^i$ *and* $C_{T+1}^i = \sum_{\tau=T+1-\omega+1}^{T} v_{T+1}^{|\tau|i} \overline{u}_{\tau}^{(\omega)i} + \varpi_{T+1}^i$, *for* $i \in N$;

$$A_k^i = \frac{(R_k^i)^2}{4c_k^i} + \delta A_{k+1}^i (1 - \lambda_k), \qquad \text{for } k \in \{1, 2, \cdots, T\} \text{ and } i \in N, \quad (4.6)$$

and C_k^i *is an expression which contains the previously executed controls* $\underline{\overline{u}}_{k-}^{(\omega)**}$.

In particular, the term involving $\underline{\overline{u}}_{k-}^{(\omega)**}$ in the C_k^i is

$$\delta A_{k+1}^i \sum_{j=1}^{n} \sum_{\tau=k-\omega+1}^{k-1} a_k^{|\tau|j} \overline{u}_{\tau}^{(\omega)j**} + \delta^2 A_{k+2}^i \sum_{j=1}^{n} \sum_{\tau=k-\omega+2}^{k-1} a_{k+1}^{|\tau|j} \overline{u}_{\tau}^{(\omega)j**}$$

$$+\delta^3 A^i_{k+3} \sum_{j=1}^{n} \sum_{\tau=k-\omega+3}^{k-1} a^{|\tau|j}_{k+2} \overline{u}^{(\omega)j**}_{\tau} + \cdots$$

$$\cdots + \delta^{\omega-1} A^i_{k+\omega-1} \sum_{j=1}^{n} \sum_{\tau=k-1}^{k-1} a^{|\tau|j}_{k+\omega-2} \overline{u}^{(\omega)j**}_{\tau} + \delta^{(T+1-k)} \sum_{\tau=T-\omega+2}^{k-1} v^{|\tau|i}_{T+1} \overline{u}^{(\omega)i**}_{\tau},$$

for $k \in \{T-\omega+2, T-\omega+3, \cdots, T\}$, where $A_k = 0$ for $k > T+1$;

$$\delta A^i_{k+1} \sum_{j=1}^{n} \sum_{\tau=k-\omega+1}^{k-1} a^{|\tau|j}_{k} \overline{u}^{(\omega)j**}_{\tau} + \delta^2 A^i_{k+2} \sum_{j=1}^{n} \sum_{\tau=k-\omega+2}^{k-1} a^{|\tau|j}_{k+1} \overline{u}^{(\omega)j**}_{\tau}$$

$$+\delta^3 A^i_{k+3} \sum_{j=1}^{n} \sum_{\tau=k-\omega+3}^{k-1} a^{|\tau|j}_{k+2} \overline{u}^{(\omega)j**}_{\tau}$$

$$\cdots + \delta^{\omega-1} A^i_{k+\omega-1} \sum_{j=1}^{n} \sum_{\tau=k-1}^{k-1} a^{|\tau|j}_{k+\omega-2} \overline{u}^{(\omega)j**}_{\tau},$$

for $k \in \{1, 2, \cdots, T-\omega+1\}$.

Proof See Appendix II. ■

Using Proposition 4.1 and Corollary 4.1 one can obtain the game equilibrium strategies in stage k as (see derivation details in Appendix II):

$$u^{i**}_k = \frac{(R^i_k)^2 x}{4(c^i_k)^2}, k \in \{1, 2, \cdots, T\} \text{ and}$$

$$\overline{u}^{(\omega)i**}_k = \frac{\sum_{\tau=k}^{k+\omega-1} \delta^{\tau-k+1} A^i_{\tau+1} a^{|k|i}_{\tau} + \delta^{(T+1-k)} v^{|k|i}_{T+1}}{2\phi^i_k},$$

for $k \in \{T-\omega+2, T-\omega+3, \cdots, T\}$ and $A_\tau = 0$ for $\tau > T+1$,

$$\overline{u}^{(\omega)i**}_k = \frac{\sum_{\tau=k}^{k+\omega-1} \delta^{\tau-k+1} A^i_{\tau+1} a^{|ki|}_{\tau}}{2\phi^i_k}, \quad \text{for } k \in \{1, 2, \cdots, T-\omega+1\}. \tag{4.7}$$

Substituting the game equilibrium level of capital investment $\overline{u}^{(\omega)**}_k$ from (5.27) into (4.1) yields the game equilibrium capital accumulation path as:

$$x_{k+1} = x_k + \sum_{j=1}^{n} a^{|k|j}_k \overline{u}^{(\omega)j**}_k + \sum_{j=1}^{n} \sum_{t=k-\omega+1}^{k-1} a^{|t|j}_k \overline{u}^{(\omega)j**}_t - \lambda_k x_k, \qquad x_1 = x^0_1, \tag{4.8}$$

where $\overline{u}^{(\omega)j**}_k$ and $\overline{u}^{(\omega)j**}_t$ are obtained from (4.7).

Equation (4.8) is a first order linear difference equation which can be solved by standard techniques. We use $\{x_k^{**}\}_{k=1}^{T+1}$ to denote the solution of the game equilibrium trajectory in (4.8).

5 Appendices

5.1 Appendix I: Proof of Proposition 2.1

Using (2.3) and Proposition 2.1, we have

$$A_{T+1} = Q_{T+1} \text{ and } C_{T+1} = \sum\nolimits_{\tau=T+1-\omega+1}^{T} v_{T+1}^{|\tau|}\overline{u}_{\tau}^{(\omega)} + \varpi_{T+1}. \tag{5.1}$$

Using Proposition 2.1 and (2.4), we can express the optimal strategies in stage T as:

$$u_T = \frac{(R_T)^2 x}{4(c_T)^2} \text{ and } \overline{u}_T^{(\omega)} = \frac{\delta(A_{T+1}a_T^{|T|} + v_{T+1}^{|T|})}{2\phi_T}. \tag{5.2}$$

Substituting the optimal strategies in (5.2) into the stage T equation of (2.4) we can obtain $V(T, x; \overline{u}_{T-}^{(\omega)})$ through

$$\begin{aligned}(A_T x + C_T)\delta^{T-1} &= \left(\frac{(R_T)^2 x}{4c_T} - \frac{(\delta(A_{T+1}a_T^{|T|} + v_{T+1}^{|T|}))^2}{4\phi_T}\right)\delta^{T-1} \\ &+ \left[A_{T+1}\left(x + a_T^{|T|}\frac{\delta(A_{T+1}a_T^{|T|} + v_{T+1}^{|T|})}{2\phi_T} + \sum_{\tau=T-\omega+1}^{T-1} a_T^{|\tau|}\overline{u}_\tau^{(\omega)} - \lambda_T x\right)\right. \\ &+ \left. v_{T+1}^{|T|}\frac{\delta(A_{T+1}a_T^{|T|} + v_{T+1}^{|T|})}{2\phi_T} + \sum_{\tau=T+1-\omega_3+1}^{T-1} v_{T+1}^{|\tau|}\overline{u}_\tau^{(\omega)} + \varpi_{T+1}\right]\delta^T.\end{aligned} \tag{5.3}$$

The right-hand-side and the left-hand-side of (5.3) are linear functions of x. For (5.3) to hold, it is required:

$$A_T = \frac{(R_T)^2}{4c_T} + \delta A_{T+1}(1 - \lambda_T). \tag{5.4}$$

In addition,

$$\begin{aligned}C_T &= -\frac{(\delta(A_{T+1}^i a_T^{|T|} + v_{T+1}^{|T|}))^2}{4\phi_T} \\ &+ \left[A_{T+1}\left(a_T^{|T|}\frac{\delta(A_{T+1}a_T^{|T|} + v_{T+1}^{|T|})}{2\phi_T} + \sum_{\tau=T-\omega+1}^{T-1} a_T^{|\tau|}\overline{u}_\tau^{(\omega)}\right)\right.\end{aligned}$$

$$+ v_{T+1}^{|T|}\frac{\delta(A_{T+1}a_T^{|T|}+v_{T+1}^{|T|})}{2\phi_T} + \sum_{\tau=T+1-\omega_3+1}^{T-1} v_{T+1}^{|\tau|}\overline{u}_\tau^{(\omega)} + \varpi_{T+1}\Bigg]\delta. \quad (5.5)$$

which is a function of previously executed controls $\overline{u}_{T-}^{(\omega)}$.

Then we move to stage $T-1$.

Using $V(T,x;\overline{u}_{T-}^{(\omega)}) = (A_T x + C_T)\delta^{T-1}$ in (5.3)–(5.5) and the stage $T-1$ equation in (2.4), the optimal strategies in stage $T-1$ can be obtained as:

$$u_{T-1} = \frac{(R_{T-1})^2 x}{4(c_{T-1})^2} \text{ and } \overline{u}_{T-1}^{(\omega)} = \frac{\delta A_T a_{T-1}^{|T-1|} + \delta^2(A_{T+1}a_T^{|T-1|} + v_{T+1}^{|T-1|})}{2\phi_{T-1}}. \quad (5.6)$$

Invoking Proposition 2.1. and substituting the optimal strategies in (5.6) into the stage $T-1$ equation of (2.4) we obtain

$$\begin{aligned}
&(A_{T-1}x + C_{T-1})\delta^{T-2}\\
&= \left(\frac{(R_{T-1})^2 x}{4c_{T-1}} - \frac{(\delta A_T a_{T-1}^{|T-1|} + \delta^2(A_{T+1}a_T^{|T-1|} + v_{T+1}^{|T-1|}))^2}{\phi}\right)\delta^{T-2}\\
&\quad + A_T\left(x + a_{T-1}^{|T-1|}\frac{\delta A_T a_{T-1}^{|T-1|} + \delta^2(A_{T+1}a_T^{|T-1|} + v_{T+1}^{|T-1|})}{2\phi_{T-1}}\right.\\
&\quad \left. + \sum_{\tau=T-\omega}^{T-2} a_{T-1}^{|\tau|}\bar{u}_\tau^{(\omega)} - \lambda_{T-1}x\right)\\
&\quad \delta^{T-1} + \left[A_{T+1}\left(a_T^{|T|}\frac{\delta(A_{T+1}a_T^{|T|}+v_{T+1}^{|T|})}{2\phi_T} + a_T^{|T-1|}\right.\right.\\
&\quad \left.\frac{\delta A_T a_{T-1}^{|T-1|} + \delta^2(A_{T+1}a_T^{|T-1|} + v_{T+1}^{|T-1|})}{2\phi_{T-1}} + \sum_{\tau=T-\omega+1}^{T-2} a_T^{|\tau|}\bar{u}_\tau^{(\omega)}\right)\\
&\quad - \frac{\delta^2(A_{T+1}a_T^{|T|}+v_{T+1}^{|T|})^2}{4\phi_T}\\
&\quad + v_{T+1}^{|T-1|}\frac{\delta A_T a_{T-1}^{|T-1|} + \delta^2(A_{T+1}a_T^{|T-1|} + v_{T+1}^{|T-1|})}{2\phi_{T-1}}\\
&\quad \left. + \sum_{\tau=T+1-\omega_3+1}^{T-2} v_{T+1}^{|\tau|}\bar{u}_\tau^{(\omega)} + \varpi_{T+1}\right]\delta^T.
\end{aligned} \quad (5.7)$$

The right-hand-side and the left-hand-side of (5.7) are linear functions of x. For (5.7) to hold, it is required:

$$A_{T-1} = \frac{(R_{T-1})^2}{4c_{T-1}} + \delta A_T(1-\lambda_{T-1}). \quad (5.8)$$

In addition,

$$C_{T-1} = -\frac{(\delta A_T a_{T-1}^{|T-1|} + \delta^2(A_{T+1}a_T^{|T-1|} + v_{T+1}^{|T-1|}))^2}{4\phi_{T-1}}\Bigg)$$

$$+ A_T\left(a_{T-1}^{|T-1|}\frac{\delta A_T a_{T-1}^{|T-1|} + \delta^2\left(A_{T+1}a_T^{|T-1|} + v_{T+1}^{|T-1|}\right)}{2\phi_{T-1}} + \sum_{\tau=T-\omega}^{T-2} a_{T-1}^{|\tau|}\overline{u}_\tau^{(\omega)}\right)\delta$$

$$+ \Bigg[A_{T+1}\Bigg(a_T^{|T|}\frac{\delta(A_{T+1}a_T^{|T|} + v_{T+1}^{|T|})}{2\phi_T}$$

$$+a_T^{|T-1|}\frac{\delta A_T a_{T-1}^{|T-1|} + \delta^2(A_{T+1}a_T^{|T-1|} + v_{T+1}^{|T-1|})}{2\phi_{T-1}} + \sum_{\tau=T-\omega+1}^{T-2} a_T^{|\tau|}\overline{u}_\tau^{(\omega)}\Bigg)$$

$$- \frac{(\delta(A_{T+1}a_T^{|T|} + v_{T+1}^{|T|}))^2}{4\phi_T} + v_{T+1}^{|T-1|}\frac{\delta A_T a_{T-1}^{|T-1|} + \delta^2(A_{T+1}a_T^{|T-1|} + v_{T+1}^{|T-1|})}{2\phi_{T-1}}$$

$$+ \sum_{\tau=T+1-\omega_3+1}^{T-2} v_{T+1}^{|\tau|}\overline{u}_\tau^{(\omega)} + \varpi_{T+1}\Bigg]\delta^2, \tag{5.9}$$

which is a function of previously executed controls $\overline{u}_{(T-1)-}^{(\omega)}$.

Then we move to stage $T-2$.

Using $V(T-1, x; \overline{u}_{(T-1)-}^{(\omega)}) = (A_{T-1}x + C_{T-1})\delta^{T-2}$ in (5.7)–(5.9) and the stage $T-2$ equation in (2.4), the optimal strategies in stage $T-2$ can be obtained as:

$$u_{T-2} = \frac{(R_{T-2})^2 x}{4(c_{T-2})^2} \text{ and}$$

$$\overline{u}_{T-2}^{(\omega)} = \frac{\delta A_{T-1}a_{T-2}^{|T-2|} + \delta^2 A_T a_{T-1}^{|T-2|} + \delta^3(A_{T+1}a_T^{|T-2|} + v_{T+1}^{|T-2|})}{2\phi_{T-2}}. \tag{5.10}$$

Substituting the optimal strategies in (5.10) into the stage $T-2$ equation of (2.4), we obtain an equation which right-hand-side and left-hand-side are linear functions of x. For the equation to hold, it is required:

$$A_{T-2} = \frac{(P_{T-2})^2}{4c_{T-2}} + \delta A_{T-1}(1 - \lambda_{T-2}). \tag{5.11}$$

In addition, C_{T-2} is an expression which contains all the terms in the right-hand-side of the equation which do not involve x including the previously executed controls $\overline{u}_{(T-2)-}^{(\omega)}$.

Following the above analysis for stage $k \in \{1, 2, \cdots, T\}$ the optimal strategies in stage k can be expressed as:

$$u_k = \frac{(R_k)^2 x}{4(c_k)^2}, \quad \text{for } k \in \{1, 2, \cdots, T\}; \text{ and}$$

$$\overline{u}_k^{(\omega)} = \frac{\sum_{\tau=k}^{k+\omega-1} \delta^{\tau-k+1} A_{\tau+1} a_\tau^{|k|} + \delta^{(T+1-k)} v_{T+1}^{|k|}}{2\phi_k},$$

for $k \in \{T - \omega + 2, T - \omega + 3, \cdots, T\}$ and $A_\tau = 0$ for $\tau > T + 1$,

$$\overline{u}_k^{(\omega)} = \frac{\sum_{\tau=k}^{k+\omega-1} \delta^{\tau-k+1} A_{\tau+1} a_\tau^{|k|}}{2\phi_k}, \qquad \text{for } k \in \{1, 2, \cdots, T - \omega + 1\}. \tag{5.12}$$

Substituting the optimal strategies in (5.12) into the stage k equation of (2.4) we obtain

$$V(k, x; \overline{u}_{k-}^{(\omega)}) = (A_k x + C_k)\delta^{k-1}, \tag{5.13}$$

where

$$A_k = \frac{(R_k)^2}{4c_k} + \delta A_{k+1}(1 - \lambda_k), \tag{5.14}$$

and C_k is an expression which contains the previously executed controls $\overline{u}_{k-}^{(\omega)}$.With $A_{T+1} = Q_{T+1}$, one can use (5.14) to obtain the values of A_k, for $k \in \{T, T - 1, \cdots, 1\}$. By definition, $A_k = 0$, for $k > T + 1$. With (5.12), the optimal levels of capital investment $\overline{u}_k^{(\omega)}$, for $k \in \{T, T - 1, \cdots, 1\}$, can be obtained. Substituting the optimal level of capital investment $\overline{u}_k^{(\omega)}$ into (2.1), the optimal capital accumulation path can be obtained as:

$$x_{k+1} = x_k + a_k^{|k|} \overline{u}_k^{(\omega)} + \sum_{t=k-\omega+1}^{k-1} a_k^{|t|} \overline{u}_t^{(\omega)} - \lambda_k x_k, \qquad x_1 = x_1^0, \tag{5.15}$$

where $\overline{u}_k^{(\omega)}$, for $k \in \{1, 2, \cdots, T\}$, are given in (5.12).

Equation (5.15) is a first order linear difference equation which can be solved by standard techniques. Using the solution $\{x_k\}_{t=1}^{T+1}$ from (5.15), the optimal non-durable input can be obtained from $u_k = \frac{(R_k)^2 x_k}{4(c_k)^2}$, for $k \in \{1, 2, \cdots, T\}$.

Moreover, the term involving $\overline{u}_{k-}^{(\omega)}$ in the C_k can be obtained as follows.

For $k \in \{T - \omega + 2, T - \omega + 3, \cdots, T\}$, it is

$$\delta A_{k+1} \sum_{\tau=k-\omega+1}^{k-1} a_k^{|\tau|} \overline{u}_\tau^{(\omega)} + \delta^2 A_{k+2} \sum_{\tau=k-\omega+2}^{k-1} a_{k+1}^{|\tau|} \overline{u}_\tau^{(\omega)}$$

$$+ \delta^3 A_{k+3} \sum_{\tau=k-\omega+3}^{k-1} a_{k+2}^{|\tau|} \overline{u}_\tau^{(\omega)}$$

$$+\,\delta^4 A_{k+4}\sum_{\tau=k-\omega+4}^{k-1} a_{k+3}^{|\tau|}\overline{u}_\tau^{(\omega)} + \cdots + \delta^{\omega-1} A_{k+\omega-1}\sum_{\tau=k-1}^{k-1} a_{k+\omega-2}^{|\tau|}\overline{u}_\tau^{(\omega)}$$

$$+\,\delta^{(T+1-k)}\sum_{\tau=T-\omega+2}^{k-1} v_{T+1}^{|\tau|}\overline{u}_\tau^{(\omega)},$$

where $A_k = 0$ for $k > T+1$.

For $k \in \{1, 2, \cdots, T-\omega+1\}$, it is

$$\delta A_{k+1}\sum_{\tau=k-\omega+1}^{k-1} a_k^{|\tau|}\overline{u}_\tau^{(\omega)} + \delta^2 A_{k+2}\sum_{\tau=k-\omega+2}^{k-1} a_{k+1}^{|\tau|}\overline{u}_\tau^{(\omega)} + \delta^3 A_{k+3}\sum_{\tau=k-\omega+3}^{k-1} a_{k+2}^{|\tau|}\overline{u}_\tau^{(\omega)}$$

$$+\,\delta^4 A_{k+4}\sum_{\tau=k-\omega+4}^{k-1} a_{k+3}^{|\tau|}\overline{u}_\tau^{(\omega)} + \cdots + \delta^{\omega-1} A_{k+\omega-1}\sum_{\tau=k-1}^{k-1} a_{k+\omega-2}^{|\tau|}\overline{u}_\tau^{(\omega)}.$$

■

5.2 Appendix II: Proof of Proposition 4.1

Using (4.3) and Proposition 4.1, we have

$$A_{T+1}^i = Q_{T+1}^i \text{ and } C_{T+1}^i = \sum_{\tau=T+1-\omega+1}^{T} v_{T+1}^{|\tau|i}\overline{u}_\tau^{(\omega)i**} + \varpi_{T+1}^i, \quad \text{for } i \in N. \tag{5.16}$$

Using Proposition 4.1 and (4.4), we can express the game equilibrium strategies in stage T as:

$$u_T^{i**} = \frac{\left(R_T^i\right)^2 x}{4\left(c_T^i\right)^2} \text{ and } \overline{u}_T^{(\omega)i**} = \frac{\delta\left(A_{T+1}^i a_T^{|T|i} + v_{T+1}^{|T|_i}\right)}{2\phi_T^i}, \quad \text{for } i \in N. \tag{5.17}$$

Substituting the game equilibrium strategies in (5.17) into the stage T equations of (4.4) we can obtain

$$(A_T^i x + C_T^i)\delta^{T-1} = \left(\frac{(R_T^i)^2 x}{4c_T^i} - \frac{(\delta(A_{T+1}^i a_T^{|T|i} + v_{T+1}^{|T|i}))^2}{4\phi_T^i}\right)\delta^{T-1}$$

$$+\left[A_{T+1}^i\left(x + \sum_{j=1}^{n} a_T^{|T|j}\frac{\delta(A_{T+1}^j a_T^{|T|j} + v_{T+1}^{|T|j})}{2\phi_T^j}\right.\right.$$

$$+\sum_{j=1}^{n}\sum_{\tau=T-\omega+1}^{T-1} a_T^{|\tau|j}\bar{u}_\tau^{(\omega)j**} - \lambda_T x\Bigg)$$

$$+ v_{T+1}^{|T|i}\frac{\delta(A_{T+1}^i a_T^{|T|i} + v_{T+1}^{|T|i})}{2\phi_T^i} + \sum_{\tau=T+1-\omega+1}^{T-1} v_{T+1}^{|\tau|i}\bar{u}_\tau^{(\omega)i**} + \varpi_{T+1}^i\Bigg]\delta^T, \text{ for } i \in N. \tag{5.18}$$

The right-hand-side and the left-hand-side of the equations in (5.18) are linear functions of x. For (5.18) to hold, it is required:

$$A_T^i = \frac{(R_T^i)^2}{4c_T^i} + \delta A_{T+1}^i(1-\lambda_T), \quad \text{for } i \in N. \tag{5.19}$$

In addition,

$$C_T^i = -\frac{(\delta(A_{T+1}^i a_T^{|T|i} + v_{T+1}^{|T|i}))^2}{4\phi_T^i} + \Bigg[A_{T+1}^i\Bigg(\sum_{j=1}^{n} a_T^{|T|j}\frac{\delta(A_{T+1}^j a_T^{|T|j} + v_{T+1}^{|T|j})}{2\phi_T^j}$$

$$+\sum_{j=1}^{n}\sum_{\tau=T-\omega+1}^{T-1} a_T^{|\tau|j}\overline{u}_\tau^{(\omega)j**}\Bigg) + v_{T+1}^{|T|i}\frac{\delta(A_{T+1}^i a_T^{|T|i} + v_{T+1}^{|T|i})}{2\phi_T^i}$$

$$+ \sum_{\tau=T+1-\omega+1}^{T-1} v_{T+1}^{|\tau|i}\overline{u}_\tau^{(\omega)i**} + \varpi_{T+1}^i\Bigg]\delta, \tag{5.20}$$

which is a function of previously executed controls $\underline{u}_{T-}^{**}$.

Then we move to stage $T-1$.

Using $(A_T^i x + C_T^i)\delta^{T-1}$ in (5.18)–(5.20) and the stage $T-1$ equation in (4.4), the game equilibrium strategies in stage $T-1$ can be obtained as:

$$u_{T-1}^{i**} = \frac{(R_{T-1}^i)^2 x}{4(c_{T-1}^i)^2} \text{ and}$$

$$\overline{u}_{T-1}^{(\omega)i**} = \frac{\delta A_T^i a_{T-1}^{|T-1|i} + \delta^2(A_{T+1}^i a_T^{|T-1|i} + v_{T+1}^{|T-1|i})}{2\phi_{T-1}^i}, \quad \text{for } i \in N. \tag{5.21}$$

Invoking Proposition 4.1. and substituting the game equilibrium strategies in (5.21) into the stage $T-1$ equation of (4.4) we obtain

$$(A_{T-1}^i x + C_{T-1}^i)\delta^{T-2} = \left(\frac{(R_{T-1}^i)^2 x}{4c_{T-1}^i} - \frac{(\delta A_T^i a_{T-1}^{|T-1|i} + \delta^2(A_{T+1}^i a_T^{|T-1|i} + v_{T+1}^{|T-1|i}))^2}{4\phi_{T-1}^i}\right)\delta^{T-2}$$

$$+\Bigg[A_T^i\Bigg(x + \sum_{j=1}^{n} a_{T-1}^{|T-1|j}\frac{\delta A_T^j a_{T-1}^{|T-1|j} + \delta^2(A_{T+1}^j a_T^{|T-1|j} + v_{T+1}^{|T-1|j})}{2\phi_{T-1}^j}$$

$$+\sum_{j=1}^{n}\sum_{\tau=T-\omega}^{T-2} a_{T-1}^{|\tau|j}\overline{u}_{\tau}^{(\omega)j**} - \lambda_{T-1}x\Bigg) - \frac{(\delta(A_{T+1}^{i}a_{T}^{|T|i} + v_{T+1}^{|T|i})^2}{4\phi_T^i}$$

$$+\delta A_{T+1}^{i}\left(\sum_{j=1}^{n} a_{T}^{|T|j}\frac{\delta(A_{T+1}^{j}a_{T}^{|T|j} + v_{T+1}^{|T|j})}{2\phi_T^j}\right.$$

$$+\sum_{j=1}^{n} a_{T}^{|T-1|j}\frac{\delta A_{T}^{j}a_{T-1}^{|T-1|j} + \delta^2(A_{T+1}^{j}a_{T}^{|T-1|j} + v_{T+1}^{|T-1|j})}{2\phi_{T-1}^{j}}$$

$$\left.+\sum_{j=1}^{n}\sum_{\tau=T-\omega+1}^{T-2} a_{T}^{|\tau|j}\overline{u}_{\tau}^{(\omega)j**}\right) + v_{T+1}^{|T|i}\frac{\delta(A_{T+1}^{i}a_{T}^{|T|i} + v_{T+1}^{|T|i})}{2\phi_T^i}$$

$$+v_{T+1}^{|T-1|i}\frac{\delta A_{T}^{i}a_{T-1}^{|T-1|i} + \delta^2(A_{T+1}^{i}a_{T}^{|T-1|i} + v_{T+1}^{|T-1|i})}{2\phi_{T-1}^{i}}$$

$$+\sum_{\tau=T+1-\omega+1}^{T-2} v_{T+1}^{|\tau|i}\overline{u}_{\tau}^{(\omega)i**} + \delta\varpi_{T+1}^{i}\Bigg]\delta^{T-1}, \qquad \text{for } i \in N. \tag{5.22}$$

The right-hand-side and the left-hand-side of the equations in (5.22) are linear functions of x. For (5.22) to hold, it is required:

$$A_{T-1}^{i} = \frac{(P_{T-1}^{i})^2}{4c_{T-1}^{i}} + \delta A_{T}^{i}(1-\lambda_{T-1}), \quad \text{for } i \in N. \tag{5.23}$$

In addition,

$$C_{T-1}^{i} = -\frac{(\delta A_{T}^{i}a_{T-1}^{|T-1|i} + \delta^2(A_{T+1}^{i}a_{T}^{|T-1|i} + v_{T+1}^{|T-1|i}))^2}{4\phi_{T-1}^{i}}$$

$$+\left[A_{T}^{i}\left(\sum_{j=1}^{n} a_{T-1}^{|T-1|j}\frac{\delta A_{T}^{j}a_{T-1}^{|T-1|j} + \delta^2(A_{T+1}^{j}a_{T}^{|T-1|j} + v_{T+1}^{|T-1|j})}{2\phi_{T-1}^{j}}\right.\right.$$

$$\left.+\sum_{j=1}^{n}\sum_{\tau=T-\omega}^{T-2} a_{T-1}^{|\tau|j}\overline{u}_{\tau}^{(\omega)j**}\right)$$

$$\left.-\frac{(\delta(A_{T+1}^{i}a_{T}^{|T|i} + v_{T+1}^{|T|i})^2}{4\phi_T^i}\right]\delta + \left[A_{T+1}^{i}\left(\sum_{j=1}^{n} a_{T}^{|T|j}\frac{\delta(A_{T+1}^{j}a_{T}^{|T|j} + v_{T+1}^{|T|j}}{2\phi_T^j}\right.\right.$$

$$+\sum_{j=1}^{n} a_{T}^{|T-1|j}\frac{\delta A_{T}^{j}a_{T-1}^{|T-1|j} + \delta^2(A_{T+1}^{j}a_{T}^{|T-1|j} + v_{T+1}^{|T-1|j})}{2\phi_{T-1}^{j}}$$

$$\left.+\sum_{j=1}^{n}\sum_{\tau=T-\omega+1}^{T-2} a_{T}^{|\tau|j}\overline{u}_{\tau}^{(\omega)j**}\right)$$

$$+v_{T+1}^{|T|i}\frac{\delta(A_{T+1}^{i}a_{T}^{|T|i} + v_{T+1}^{|T|i})}{2\phi_T^i} + v_{T+1}^{|T-1|i}\frac{\delta A_{T}^{i}a_{T-1}^{|T-1|i} + \delta^2(A_{T+1}^{i}a_{T}^{|T-1|i} + v_{T+1}^{|T-1|i})}{2\phi_{T-1}^{i}}$$

$$\left.+\sum_{\tau=T+1-\omega+1}^{T-2} v_{T+1}^{|\tau|i}\overline{u}_{\tau}^{(\omega)i**} + \varpi_{T+1}^{i}\right]\delta^2, \tag{5.24}$$

which is a function of previously executed controls $\underline{\overline{u}}_{(T-1)-}^{(\omega)**}$.

Then we move to stage $T - 2$.

Using $(A^i_{T-1}x + C^i_{T-1})\delta^{T-2}$ in (5.22)–(5.24) and the stage $T - 2$ equation in (4.4), the game equilibrium strategies in stage $T - 2$ can be obtained as:

$$u^{i**}_{T-2} = \frac{(R^i_{T-2})^2 x}{4(c^i_{T-2})^2} \text{ and}$$

$$\overline{u}^{(\omega)i**}_{T-2} = \frac{\delta A^i_{T-1} a^{|T-2|i}_{T-2} + \delta^2 A^i_T a^{|T-2|i}_{T-1} + \delta^3 (A_{T+1} a^{|T-2|i}_T + v^{|T-2|i}_{T+1})}{2\phi^i_{T-2}}, \tag{5.25}$$

for $i \in N$.

Substituting the game equilibrium strategies in (5.25) into the stage $T-2$ equations of (4.4), we obtain a system of equations which right-hand-side and left-hand-side are linear functions of x. For the equation to hold, it is required:

$$A^i_{T-2} = \frac{(R^i_{T-2})^2}{4c^i_{T-2}} + \delta A^i_{T-1}(1 - \lambda_{T-2}). \tag{5.26}$$

In addition, C^i_{T-2} is an expression which contains all the terms in the right-hand-side of the equations which do not involve x including the previously executed controls $\underline{\overline{u}}^{(\omega)*}_{(T-2)-}$.

Following the above analysis for stage $k \in \{1, 2, \cdots, T\}$ the game equilibrium strategies in stage k can be expressed as:

$$u^{i**}_k = \frac{(R^i_k)^2 x}{4(c^i_k)^2}, \text{ for } k \in \{1, 2, \cdots, T\}; \text{ and}$$

$$\overline{u}^{(\omega)i**}_k = \frac{\sum_{\tau=k}^{k+\omega-1} \delta^{\tau-k+1} A^i_{\tau+1} a^{|k|i}_\tau + \delta^{(T+1-k)} v^{|k|i}_{T+1}}{2\phi^i_k},$$

for $k \in \{T - \omega + 2, T - \omega + 3, \cdots, T\}$ and $A_\tau = 0$ for $\tau > T + 1$,

$$\overline{u}^{(\omega)i**}_k = \frac{\sum_{\tau=k}^{k+\omega-1} \delta^{\tau-k+1} A^i_{\tau+1} a^{|ki|}_\tau}{2\phi^i_k}, \quad \text{for } k \in \{1, 2, \cdots, T - \omega + 1\}. \tag{5.27}$$

Substituting the game equilibrium strategies in (5.27) into the stage k equation of (4.4) we can obtain the game equilibrium payoff of firm i as

$$(A^i_k x + C^i_k)\delta^{k-1}, \quad \text{for } i \in N, \tag{5.28}$$

where

$$A^i_k = \frac{(R^i_k)^2}{4c^i_k} + \delta A^i_{k+1}(1 - \lambda_k), \tag{5.29}$$

and C_k^i is an expression which contains the previously executed controls $\overline{u}_{k-}^{(\omega)**}$.

With $A_{T+1}^i = Q_{T+1}^i$, one can use (5.29) to obtain the values of A_k^i, for $k \in \{T, T-1, \cdots, 1\}$. Invoking (5.27), the game equilibrium level of capital investment $\overline{u}_k^{(\omega)i**}$, for $k \in \{T, T-1, \cdots, 1\}$, can be obtained. Substituting the game equilibrium level of capital investment $\overline{u}_k^{(\omega)**}$ from (5.27) into (4.1) yields the game equilibrium capital accumulation path as:

$$x_{k+1} = x_k + \sum_{j=1}^{n} a_k^{|k|j} \overline{u}_k^{(\omega)j**} + \sum_{j=1}^{n} \sum_{t=k-\omega+1}^{k-1} a_k^{|t|j} \overline{u}_t^{(\omega)j**} - \lambda_k x_k,$$
$$x_1 = x_1^0 \tag{5.30}$$

Equation (5.30) is a first order linear difference equation which can be solved by standard techniques. Using the solution $\{x_k\}_{t=1}^{T+1}$ from (5.30), the game equilibrium non-durable input can be obtained as $u_k^i = \frac{(R_k^i)^2 x_k}{4(c_k^i)^2}$, for $k \in \{1, 2, \cdots, T\}$ and $i \in N$.

Moreover, the term involving $\overline{u}_{k-}^{(\omega)}$ in the C_k^i can be obtained as follows.

For $k \in \{T - \omega + 2, T - \omega + 3, \cdots, T\}$, it is

$$\begin{aligned}
&\delta A_{k+1}^i \sum_{j=1}^{n} \sum_{\tau=k-\omega+1}^{k-1} a_k^{|\tau|j} \overline{u}_\tau^{(\omega)j**} + \delta^2 A_{k+2}^i \sum_{j=1}^{n} \sum_{\tau=k-\omega+2}^{k-1} a_{k+1}^{|\tau|j} \overline{u}_\tau^{(\omega)j**} \\
&+ \delta^3 A_{k+3}^i \sum_{j=1}^{n} \sum_{\tau=k-\omega+3}^{k-1} a_{k+2}^{|\tau|j} \overline{u}_\tau^{(\omega)j**} + \cdots \\
&\cdots + \delta^{\omega-1} A_{k+\omega-1}^i \sum_{j=1}^{n} \sum_{\tau=k-1}^{k-1} a_{k+\omega-2}^{|\tau|j} \overline{u}_\tau^{(\omega)j**} \\
&+ \delta^{(T+1-k)} \sum_{\tau=T-\omega+2}^{k-1} v_{T+1}^{|\tau|i} \overline{u}_\tau^{(\omega)i**},
\end{aligned} \tag{5.31}$$

where $A_k = 0$ for $k > T + 1$.

For $k \in \{1, 2, \cdots, T - \omega + 1\}$, it is

$$\begin{aligned}
&\delta A_{k+1}^i \sum_{j=1}^{n} \sum_{\tau=k-\omega+1}^{k-1} a_k^{|\tau|j} \overline{u}_\tau^{(\omega)j**} + \delta^2 A_{k+2}^i \sum_{j=1}^{n} \sum_{\tau=k-\omega+2}^{k-1} a_{k+1}^{|\tau|j} \overline{u}_\tau^{(\omega)j**} \\
&+ \delta^3 A_{k+3}^i \sum_{j=1}^{n} \sum_{\tau=k-\omega+3}^{k-1} a_{k+2}^{|\tau|j} \overline{u}_\tau^{(\omega)j**} + \cdots \\
&\cdots + \delta^{\omega-1} A_{k+\omega-1}^i \sum_{j=1}^{n} \sum_{\tau=k-1}^{k-1} a_{k+\omega-2}^{|\tau|j} \overline{u}_\tau^{(\omega)j**} + \delta^{\omega-1} \sum_{\tau=T-\omega+2}^{k-1} v_{T+1}^{|\tau|i} \overline{u}_\tau^{(\omega)i}.
\end{aligned} \tag{5.32}$$

■

6 Chapter Notes

Durable controls that have influences lasting over a certain period of time are frequently observed in real-life situations. In the presence of durable controls, modification of the dynamic optimization techniques has to be made to accommodate the lagged effects of the controls. Most of the existing works on lagged control optimization involved lags in the state dynamics (see Bellman (1957), Arthur (1977), Burdet and Sethi (1976) and Hartl and Sethi (1984), Brandt-Pollmann et al. (2008), Huschto et al. (2011), Sethi and McGuire (1977) and Winkler et al. (2003)). Bokov (2011) considered continuous-time optimal control with control lags in the payoff and time lags in the state dynamics. In reality, both the decision-maker's payoffs and the evolution of the state dynamics can be subjected to effects from durable controls. For instance, durable goods, vehicles, factories, real estate, revenue generating investments, toxic waste disposal, business contracts and regulations affect the decision-maker's payoffs for more than one stage. At the same time, tariffs have delayed impacts on the economy's growth dynamics, knowledge diffusion takes time in building up knowledge-based capital, conversion of investments into physical capital stock involves a certain number of stages and advertisement enhances the firm's reputation gradually over time. Advertising games with promotion lags in which reputation takes time to establish and time lags often occur in advertising efforts (see Mela et al. (1997), Tellis (2006) and Ruffino (2008)). Sarkar and Zhang (2015) and Agliardi and Koussisc (2013) showed the time-to-build characteristics in investment and capital formation. Tsuruga and Shota (2019) and Sarkar and Zhang (2013) studied implementation lags which generates lag effects fiscal policy and investment decision. Rigidities is another source of delay in policy implementation (see Christiano et al. (2005) and Shibata et al. (2019)).

Based on the work of Petrosyan and Yeung (2020a), this Chapter presents a new class of dynamic games with durable controls of different lag durations affecting both the players' payoffs and the state dynamics. Worth noting is that this class of games cannot be handled by the standard approach of dynamic programming. A common supposition is that the problem involving durable controls in the payoff functions can be converted into standard dynamic programming problem by introducing artificial states to represent the effects of lagged controls. This works only in a small number of simple cases, like the durable control being the purchase of a durable good which can add to the stock of the good. The lagged effects of durable controls executed in one stage may be different from (and cannot be mixed with) the lagged effects of the same controls executed in another stage. Even in some special case where each durable control affecting the payoff can be treated as a state variable with an expressible state dynamic, there would be a large number of states – one for each durable control executed in every stage throughout the game. Burdet and Sethi (1976) have derived the maximum principle for solving optimal control problems with a specific class of discrete-time dynamics with lagged controls by converting the problem into a problem without lags with a larger number of states. Moreover, durable controls affecting the payoff may appear in irregular patterns of future stages with haphazard degrees of impacts. That makes the formulation of the corresponding state dynamics impracticable. Concerning durable controls in the state dynamics, it is well-known that solving control lags in continuous-time state dynamics requires more complicated control techniques than those in standard dynamic programming (see for examples, Hartl and Sethi (1984), Wong et al. (1985), Brandt-Pollmann et al. (2008) and Bokov (2011)). In general, standard dynamic programming technique cannot be applied to solve problems with durable controls affecting the payoff and the state dynamics.

7 Problems

1. (Warmup exercise) Consider a 5-stage optimization problem in which the decision-maker uses a durable strategy $\overline{u}_k^{(2)}$ which has effect in two stages.
   The dynamic equation for the state variable is

$$x_{k+1} = x_k + \overline{u}_k^{(2)} + 0.2\overline{u}_{k-1}^{(2)} - 0.1x_k, \qquad x_1 = 100,$$

   for $k \in \{1, 2, \cdots, 5\}$ and $\overline{u}_k^{(2)} = 0$.
   The payoff is

$$\sum_{k=1}^{5} \left(x_k - (\overline{u}_k^{(2)})^2\right)(0.9)^{k-1} + \left(3x_6(0.8)^5 + 30\right)(0.8)^5.$$

   Derive the optimal strategy $\overline{u}_k^{(2)}$ and the payoff of the decision-maker in stage $k \in \{1, 2, \cdots, 5\}$. (Hints: Follow the computational illustration in Sect. 2. Note that the terminal payoff in stage 6 contains previously executed durable strategies which are executed in different stages, due to

$$x_6 = x_6 + \overline{u}_5^{(2)} + 0.2\overline{u}_4^{(2)} - 0.1x_5.)$$

2. Consider a monopolistic firm which planning horizon involves 6 stages. It uses a knowledge-based capital $x_k \in X \subset R^+$ to produce its output. Capital investment takes three stages to be built up into the capital stock. The accumulation process of the capital stock is governed by the dynamics

$$x_{k+1} = x_k + \overline{u}_k^{(3)} + \sum_{\tau=k-2}^{k-1} 0.5\overline{u}_\tau^{(3)} - 0.05x_k, \qquad x_1 = 80,$$

   where $\overline{u}_k^{(3)}$ is the investment in the knowledge-based capital in stage k.

The revenue of the firm is related to the capital stock and yields a revenue $4x_k$ and the cost of capital investment is $(\overline{u}_k^{(3)})^2$.

The payoff of the firm is

$$\sum_{k=1}^{6}\left(4x_k - \left(\overline{u}_k^{(3)}\right)^2\right)(0.9)^{k-1} + (3x_7 + 5)(0.9)^5.$$

Derive (i) the optimal investment strategy, (ii) the payoff of firm in stage k and (iii) the optimal capital accumulation trajectory.

3. Consider an oligopoly with n firms. The planning horizon of these firms involves T stages. The firms use a knowledge-based public capital $x_k \in X \subset R^+$ to produce its output. Capital investment takes time to be built up into the capital stock. The accumulation process of the capital stock is governed by the dynamics

$$x_{k+1} = x_k + \sum_{j=1}^{n} a_k^{|k|j}\overline{u}_k^{(\omega)j} + \sum_{j=1}^{n}\sum_{\tau=k-\omega+1}^{k-1} a_k^{|\tau|j}\overline{u}_\tau^{(\omega)j} - \lambda_k x_k, \qquad x_1 = x_1^0,$$

where $\overline{u}_k^{(\omega)j}$ is the investment in the knowledge-based capital by firm j in stage k and it takes ω stages for $\overline{u}_k^{(\omega)j}$ to complete its conversion into capital.

Firm i's revenue of the firm is related to the capital stock and yields a net revenue $R_k^i x_k$ and the cost of capital investment is $\phi_k^i(\overline{u}_k^{(\omega)})^2$. The payoff of the firm i is

$$\sum_{k=1}^{T}\left(\left[R_k^j x_k - \vartheta_k^j\left(\overline{u}_k^{(\omega)i}\right)^2\right]\right)\delta^{k-1} + (Q_{T+1}^i x_{T+1} + \varpi_{T+1}^i)\delta^T,$$

where $\left(Q_{T+1}^i x_{T+1} + \varpi_{T+1}^i\right)$ is the salvage value of firm i in stage $T+1$.

The above game is a durable-strategies version of the game in Yeung and Petrosyan (2013) and a discrete-time durable-strategies version of the games Fershtman and Nitzan (1991) and Wirl (1996).

Derive (i) the equilibrium investment strategies of the game, (ii) the payoff of firms in stage k and (iii) the game equilibrium capital accumulation trajectory.

References

Agliardi E, Koussisc N (2013) Optimal capital structure and the impact of time-to-build. Financ Res Lett 10:124–130

Arthur WB (1977) Control of linear processes with distributed lags using dynamic programming from first principles. J Optim Theory Appl 23:429–443

Başar T, Zaccour G (2018) Handbook of dynamic game theory, vol 1 and 2. Springer

Bellman R (1957) Terminal control, time lags, and dynamic programming. Proc Natl Acad Sci USA 43:927–930
Bokov GV (2011) Pontryagin's maximum principle of optimal control problems with time-delay. J Math Sci 172:623–634
Brandt-Pollmann U, Winkler R, Sager S, Moslener U, Schlöder J (2008) Numerical solution of optimal control problems with constant control delays. Comput Econ 31:181–206
Burdet CA, Sethi SP (1976) On the maximum principle for a class of discrete dynamical systems with lags. J Optim Theory Appl 19:445–454
Christiano LJ, Eichenbaum M, Evans CL (2005) Nominal rigidities and the dynamic effects of a shock to monetary policy. J Polit Econ 113:1–45
Fershtman C, Nitzan S (1991) Dynamic voluntary provision of public goods. Eur Econ Rev 35:1057–1067
Hartl RF, Sethi SP (1984) Optimal control of a class of systems with continuous lags: dynamic programming approach and economic interpretations. J Optim Theory Appl 43(1):73–88
Huschto T, Feichtinger G, Hartl RF, Kort PM, Sager S (2011) Numerical solution of a conspicuous consumption model with constant control delay. Automatica 47:1868–1877
Jørgensen S, Zaccour G (2004) Differential games in marketing. Kluwer Academic Publishers
Long NV (2010) A survey of dynamic games in economics. World Scientific
Mela CF, Gupta S, Lehmann DR (1997) The long-term impact of promotion and advertising on consumer brand choice. J Mark Res 34:248–261
Petrosyan LA, Yeung DWK (2020) Cooperative dynamic games with durable controls: theory and application. Dyn Games Appl 10:872–896. https://doi.org/10.1007/s13235-019-00336-w
Ruffino CC (2008) Lagged effects of TV advertising on sales of an intermittently advertised product. Bus Econ Rev 18:1–12
Sarkar S, Zhang H (2013) Implementation lag and the investment decision. Econ Lett 119:136–140
Sarkar S, Zhang C (2015) Investment policy with time-to-build. J Bank Finance 55:142–156
Sethi SP, Mcguire TW (1977) Optimal skill mix: an application of the maximum principle for systems with retarded controls. J Optim Theory Appl 23:245–275
Shibata A, Shintani M, Tsurugac T (2019) Current account dynamics under information rigidity and imperfect capital mobility. J Int Money Financ 92:153–167
Tellis GJ (2006) Optimal data interval for estimating advertising response. Mark Sci 25:217–229
Tsuruga T, Shota W (2019) Money-financed fiscal stimulus: the effects of implementation lag. J Econ Dyn Control 104:132–151
Winkler R, Brandt-Pollmann U, Moslener U, Schlöder J (2003) Time-lags in capital accumulation. In: Ahr D, Fahrion R, Oswald M, Reinelt G (eds) Operations research proceedings, pp 451–458
Wirl F (1996) Dynamic voluntary provision of public goods: extension to nonlinear strategies. Eur J Polit Econ 12:555–560
Wong KH, Clements DJ, Teo KL (1985) Optimal control computation for nonlinear time-lag systems. J Optim Theory Appl 47:91–107
Yeung DWK, Petrosyan LA (2012a) Subgame consistent economic optimization: an advanced cooperative dynamic game analysis. Boston, Birkhäuser, pp 395. ISBN 978-0-8176-8261-3
Yeung DWK, Petrosyan LA (2013) Subgame-consistent cooperative solutions in randomly furcating stochastic dynamic games. Math Comput Model 57(3–4):976–991

Chapter 3
Durable-Strategies Dynamic Games of Investments

Durable strategies are prevalent in business investments. This Chapter presents a class of dynamic games of firms with durable interactive investments. Four major types of investments in a firm's portfolio which have durable effects on the firm's profits and capital are considered. These include the firm's investments in private capitals and in public capitals accessible to all firms, revenue generating, revenue generating investments and debt-financed investments. The Chapter is organized as follows. Section 1 discusses durable interactive strategies available to business firms. Section 2 presents the formulation and equilibrium solution of a generic class of dynamic games of durable interactive investments. An illustrative game with explicit functional forms is provided in Sect. 3. A novel dynamic theory of the valuation of the firm is provided in Sect. 4. Section 5 contains a mathematical appendix and Chapter notes are given in Sect. 6.

1 Durable Interactive Investment Strategies

Durable strategies are far more common than non-durable strategies in business investments. We consider four major and common types of investments in a firm's portfolio which have durable effects on the firm's profits and capital stocks. The first are investments in private capital owned by the firm, the second are investments in public capitals which are accessible to the other firms, the third is revenue generating investments which yield returns in a number of stages and the fourth is debt-financed investments which costs are paid by instalment over a number of periods.

(i) *Investments in Private Capitals*

Private capitals are solely owned by the firm holding them. This type of capital can be classified into two major categories—physical capitals and knowledge capitals. Private physical capitals include man-made items or products that make the manufacturing process possible or enable it to run smoothly. Many physical capitals

D. W. K. Yeung and L. A. Petrosyan, *Durable-Strategies Dynamic Games*,
Theory and Decision Library C 50, https://doi.org/10.1007/978-3-030-92742-4_3

are immobile fixed structures, like corporate headquarters, campus of office buildings, production plants, scientific laboratories, train and bus terminals, container ports, factories, nuclear plants, water dams, warehouses and pharmaceutical laboratories. In addition, large scale computer networks, communication systems, specific manufacturing equipment, airline fleets, container vessels, transmission towers and railway tracks are also examples of major physical capitals. Physical capitals have long-term values, but these values can change over time. Usually, they decline due to depreciation and obsolescence but the costs of investments in physical capital may increase.

Private knowledge-based capitals of an organization consist of the firm's knowledge, technical knowhow, relationships, learned techniques, operation procedures and innovation skills. Many of these are referred to as intellectual capitals, some of these are intangible and they contribute to the productivity of the firm owning it. This type of capital can be classified into three categories: scientific and technical knowhow, relationship capital and structural capital. Scientific and technical knowhow increase the productivity of the firm and contribute to its productivity. Relationship capital including client base, trademarks, company goodwill, databases, information assets, intellectual property and reputation increase the operational efficiency of the firm and give the firm cost advantages. Structural capitals include, for example, patents, franchises, proprietary processes and trademarks which enhance the firm's profits with products under its property rights. Often, physical and knowledge capitals are fused together in the firm's production and operation processes.

While trade is an essential operation of firms, trading activities have been considered to be a major channel of technology spillover between trading partners (see Maskus 2004; Coe and Helpman, 1995). Lichtenberg and Pottelsberghe (1998) showed that the more open the country is to trade, the more likely the country is to benefit from foreign R&D through technology spillover. Parrado and Cian (2014) found significant effects of trade spillovers through the transmission mechanisms underlying imports of machinery and equipment. Madsen (2007) showed that knowledge has been transmitted internationally through trade activities. The increment of a firm's private capital can experience positive impacts from its trading partners' technology.

Worth noting is that the building up of the physical capital and the accumulation of knowledge capital involve certain periods of time or through a number of stages. The building of physical capitals, such as corporate headquarters, offices, production plants, laboratories, terminals, ports, specific manufacturing equipment, railway tracks, energy plants and warehouses, will take a considerable amount of time and cannot be realized instantaneous. The accumulation of knowledge capitals, such as the design of large scale computer networks, communication systems, the development of patents, proprietary processes, software and the establishment of company reputation and goodwill, are time consuming endeavours. Since the conversions of investments into physical and knowledge capitals have to take time, investments in the firm's private capitals can be viewed as durable strategies.

(ii) *Investments in Public Capital*

Firms are also facing public capitals which are non-excludable and non-rivalrous. Public infrastructure, pollution abatement schemes, decease control centres, highway systems, mass-transit systems, school systems, bridges, public sports facilities, airports, weather warning systems and the classic example of lighthouses are public physical capitals that could enhance the productivity and profitability of a firm. Scientific knowledge such as public information, TV and radio signals, internet, public health information dissemination, news and public communication networks are examples of public knowledge capitals.

Similar to the case of private capitals, the building up of public capitals involves certain period of time or through a number of stages. The building of public infrastructure, pollution abatement schemes, decease control centres, highway systems, mass-transit systems, school systems, bridges, public sports facilities, airports and weather warning systems take time and many of these projects may take years to complete. The diffusion of knowledge in science, public information, internet, public health information distribution and forming of public communication networks also take time.

Again, the conversion of investments into public capitals have to take time. Hence, investments in public capital can be viewed as durable strategies which impact public capital build up in a number of stages.

(iii) *Revenue Generating Investments*

Revenue generating investments target income producing assets that generate consistent and recurring returns. The returns can be in the form of dividends, interest payments or other cash distributions that are recurring over a certain time period to the investing firm. The advantages of investing in income producing assets include the possibility to reinvest frequently and relatively high predictability of the returns. There is a diversity of choices of income producing assets—bonds and debentures, syndicated mortgages, dividend stocks, annuities, loans, bank deposits, investment funds, P2P lending, real estate investment trusts, certificates of deposit and guaranteed investment certificates. In fact, revenue generating investments are among the safest incoming producing instruments in a firm's portfolio. They also appear in almost all firms' lists of investments. Revenue generating investments are durable strategies because they generate returns in multiple stages over time.

(iv) *Debt-financed Investments*

Debt financing investments occur when a firm raises money for working capital or capital expenditures by selling debt instruments to individuals and/or institutional investors. One way of handling the debt is through instalment debt. An instalment debt is a liability with a repayment schedule that spans over a certain period of time. However, it is generally repaid in equal periodic instalments that include interest and a portion of the principal. It is a rule rather than an exception that business firms would involve debt financing. Debt-financed investments are durable strategies that produce impacts on the firm in the future.

2 Dynamic Games of Durable Interactive Investments

In this section, we present an analytical framework for formulating a general class of dynamic games under durable interactive investments. Then, we characterize the game equilibrium solution.

2.1 Game Formulation

Consider a $n-$ firm $T-$ stage dynamic game in which firm (economic agent) $i \in N$ faces:

(i) a vector of private production capitals $x_k^i \in X^i \subset R^{m_i}$ which investments, $\overline{u}_k^{(\omega_1)i} \in \overline{U}^{(\omega_1)i} \subset R^{w_1^i}$, take ω_1 stages to build up into the firm's capitals. There is spillover through trade so that the private capitals of its trading partners can positively affect the increments of its private capitals;

(ii) a vector of public capitals $y_k \in Y \subset R^m$ which investments, $\overline{u}_k^{(\omega_2)i} \in \overline{U}^{(\omega_2)i} \subset R^{w_2^i}$, take ω_2 stages to be fully transformed into public capital stocks;

(iii) a set of revenue generating investments, $\overline{u}_k^{(\omega_3)i} \in \overline{U}^{(\omega_3)i} \subset R^{w_3^i}$, which yields an income stream over ω_3 stages and

(iv) a set of debt-financed investments, $\overline{u}_k^{(\omega_4)i} \in \overline{U}^{(\omega_4)i} \subset R^{w_4^i}$, which costs are paid by instalments in ω_4 stages.

The accumulation process of the private capital stocks of firm i is governed by the dynamics

$$x_{k+1}^i = x_k^i + \varepsilon_k^{|k|i}(\overline{u}_k^{(\omega_1)i}) + \sum_{\tau=k-\omega_1+1}^{k-1} \varepsilon_k^{|\tau|i}(\overline{u}_\tau^{(\omega_1)i}) - \lambda_k^i(x_k^i) + \sum_{j \in K(i)} \gamma_k^{(i)j}(x_k^j),$$
$$x_1^i = x_1^{i(0)}, \quad i \in N, \tag{2.1}$$

where $\varepsilon_k^{|k|i}(\overline{u}_k^{(\omega_1)i})$ are the increments in the private capitals of firm $i \in N$ in stage k brought about by $\overline{u}_k^{(\omega_1)i}$ in stage k, $\varepsilon_k^{|\tau|i}(\overline{u}_\tau^{(\omega_1)i})$ are the increments in firm i's private capitals in stage k brought about by firm i's private capital building investments $\overline{u}_\tau^{(\omega_1)i}$ in stage $\tau \in \{k-1, k-2, \cdots, k-\omega_1+1\}$, $\lambda_k^i(x_k^i)$ are the of depreciations of capitals x_k^i, $K(i)$ is the set of trading partners of firm i and $\sum_{j \in K(i)} \gamma_k^{(i)j}(x_k^j)$ are the technology spillover of firm i's trading partners to the increments of firm i's private capital stocks. The costs of investments in the private capitals are $\phi_k^{(x)i}(\overline{u}_\tau^{(\omega_2)i})$.

The accumulation process of the public capital accessible to all firms is

$$y_{k+1} = y_k + \sum_{j=1}^{n} a_k^{|k|j}(\overline{u}_k^{(\omega_2)j}) + \sum_{j=1}^{n} \sum_{\tau=k-\omega_2+1}^{k-1} a_k^{|\tau|j}(\overline{u}_\tau^{(\omega_2)j}) - \lambda_k(y_k), \quad y_1 = y_1^0, \tag{2.2}$$

where $a_k^{|k|j}(\overline{u}_k^{(\omega_2)j})$ are the additions to public capitals in stage k brought about by firm j's capital building investments $\overline{u}_k^{(\omega_2)j}$ in stage k, $a_k^{|\tau|j}(\overline{u}_\tau^{(\omega_2)j})$ are the additions to public capitals in stage k by firm j's investments $\overline{u}_\tau^{(\omega_2)j}$ in stage $\tau \in \{k-1, k-2, \cdots, k-\omega_2+1\}$ and $\lambda_k(y_k)$ is the obsolescence of the public capitals in stage k. The costs of investments in the public capitals are $\phi_k^{(y)i}(\overline{u}_\tau^{(\omega_2)i})$.

Firm $i \in N$ uses its private and public capital together with non-durable productive inputs u_k^{Ii} and u_k^{IIi} to produce outputs. The gross revenue from the outputs produced is $R_k^i[u_k^{Ii}, u_k^{IIi}, x_k^i, y_k]$. The non-durable inputs u_k^{Ii} are used for production related to the private capitals and the non-durable inputs u_k^{IIi} are used for production related to the public capitals. The input costs of u_k^{Ii} are $c_k^{Ii}(u_k^{Ii})$ and the input costs of u_k^{IIi} are $c_k^{IIi}(u_k^{IIi})$.

Firm i's revenue generating investments $\overline{u}_k^{(\omega_3)i}$ yield incomes in ω_3 stages. The income generated by $\overline{u}_k^{(\omega_3)i}$ in stage $\tau \in \{k, k+1, \cdots, k+\omega_3-1\}$ is $p_\tau^{|k|i}(\overline{u}_k^{(\omega_3)i})$ and the costs of the revenue generating investments are $\varphi_k^i(\overline{u}_k^{(\omega_3)i})$.

Finally, firm i's debt-financed investments $\overline{u}_k^{(\omega_4)i}$ generate a gain $\chi_k^i(\overline{u}_k^{(\omega_4)i})$ in stage k and the project costs are paid in ω_4 instalments—with a cost payment $q_\tau^{|k|i}(\overline{u}_k^{(\omega_4)i})$ in stage $\tau \in \{k, k+1, \cdots, k+\omega_4-1\}$.

The payoff of firm $i \in N$ is

$$\sum_{k=1}^{T}\Bigg(R_k^i[u_k^{Ii}, u_k^{IIi}, x_k^i, y_k] - c_k^{Ii}(u_k^{Ii}) - c_k^{IIi}(u_k^{IIi}) - \phi_k^{(x)i}(\overline{u}_\tau^{(\omega_2)i}) - \phi_k^{(y)i}(\overline{u}_\tau^{(\omega_2)i})$$
$$+\left[\sum_{\tau=k-\omega_3+1}^{k} p_k^{|\tau|i}(\overline{u}_k^{(\omega_3)i}) - \varphi_k^i(\overline{u}_k^{(\omega_3)i})\right] + \left[\chi_k^i(\overline{u}_k^{(\omega_4)i}) - \sum_{\tau=k-\omega_4+1}^{k} q_k^{|\tau|i}(\overline{u}_\tau^{(\omega_4)i})\right]\Bigg)\delta_1^k$$
$$+\left(Q_{T+1}^{(x)i}(x_{T+1}^i) + Q_{T+1}^{(y)i}(y_{T+1}) + \sum_{\tau=T+1-\omega_3+1}^{T} p_{T+1}^{|\tau|i}(\overline{u}_\tau^{(\omega_3)i})\right.$$
$$\left. - \sum_{\tau=T+1-\omega_4+1}^{T} q_{T+1}^{|\tau|i}(\overline{u}_\tau^{(\omega_4)i}) + \varpi_{T+1}^i\right)\delta_1^{T+1}, \tag{2.3}$$

where $\left(Q_{T+1}^{(x)i}(x_{T+1}^i) + Q_{T+1}^{(y)i}(y_{T+1}) + \sum_{\tau=T+1-\omega_3+1}^{T} p_{T+1}^{|\tau|i}(\overline{u}_\tau^{(\omega_3)i}) - \sum_{\tau=T+1-\omega_4+1}^{T} q_{T+1}^{|\tau|i}(\overline{u}_\tau^{(\omega_4)i}) + \varpi_{T+1}^i\right)$ is the value of firm i's in stage $T+1$ and $\delta = (1+r)^{-1}$ is the discount factor.

In particular, $Q_{T+1}^{(x)i}(x_{T+1}^i)$ is the value of the private capital x_{T+1}^i. The term $Q_{T+1}^{(y)i}(y_{T+1})$ is the value of the public capital y_{T+1} of firm i. The term $\sum_{\tau=T+1-\omega_3+1}^{T} p_{T+1}^{|\tau|i}(\overline{u}_\tau^{(\omega_3)i})$ is the value of the returns from previously procured revenue generating investments. Finally, $\sum_{\tau=T+1-\omega_4+1}^{T} q_{T+1}^{|\tau|i}(\overline{u}_\tau^{(\omega_4)i})$ are the payment obligations of previously engaged debt-financed investments.

The value of T can is arbitrarily large. If the firms plan to re-evaluate its operation in stage $T+1$ and then continue their operation from stage $T+1$ onwards, the term

$$\left(Q_{T+1}^{(x)i}(x_{T+1}^{i}) + Q_{T+1}^{(y)i}(y_{T+1}) + \sum_{\tau=T+1-\omega_3+1}^{T} p_{T+1}^{|\tau|i}(\overline{u}_{\tau}^{(\omega_3)i}) - \sum_{\tau=T+1-\omega_4+1}^{T} q_{T+1}^{|\tau|i}(\overline{u}_{\tau}^{(\omega_4)i}) + \varpi_{T+1}^{i} \right)$$

is the known value of the firm at stage $T+1$ based on which another planning scheme from stage $T+1$ onwards would be made.

If the firms expect to terminate their operation after stage T, the term

$$\left(Q_{T+1}^{(x)i}(x_{T+1}^{i}) + Q_{T+1}^{(y)i}(y_{T+1}) + \sum_{\tau=T+1-\omega_3+1}^{T} p_{T+1}^{|\tau|i}(\overline{u}_{\tau}^{(\omega_3)i}) - \sum_{\tau=T+1-\omega_4+1}^{T} q_{T+1}^{|\tau|i}(\overline{u}_{\tau}^{(\omega_4)i}) + \varpi_{T+1}^{i} \right)$$

is the realizable salvage value of the firm in stage $T+1$.

2.2 Game Equilibrium

We use x_k to stand for $(x_k^1, x_k^2, \cdots, x_k^n)$ and we use $\underline{u}_k^{I}, \underline{u}_k^{II}, \underline{\overline{u}}_k^{(\omega_1)}, \underline{\overline{u}}_k^{(\omega_2)}, \underline{\overline{u}}_k^{(\omega_3)}, \underline{\overline{u}}_k^{(\omega_4)}$ to denote the different sets of strategies of all the players. Invoking Theorem 3.1 in Chap. 2 we characterize the equilibrium of the game (2.1–2.3) in the following corollary.

Corollary 2.1 *Let $(\underline{u}_k^{I**}, \underline{u}_k^{II**}, \underline{\overline{u}}_k^{(\omega_1)**}, \underline{\overline{u}}_k^{(\omega_2)**}, \underline{\overline{u}}_k^{(\omega_3)**}, \underline{\overline{u}}_k^{(\omega_4)**})$ be the set of feedback Nash equilibrium strategies and $V^i(k, x, y; \underline{\overline{u}}_{k-}^{(\omega_1)**}, \underline{\overline{u}}_{k-}^{(\omega_2)**}, \underline{\overline{u}}_{k-}^{(\omega_3)**}, \underline{\overline{u}}_{k-}^{(\omega_4)**})$ be the feedback Nash equilibrium payoff of firm i at stage k in the durable-strategies non-cooperative dynamic game(2.1–2.3), the functions $V^i(k, x, y; \underline{\overline{u}}_{k-}^{(\omega_1)**}, \underline{\overline{u}}_{k-}^{(\omega_2)**}, \underline{\overline{u}}_{k-}^{(\omega_3)**}, \underline{\overline{u}}_{k-}^{(\omega_4)**})$, for $k \in \{1, 2, \cdots, T+1\}$ and $i \in N$, satisfy the following recursive equations:*

$$V^i(T+1, x, y; \underline{\overline{u}}_{(T+1)-}^{(\omega_1)**}, \underline{\overline{u}}_{(T+1)-}^{(\omega_2)**}, \underline{\overline{u}}_{(T+1)-}^{(\omega_3)**}, \underline{\overline{u}}_{(T+1)-}^{(\omega_4)**})$$
$$= \left(Q_{T+1}^{(x)i}(x^{i}) + Q_{T+1}^{(y)i}(y) + \sum_{\tau=T+1-\omega_3+1}^{T} p_{T+1}^{|\tau|i}(\overline{u}_{\tau}^{(\omega_3)i}) - \sum_{\tau=T+1-\omega_4+1}^{T} q_{T+1}^{|\tau|i}(\overline{u}_{\tau}^{(\omega_4)i}) + \varpi_{T+1}^{i} \right) \delta_1^{T+1}; \quad (2.4)$$

$$V^i(k,x,y;\underline{\overline{u}}_{k-}^{(\omega_1)**},\underline{\overline{u}}_{k-}^{(\omega_2)**},\underline{\overline{u}}_{k-}^{(\omega_3)**},\underline{\overline{u}}_{k-}^{(\omega_4)**})$$

$$= \max_{\substack{u_k^{Ii},u_k^{IIi},\overline{u}_k^{(\omega_1)i}\\ \overline{u}_k^{(\omega_2)i},\overline{u}_k^{(\omega_3)i},\overline{u}_k^{(\omega_4)i}}} \Big\{ \Big(R_k^i[u_k^{Ii},u_k^{IIi},x^i,y] - c_k^{Ii}(u_k^{Ii})$$

$$- c_k^{IIi}(u_k^{IIi}) - c_k^{(x)i}(\overline{u}_\tau^{(\omega_2)i}) - \phi_k^{(y)i}(\overline{u}_\tau^{(\omega_2)i})$$

$$+ \left[\sum_{\tau=k-\omega_3+1}^{k} p_k^{|\tau|i}(\overline{u}_k^{(\omega_3)i}) - \varphi_k^i(\overline{u}_k^{(\omega_3)i}) \right] + [\chi_k^i(\overline{u}_k^{(\omega_4)i})$$

$$- \sum_{\tau=k-\omega_4+1}^{k} q_k^{|\tau|i}(\overline{u}_\tau^{(\omega_4)i})] \Big) \delta_1^k$$

$$+ V^i(k+1, x_{k+1}^{***}, y_{k+1}^{***}; \underline{\overline{u}}_{(k+1)-}^{(\omega_1)***}, \underline{\overline{u}}_{(k+1)-}^{(\omega_2)***}, \underline{\overline{u}}_{(k+1)-}^{(\omega_3)***}, \underline{\overline{u}}_{(k+1)-}^{(\omega_4)***}) \Big\}, \qquad (2.5)$$

for $k \in \{1, 2, \cdots, T\}$ and $i \in N$,
where x_{k+1}^{***} is a vector with.

$$x_{k+1}^{***i} = x^i + \varepsilon_k^{|k|i}(\bar{u}_k^{(\omega_1)i}) + \sum_{\tau=k-\omega_1+1}^{k-1} \varepsilon_k^{|\tau|i}(\bar{u}_\tau^{(\omega_1)i**}) - \lambda_k^{(i)}(x^i) + \sum_{j\in K(i)} \gamma_k^{(i)j}(x^j), \text{ and}$$

$$x_{k+1}^{***\ell} == x^\ell + \varepsilon_k^{|k|\ell}(\bar{u}_k^{(\omega_1)\ell**}) + \sum_{\tau=k-\omega_1+1}^{k-1} \varepsilon_k^{|\tau|\ell}(\vec{u}_\tau^{\ell**}) - \lambda_k^{(\ell)}(x^\ell) + \sum_{j\in K(\ell)} \gamma_k^{(\ell)j}(x^j),$$

for $\ell \in N$ and $\ell \neq i$;

$$y_{k+1}^{***} = y_k + a_k^{|k|i}(\bar{u}_k^{(\omega_2)i}) + \sum\nolimits_{\substack{j=1\\ j\neq i}}^{n} a_k^{|k|j}(\bar{u}_k^{(\omega_2)j**}) + \sum\nolimits_{j=1}^{n} \sum\nolimits_{\tau=k-\omega_2+1}^{k-1} a_k^{|\tau|j}(\bar{u}_\tau^{(\omega_2)j**}) - \lambda_k(y),$$

$$\underline{\bar{u}}_{(k+1)-}^{(\omega_1)***} = (\bar{u}_k^{(\omega_1)i}, \bar{u}_k^{(\omega_1)(\neq i)**}, \underline{\bar{u}}_{(k+1)-}^{(\omega_1)**} \cap \underline{\bar{u}}_{k-}^{(\omega_1)**}),$$

$$\underline{\bar{u}}_{(k+1)-}^{(\omega_2)***} = (\bar{u}_k^{(\omega_2)i}, \bar{u}_k^{(\omega_2)(\neq i)**}, \underline{\bar{u}}_{(k+1)-}^{(\omega_2)**} \cap \underline{\bar{u}}_{k-}^{(\omega_2)**}),$$

$$\underline{\bar{u}}_{(k+1)-}^{(\omega_3)***} = (\bar{u}_k^{(\omega_3)i}, \bar{u}_k^{(\omega_3)(\neq i)**}, \underline{\bar{u}}_{(k+1)-}^{(\omega_3)**} \cap \underline{\bar{u}}_{k-}^{(\omega_3)**}) \text{ and}$$

$$\underline{\overline{u}}_{(k+1)-}^{(\omega_4)***} = \left(\overline{u}_k^{(\omega_4)i}, \overline{u}_k^{(\omega_4)(\neq i)**}, \underline{\overline{u}}_{(k+1)-}^{(\omega_4)**} \cap \underline{\overline{u}}_{k-}^{(\omega_4)**}\right).$$

■

Performing the indicated maximization operator in Corollary 2.1, we obtain the following game equilibrium conditions.

(i) The non-durable inputs for production related to the private capitals are employed up to the levels where the marginal revenues equal the marginal costs, that is

$$\frac{\partial}{\partial u_k^{Ii}} R_k^i[u_k^{Ii}, u_k^{IIi}, x^i, y] = \frac{\partial}{\partial u_k^{Ii}} c_k^{Ii}(u_k^{Ii}), \quad \text{for } i \in N, \qquad (2.6)$$

(ii) The non-durable inputs for production related to the public capitals are employed up to the levels where the marginal revenues equal the marginal costs, that is

$$\frac{\partial}{\partial u_k^{IIi}} R_k^i[u_k^{Ii}, u_k^{IIi}, x^i, y] = \frac{\partial}{\partial u_k^{IIi}} c_k^{IIi}(u_k^{IIi}), \quad \text{for } i \in N. \tag{2.7}$$

(iii) Investments in private capitals by firm i in stage k will generate benefits to the increment of the capitals in stage $k+1$ plus the benefits in capital accumulation subsequently in stages $k+2$ to $k+\omega_1-1$. The marginal benefits of investments in private capital can be expressed as

$$\frac{\partial V_{k+1}^i}{\partial x_{k+1}^{***i}} \frac{\partial x_{k+1}^{***i}}{\partial \overline{u}_k^{(\omega_1)i}} + \frac{\partial V_{k+1}^i}{\partial \overline{u}_k^{(\omega_1)i}}, \quad \text{for } i \in N,$$

where V_{k+1}^i is the short form of

$$V^i(k+1, x_{k+1}^{***}, y_{k+1}^{***}; \underline{\overline{u}}_{(k+1)-}^{(\omega_1)***}, \underline{\overline{u}}_{(k+1)-}^{(\omega_2)***}, \underline{\overline{u}}_{(k+1)-}^{(\omega_3)***}, \underline{\overline{u}}_{(k+1)-}^{(\omega_4)***}).$$

The marginal costs of investments in private capitals can be expressed as

$$\delta_1^k \frac{\partial \phi_k^{(x)i}(\overline{u}_k^{(\omega_1)i})}{\partial \overline{u}_k^{(\omega_1)i}}, \quad \text{for } i \in N.$$

In a game equilibrium, the marginal benefits of investments equal the marginal costs of investments in private capitals, that is

$$\frac{\partial V_{k+1}^i}{\partial x_{k+1}^{***i}} \frac{\partial x_{k+1}^{***i}}{\partial \overline{u}_k^{(\omega_1)i}} + \frac{\partial V_{k+1}^i}{\partial \overline{u}_k^{(\omega_1)i}} = \delta_1^k \frac{\partial \phi_k^{(x)i}(\overline{u}_k^{(\omega_1)i})}{\partial \overline{u}_k^{(\omega_1)i}}, \quad \text{for } i \in N. \tag{2.8}$$

(iv) Investments in public capitals by firm i in stage k will generate benefits to the increment of the capitals in stage $k+1$ plus the benefits in capital accumulation subsequently in stages $k+2$ to $k+\omega_2-1$. The marginal benefits of investments in public capital can be expressed as

$$\frac{\partial V_{k+1}^i}{\partial y_{k+1}^{***}} \frac{\partial y_{k+1}^{***}}{\partial \overline{u}_k^{(\omega_2)i}} + \frac{\partial V_{k+1}^i}{\partial \overline{u}_k^{(\omega_2)i}}, \quad \text{for } i \in N.$$

The marginal costs of investments in public capitals can be expressed as

$$\delta_1^k \frac{\partial \phi_k^{(y)i}(\overline{u}_k^{(\omega_2)i})}{\partial \overline{u}_k^{(\omega_2)i}}, \quad \text{for } i \in N.$$

In a game equilibrium, the marginal benefits of investments equal the marginal costs of investments in public capitals, that is

$$\frac{\partial V_{k+1}^i}{\partial y_{k+1}^{***}} \frac{\partial y_{k+1}^{***}}{\partial \overline{u}_k^{(\omega_2)i}} + \frac{\partial V_{k+1}^i}{\partial \overline{u}_k^{(\omega_2)i}} = \delta_1^k \frac{\partial \phi_k^{(y)i}(\overline{u}_k^{(\omega_2)i})}{\partial \overline{u}_k^{(\omega_2)i}}, \quad \text{for } i \in N. \tag{2.9}$$

(v) The marginal gains from the revenue generating investments over stage k to stage $k + \omega_3 - 1$ can be obtained as:

$$\frac{\partial p_k^{|k|i}(\overline{u}_k^{(\omega_3)i})}{\partial \overline{u}_k^{(\omega_3)i}} \delta_1^k + \frac{\partial V_{k+1}^i}{\partial \overline{u}_k^{(\omega_3)i}}, \quad \text{for } i \in N.$$

The marginal costs of the revenue generating investments are

$$\frac{\partial \varphi_k^i(\overline{u}_k^{(\omega_3)i})}{\partial \overline{u}_k^{(\omega_3)i}} \delta_1^k, \quad i \in N.$$

In a game equilibrium, the marginal gains from the revenue generating investments equal the marginal costs, that is

$$\frac{\partial p_k^{|k|i}(\overline{u}_k^{(\omega_3)i})}{\partial \overline{u}_k^{(\omega_3)i}} \delta_1^k + \frac{\partial V_{k+1}^i}{\partial \overline{u}_k^{(\omega_3)i}} = \frac{\partial \phi_k^i(\overline{u}_k^{(\omega_3)i})}{\partial \overline{u}_k^{(\omega_3)i}} \delta_1^k, \quad \text{for } i \in N. \tag{2.10}$$

(vi) The marginal gains from the debt-financed investments are

$$\frac{\partial \chi_k^i(\overline{u}_k^{(\omega_4)i})}{\partial \overline{u}_k^{(\omega_4)i}} \delta_1^k, \quad \text{for } i \in N.$$

The marginal costs of the debt-financed investments (through installed payments from stage k to stage $k + \omega_4 - 1$) is reflected by

$$\frac{\partial q_k^{|k|i}(\overline{u}_k^{(\omega_4)i})}{\partial \overline{u}_k^{(\omega_4)i}} \delta_1^k + \frac{\partial V_{k+1}^i}{\partial \overline{u}_k^{(\omega_4)i}}, \quad \text{for } i \in N.$$

In a game equilibrium, the marginal gains from debt-financed investments equal the investments' marginal costs, that is

$$\frac{\partial \chi_k^i(\overline{u}_k^{(\omega_4)i})}{\partial \overline{u}_k^{(\omega_4)i}}\delta_1^k = \frac{\partial q_k^{|k|i}(\overline{u}_k^{(\omega_4)i})}{\partial \overline{u}_k^{(\omega_4)i}}\delta_1^k + \frac{\partial V_{k+1}^i}{\partial \overline{u}_k^{(\omega_4)i}}, \quad \text{for } i \in N. \tag{2.11}$$

If the game equilibrium conditions (2.6–2.11) satisfy the implicit function theorem, the game equilibrium strategies $(u_k^{I**}, u_k^{II**}, \overline{u}_k^{(\omega_1)**}, \overline{u}_k^{(\omega_2)**}, \overline{u}_k^{(\omega_3)**}, \overline{u}_k^{(\omega_4)**})$, for $i \in N$, can be obtained as explicit functions of $(x, y; \overline{u}_{k-}^{(\omega_1)**}, \overline{u}_{k-}^{(\omega_2)**}, \overline{u}_{k-}^{(\omega_3)**}, \overline{u}_{k-}^{(\omega_4)**})$. Substituting these functions into the stage k equations of (2.5), we can obtain $V^i(k, x, y; \overline{u}_{k-}^{(\omega_1)**}, \overline{u}_{k-}^{(\omega_2)**}, \overline{u}_{k-}^{(\omega_3)**}, \overline{u}_{k-}^{(\omega_4)**})$, for $i \in N$ and $k \in \{1, 2, \cdots, T\}$. In particular, $V^i(k, x, y; \overline{u}_{k-}^{(\omega_1)**}, \overline{u}_{k-}^{(\omega_2)**}, \overline{u}_{k-}^{(\omega_3)**}, \overline{u}_{k-}^{(\omega_4)**})$ reflects the value of firm i under durable interactive investments.

3 A Game Example

In this section, we consider a representative dynamic game of firms under interactive durable investments with explicit functional forms as a computational illustration.

3.1 *Formulation of the Game*

In the $n-$firm, $T-$stage dynamic game, firm $i \in N$ faces:

(i) a private production capital $x_k^i \in X^i \subset R^+$ which investment, $\overline{u}_k^{(\omega_1)i} \in \overline{U}^{(\omega_1)i} \subset R$, takes ω_1 stages to build up into the firm's capitals. There is spillover through trade so that the private capitals of its trading partners can positively affect the increments of its private capital;

(ii) a public capital $y_k \in Y \subset R^+$ which investment, $\overline{u}_k^{(\omega_2)i} \in \overline{U}^{(\omega_2)i} \subset R$, takes ω_2 stages to be fully transformed into the public capital stock;

(iii) a revenue generating investment, $\overline{u}_k^{(\omega_3)i} \in \overline{U}^{(\omega_3)i} \subset R$, which yields an income stream over ω_3 stages and

(iv) a set of debt-financed investments, $\overline{u}_k^{(\omega_4)i} \in \overline{U}^{(\omega_4)i} \subset R$, which costs are paid by instalments in ω_4 stages.

The accumulation process of the private capital stock of firm i is governed by the dynamics

$$x_{k+1}^i = x_k^i + \varepsilon_k^{|k|i}\overline{u}_k^{(\omega_1)i} + \sum_{\tau=k-\omega_1+1}^{k-1} \varepsilon_k^{|\tau|i}\overline{u}_\tau^{(\omega_1)i} - \lambda_k^i x_k^i + \sum_{j \in K(i)} \gamma_k^{(i)j} x_k^j, \quad x_1^i = x_1^{i(0)},$$
$$i \in N, \tag{3.1}$$

where $\varepsilon_k^{|k|i}\overline{u}_k^{(\omega_1)i}$ is the increments in the private capital of firm $i \in N$ in stage k brought about by $u_k^{(\omega_1)i}$ in stage k, $\varepsilon_k^{|\tau|i}\overline{u}_\tau^{(\omega_1)i}$ is the increments in firm i's private capitals in stage k brought about by firm i's private capital building investments $\overline{u}_\tau^{(\omega_1)i}$ in stage $\tau \in \{k-1, k-2, \cdots, k-\omega_1+1\}$, $\lambda_k^i x_k^i$ is the of depreciation of capital x_k^i, $K(i)$ is the set of trading partners of firm i and $\sum_{j\in K(i)} \gamma_k^{(i)j} x_k^j$ are the technology spillover of firm i's trading partners to the increments of firm i's private capital stocks. k. The costs of investments in the private capitals are $\varphi_k^{(x)i}(\overline{u}_\tau^{(\omega_2)i})^2$. The terms $\varepsilon_k^{|\tau|i}$, λ_k^i, $\gamma_k^{(i)j}$ and $\varphi_k^{(x)i}$ are positive parameters.

The accumulation process of the public capital accessible to all firms is

$$y_{k+1} = y_k + \sum_{j=1}^{n} a_k^{|k|j}\overline{u}_k^{(\omega_2)j} + \sum_{j=1}^{n}\sum_{\tau=k-\omega_2+1}^{k-1} a_k^{|\tau|j}\overline{u}_\tau^{(\omega_2)j} - \lambda_k y_k, \quad y_1 = y_1^0, \tag{3.2}$$

where $a_k^{|k|j}\overline{u}_k^{(\omega_2)j}$ is the addition to the public capital in stage k brought about by firm j's capital building investments $\overline{u}_k^{(\omega_2)j}$ in stage k, $a_k^{|\tau|j}\overline{u}_\tau^{(\omega_2)j}$ is the addition to the public capital in stage k by firm j's investments $\overline{u}_\tau^{(\omega_2)j}$ in stage $\tau \in \{k-1, k-2, \cdots, k-\omega_2+1\}$ and $\lambda_k y_k$ is the obsolescence of the public capitals in stage k. The costs of investments in the public capitals are $\varphi_k^{(y)i}(\overline{u}_\tau^{(\omega_2)i})^2$. The terms $a_k^{|\tau|i}$, λ_k and $\varphi_k^{(y)i}$ are positive parameters.

Firm $i \in N$ uses its private and public capital together with non-durable productive inputs u_k^{Ii} and u_k^{IIi} to produce outputs. The gross revenue from the outputs produced is $R_k^{Ii}(u_k^{Ii}x_k^i)^{1/2} + R_k^{IIi}(u_k^{IIi}y_k)^{1/2}$. The input cost of u_k^{Ii} is $c_k^{Ii}u_k^{Ii}$ and the input cost of u_k^{IIi} is $c_k^{IIi}u_k^{IIi}$. The terms R_k^{Ii}, R_k^{IIi}, c_k^{Ii} and c_k^{IIi} are positive parameters.

Firm i's revenue generating investment $u_k^{(\omega_3)i}$ yields incomes in ω_3 stages. The income generated in stage $\tau \in \{k, k+1, k+2, \cdots, k+\omega_3-1\}$ is $p_\tau^{|k|i}\overline{u}_k^{(\omega_3)i}$ and the cost of the revenue generating investments is $\phi_k^i(\overline{u}_k^{(\omega_3)i})^2$. The terms $p_\tau^{|k|i}$ and ϕ_k^i are positive parameters.

Finally, firm i's debt-financed investment $\overline{u}_k^{(\omega_4)i}$ generates a gain $\chi_k^i(\overline{u}_k^{(\omega_4)i})^{1/2}$ in stage k and the project costs are paid in ω_4 instalments—with a cost payment $q_\tau^{|k|i}\overline{u}_k^{(\omega_4)i}$ in stage $\tau \in \{k, k+1, k+2, \cdots, k+\omega_4-1\}$. The terms χ_k^i and $q_\tau^{|k|i}$ are positive parameters.

The payoff of firm $i \in N$ is

$$\sum_{k=1}^{T}\Big(R_k^{Ii}(u_k^{Ii}x_k^i)^{1/2} + R_k^{IIi}(u_k^{IIi}y_k)^{1/2} - c_k^{Ii}u_k^{Ii} - c_k^{IIi}u_k^{IIi} - \phi_k^{(x)i}(\overline{u}_\tau^{(\omega_2)i})^2$$
$$- \phi_k^{(y)i}(\overline{u}_\tau^{(\omega_2)i})^2 + \left[\sum_{\tau=k-\omega_3+1}^{k} p_k^{|\tau|i}\overline{u}_k^{(\omega_3)i} - \varphi_k^i(\overline{u}_k^{(\omega_3)i})^2\right]$$
$$+ \left[\chi_k^i(\overline{u}_k^{(\omega_4)i})^{1/2} - \sum_{\tau=k-\omega_4+1}^{k} q_k^{|\tau|i}\overline{u}_\tau^{(\omega_4)i}\right]\Big)\delta^{k-1}$$

$$+\left(Q_{T+1}^{(x)i}x_{T+1}^i+Q_{T+1}^{(y)i}y_{T+1}+\sum_{\tau=T+1-\omega_3+1}^{T}p_{T+1}^{|\tau|i}\overline{u}_\tau^{(\omega_3)i}\right.$$
$$\left.-\sum_{\tau=T+1-\omega_4+1}^{T}q_{T+1}^{|\tau|i}\overline{u}_\tau^{(\omega_4)i}+\varpi_{T+1}^i\right)\delta^T, \tag{3.3}$$

where $\delta=(1+r)^{-1}$ is the discount factor and $Q_{T+1}^{(x)i}$ and $Q_{T+1}^{(y)i}$ are positive parameters.

3.2 Game Equilibrium Outcome

Invoking Corollary 2.1, we can characterize the game equilibrium of the game in (3.1–3.3) as follows.

Corollary 3.1 *Let* $(u_k^{I**},u_k^{II**},\underline{\overline{u}}_k^{(\omega_1)**},\underline{\overline{u}}_k^{(\omega_2)**},\underline{\overline{u}}_k^{(\omega_3)**},\underline{\overline{u}}_k^{(\omega_4)**})$ *be the set of feedback Nash equilibrium strategies and* $V^i(k,x,y;\underline{\overline{u}}_{k-}^{(\omega_1)**},\underline{\overline{u}}_{k-}^{(\omega_2)**},\underline{\overline{u}}_{k-}^{(\omega_3)**},\underline{\overline{u}}_{k-}^{(\omega_4)**})$ *be the feedback Nash equilibrium payoff of firm i at stage k in the durable-strategies non-cooperative dynamic game (3.1–3.3), the functions* $V^i(k,x,y;\underline{\overline{u}}_{k-}^{(\omega_1)**},\underline{\overline{u}}_{k-}^{(\omega_2)**},\underline{\overline{u}}_{k-}^{(\omega_3)**},\underline{\overline{u}}_{k-}^{(\omega_4)**})$*, for* $k\in\{1,2,\cdots,T+1\}$ *and* $i\in N$*, satisfy the following recursive equations:*

$$V^i(T+1,x,y;\underline{\overline{u}}_{(T+1)-}^{(\omega_1)**},\underline{\overline{u}}_{(T+1)-}^{(\omega_2)**},\underline{\overline{u}}_{(T+1)-}^{(\omega_3)**},\underline{\overline{u}}_{(T+1)-}^{(\omega_4)**})$$
$$=\left(Q_{T+1}^{(x)i}x^i+Q_{T+1}^{(y)i}y+\sum_{\tau=T+1-\omega_3+1}^{T}p_{T+1}^{|\tau|i}\overline{u}_\tau^{(\omega_3)i}\right.$$
$$\left.-\sum_{\tau=T+1-\omega_4+1}^{T}q_{T+1}^{|\tau|i}\overline{u}_\tau^{(\omega_4)i}+\varpi_{T+1}^i\right)\delta_1^{T+1}; \tag{3.4}$$

$$V^i(k,x,y;\underline{\overline{u}}_{k-}^{(\omega_1)**},\underline{\overline{u}}_{k-}^{(\omega_2)**},\underline{\overline{u}}_{k-}^{(\omega_3)**},\underline{\overline{u}}_{k-}^{(\omega_4)**})$$
$$=\max_{\substack{u_k^{Ii},u_k^{IIi},\overline{u}_k^{(\omega_1)i}\\ \overline{u}_k^{(\omega_2)i},\overline{u}_k^{(\omega_3)i},\overline{u}_k^{(\omega_4)i}}}\left\{\left(R_k^{Ii}(u_k^{Ii}x_k^i)^{1/2}\right.\right.$$
$$+R_k^{IIi}(u_k^{IIi}y_k)^{1/2}-c_k^{Ii}u_k^{Ii}-c_k^{IIi}u_k^{IIi}-\varphi_k^{(x)i}(\overline{u}_\tau^{(\omega_2)i})^2-\varphi_k^{(y)i}(\overline{u}_\tau^{(\omega_2)i})^2$$
$$+\left[\sum_{\tau=k-\omega_3+1}^{k}p_k^{|\tau|i}\overline{u}_k^{(\omega_3)i}-\phi_k^i(\overline{u}_k^{(\omega_3)i})^2\right]$$
$$\left.+\left[\chi_k^i(\overline{u}_k^{(\omega_4)i})^{1/2}-\sum_{\tau=k-\omega_4+1}^{k}q_k^{|\tau|i}\overline{u}_\tau^{(\omega_4)i}\right]\right)\delta^{k-1}$$
$$\left.+V^i(k+1,x_{k+1}^{***},y_{k+1}^{***};\underline{\overline{u}}_{(k+1)-}^{(\omega_1)***},\underline{\overline{u}}_{(k+1)-}^{(\omega_2)***},\underline{\overline{u}}_{(k+1)-}^{(\omega_3)***},\underline{\overline{u}}_{(k+1)-}^{(\omega_4)***})\right\}, \tag{3.5}$$

for $k \in \{1, 2, \cdots, T\}$ *and* $i \in N$,
where x_{k+1}^{***} *is a vector with*

$$x_{k+1}^{***i} = x^i + \varepsilon_k^{|k|i}\bar{u}_k^{(\omega_1)i} + \sum_{\tau=k-\omega_1+1}^{k-1} \varepsilon_k^{|\tau|i}\bar{u}_\tau^{(\omega_1)i**} - \lambda_k^{(i)}x^i + \sum_{j \in K(i)} \gamma_k^{(i)j}x^j, \text{ and}$$

$$x_{k+1}^{***\ell} = x^\ell + \varepsilon_k^{|k|\ell}\bar{u}_k^{(\omega_1)\ell**} + \sum_{\tau=k-\omega_1+1}^{k-1} \varepsilon_k^{|\tau|\ell}\bar{u}_\tau^{\ell**} - \lambda_k^{(\ell)}x^\ell + \sum_{j \in K(\ell)} \gamma_k^{(\ell)j}x^j,$$

for $\ell \in N$ *and* $\ell \neq i$;

$$y_{k+1}^{***} = y_k + a_k^{|k|i}(\bar{u}_k^{(\omega_2)i}) + \sum_{\substack{j=1 \\ j \neq i}}^{n} a_k^{|k|j}(\bar{u}_k^{(\omega_2)j**}) + \sum_{j=1}^{n}\sum_{\tau=k-\omega_2+1}^{k-1} a_k^{|\tau|j}(\bar{u}_\tau^{(\omega_2)j**}) - \lambda_k(y),$$

$$\underline{\bar{u}}_{(k+1)-}^{(\omega_1)***} = (\bar{u}_k^{(\omega_1)i}, \bar{u}_k^{(\omega_1)(\neq i)**}, \underline{\bar{u}}_{(k+1)-}^{(\omega_1)**} \cap \underline{\bar{u}}_{k-}^{(\omega_1)**}),$$
$$\underline{\bar{u}}_{(k+1)-}^{(\omega_2)***} = (\bar{u}_k^{(\omega_2)i}, \bar{u}_k^{(\omega_2)(\neq i)**}, \underline{\bar{u}}_{(k+1)-}^{(\omega_2)**} \cap \underline{\bar{u}}_{k-}^{(\omega_2)**}),$$
$$\underline{\bar{u}}_{(k+1)-}^{(\omega_3)***} = (\bar{u}_k^{(\omega_3)i}, \bar{u}_k^{(\omega_3)(\neq i)**}, \underline{\bar{u}}_{(k+1)-}^{(\omega_3)**} \cap \underline{\bar{u}}_{k-}^{(\omega_3)**}) \text{ and}$$
$$\underline{\overline{u}}_{(k+1)-}^{(\omega_4)***} = \left(\overline{u}_k^{(\omega_4)i}, \overline{u}_k^{(\omega_4)(\neq i)**}, \underline{\overline{u}}_{(k+1)-}^{(\omega_4)**} \cap \underline{\overline{u}}_{k-}^{(\omega_4)**}\right).$$

■

Performing the indicated maximization operator in Corollary 3.1, we obtain the following game equilibrium conditions.

(i) Equating the marginal revenue with the marginal cost of the non-durable input for production related to the private capital yields the game equilibrium level of the input as

$$u_k^{Ii} = \frac{(R_k^{Ii})^2 x^i}{4(c_k^{Ii})^2}, \quad \text{for } i \in N. \tag{3.6}$$

(ii) Equating the marginal revenue with the marginal cost of the non-durable input for production related to the public capital yields the game equilibrium level of the input as

$$u_k^{IIi} = \frac{(R_k^{IIi})^2 y}{4(c_k^{IIi})^2}, \quad \text{for } i \in N. \tag{3.7}$$

(iii) Investment in private capital by firm i in stage k will generate benefits to the increment of the capital in stage $k+1$ plus the benefits in capital accumulation in stages $k+2$ to $k+\omega_1-1$. In a game equilibrium, the marginal benefit of investment equals the marginal cost of investment in private capital, that is

$$\frac{\partial V_{k+1}^i}{\partial x_{k+1}^{***i}}\frac{\partial x_{k+1}^i}{\partial u_k^{(\omega_1)i}}+\frac{\partial V_{k+1}^i}{\partial u_k^{(\omega_1)i}}=2\varphi_k^{(x)i}\overline{u}_k^{(\omega_1)i}\delta^{k-1}, \text{ for } i\in N,$$

which implies

$$u_k^{(\omega_1)i}=\left(\frac{\partial V_{k+1}^i}{\partial x_{k+1}^i}\varepsilon_k^{|k|i}+\frac{\partial V_{k+1}^i}{\partial \overline{u}_k^{(\omega_1)i}}\right)\div 2\varphi_k^{(x)i}\delta^{k-1}, \quad \text{for } i\in N. \tag{3.8}$$

(iv) Investment in public capital by firm i in stage k will generate benefits to the increment of the capital in stage $k+1$ plus the benefits in capital accumulation in stages $k+2$ to $k+\omega_2-1$. In a game equilibrium, the marginal benefit of investment equals the marginal cost of investment in public capital, that is

$$\frac{\partial V_{k+1}^i}{\partial y_{k+1}^{***}}\frac{\partial y_{k+1}^{***}}{\partial \overline{u}_k^{(\omega_2)i}}+\frac{\partial V_{k+1}^i}{\partial u_k^{(\omega_2)i}}=2\varphi_k^{(y)i}u_k^{(\omega_2)i}\delta^{k-1}, \quad \text{for } i\in N,$$

which implies

$$\overline{u}_k^{(\omega_2)i}=\left(\frac{\partial V_{k+1}^i}{\partial y_{k+1}^{***}}a_k^{|k|i}+\frac{\partial V_{k+1}^i}{\partial u_k^{(\omega_2)i}}\right)\div 2\varphi_k^{(y)i}\delta^{k-1}, \quad \text{for } i\in N. \tag{3.9}$$

(v) In a game equilibrium, the marginal gain from the revenue generating investment equals the marginal cost, that is

$$p_k^{|k|i}\delta^{k-1}+\frac{\partial V_{k+1}^i}{\partial u_k^{(\omega_3)i}}=2\phi_k^i\overline{u}_k^{(\omega_3)i}\delta^{k-1}, \quad \text{for } i\in N,$$

which implies

$$\overline{u}_k^{(\omega_3)i}=\left(p_k^{|k|}\delta^{k-1}+\frac{\partial V_{k+1}^i}{\partial \overline{u}_k^{(\omega_3)i}}\right)\div 2\phi_k^i\delta^{k-1}, \quad \text{for } i\in N. \tag{3.10}$$

(vi) In a game equilibrium, the marginal gain from debt-financed investment equals the investment's marginal cost, that is

$$\frac{\chi_k^i}{2(\overline{u}_k^{(\omega_4)i})^{1/2}}\delta^{k-1} = q_k^{|k|i}\delta^{k-1} + \frac{\partial V_{k+1}^i}{\partial \overline{u}_k^{(\omega_4)i}}, \quad \text{for } i \in N,$$

which implies

$$\overline{u}_k^{(\omega_4)i} = (\chi_k^i \delta^{k-1})^2 \div 4\left(q_k^{|k|i}\delta^{k-1} + \frac{\partial V_{k+1}^i}{\partial \overline{u}_k^{(\omega_4)i}}\right)^2, \quad \text{for } i \in N. \tag{3.11}$$

A feedback Nash equilibrium solution of the game (3.1–3.3) can be solved as:

Proposition 3.1 *System (3.4–3.5)admits a solution with the game equilibrium payoff of firm i being.*

$$\begin{aligned} &V^i(k, x, y; \overline{u}_{k-}^{(\omega_1)**}, \overline{u}_{k-}^{(\omega_2)**}, \overline{u}_{k-}^{(\omega_3)**}, \overline{u}_{k-}^{(\omega_4)**}) \\ &= \left(A_k^{(i)i}x^i + \sum\nolimits_{j \in N, j \neq i} A_k^{(i)j}x^j + B_k^i y + C_k^i\right)\delta^{k-1}, \end{aligned} \tag{3.12}$$

where $A_{T+1}^{(i)i} = Q_{T+1}^{(x)i}$, $B_{T+1}^i = Q_{T+1}^{(y)i}$ *and* $A_{T+1}^{(i)j} = 0$, *for* $i, j \in N$ *and* $j \neq i$;

$$\begin{aligned} C_{T+1}^i &= \left(\sum\nolimits_{\tau=T+2-\omega_3}^{T} p_{T+1}^{|\tau|i}\overline{u}_\tau^{(\omega_3)i} - \sum\nolimits_{\tau=T+2-\omega_4}^{T} q_{T+1}^{|\tau|i}\overline{u}_\tau^{(\omega_4)i} + \varpi_{T+1}^i\right); \\ A_T^{(i)i} &= \frac{(R_T^{Ii})^2}{4c_T^{Ii}} + \delta A_{T+1}^{(i)i}(1 - \lambda_T^i), \\ A_T^{(i)j} &= \delta A_{T+1}^{(i)i}\gamma_T^{(i)j}, \quad j \in K(i), \\ A_T^{(i)\ell} &= 0, \quad \ell \in N\backslash K(i) \text{ and } \ell \neq i, \end{aligned}$$

$B_T^i = \frac{(R_T^{IIi})^2}{4c_T^{IIi}} + \delta B_{T+1}^i(1-\lambda_T)$ and C_T^i is an expression that contains the previously executed strategies $(\overline{u}_{T-}^{(\omega_1)**}, \overline{u}_{T-}^{(\omega_2)**}, \overline{u}_{T-}^{(\omega_3)**}, \overline{u}_{T-}^{(\omega_4)**})$;

$$\begin{aligned} A_{T-1}^{(i)i} &= \frac{(R_{T-1}^{Ii})^2}{4c_{T-1}^{Ii}} + \delta A_{T+1}^{(i)i}(1 - \lambda_{T-1}^i) + \delta \sum_{j \in K(i)} A_T^{(i)j}\gamma_{T-1}^{(j)i}, \\ A_{T-1}^{(i)j} &= \delta A_T^{(i)i}\gamma_{T-1}^{(i)j} + \delta A_T^{(i)j}\left(1 - \lambda_{T-1}^j\right) + \delta \sum_{\ell \in K(j)} A_T^{(i)\ell}\gamma_{T-1}^{(\ell)j}, \quad j \in K(i), \\ A_{T-1}^{(i)\ell} &= \delta \sum_{\ell \in K(j)} A_T^{(i)j}\gamma_{T-1}^{(j)\ell} + \delta A_T^{(i)\ell}\left(1 - \lambda_{T-1}^\ell\right), \quad \ell \in K(j) \\ &\text{for } \ j \in K(i) \ \text{ and } \ \ell \notin K(i) \\ A_{T-1}^{(i)h} &= 0, h \notin K(i), h \notin K(j) \ \text{ for } \ j \in K(i) \\ B_{T-1}^i &= \frac{\left(R_{T-1}^{IIi}\right)^2}{4c_{T-1}^{IIi}} + \delta B_T^i(1 - \lambda_{T-1}), \end{aligned}$$

and C^i_{T-1} is an expression that contains the previously executed strategies

$$(\underline{\overline{u}}^{(\omega_1)**}_{(T-1)-}, \underline{\overline{u}}^{(\omega_2)**}_{(T-1)-}, \underline{\overline{u}}^{(\omega_3)**}_{(T-1)-}, \underline{\overline{u}}^{(\omega_4)**}_{(T-1)-}).$$

For $k \in \{T-2, T-3, \cdots, 1\}$,

$$A^{(i)i}_k = \frac{(R^{Ii}_k)^2}{4c^{Ii}_k} + \delta A^{(i)i}_{k+1}(1-\lambda^i_k) + \delta \sum_{j\in K(i)} A^{(i)j}_{k+1}\gamma^{(j)i}_k,$$

$$A^{(i)j}_k = \delta A^{(i)i}_{k+1}\gamma^{(i)j}_k + \delta A^{(i)j}_{k+1}\left(1-\lambda^j_k\right) + \delta \sum_{\ell\in K(j)} A^{(i)\ell}_{k+1}\gamma^{(\ell)j}_k, \quad j \in K(i),$$

$$A^{(i)\ell}_k = \delta \sum_{\ell\in K(j)} A^{(i)j}_{k+1}\gamma^{(j)\ell}_k + \delta A^{(i)\ell}_{k+1}(1-\lambda^\ell_k),$$

$\ell \in K(j)$ for $j \in K(i)$ and $\ell \notin K(i)$,

$$A^{(i)h}_k = 0, \; h \notin K(i), \; h \notin K(j) \text{ for } j \in K(i),$$

$$B^i_k = \frac{(R^{IIi}_k)^2}{4c^{IIi}_k} + \delta B^i_{k+1}(1-\lambda_k),$$

and C^i_k is an expression that contains the previously executed strategies

$$(\underline{\overline{u}}^{(\omega_1)**}_{k-}, \underline{\overline{u}}^{(\omega_2)**}_{k-}, \underline{\overline{u}}^{(\omega_3)**}_{k-}, \underline{\overline{u}}^{(\omega_4)**}_{k-}).$$

Proof See Appendix in Sect. 5. ■

Invoking Proposition 3.1, the game equilibrium strategies in (3.6–3.11) can be obtained as (see derivation details in the Appendix in Sect. 5):

$$u^{Ii**}_k = \frac{(R^{(x)i}_k)^2 x^i}{4(c^{Ii}_k)^2}, \; u^{IIi**}_k = \frac{(R^{IIi}_k)^2 y}{4(c^{Ii}_k)^2};$$

$$\overline{u}^{(\omega_1)i**}_k = \frac{\sum_{\tau=k}^{k+\omega_1-1} \delta^{\tau-k+1} A^{(i)i}_{\tau+1}\varepsilon^{|k|i}_\tau}{2c^{(x)i}_k}, \quad A^{(i)i}_\tau = 0 \text{ for } \tau > T+1;$$

$$\overline{u}^{(\omega_2)i**}_k = \frac{\sum_{\tau=k}^{k+\omega_2-1} \delta^{\tau-k+1} B^i_{\tau+1} a^{|k|i}_\tau}{2\varphi^{(y)i}_k}, \quad B^i_\tau = 0 \text{ for } \tau > T+1;$$

$$\overline{u}^{(\omega_3)i**}_k = \frac{\sum_{\tau=k}^{k+\omega_3-1} \delta^{\tau-k} p^{|k|i}_\tau}{2c^{(w)i}_k}, \quad p^{|k|i}_\tau = p^{|k|i}_{T+1} \text{ for } \tau > T+1;$$

$$\overline{u}_k^{(\omega_4)i**} = \frac{(\chi_k^i)^2}{4\left(\sum_{\tau=k}^{k+\omega_4-1} \delta^{\tau-k} q_\tau^{|k|i}\right)^2}, \quad q_\tau^{|k|i} = q_{T+1}^{|k|i} \text{ for } \tau > T+1;$$

$$\text{for } i \in N \text{ and } k \in \{1, 2, \cdots, T\}. \tag{3.13}$$

Substituting the game equilibrium investment strategies from (3.13) into the state dynamics (3.1–3.2), we obtain the dynamics of private and public capitals. In particular, the game equilibrium accumulation process of the private capital stock of firm i is

$$x_{k+1}^i = x_k^i + \varepsilon_k^{|k|i} \frac{\sum_{\tau=k}^{k+\omega_1-1} \delta^{\tau-k+1} A_{\tau+1}^{(i)i} \varepsilon_\tau^{|k|i}}{2\varphi_k^{(x)i}} + \sum_{\tau=k-\omega_1+1}^{k-1} \varepsilon_k^{|\tau|i} \overline{u}_\tau^{(\omega_1)i**} - \lambda_k^i x_k^i + \sum_{j \in K(i)} \gamma_k^{(i)j} x_k^j,$$

$$x_1^i = x_1^{i(0)}, \quad i \in N. \tag{3.14}$$

In initial stage 1, $u_\tau^{(\omega_1)i} = 0$, for $i \in N$ and $\tau \in \{0, -1, -2, \cdots, -\omega_1 + 1\}$ and we can obtain x_2^i, for $i \in N$. Then with x_2^i, for $i \in N$ and (3.14), we can obtain x_3^i, for $i \in N$. Repeating the process, we can obtain the game equilibrium path of private capitals $\{x_k^1, x_k^2, \cdots, x_k^n\}_{k=1}^{T+1}$ can be obtained.

The game equilibrium accumulation process of the public capital is

$$y_{k+1} = y_k + \sum_{j=1}^n a_k^{|k|j} \frac{\sum_{\tau=k}^{k+\omega_2-1} \delta^{\tau-k+1} B_{\tau+1}^j a_\tau^{|k|j}}{2\varphi_k^{(y)j}}$$

$$+ \sum_{j=1}^n \sum_{\tau=k-\omega_2+1}^{k-1} a_k^{|\tau|j} u_\tau^{(\omega_2)j**} - \lambda_k y_k, \quad y_1 = y_1^0, \tag{3.15}$$

In initial stage 1, $u_\tau^{(\omega_2)i} = 0$, for $i \in N$ and $\tau \in \{0, -1, -2, \cdots, -\omega_2 + 1\}$ and we can obtain y_2. Then with y_2 and (3.15), we can obtain y_3. Repeating the process, the game equilibrium path of public capital $\{y_k\}_{k=1}^{T+1}$ can be obtained.

4 A Novel Dynamic Theory of the Firm

The dynamic capital theory originated by Jorgenson (1967, 1971, 1963) laid down the foundation of a formal dynamic analytical framework of the value of the firm. In his seminal work and subsequent analyses (see, for example, Lucas (1967), Gould (1968) and Treadway (1969), Hayashi (1982), Loon (1983) and Plourde and Yeung (1988)), investments are confined to investments that build up the firm's productive capital stocks. Jorgenson's analysis considers a continuous-time infinite horizon model with time-invariant parameters and constant discount rates. For the sake of comparison, a discrete-time counterpart of the Jorgenson continuous-time paradigm can be expressed as maximizing the firm's payoff

$$\sum_{k=1}^{\infty}[R(u_k, x_k) - c(u_k) - c(v_k, x_k)]\delta^k \tag{4.1}$$

subject to the capital accumulation dynamics

$$x_{k+1} = x_k + f(v_k) - \lambda x_k, \tag{4.2}$$

where u_k is the input for production of output, $R(u_k, x_k)$ is gross revenue from the output which is produced by using u_k and the capital stock x_k, $c(u_k)$ is the cost of input, v_k is the investment in building up the capital stock, $c(v_k, x_k)$ is the installation cost of capital investment, $f(v_k)$ is the increment in capital brought about investment v_k and δ^k is the discount factor.

The firm maximization of payoff (4.1) subject to the capital stock dynamics (4.2) yields a steady-state solution which reflects the value of the firm as:

$$V(x) = \sum_{k=1}^{\infty}[R(u_k^*, x_k^*) - c(u_k^*) - c(v_k^*, x_k^*)]\delta^k, \tag{4.3}$$

where u_k^* and v_k^* are the steady-state levels of input and investment and x_k^* is the steady-state level of capital stock that satisfies

$$x_k^* = \frac{f(v_k^*)}{\lambda}.$$

In particular, the value of the firm, $V(x)$, depends solely on the firm's private capital stock.

4.1 Firm's Value Under Interactive Durable Strategies

The analysis in this chapter complements the conventional theory of the value of the firm with a comprehensive investment portfolio. Modern investment portfolio of firms normally includes revenue generating investment assets such as incoming earning assets, securities, real estate, bonds and mutual funds which would yield future returns. At the same time, firms also acquire debt-financed investments which yield current gains and incur a series of payment obligations in the future. In the current knowledge-based economy, public capital freely accessible to economic agents is not uncommon. Public capital could serve as important capital inputs in the firm's productive process. The diffusion of investment in knowledge-based public capital often takes time to be fully realized. The investment to build up private capital would often take a number of stages to be fully converted into physical capital stock. Moreover, technology spillover through trade affecting the private technology/capital of firms has been observed widely. Incorporating all these factors, we present a practically relevant theory of the value of the firm with durable strategies dynamic interactive investments.

Invoking the results in Sects. 2 and 3, a number of novel characteristics of the value of the firm are developed. Note that the firm's value $V^i(k, x, y; \underline{\overline{u}}_{k-}^{(\omega_1)**}, \underline{\overline{u}}_{k-}^{(\omega_2)**}, \underline{\overline{u}}_{k-}^{(\omega_3)**}, \underline{\overline{u}}_{k-}^{(\omega_4)**})$ in stage k depends on the capital state vectors (x, y) and the durable strategies $(\underline{\overline{u}}_{k-}^{(\omega_1)**}, \underline{\overline{u}}_{k-}^{(\omega_2)**}, \underline{\overline{u}}_{k-}^{(\omega_3)**}, \underline{\overline{u}}_{k-}^{(\omega_4)**})$, for $k \in \{1, 2, \cdots, T+1\}$. This differs significantly from the conventional dynamic theory of the firm which value depends only on its own private capital x^i.

First, $V^i(k, x, y; \underline{\overline{u}}_{k-}^{(\omega_1)**}, \underline{\overline{u}}_{k-}^{(\omega_2)**}, \underline{\overline{u}}_{k-}^{(\omega_3)**}, \underline{\overline{u}}_{k-}^{(\omega_4)**})$ depends on the private capital of other firms. These private capitals include (i) the firm's own private capital, that is x^i; (ii) the private capital of the firm's trading partners, that is x^j for $j \in K(i)$; and (iii) the private capital of the trading partners of firm $j \in K(i)$, that is x^ℓ for $\ell \in K(j)$, $j \in K(i)$ and $\ell \notin K(i)$. Proposition 3.1 demonstrates this outcome explicitly with

$$V^i(k, x, y; \underline{\overline{u}}_{k-}^{(\omega_1)**}, \underline{\overline{u}}_{k-}^{(\omega_2)**}, \underline{\overline{u}}_{k-}^{(\omega_3)**}, \underline{\overline{u}}_{k-}^{(\omega_4)**})$$
$$= \left(A_k^{(i)i} x^i + \sum_{j \in N, j \neq i} A_k^{(i)j} x^j + B_k^i y + C_k^i \right) \delta^{k-1}.$$

In addition, $V^i(k, x, y; \underline{\overline{u}}_{k-}^{(\omega_1)**}, \underline{\overline{u}}_{k-}^{(\omega_2)**}, \underline{\overline{u}}_{k-}^{(\omega_3)**}, \underline{\overline{u}}_{k-}^{(\omega_4)**})$ also depends on previously executed investments in private capital $\underline{\overline{u}}_{k-}^{(\omega_1)**}$ by firm i and by firms directly or indirectly related in trade. As shown in Eq. (5.18) of the Appendix, the terms involving $\underline{\overline{u}}_{k-}^{(\omega_1)**}$ appear in the value of firm i in Sect. 3 as

$$\delta \left(\begin{array}{l} A_{k+1}^{(i)i} \displaystyle\sum_{\tau=k-\omega_1+1}^{k-1} \varepsilon_k^{|\tau|i} \overline{u}_\tau^{(\omega_1)i**} + \sum_{j \in K(i)} A_{k+1}^{(i)j} \sum_{\tau=k-\omega_1+1}^{k-1} \varepsilon_k^{|\tau|j} \overline{u}_\tau^{(\omega_1)j**} \\ + \displaystyle\sum_{\substack{\ell \in K(j) \\ j \in K(i) \\ \ell \notin K(i)}} A_{k+1}^{(i)\ell} \sum_{\tau=k-\omega_1+1}^{k-1} \varepsilon_k^{|\tau|\ell} \overline{u}_\tau^{(\omega_1)\ell**} \end{array} \right)$$
$$+ \delta^2 \left(\begin{array}{l} A_{k+2}^{(i)i} \displaystyle\sum_{\tau=k-\omega_1+2}^{k-1} \varepsilon_{k+1}^{|\tau|i} \overline{u}_\tau^{(\omega_1)i**} + \sum_{j \in K(i)} A_{k+2}^{(i)j} \sum_{\tau=k-\omega_1+2}^{k-1} \varepsilon_{k+1}^{|\tau|j} \overline{u}_\tau^{(\omega_1)j**} \\ + \displaystyle\sum_{\substack{\ell \in K(j) \\ j \in K(i) \\ \ell \notin K(i)}} A_{k+2}^{(i)\ell} \sum_{\tau=k-\omega_1+2}^{k-1} \varepsilon_{k+1}^{|\tau|\ell} \overline{u}_\tau^{(\omega_1)\ell**} \end{array} \right)$$

$$+\delta^3\left(\begin{array}{l}A_{k+3}^{(i)i}\sum_{\tau=k-\omega_1+3}^{k-1}\varepsilon_{k+2}^{|\tau|i}\overline{u}_\tau^{(\omega_1)i**}+\sum_{j\in K(i)}A_{k+3}^{(i)j}\sum_{\tau=k-\omega_1+3}^{k-1}\varepsilon_{k+2}^{|\tau|j}\overline{u}_\tau^{(\omega_1)j**}\\ +\sum_{\substack{\ell\in K(j)\\ j\in K(i)\\ \ell\notin K(i)}}A_{k+3}^{(i)\ell}\sum_{\tau=k-\omega_1+3}^{k-1}\varepsilon_{k+2}^{|\tau|\ell}\overline{u}_\tau^{(\omega_1)\ell**}\end{array}\right)$$

$$+\cdots+\delta^{\omega_1-1}\left(\begin{array}{l}A_{k+\omega_1-1}^{(i)i}\sum_{\tau=k-1}^{k-1}\varepsilon_{k+\omega_1-2}^{|\tau|i}\overline{u}_\tau^{(\omega_1)i**}\\ +\sum_{j\in K(i)}A_{k+\omega_1-1}^{(i)j}\sum_{\tau=k-1}^{k-1}\varepsilon_{k+\omega_1-2}^{|\tau|j}\overline{u}_\tau^{(\omega_3)j**}\\ +\sum_{\substack{\ell\in K(j)\\ j\in K(i)\\ \ell\notin K(i)}}A_{k+\omega_1-1}^{(i)\ell}\sum_{\tau=k-1}^{k-1}\varepsilon_{k+\omega_1-2}^{|\tau|\ell}\overline{u}_\tau^{(\omega_1)\ell**}\end{array}\right), \tag{4.4}$$

where $A_t^{(i)i}=A_t^{(i)j}=A_t^{(i)\ell}=0$ for $t>T+1$.

The term in (4.4) is the sum of the contributions to the value of firm i by the previously executed durable private capital investment strategies by all the relevant firms including those from its trading partners and indirectly related firms. This is the first time that previously executed investments in private capital by firms appear as a factor affecting the value of the firm.

Second, the level of public capital stocks y appears in $V^i(k,x,y;\underline{\overline{u}}_{k-}^{(\omega_1)**},\underline{\overline{u}}_{k-}^{(\omega_2)**},\underline{\overline{u}}_{k-}^{(\omega_3)**},\underline{\overline{u}}_{k-}^{(\omega_4)**})$. The prevalence of accessible public capital in our knowledge-based economy makes public capital important capital factor inputs in the firm's production process.

In addition, $V^i(k,x,y;\underline{\overline{u}}_{k-}^{(\omega_1)**},\underline{\overline{u}}_{k-}^{(\omega_2)**},\underline{\overline{u}}_{k-}^{(\omega_3)**},\underline{\overline{u}}_{k-}^{(\omega_4)**})$ also depends on previously executed investments in public capital $\underline{\overline{u}}_{k-}^{(\omega_2)**}$ by all firms. As shown in Eq. (5.17) of the Appendix, the terms involving $\underline{\overline{u}}_{k-}^{(\omega_2)**}$ appear in the value of firm i in Sect. 3 as

$$\delta B_{k+1}^i\sum_{j=1}^{n}\sum_{\tau=k-\omega_2+1}^{k-1}a_k^{|\tau|j}\overline{u}_\tau^{(\omega_2)j**}+\delta^2 B_{k+2}^i\sum_{j=1}^{n}\sum_{\tau=k-\omega_2+2}^{k-1}a_{k+1}^{|\tau|j}\overline{u}_\tau^{(\omega_2)j**}$$

$$+\delta^3 B_{k+3}^i\sum_{j=1}^{n}\sum_{\tau=k-\omega_2+3}^{k-1}a_{k+2}^{|\tau|j}\overline{u}_\tau^{(\omega_2)j**}$$

$$+\delta^4 B_{k+4}^i\sum_{j=1}^{n}\sum_{\tau=k-\omega_2+4}^{k-1}a_{k+3}^{|\tau|j}\overline{u}_\tau^{(\omega_2)j**}+\cdots$$

$$+ \delta^{\omega_2 - 1} B^i_{k+\omega_2-1} \sum_{j=1}^{n} \sum_{\tau=k-1}^{k-1} a^{|\tau|j}_{k+\omega_2-2} \overline{u}^{(\omega_2)j**}_{\tau}, \tag{4.5}$$

where $B^i_t = 0$ for $t > T + 1$.

The term in (4.5) is the sum of the contributions to the value of firm i by the previously executed durable public capital investment strategies by all firms. This is the first time that previously executed investments in public capital by all firms appear as a factor affecting the value of the firm.

Third, previously executed revenue generating investments $\underline{\overline{u}}^{(\omega_3)i**}_{k-}$ by firm i appear in the value of the firm $V^i(k, x, y; \underline{\overline{u}}^{(\omega_1)**}_{k-}, \underline{\overline{u}}^{(\omega_2)**}_{k-}, \underline{\overline{u}}^{(\omega_3)**}_{k-}, \underline{\overline{u}}^{(\omega_4)**}_{k-})$. As shown in Eq. (5.15) of the Appendix, the terms involving $\underline{u}^{(\omega_3)i**}_{k-}$ appear in the value of the firm in Sect. 3 as

$$\sum_{\tau=k-\omega_3+1}^{k-1} p^{|\tau|i}_{k} u^{(\omega_3)i**}_{\tau} + \delta \sum_{\tau=k-\omega_3+2}^{k-1} p^{|\tau|i}_{k+1} \overline{u}^{(\omega_3)i**}_{\tau}$$
$$+ \delta^2 \sum_{\tau=k-\omega_3+3}^{k-1} p^{|\tau|i}_{k+2} \overline{u}^{(\omega_3)i**}_{\tau} + \delta^3 \sum_{\tau=k-\omega_3+4}^{k-1} p^{|\tau|i}_{k+3} \overline{u}^{(\omega_3)i**}_{\tau}$$
$$+ \cdots + \delta^{\omega_3 - 2} \sum_{\tau=k-1}^{k-1} p^{|\tau|i}_{k+\omega_3-2} \overline{u}^{(\omega_3)i**}_{\tau}, \tag{4.6}$$

where $p^{|\tau|i}_t = p^{|\tau|i}_{T+1}$ for $t > T + 1$.

The term in (4.6) is the present value of the sum of the returns from firm i's revenue generating investments executed before stage k. This is the first time that previously executed revenue generating investments by firm i appear as a factor affecting the value of the firm.

Fourth, previously procured debt-financed investments $\underline{\overline{u}}^{(\omega_4)i**}_{k-}$ by firm i appear in the value of the firm $V^i(k, x, y; \underline{\overline{u}}^{(\omega_1)**}_{k-}, \underline{\overline{u}}^{(\omega_2)**}_{k-}, \underline{\overline{u}}^{(\omega_3)**}_{k-}, \underline{\overline{u}}^{(\omega_4)**}_{k-})$. As shown in Eq. (5.16) of the Appendix, the terms involving $\underline{\overline{u}}^{(\omega_4)i**}_{k-}$ appear in the value of the firm in Sect. 3 as

$$- \sum_{\tau=k-\omega_4+1}^{k-1} q^{|\tau|i}_{k} \overline{u}^{(\omega_4)i**}_{\tau} - \delta \sum_{\tau=k-\omega_4+2}^{k-1} q^{|\tau|i}_{k+1} \overline{u}^{(\omega_4)i**}_{\tau}$$
$$- \delta^2 \sum_{\tau=k-\omega_4+3}^{k-1} q^{|\tau|i}_{k+2} \overline{u}^{(\omega_4)i**}_{\tau} - \delta^3 \sum_{\tau=k-\omega_4+4}^{k-1} q^{|\tau|i}_{k+3} \overline{u}^{(\omega_4)i**}_{\tau}$$
$$- \cdots - \delta^{\omega_4 - 2} \sum_{\tau=k-1}^{k-1} q^{|\tau|i}_{k+\omega_4-2} \overline{u}^{(\omega_4)i**}_{\tau}, \tag{4.7}$$

where $q^{|\tau|i}_t = q^{|\tau|i}_{T+1}$ for $t > T + 1$.

The term in (4.7) is the present value of the debt obligations from the previously executed investment projects $\underline{\overline{u}}_{k-}^{(\omega_4)**}$. This is the first time that debt obligations from previously executed project investments by firm i appear as a factor affecting the value of the firm.

The first item to the fourth item mentioned above is novel features in the dynamic theory of the firm.

4.2 Non-interactive Standard Theory Under Durable Strategies

The value of the firm without interaction between firms in the traditional perspective can be derived from the analysis in Sect. 4.1. In particular, there is no technology spillover in the accumulation process of the firm's private capital and there is public capital.

The accumulation process of the private capital stock of the firm is governed by the dynamics

$$x_{k+1} = x_k + \varepsilon_k^{|k|} \overline{u}_k^{(\omega_1)} + \sum_{\tau=k-\omega_1+1}^{k-1} \varepsilon_k^{|\tau|} \overline{u}_\tau^{(\omega_1)} - \lambda_k x_k, \quad x_1 = x_1^0. \tag{4.8}$$

Following previous analysis, the value of the firm becomes $V(k, x; \overline{u}_{k-}^{(\omega_1)}, \overline{u}_{k-}^{(\omega_3)}, \overline{u}_{k-}^{(\omega_4)})$. The value of the firm without spillover interaction can be obtained as:

Proposition 4.1 *The value of the firm without interaction in* Sect. 3 *can be obtained as:*

$$V(k, x; \overline{u}_{k-}^{(\omega_1)}, \overline{u}_{k-}^{(\omega_3)}, \overline{u}_{k-}^{(\omega_4)}) = (A_k x + C_k)\delta^{k-1}, \tag{4.9}$$

where $A_{T+1} = Q_{T+1}^{(x)}$;

$$C_{T+1} = \left(\sum_{\tau=T+2-\omega_3}^{T} p_{T+1}^{|\tau|} \bar{u}_\tau^{(\omega_3)} - \sum_{\tau=T+2-\omega_4}^{T} q_{T+1}^{|\tau|} \bar{u}_\tau^{(\omega_4)} + \varpi_{T+1}\right);$$

$$A_k = \frac{(R_k^I)^2}{4c_k^I} + \delta A_k(1 - \lambda_k),$$

$B_k = \dfrac{(R_k^{II})^2}{4c_k^{II}} + \delta B_k(1 - \lambda_k)$ *and* C_k *is an expression that contains the previously executed controls* $(\overline{u}_{k-}^{(\omega_1)}, \overline{u}_{k-}^{(\omega_3)}, \overline{u}_{k-}^{(\omega_4)})$.

Proof Follow the proof of Proposition 3.1. ■

Note that the value of the firm $V(k, x; \overline{u}_{k-}^{(\omega_1)}, \overline{u}_{k-}^{(\omega_3)}, \overline{u}_{k-}^{(\omega_4)})$ depends on the firm's private capital. In addition, $V(k, x; \overline{u}_{k-}^{(\omega_1)}, \overline{u}_{k-}^{(\omega_3)}, \overline{u}_{k-}^{(\omega_4)})$ also depends on firm i's previously executed investments in private capital $\overline{u}_{k-}^{(\omega_1)}$. The terms involving $\overline{u}_{k-}^{(\omega_1)}$ appearing in the value of firm can be derived from (4.4) as

$$\delta A_{k+1} \sum_{\tau=k-\omega_1+1}^{k-1} \varepsilon_k^{|\tau|} u_\tau^{(\omega_1)} + \delta^2 A_{k+2} \sum_{\tau=k+2-\omega_1}^{k-1} \varepsilon_{k+1}^{|\tau|} u_\tau^{(\omega_1)} + \delta^3 A_{k+3} \sum_{\tau=k+3-\omega_1}^{k-1} \varepsilon_{k+2}^{|\tau|} u_\tau^{(\omega_1)}$$
$$+ \delta^4 A_{k+4} \sum_{\tau=k+4-\omega_1}^{k-1} \varepsilon_{k+3}^{|\tau|} u_\tau^{(\omega_1)} + \cdots + \delta^{\omega_1-1} A_{k+\omega_1-1} \sum_{\tau=k-1}^{k-1} \varepsilon_{k+\omega_1-2}^{|\tau|} u_\tau^{(\omega_1)}. \quad (4.10)$$

The term in (4.10) is the sum of the contributions to the value of the firm by its previously executed durable private capital investment strategies.

Third, previously executed revenue generating investments $\overline{u}_{k-}^{(\omega_3)}$ appear in the value of the firm $V(k, x; \overline{u}_{k-}^{(\omega_1)}, \overline{u}_{k-}^{(\omega_3)}, \overline{u}_{k-}^{(\omega_4)})$. The terms involving $u_{k-}^{(\omega_3)}$ appearing in the value of firm can be derived from (4.6) as

$$\sum_{\tau=k-\omega_3+1}^{k-1} p_k^{|\tau|} \overline{u}_\tau^{(\omega_3)} + \delta \sum_{\tau=k-\omega_3+2}^{k-1} p_{k+1}^{|\tau|} \overline{u}_\tau^{(\omega_3)}$$
$$+ \delta^2 \sum_{\tau=k-\omega_3+3}^{k-1} p_{k+2}^{|\tau|} \overline{u}_\tau^{(\omega_3)} + \delta^3 \sum_{\tau=k-\omega_3+4}^{k-1} p_{k+3}^{|\tau|} \overline{u}_\tau^{(\omega_3)}$$
$$+ \cdots + \delta^{\omega_3-2} \sum_{\tau=k-1}^{k-1} p_{k+\omega_3-3}^{|\tau|} \overline{u}_\tau^{(\omega_3)}. \quad (4.11)$$

The term in (4.11) is the present value of the sum of the returns from the firm's revenue generating investments executed before stage k.

Fourth, previously procured debt-financed investments $\overline{u}_{k-}^{(\omega_4)}$ by the firm appear in the value of the firm $V(k, x; \overline{u}_{k-}^{(\omega_1)}, \overline{u}_{k-}^{(\omega_3)}, \overline{u}_{k-}^{(\omega_4)})$. The terms involving $\overline{u}_{k-}^{(\omega_4)}$ appearing in the value of firm can be derived from (4.7) as

$$- \sum_{\tau=k-\omega_4+1}^{k-1} q_k^{|\tau|} \overline{u}_\tau^{(\omega_4)} - \delta \sum_{\tau=k+2-\omega_4}^{k-1} q_{k+1}^{|\tau|} \overline{u}_\tau^{(\omega_4)}$$
$$- \delta^2 \sum_{\tau=k+3-\omega_4}^{k-1} q_{k+2}^{|\tau|} \overline{u}_\tau^{(\omega_4)} - \delta^3 \sum_{\tau=k+4-\omega_4}^{k-1} q_{k+3}^{|\tau|} \overline{u}_\tau^{(\omega_4)}$$
$$- \cdots - \delta^{\omega_4-2} \sum_{\tau=k-1}^{k-1} q_{k+\omega_4-2}^{|\tau|} \overline{u}_\tau^{(\omega_4)}. \quad (4.12)$$

The term in (4.12) is the present value of the debt obligations from the previously executed investment projects $\overline{u}_{k-}^{(\omega_4)}$.

5 Appendices

5.1 Appendix I: Proof of Proposition 3.1

Using (3.4), we have
$A^{(i)i}_{T+1} = Q^{(x)i}_{T+1}$, $B^{i}_{T+1} = Q^{(y)i}_{T+1}$ and $A^{(i)j}_{T+1} = 0$, for $i, j \in N$ and $j \neq i$;

$$C^{i}_{T+1} = \left(\sum_{\tau=T+2-\omega_3}^{T} p^{|\tau|i}_{T+1} \overline{u}^{(\omega_3)i}_{\tau} - \sum_{\tau=T+2-\omega_4}^{T} q^{|\tau|i}_{T+1} \overline{u}^{(\omega_4)i}_{\tau} + \varpi^{i}_{T+1} \right). \tag{5.1}$$

Invoking the technique of backward induction, we consider first the last operational stage T. Using Proposition 3.1 and (3.5), we can express the game equilibrium strategies in stage T as:

$$u^{Ii**}_{T} = \frac{(R^{Ii}_{T})^2 x^i}{4(c^{Ii}_{T})^2}, u^{IIi**}_{T} = \frac{(R^{IIi}_{T})^2 y}{4(c^{IIi}_{T})^2}, \overline{u}^{(\omega_1)i**}_{T} = \frac{\delta A^{(i)i}_{T+1} \varepsilon^{(T)i}_{T}}{2\varphi^{(x)i}_{T}}, \overline{u}^{(\omega_2)i**}_{T} = \frac{\delta B^{i}_{T+1} a^{|T|i}_{T}}{2\varphi^{(y)i}_{T}},$$

$$\overline{u}^{(\omega_3)i**}_{T} = \frac{p^{|T|i}_{T} + \delta p^{|T|i}_{T+1}}{2\phi^{i}_{T}}, \overline{u}^{(\omega_4)i**}_{T} = \frac{(\chi^{i}_{T})^2}{4(q^{|T|i}_{T} + \delta q^{|T|i}_{T+1})^2}, \quad \text{for } i \in N. \tag{5.2}$$

Substituting the optimal strategies in (5.2) into the stage T equation of (3.5) we obtain

$$\begin{aligned}
&V^{i}(T, x, y; \overline{u}^{(\omega_1)**}_{T-}, \overline{u}^{(\omega_2)**}_{T-}, \overline{u}^{(\omega_3)**}_{T-}, \overline{u}^{(\omega_4)**}_{T-}) \\
&= \left(A^{(i)i}_{T} x^i + \sum_{j \in N, j \neq i} A^{(i)j}_{T} x^j + B^{i}_{T} y + C^{i}_{T} \right) \delta^{T-1}, \\
&= \left[\frac{(R^{Ii}_{T})^2 x^i}{4c^{Ii}_{T}} - \frac{(\delta A^{(i)i}_{T+1} \varepsilon^{|T|i}_{T})^2}{4\varphi^{(x)i}_{T}} + \frac{(R^{IIi}_{T})^2 y}{4c^{IIi}_{T}} - \frac{(\delta B^{i}_{T+1} a^{|T|i}_{T})^2}{4\varphi^{(y)i}_{T}} \right. \\
&\quad + p^{|T|i}_{T} \frac{p^{|T|i}_{T} + \delta p^{|T|i}_{T+1}}{2\phi^{i}_{T}} + \sum_{\tau=T-\omega_3+1}^{T-1} p^{|\tau|i}_{T} \overline{u}^{(\omega_3)i**}_{\tau} - \frac{(p^{|T|i}_{T} + \delta p^{|T|i}_{T+1})^2}{4\phi^{i}_{T}} \\
&\quad \left. + \frac{(\chi^{i}_{T})^2}{2(q^{|T|i}_{T} + \delta q^{|T|i}_{T+1})} - q^{|T|i}_{T} \frac{(\chi^{i}_{T})^2}{4(q^{|T|i}_{T} + \delta q^{|T|i}_{T+1})^2} - \sum_{\tau=T-\omega_4+1}^{T-1} q^{|\tau|i}_{T} \overline{u}^{(\omega_4)i**}_{\tau} \right] \delta^{T-1} \\
&\quad + \left[A^{(i)i}_{T+1} x^{***i}_{T+1} + B^{i}_{T+1} y^{***}_{T+1} + \left(p^{|T|i}_{T+1} \frac{p^{|T|i}_{T} + \delta p^{|T|i}_{T+1}}{2\varphi^{i}_{T}} + \sum_{\tau=T+2-\omega_3}^{T-1} p^{|\tau|i}_{T+1} \overline{u}^{(\omega_3)i**}_{\tau} \right.\right. \\
&\quad \left.\left. - q^{|T|i}_{T+1} \frac{(\chi^{i}_{T})^2}{4(q^{|T|i}_{T} + \delta q^{|T|i}_{T+1})^2} - \sum_{\tau=T+2-\omega_4}^{T-1} \overline{q}^{|\tau|i}_{T+1} u^{(\omega_4)i**}_{\tau} + \varpi^{i}_{T+1} \right) \right] \delta^{T}, \quad i \in N,
\end{aligned} \tag{5.3}$$

where

$$x_{T+1}^{***i} = x^i + \varepsilon_T^{|T|i} \frac{\delta A_{T+1}^{(i)i} \varepsilon_T^{|T|i}}{2\varphi_T^{(x)i}} + \sum\nolimits_{\tau=T-\omega_1+1}^{T-1} \varepsilon_T^{|\tau|i} \overline{u}_\tau^{(\omega_1)i**} - \lambda_T^i x^i + \sum\nolimits_{j\in K(i)} \gamma_T^{x(i)j} x^j,$$

$$y_{T+1}^{***} = y + \sum\nolimits_{j=1}^{n} a_T^{|T|j} \frac{\delta B_{T+1}^j a_T^{|T|j}}{2\varphi_T^{(y)j}} + \sum\nolimits_{j=1}^{n} \sum\nolimits_{\tau=T-\omega_2+1}^{T-1} a_T^{|\tau|j} \overline{u}_\tau^{(\omega_2)j**} - \lambda_T y.$$

System (5.3) yields equations in which right-hand-side and left-hand-side are linear functions of $x^1, x^2, \cdots, x^n$ and y, for $i \in N$. For (5.3) to hold, it is required that:

$$\begin{aligned}
A_T^{(i)i} x^i &= \frac{\left(R_T^{Ii}\right)^2 x^i}{4c_T^{Ii}} + \delta A_{T+1}^{(i)i}(1-\lambda_T^i)x^i, \\
A_T^{(i)j} x^j &= \delta A_{T+1}^{(i)i} \gamma_T^{(i)j} x^j, \quad j \in K(i), \\
A_T^{(i)\ell} &= 0, \ldots \ell \in N \backslash K(i) \text{ and } \ell \neq i, \\
B_T^i y &= \frac{\left(R_T^{IIi}\right)^2 y}{4c_T^{IIi}} + \delta B_{T+1}^i (1-\lambda_T) y.
\end{aligned} \tag{5.4}$$

From (5.4) we obtain:

$$\begin{aligned}
A_T^{(i)i} &= \frac{(R_T^{Ii})^2}{4c_T^{Ii}} + \delta A_{T+1}^{(i)i}(1-\lambda_T^i), \\
A_T^{(i)j} &= \delta A_{T+1}^{(i)i} \gamma_T^{(i)j}, \quad j \in K(i), \\
A_T^{(i)\ell} &= 0, \ldots \ell \in N \backslash K(i) \text{and} \ell \neq i, \\
B_T^i &= \frac{(R_T^{IIi})^2}{4c_T^{IIi}} + \delta B_{T+1}^i (1-\lambda_T).
\end{aligned} \tag{5.5}$$

In addition,

$$\begin{aligned}
C_T^i = \Bigg[&- \frac{\left(\delta A_{T+1}^{(i)i} \varepsilon_T^{|T|i}\right)^2}{4\varphi_T^{(x)i}} - \frac{\left(\delta B_{T+1}^i a_T^{|T|i}\right)^2}{4\varphi_T^{(y)i}} \\
&+ p_T^{|T|i} \frac{p_T^{|T|i} + \delta p_{T+1}^{(T)i}}{2\phi_T^i} + \sum_{\tau=T-\omega_3+1}^{T-1} p_T^{|\tau|i} \overline{u}_\tau^{(\omega_3)i**} - \frac{(p_T^{|T|i} + \delta p_{T+1}^{|T|i})^2}{4\phi_T^i} \\
&+ \frac{(\chi_T^i)^2}{2(q_T^{|T|i} + \delta q_{T+1}^{|T|i})} - q_T^{|T|i} \frac{(\chi_T^i)^2}{4(q_T^{|T|i} + \delta q_{T+1}^{|T|i})^2} - \sum_{\tau=T-\omega_4+1}^{T-1} q_T^{|\tau|i} \overline{u}_\tau^{(\omega_4)i**} \Bigg] \\
&+ \Bigg[A_{T+1}^{(i)i} \left(\varepsilon_T^{|T|i} \frac{\delta A_{T+1}^{(i)i} \varepsilon_T^{|T|i}}{2\varphi_T^{(x)i}} + \sum_{\tau=T-\omega_1+1}^{T-1} \varepsilon_T^{|\tau|i} \overline{u}_\tau^{(\omega_1)i**} \right)
\end{aligned}$$

$$+ B^i_{T+1}\left(\sum_{j=1}^{n} a_T^{|T|j}\frac{\delta B^j_{T+1}a_T^{|T|j}}{2\varphi_T^{(y)j}} + \sum_{j=1}^{n}\sum_{\tau=T-\omega_2+1}^{T-1} a_T^{|\tau|j}\overline{u}_\tau^{(\omega_2)j**}\right)$$

$$+\left(p_T^{|T|i}\frac{p_T^{|T|i}+\delta p_{T+1}^{|T|i}}{2\phi_T^i} + \sum_{\tau=T-\omega_3+1}^{T-1} p_{T+1}^{|\tau|i}\overline{u}_\tau^{(\omega_3)i**}\right.$$

$$\left.\left. - q_T^{|T|i}\frac{(\chi_T^i)^2}{4(q_T^{|T|i}+\delta q_{T+1}^{|T|i})^2} - \sum_{\tau=T-\omega_4+1}^{T-1} q_{T+1}^{|\tau|i}\overline{u}_\tau^{(\omega_4)i**} + \varpi_{T+1}^i\right)\right]\delta, \tag{5.6}$$

which is a function of previously executed controls $(\underline{\overline{u}}_{T-}^{(\omega_1)**}, \underline{\overline{u}}_{T-}^{(\omega_2)**}, \underline{\overline{u}}_{T-}^{(\omega_3)**}, \underline{\overline{u}}_{T-}^{(\omega_4)**})$.

Then we move to stage $T-1$.

Using $V^i(T,x,y;\underline{\overline{u}}_{T-}^{(\omega_1)**}, \underline{\overline{u}}_{T-}^{(\omega_2)**}, \underline{\overline{u}}_{T-}^{(\omega_3)**}, \underline{\overline{u}}_{T-}^{(\omega_4)**}) = (A_T^{(i)i}x^i + \sum_{j\in N, j\neq i} A_T^{(i)j}x^j + B_T^i y + C_T^i)\delta^{T-1}$ derived in (5.5–5.6) and the stage $T-1$ equation in (3.5), the game equilibrium strategies in stage $T-1$ can be expressed as:

$$u_{T-1}^{Ii**} = \frac{(R_{T-1}^{Ii})^2 x^i}{4(c_{T-1}^{Ii})^2},\ u_{T-1}^{IIi**} = \frac{(R_{T-1}^{IIi})^2 y}{4(c_{T-1}^{IIi})^2},\ \overline{u}_{T-1}^{(\omega_1)i**} = \frac{\delta A_T^{(i)i}\varepsilon_{T-1}^{|T-1|i} + \delta^2 A_{T+1}^{(i)i}\varepsilon_T^{|T-1|i}}{2c_{T-1}^{(x)i}},$$

$$\overline{u}_{T-1}^{(\omega_2)i**} = \frac{\delta B_T^i a_{T-1}^{|T-1|i} + \delta^2 B_{T+1}^i a_T^{|T-1|i}}{2c_{T-1}^{(y)i}},\ \overline{u}_{T-1}^{(\omega_3)i**} = \frac{p_{T-1}^{|T-1|i} + \delta p_T^{|T-1|i} + \delta^2 p_{T+1}^{|T-1|i}}{2\phi_{T-1}^i},$$

$$\overline{u}_{T-1}^{(\omega_4)i**} = \frac{(\chi_{T-1}^i)^2}{4(q_{T-1}^{|T-1|i} + \delta q_T^{|T-1|i} + \delta^2 q_{T+1}^{|T-1|i})^2},\quad \text{for } i \in N. \tag{5.7}$$

Substituting the optimal strategies in (5.7) into the stage $T-1$ equation of (3.5) we obtain

$$V^i(T-1,x,y;\underline{\overline{u}}_{(T-1)-}^{(\omega_1)**}, \underline{\overline{u}}_{(T-1)-}^{(\omega_2)**}, \underline{\overline{u}}_{(T-1)-}^{(\omega_3)**}, \underline{\overline{u}}_{(T-1)-}^{(\omega_4)**})$$

$$= \left(A_{T-1}^{(i)i}x^i + \sum_{j\in N, j\neq i} A_{T-1}^{(i)j}x^j + B_{T-1}^i y + C_{T-1}^i\right)\delta^{T-2}$$

$$= \left[\frac{(R_{T-1}^{Ii})^2 x^i}{4c_{T-1}^{Ii}} + \frac{(R_{T-1}^{IIi})^2 y}{4c_{T-1}^{IIi}} - \frac{(\delta A_T^{(i)i}\varepsilon_{T-1}^{|T-1|i} + \delta^2 A_{T+1}^{(i)i}\varepsilon_T^{|T-1|i})^2}{4\varphi_{T-1}^{(x)i}}\right.$$

$$- \frac{(\delta B_T^i a_{T-1}^{|T-1|i} + \delta^2 B_{T+1}^i a_T^{|T-1|i})^2}{4\varphi_{T-1}^{(y)i}}$$

$$+ p_{T-1}^{|T-1|i}\frac{p_{T-1}^{|T-1|i} + \delta p_T^{|T-1|i} + \delta^2 p_{T+1}^{|T-1|i}}{2\phi_{T-1}^i} + \sum_{\tau=T-\omega_3}^{T-2} p_{T-1}^{|\tau|i}\overline{u}_\tau^{(\omega_3)i**}$$

$$
-\frac{(p_{T-1}^{|T-1|i}+\delta p_{T}^{|T-1|i}+\delta^2 p_{T+1}^{|T-1|i})^2}{4\phi_{T-1}^{i}}+\frac{(\chi_{T-1}^{i})^2}{2(q_{T-1}^{|T-1|i}+\delta q_{T}^{|T-1|i}+\delta^2 q_{T+1}^{|T-1|i})}
$$

$$
\left.-q_{T-1}^{|T-1|i}\frac{(\chi_{T-1}^{i})^2}{4(q_{T-1}^{|T-1|i}+\delta q_{T}^{|T-1|i}+\delta^2 q_{T+1}^{|T-1|i})^2}-\sum_{\tau=T-\omega_4}^{T-2} q_{T-1}^{|\tau|i}\overline{u}_{\tau}^{(\omega_4)i**}\right]\delta^{T-2}
$$

$$
+\left[A_T^{(i)i}x_T^{i***}+\sum_{j\in K(i)}A_T^{(i)j}x_T^{j***}+B_T^i y_T^{***}+C_T^{i***}\right]\delta^{T-1},\quad i\in N, \tag{5.8}
$$

where

$$
x_T^{***i}=x^i+\varepsilon_{T-1}^{|T-1|i}\frac{\delta A_T^{(i)i}\varepsilon_{T-1}^{|T-1|i}+\delta^2 A_{T+1}^{(i)i}\varepsilon_T^{|T-1|i}}{2\varphi_{T-1}^{(x)i}}
$$

$$
+\sum_{\tau=T-\omega_1+1}^{T-2}\varepsilon_{T-1}^{|\tau|i}\overline{u}_{\tau}^{(\omega_1)i**}-\lambda_{T-1}^{(i)}x^i+\sum_{j\in K(i)}\gamma_{T-1}^{(i)j}x^j,
$$

$$
x_T^{***j}=x^j+\varepsilon_{T-1}^{|T-1|j}\frac{\delta A_T^{(j)j}\varepsilon_{T-1}^{|T-1|j}+\delta^2 A_{T+1}^{(j)j}\varepsilon_T^{|T-1|j}}{2\varphi_{T-1}^{(x)j}}
$$

$$
+\sum_{\tau=T-\omega_1+1}^{T-2}\varepsilon_{T-1}^{|\tau|j}\overline{u}_{\tau}^{(\omega_1)j**}-\lambda_{T-1}^{j}x^j+\sum_{\ell\in K(j)}\gamma_{T-1}^{(j)\ell}x^{\ell},
$$

for $j\in K(i)$ and

$$
y_T^{***}=y+\sum_{j=1}^{n}a_{T-1}^{|T-1|j}\frac{\delta B_T^i a_{T-1}^{|T-1|i}+\delta^2 B_{T+1}^i a_T^{|T-1|i}}{2\varphi_{T-1}^{(y)i}}
$$

$$
+\sum_{j=1}^{n}\sum_{\tau=T-\omega_2+1}^{T-2}a_{T-1}^{|\tau|j}\overline{u}_{\tau}^{(\omega_2)j**}-\lambda_{T-1}y,
$$

C_T^{i***} is the C_T^i in (5.6) with

$$
\sum_{\tau=T-\omega_3}^{T-1}p_T^{|\tau|i}\overline{u}_{\tau}^{(\omega_3)i**}=p_T^{|T-1|i}\frac{p_{T-1}^{|T-1|i}+\delta p_T^{|T-1|i}+\delta^2 p_{T+1}^{|T-1|i}}{2\phi_{T-1}^{i}}
$$

$$
+\sum_{\tau=T-\omega_3+1}^{T-2}p_T^{|\tau|i}\overline{u}_{\tau}^{(\omega_3)i**},\quad \sum_{\tau=T-\omega_4}^{T-1}q_T^{|\tau|i}\overline{u}_{\tau}^{(\omega_4)i**}
$$

$$
=q_T^{|T-1|i}\frac{(\chi_{T-1}^{i})^2}{4(q_{T-1}^{|T-1|i}+\delta q_T^{|T-1|i}+\delta^2 q_{T+1}^{|T-1|i})^2}
$$

$$+\sum_{\tau=T-\omega_4+1}^{T-2} q_T^{(\tau)i}\hat{z}_\tau^{i**},\ \sum_{\tau=T+1-\omega_1}^{T-1} \varepsilon_T^{|\tau|i} u_\tau^{(\omega_1)i**}$$

$$= \varepsilon_T^{|T-1|i} \frac{\delta A_T^{(i)i}\varepsilon_{T-1}^{|T-1|i} + \delta^2 A_{T+1}^{(i)i}\varepsilon_T^{|T-1|i}}{2\varphi_{T-1}^{(x)i}}$$

$$+\sum_{\tau=T-\omega_1+1}^{T-2} \varepsilon_T^{|\tau|i} u_\tau^{(\omega_1)i**},\ \sum_{j=1}^{n}\sum_{\tau=T+1-\omega_2}^{T-1} a_T^{|\tau|j}\bar{u}_\tau^{(\omega_2)j**}$$

$$= \sum_{j=1}^{n} a_T^{|T-1|j} \frac{\delta B_T^j a_{T-1}^{|T-1|j} + \delta^2 B_{T+1}^j a_T^{|T-1|j}}{2\varphi_{T-1}^{(y)j}}$$

$$+\sum_{j=1}^{n}\sum_{\tau=T-\omega_2+1}^{T-2} a_T^{|\tau|j}\bar{u}_\tau^{(\omega_2)j**},$$

$$\sum_{\tau=T+1-\omega_3}^{T-1} p_{T+1}^{|\tau|i}\bar{u}_\tau^{(\omega_3)i**} = p_{T+1}^{|T-1|i} \frac{p_{T-1}^{|T-1|i} + \delta p_T^{|T-1|i} + \delta^2 p_{T+1}^{|T-1|i}}{2\phi_{T-1}^i}$$

$$+\sum_{\tau=T-\omega_3+1}^{T-2} p_{T+1}^{|\tau|i}\bar{u}_\tau^{(\omega_3)i**},\ \sum_{\tau=T+1-\omega_4}^{T-1} q_{T+1}^{|\tau|i}\bar{u}_\tau^{(\omega_4)i**}$$

$$= q_{T+1}^{|T-1|i} \frac{(\chi_{T-1}^i)^2}{4(q_{T-1}^{|T-1|i} + \delta q_T^{|T-1|i} + \delta^2 q_{T+1}^{|T-1|i})^2} + \sum_{\tau=T-\omega_2+1}^{T-2} q_{T+1}^{|\tau|i}\bar{u}_\tau^{(\omega_4)i**}.$$

System (5.8) yields equations in which right-hand-side and left-hand-side are linear functions of $x^1, x^2, \cdots, x^n$ and y, for $i \in N$. For (5.8) to hold, it is required:

$$A_{T-1}^{(i)i}x^i = \frac{\left(R_{T-1}^{Ii}\right)^2 x^i}{4c_{T-1}^{Ii}} + \delta A_T^{(i)i}\left(1-\lambda_{T-1}^i\right)x^i$$
$$+\delta\sum_{j\in K(i)} A_T^{(i)j}\gamma_{T-1}^{(j)i}x^i,$$
$$A_{T-1}^{(i)j}x^j = \delta A_T^{(i)i}\gamma_{T-1}^{(i)j}x^j + \delta A_T^{(i)j}\left(1-\lambda_{T-1}^j\right)x^j$$
$$+\delta\sum_{\ell\in K(j)} A_T^{(i)\ell}\gamma_{T-1}^{(\ell)j}x^j,\quad j\in K(i),$$
$$A_{T-1}^{(i)\ell}x^\ell = \delta\sum_{\ell\in K(j)} A_T^{(i)j}\gamma_{T-1}^{(j)\ell}x^\ell + \delta A_T^{(i)\ell}$$
$$(1-\lambda_{T-1}^\ell)x^\ell, \ldots \ell\in K(j) \text{ for } j\in K(i) \text{ and } \ell\notin K(i)$$
$$A_{T-1}^{(i)h} = 0,\ h\notin K(i), h\notin K(j) \text{ for } j\in K(i),$$
$$B_{T-1}^i y = \frac{(R_{T-1}^{IIi})^2 y}{4c_{T-1}^{IIi}} + \delta B_T^i(1-\lambda_{T-1})y. \tag{5.9}$$

From (5.9) we obtain:

$$A_{T-1}^{(i)i} = \frac{\left(R_{T-1}^{Ii}\right)^2}{4c_{T-1}^{Ii}} + \delta A_T^{(i)i}(1-\lambda_{T-1}^i) + \delta \sum\nolimits_{j\in K(i)} A_T^{(i)j}\gamma_{T-1}^{(j)i},$$
$$A_{T-1}^{(i)j} = \delta A_T^{(i)i}\gamma_{T-1}^{(i)j} + \delta A_T^{(i)j}\left(1-\lambda_{T-1}^j\right)$$
$$+\delta \sum\nolimits_{\ell\in K(j)} A_T^{(i)\ell}\gamma_{T-1}^{(\ell)j}, \quad j \in K(i),$$
$$A_{T-1}^{(i)\ell} = \delta \sum\nolimits_{\ell\in K(j)} A_T^{(i)j}\gamma_{T-1}^{(j)\ell} + \delta A_T^{(i)\ell}(1-\lambda_{T-1}^\ell), \ldots$$
$$\ell \in K(j) \text{ for } j \in K(i) \text{ and } \ell \notin K(i),$$
$$A_{T-1}^{(i)h} = 0,\ h \notin K(i),\ h \notin K(j) \text{ for } j \in K(i),$$
$$B_{T-1}^i = \frac{(R_{T-1}^{IIi})^2}{4c_{T-1}^{IIi}} + \delta B_T^i(1-\lambda_{T-1}). \tag{5.10}$$

In addition, C_{T-1}^i is an expression that contains all the terms in the right-hand-side of the system of (5.8) which do not involve $x^1, x^2, \cdots, x^n, y$ including the previously executed controls $(\underline{u}_{(T-1)-}^{(\omega_1)**}, \underline{u}_{(T-1)-}^{(\omega_2)**}, \underline{u}_{(T-1)-}^{(\omega_3)**}, \underline{u}_{(T-1)-}^{(\omega_4)**})$.

Then, we move to stage $T-2$.

Using $V^i(T-1, x, y; \underline{u}_{(T-1)-}^{(\omega_1)**}, \underline{u}_{(T-1)-}^{(\omega_2)**}, \underline{u}_{(T-1)-}^{(\omega_3)**}, \underline{u}_{(T-1)-}^{(\omega_4)**})$ equals. $\left(A_{T-1}^{(i)i}x^i + \sum_{j\in N, j\neq i} A_{T-1}^{(i)j}x^j + B_{T-1}^i y + C_{T-1}^i\right)\delta^{T-2}$ in (5.10) and the stage $T-2$ equation in (3.5), the game equilibrium strategies in stage $T-2$ can be expressed as:

$$u_{T-2}^{Ii**} = \frac{(R_{T-2}^{(x)i})^2 x^i}{4(c_{T-2}^{Ii})^2},\ u_{T-2}^{IIi**} = \frac{(R_{T-2}^{(y)i})^2 y}{4(c_{T-2}^{IIi})^2},$$
$$\overline{u}_{T-2}^{(\omega_1)i**} = \frac{\delta A_{T-1}^{(i)i}\varepsilon_{T-2}^{|T-2)|} + \delta^2 A_T^{(i)i}\varepsilon_{T-1}^{|T-2|i} + \delta^3 A_{T+1}^{(i)i}\varepsilon_T^{|T-2|i}}{2\varphi_{T-2}^{(x)i}},$$
$$\overline{u}_{T-2}^{(\omega_2)i**} = \frac{\delta B_{T-1}^i a_{T-2}^{I|T-2|i} + \delta^2 B_T^i a_{T-1}^{|T-2|i} + \delta^3 B_{T+1}^i a_T^{|T-2|i}}{2\varphi_{T-2}^{(y)i}},$$
$$\overline{u}_{T-2}^{(\omega_3)i**} = \frac{p_{T-2}^{|T-2|i} + \delta p_{T-1}^{|T-2|i} + \delta^2 p_T^{|T-2|i} + \delta^3 p_{T+1}^{|T-2|i}}{2\phi_{T-2}^i}$$
$$\overline{u}_{T-2}^{(\omega_4)i**} = \frac{(\chi_{T-2}^i)^2}{4(q_{T-2}^{|T-2|i} + \delta q_{T-1}^{|T-2|i} + \delta^2 q_T^{|T-2|i} + \delta^3 q_{T+1}^{|T-2|i})^2}, \quad \text{for } i \in N. \tag{5.11}$$

Substituting the optimal strategies in (5.11) into the stage $T-2$ equation of (3.5) we obtain a system of equations in which right-hand-side and left-hand-side are linear functions of $x^1, x^2, \cdots, x^n$ and y, for $i \in N$. For the system to hold, it is required:

$$A_{T-2}^{(i)i} = \frac{\left(R_{T-2}^{Ii}\right)^2}{4c_{T-2}^{Ii}} + \delta A_{T-1}^{(i)i}(1 - \lambda_{T-2}^{i}) + \delta \sum\nolimits_{j\in K(i)} A_{T-1}^{(i)j}\gamma_{T-2}^{(j)i},$$

$$A_{T-2}^{(i)j} = \delta A_{T-1}^{(i)i}\gamma_{T-2}^{(i)j} + \delta A_{T-1}^{(i)j}\left(1 - \lambda_{T-2}^{j}\right) + \delta \sum\nolimits_{\ell\in K(j)} A_{T-1}^{(i)\ell}\gamma_{T-2}^{(\ell)j}, \quad j \in K(i),$$

$$A_{T-2}^{(i)\ell} = \delta \sum\nolimits_{\ell\in K(j)} A_{T-1}^{(i)j}\gamma_{T-2}^{(j)\ell} + \delta A_{T-1}^{(i)\ell}(1 - \lambda_{T-2}^{\ell}), \ldots$$

$\ell \in K(j)$ for $j \in K(i)$ and $\ell \notin K(i)$,

$$A_{T-2}^{(i)h} = 0,\ h \notin K(i),\ h \notin K(j) \text{ for } j \in K(i),$$

$$B_{T-2}^{i} = \frac{(R_{T-2}^{IIi})^2}{4c_{T-2}^{IIi}} + \delta B_{T-1}^{i}(1 - \lambda_{T-2}). \tag{5.12}$$

In addition, C_{T-2}^{i} is an expression that contains the previously executed controls $(\overline{u}_{(T-2)-}^{(\omega_1)**}, \overline{u}_{(T-2)-}^{(\omega_2)**}, \overline{u}_{(T-2)-}^{(\omega_3)**}, \overline{u}_{(T-2)-}^{(\omega_4)**})$.

Following the above analysis for stage $k \in \{T-3, T-4, \cdots, 1\}$ the game equilibrium strategies in stage k can be expressed as:

$$u_k^{Ii**} = \frac{(R_k^{(x)i})^2 x^i}{4(c_k^{Ii})^2},\ u_k^{IIi**} = \frac{(R_k^{IIi})^2 y}{4(c_k^{IIi})^2},$$

$$\overline{u}_k^{(\omega_1)i**} = \frac{\sum_{\tau=k}^{k+\omega_1-1} \delta^{\tau-k+1} A_{\tau+1}^{(i)i}\varepsilon_{\tau}^{|k|i}}{2\varphi_k^{(x)i}}, \quad A_{\tau}^{(i)i} = 0 \text{ for } \tau > T+1;$$

$$\overline{u}_k^{(\omega_2)i**} = \frac{\sum_{\tau=k}^{k+\omega_2-1} \delta^{\tau-k+1} B_{\tau+1}^{i} a_{\tau}^{|k|i}}{2\varphi_k^{(y)i}}, \quad B_{\tau}^{i} = 0 \text{ for } \tau > T+1;$$

$$\overline{u}_k^{(\omega_3)i**} = \frac{\sum_{\tau=k}^{k+\omega_3-1} \delta^{\tau-k} p_{\tau}^{|k|i}}{2\phi_k^{i}}, \quad p_{\tau}^{|k|i} = p_{T+1}^{|k|i} \text{ for } \tau > T+1;$$

$$\overline{u}_k^{(\omega_4)i**} = \frac{(\chi_k^{i})^2}{4\left(\sum_{\tau=k}^{k+\omega_4-1} \delta^{\tau-k} q_{\tau}^{|k|i}\right)^2}, \quad q_{\tau}^{|k|i} = q_{T+1}^{|k|i} \text{ for } \tau > T+1;$$

$$\text{for } i \in N \text{ and } k \in \{1, 2, \cdots, T\}. \tag{5.13}$$

Substituting the optimal strategies in (5.13) into the stage k equation of (3.5) we obtain

$$V^i(k, x, y; \overline{u}_{k-}^{(\omega_1)**}, \overline{u}_{k-}^{(\omega_2)**}, \overline{u}_{k-}^{(\omega_3)**}, \overline{u}_{k-}^{(\omega_4)**})$$
$$= \left(A_k^{(i)i} x^i + \sum_{j\in N, j\neq i} A_k^{(i)j} x^j + B_k^i y + C_k^i \right) \delta^{k-1}, \tag{5.14}$$

where

$$A_k^{(i)i} = \frac{(R_k^{Ii})^2}{4c_k^{Ii}} + \delta A_{k+1}^{(i)i}(1-\lambda_k^i) + \delta \sum_{j\in K(i)} A_{k+1}^{(i)j}\gamma_k^{(j)i},$$
$$A_k^{(i)j} = \delta A_{k+1}^{(i)i}\gamma_k^{(i)j} + \delta A_{k+1}^{(i)j}\left(1-\lambda_k^j\right) + \delta \sum_{\ell\in K(j)} A_{k+1}^{(i)\ell}\gamma_k^{(\ell)j}, \quad j \in K(i),$$
$$A_k^{(i)\ell} = \delta \sum_{\ell\in K(j)} A_{k+1}^{(i)j}\gamma_k^{(j)\ell} + \delta A_{k+1}^{(i)\ell}(1-\lambda_k^\ell), \ldots \ell \in K(j)$$

for $j \in K(i)$ and $\ell \notin K(i)$,

$$A_k^{(i)h} = 0, h \notin K(i), h \notin K(j) \text{ for } j \in K(i)$$
$$B_k^i = \frac{(R_k^{IIi})^2}{4c_k^{IIi}} + \delta B_{k+1}^i(1-\lambda_k),$$

and C_k^i is an expression that contains the previously executed controls

$$(\overline{\underline{u}}_{k-}^{(\omega_1)**}, \overline{\underline{u}}_{k-}^{(\omega_2)**}, \overline{\underline{u}}_{k-}^{(\omega_3)**}, \overline{\underline{u}}_{k-}^{(\omega_4)**}).$$

In particular, the term involving $\overline{\underline{u}}_{k-}^{(\omega_3)**}$ in C_k^i can be obtained as

$$\begin{aligned}
&\sum_{\tau=k-\omega_3+1}^{k-1} p_k^{|\tau|i}\overline{u}_\tau^{(\omega_3)i**} + \delta \sum_{\tau=k-\omega_3+2}^{k-1} p_{k+1}^{|\tau|i}\overline{u}_\tau^{(\omega_3)i**} \\
&+ \delta^2 \sum_{\tau=k-\omega_3+3}^{k-1} p_{k+2}^{|\tau|i}\overline{u}_\tau^{(\omega_3)i**} + \delta^3 \sum_{\tau=k-\omega_3+4}^{k-1} p_{k+3}^{|\tau|i}\overline{u}_\tau^{(\omega_3)i**} \\
&+ \cdots + \delta^{\omega_3-2} \sum_{\tau=k-1}^{k-1} p_{k+\omega_3-2}^{|\tau|i}\overline{u}_\tau^{(\omega_3)i**},
\end{aligned} \tag{5.15}$$

where $p_t^{|\tau|i} = p_{T+1}^{|\tau|i}$ for $t > T+1$.

The term in (5.15) is the present value of the present and future revenues are generated by the revenue generating investments $\overline{\underline{u}}_{k-}^{(\omega_3)**}$.

The term involving $\overline{\underline{u}}_{k-}^{(\omega_4)**}$ in C_k^i can be obtained as

$$\begin{aligned}
&-\sum_{\tau=k-\omega_4+1}^{k-1} q_k^{|\tau|i}\overline{u}_\tau^{(\omega_4)i**} - \delta \sum_{\tau=k-\omega_4+2}^{k-1} q_{k+1}^{|\tau|i}\overline{u}_\tau^{(\omega_4)i**} \\
&- \delta^2 \sum_{\tau=k-\omega_4+3}^{k-1} q_{k+2}^{|\tau|i}\overline{u}_\tau^{(\omega_4)i**} - \delta^3 \sum_{\tau=k-\omega_4+4}^{k-1} q_{k+3}^{|\tau|i}\overline{u}_\tau^{(\omega_4)i**} \\
&- \cdots - \delta^{\omega_4-2} \sum_{\tau=k-1}^{k-1} q_{k+\omega_4-2}^{|\tau|i}\overline{u}_\tau^{(\omega_4)i**},
\end{aligned} \tag{5.16}$$

where $q_t^{|\tau|i} = q_{T+1}^{|\tau|i}$ for $t > T+1$.

The term in (5.16) is the present value of the debt obligations from the previously executed investment projects $\overline{\underline{u}}_{k-}^{(\omega_4)**}$.

The term involving $\underline{\overline{u}}_{k-}^{(\omega_2)**}$ in C_k^i is

$$\delta B_{k+1}^i \sum_{j=1}^{n} \sum_{\tau=k-\omega_2+1}^{k-1} a_k^{|\tau|j} \overline{u}_\tau^{(\omega_2)j**} + \delta^2 B_{k+2}^i \sum_{j=1}^{n} \sum_{\tau=k-\omega_2+2}^{k-1} a_{k+1}^{|\tau|j} \overline{u}_\tau^{(\omega_2)j**}$$
$$+ \delta^3 B_{k+3}^i \sum_{j=1}^{n} \sum_{\tau=k-\omega_2+3}^{k-1} a_{k+2}^{|\tau|j} \overline{u}_\tau^{(\omega_2)j**}$$
$$+ \delta^4 B_{k+4}^i \sum_{j=1}^{n} \sum_{\tau=k-\omega_2+4}^{k-1} a_{k+3}^{|\tau|j} \overline{u}_\tau^{(\omega_2)j**} + \cdots$$
$$+ \delta^{\omega_2-1} B_{k+\omega_2-1}^i \sum_{j=1}^{n} \sum_{\tau=k-1}^{k-1} a_{k+\omega_2-2}^{|\tau|j} \overline{u}_\tau^{(\omega_2)j**}, \tag{5.17}$$

where $B_t^i = 0$ for $t > T+1$.
The term involving $\underline{\overline{u}}_{k-}^{(\omega_1)**}$ in C_k^i is

$$\delta \left(\begin{array}{l} A_{k+1}^{(i)i} \displaystyle\sum_{\tau=k-\omega_1+1}^{k-1} \varepsilon_k^{|\tau|i} \overline{u}_\tau^{(\omega_1)i**} + \sum_{j\in K(i)} A_{k+1}^{(i)j} \sum_{\tau=k-\omega_1+1}^{k-1} \varepsilon_k^{|\tau|j} \overline{u}_\tau^{(\omega_1)j**} \\ + \displaystyle\sum_{\substack{\ell\in K(j)\\ j\in K(i)\\ \ell\notin K(i)}} A_{k+1}^{(i)\ell} \sum_{\tau=k-\omega_1+1}^{k-1} \varepsilon_k^{|\tau|\ell} \overline{u}_\tau^{(\omega_1)\ell**} \end{array} \right)$$
$$+ \delta^2 \left(\begin{array}{l} A_{k+2}^{(i)i} \displaystyle\sum_{\tau=k-\omega_1+2}^{k-1} \varepsilon_{k+1}^{|\tau|i} \overline{u}_\tau^{(\omega_1)i**} + \sum_{j\in K(i)} A_{k+2}^{(i)j} \sum_{\tau=k-\omega_1+2}^{k-1} \varepsilon_{k+1}^{|\tau|j} \overline{u}_\tau^{(\omega_1)j**} \\ + \displaystyle\sum_{\substack{\ell\in K(j)\\ j\in K(i)\\ \ell\notin K(i)}} A_{k+2}^{(i)\ell} \sum_{\tau=k-\omega_1+2}^{k-1} \varepsilon_{k+1}^{|\tau|\ell} \overline{u}_\tau^{(\omega_1)\ell**} \end{array} \right)$$
$$+ \delta^3 \left(\begin{array}{l} A_{k+3}^{(i)i} \displaystyle\sum_{\tau=k-\omega_{1+3}}^{k-1} \varepsilon_{k+2}^{|\tau|i} \overline{u}_\tau^{(\omega_1)i**} + \sum_{j\in K(i)} A_{k+3}^{(i)j} \sum_{\tau=k-\omega_1+3}^{k-1} \varepsilon_{k+2}^{|\tau|j} \overline{u}_\tau^{(\omega_1)j**} \\ + \displaystyle\sum_{\substack{\ell\in K(j)\\ j\in K(i)\\ \ell\notin K(i)}} A_{k+3}^{(i)\ell} \sum_{\tau=k-\omega_1+3}^{k-1} \varepsilon_{k+2}^{|\tau|\ell} \overline{u}_\tau^{(\omega_1)\ell**} \end{array} \right)$$

$$+\,\delta^4\left(\begin{array}{l} A_{k+4}^{(i)i}\displaystyle\sum_{\tau=k-\omega_1+4}^{k-1}\varepsilon_{k+3}^{|\tau|i}\overline{u}_{\tau}^{(\omega_1)i**}+\sum_{j\in K(i)}A_{k+4}^{(i)j}\sum_{\tau=k-\omega_1+4}^{k-1}\varepsilon_{k+3}^{|\tau|j}\overline{u}_{\tau}^{(\omega_1)j**}\\ +\displaystyle\sum_{\substack{\ell\in K(j)\\ j\in K(i)\\ \ell\notin K(i)}}A_{k+4}^{(i)\ell}\sum_{\tau=k-\omega_1+4}^{k-1}\varepsilon_{k+3}^{|\tau|\ell}\overline{u}_{\tau}^{(\omega_1)\ell**}\end{array}\right)$$

$$+\cdots+\delta^{\omega_1-1}\left(A_{k+\omega_1-1}^{(i)i}\sum_{\tau=k-1}^{k-1}\varepsilon_{k+\omega_1-2}^{|\tau|i}\overline{u}_{\tau}^{(\omega_1)i**}+\sum_{j\in K(i)}A_{k+\omega_1}^{(i)j}\sum_{\tau=k-1}^{k-1}\varepsilon_{k+\omega_1-2}^{|\tau|j}\overline{u}_{\tau}^{(\omega_3)j**}\right.$$

$$\left.+\sum_{\substack{\ell\in K(j)\\ j\in K(i)\\ \ell\notin K(i)}}A_{k+\omega_1}^{(i)\ell}\sum_{\tau=k-1}^{k-1}\varepsilon_{k+\omega_1-2}^{|\tau|\ell}\overline{u}_{\tau}^{(\omega_1)\ell**}\right), \tag{5.18}$$

where $A_t^{(i)i}=A_t^{(i)j}=A_t^{(i)\ell}=0$ for $t>T+1$. ■

6 Chapter Notes

This Chapter presents an innovative dynamic game of interactive investments of the firm which incorporates private capitals, public capitals, revenue generating investments and debt liabilities from investments under the effects of durable strategies. Moreover, the value of firm i at the terminal planning stage $T+1$ can be extended to include previous investments in private and public capitals as

$$\left(Q_{T+1}^{(x)i}x_{T+1}^i+Q_{T+1}^{(y)i}y_{T+1}+\sum_{\tau=T+1-\omega_3}^{T}p_{T+1}^{|\tau|i}\overline{u}_{\tau}^{(\omega_3)i}\right.$$

$$-\sum_{\tau=T+1-\omega_4}^{T}q_{T+1}^{|\tau|i}\overline{u}_{\tau}^{(\omega_4)i}+\sum_{\tau=T+2-\omega_1}^{T}p_{T+1}^{(x)|\tau|i}\overline{u}_{\tau}^{(\omega_1)i}$$

$$\left.+\sum_{\tau=T+2-\omega_1}^{T}p_{T+1}^{(y)|\tau|i}\overline{u}_{\tau}^{(\omega_2)i}+\varpi_{T+1}^i\right),\quad\text{for } i\in N, \tag{6.1}$$

where $p_{T+1}^{(x)|\tau|i}\overline{u}_{\tau}^{(\omega_1)i}$ is the terminal value of the firm's investments in private capital $\overline{u}_{\tau}^{(\omega_1)i}$ in stage $\tau\in\{T,T-1,\cdots,T-\omega_1+2\}$ and $p_{T+1}^{(y)|\tau|i}\overline{u}_{\tau}^{(\omega_2)i}$ is the value of the firm's investments in public capital $\overline{u}_{\tau}^{(\omega_2)i}$ in stage $\tau\in\{T,T-1,\cdots,T-\omega_1+2\}$.

In addition, following Yeung and Petrosian (2017) we can extend the analysis to an infinite horizon dynamic game framework through information updating. We consider the updating of future information and the evolution of the game. In initial stage 1, information about the value of the firm (6.1) $T+1$ is known. By the time stage $T+1$ is reached, information about the value of the firm in stage $T+T_1+1$ is known. The firms will face a new game in stage $T+1$ with a game horizon of T_1+1. The process can continue on as new information becomes available. Moreover, the duration of the effectiveness of same type of investments can be different. Similarly, the duration of the effectiveness of same investments can be different among the firms.

Some notable studies on dynamic games involving investments of firms are given below. Feichtinger, et al. (2016) considered R&D for green technologies in a dynamic oligopoly. Cellini and Lambertini (2002) presented a differential game approach to investment in product differentiation. Cellini and Lambertini (2003a) examined advertising in a differential oligopoly game. Dragone, et al. (2010) studied the Leitmann-Schmitendorf advertising game with n players and time discounting. In studies on dynamic games with effects spillover, Cellini and Lambertini (2003b) considered advertising with spillover effects in a differential oligopoly game with differentiated goods and Cellini and Lambertini (2009) examined dynamic R&D with spillovers.

7 Problems

1. (Warmup Exercise) Consider a 2−player 5−stage dynamic game. Firm 1 has a durable strategy $\overline{u}_k^{(2)1}$ and a non-durable strategy u_k^1.

The accumulation process of the private capital x_k^1 of firm 1 is governed by the dynamics

$$x_{k+1}^1 = x_k^1 + 2\overline{u}_k^{(2)1} + \overline{u}_{k-1}^{(2)1} - 0.1x_k^1 + 0.4x_k^2, \quad x_k^1 = 50.$$

The accumulation process of the public capital accessible to both firms is

$$y_{k+1} = y_k + \sum_{j=1}^{2} 2u_k^j - 0.2y_k.$$

The payoff of firm 1 is

$$\sum_{k=1}^{5} \left(4x_k^1 + 3y_k - \left(\overline{u}_\tau^{(2)1}\right)^2 - 2\left(u_\tau^1\right)^2\right)(0.9)^{k-1} + \left(3x_6^1 + 2y_6 + 50\right)(0.9)^5.$$

Firm 2 has a durable strategy $\overline{u}_k^{(2)2}$ and a non-durable strategy u_k^2. The accumulation process of firm 2's private capital x_k^2 is governed by the dynamics

$$x_{k+1}^2 = x_k^2 + 2\overline{u}_k^{(2)1} + \overline{u}_{k-1}^{(2)1} - 0.2x_k^2 + 0.5x_k^1, \quad x_k^2 = 60.$$

The payoff of firm 2 is

$$\sum_{k=1}^{5}\left(3x_k^2 + y_k - \left(\overline{u}_\tau^{(2)2}\right)^2 - \left(u_\tau^2\right)^2\right)(0.9)^{k-1} + \left(3x_6^2 + 2y_6 + 70\right)(0.9)^5.$$

Characterize a feedback Nash equilibrium solution with (i) the game equilibrium strategies and (ii) the game equilibrium value of the firms. (Hints: Follow the analysis in Sect. 3).

2. Consider a 2–firm 10–stage dynamic game in which firm $i \in N$ faces:

(i) a private production capital $x_k^i \in X^i \subset R^+$ which investment, $\overline{u}_k^{(2)i} \in \overline{U}^{(2)i} \subset R$, takes two stages to build up into the firm's capitals. There is spillover through trade so that the private capital of its trading partner can positively affect the increments of its private capital;
(ii) a public capital $y_k \in Y \subset R^+$ which investment, $\overline{u}_k^{(3)i} \in \overline{U}^{(3)i} \subset R$, takes three stages to be fully transformed into the public capital stock;
(iii) a revenue generating investment, $\overline{u}_k^{(4)i} \in \overline{U}^{(4)i} \subset R$, which yields an income stream over four stages and
(iv) a set of debt-financed investments, $\overline{u}_k^{(5)i} \in \overline{U}^{(5)i} \subset R$, which costs are paid by instalments in five stages.

The accumulation process of the private capital stock of firm i is governed by the dynamics

$$x_{k+1}^i = x_k^i + 0.5u_k^{(2)i} + \sum_{\tau=k-1}^{k-1} 0.5\overline{u}_\tau^{(2)i} - 0.05\,x_k^i + 0.1x_k^j, \quad x_1^i = 70, \quad i \in N.$$

The cost of investment in private capital is $(\overline{u}_\tau^{(2)i})^2$.
The accumulation process of the public capital accessible to all firms is

$$y_{k+1} = y_k + \sum\nolimits_{j=1}^{2} 0.4\overline{u}_k^{(3)j} + \sum\nolimits_{j=1}^{2} \sum\nolimits_{\tau=k-2}^{k-1} 0.2\overline{u}_\tau^{(3)j} - 0.1y_k, \quad y_1 = 50.$$

The cost of investment in public capital is $2(u_\tau^{(3)i})^2$.
Firm $i \in \{1, 2\}$ uses its private and public capital together with non-durable productive inputs u_k^{Ii} and u_k^{IIi} to produce outputs. The gross revenue from the outputs produced is $20u_k^{Ii}(x_k^i)^{1/2} + 15u_k^{IIi}(y_k)^{1/2}$. The input cost of u_k^{Ii} is $(2u_k^{Ii})^2$ and the input cost of u_k^{IIi} is $(u_k^{IIi})^2$.

Firm i's revenue generating investment $\overline{u}_k^{(4)i}$ yields incomes in four stages. The income generated in stage $\tau \in \{k, k+1, k+2, k+3\}$ is $3\overline{u}_k^{(4)i}$ and the cost of the revenue generating investments is $2(\overline{u}_k^{(4)i})^2$.

Finally, firm i's debt-financed investment $\overline{u}_k^{(5)i}$ generates a gain $50(\overline{u}_k^{(5)i})^{1/2}$ in stage k and the project costs are paid in five instalments—with a cost payment $2\overline{u}_k^{(5)i}$ in stage $\tau \in \{k+1, k+2, \cdots, k+4\}$.

The payoff of firm $i \in N$ is

$$\sum_{k=1}^{10}\Bigg(20u_k^{Ii}(x_k^i)^{1/2} + 15u_k^{IIi}(y_k)^{1/2} - \left(2u_k^{Ii}\right)^2 - \left(u_k^{IIi}\right)^2 - \left(\overline{u}_\tau^{(2)i}\right)^2 - 2\left(\overline{u}_k^{(3)i}\right)^2$$
$$+\left[\sum_{\tau=k-3}^{k} 3\overline{u}_k^{(4)i} - 2\left(\overline{u}_k^{(4)i}\right)^2\right] + \left[50\left(\overline{u}_k^{(5)i}\right)^{1/2} - \sum_{\tau=k-4}^{k} 2\overline{u}_\tau^{(5)i}\right]\Bigg)(0.9)^{k-1}$$
$$+\left(15x_{11}^i + 5y_{11} + \sum_{\tau=8}^{10} 3\overline{u}_\tau^{(4)i} - \sum_{\tau=7}^{10} 2\overline{u}_\tau^{(5)i} + \varpi_{T+1}^i\right)(0.9)^{10}.$$

Characterize a feedback Nash equilibrium solution with (i) the game equilibrium strategies and (ii) the game equilibrium value of the firms. (Note that the revenues and costs of the outputs are different from the game (3.1–3.3).

References

Cellini R, Lambertini L (2002) A differential game approach to investment in product differentiation. J Econ Dyn Control 27:51–62

Cellini R, Lambertini L (2003) Advertising in a differential oligopoly game. J Optim Theory Appl 116:61–81

Cellini R, Lambertini L (2003) Advertising with spillover effects in a differential oligopoly game with differentiated goods. CEJOR 11:409–423

Cellini R, Lambertini L (2009) Dynamic R&D with spillovers: competition vs cooperation. J Econ Dyn Control 33:568–582

Coe DT, Helpman E (1995) International R&D spillovers. Eur Econ Rev 39(5):859–887

Dragone D, Lambertini L, Palestini A (2010) The Leitmann-Schmitendorf advertising game with n players and time discounting. Appl Math Comput 217:1010–1016

Feichtinger G, Lambertini L, Leitmann G, Wrzaczek S (2016) R&D for green technologies in a dynamic oligopoly: schumpeter. Arrow and Inverted-U's 249:1131–1138

Gould JP (1968) Adjustment costs in the theory of investment of the Örm. Rev Econ Stud 35:47–56

Hayashi F (1982) Tobin's marginal q and average q: a neoclassical interpretation. Econometrica 50(1):213–224

Jorgensen DW (1967) The theory of investment behavior. In: Ferber (ed) Determinants of investment behavior. New York, NBER

Jorgensen DW (1971) Econometric studies of investment behavior: a survey. J Econ Literature 9:1111–1147

Jorgenson DW (1963) Capital theory and investment behavior. American Econ Rev 53:247–259

Lichtenberg F, Pottelsberghe B (1998) International R&D spillovers: a comment. Eur Econ Rev 42(8):1483–1491

Loon P van (1983) A dynamic theory of the firm: Production, finance and investment. Lecture notes in economics and mathematical systems. New York, Springer

Lucas RE Jr (1967) Optimal investment policy and the flexible accelerator. Int Econ Rev 8:78–85

Madsen JB (2007) Technology spillover through trade and TFP convergence: 135 years of evidence for the OECD countries. J Int Econ 72:464–480
Maskus KE (2004) Encouraging international technology transfer. International Centre for Trade and Sustainable Development, Geneva
Parrado R, Cian ED (2014) Technology spillovers embodied in international trade: intertemporal, regional and sectoral effects in a global CGE framework. Energy Econ 41:76–89
Plourde CG, Yeung DWK (1988) A note on Hayashi's neoclassical interpretation of Tobin's q. Atlantic Econ J XV I(2):80–81
Treadway A (1969) On rational entrepreneurial behaviour and the demand for investment. Rev Econ Stud 36:227–239
Yeung DWK, Petrosian O (2017) Infinite horizon dynamic games: a new approach via information updating. Int Game Theory Rev 19(4):1750026. https://doi.org/10.1142/S0219198917500268

Chapter 4
Dynamic Environmental Games: Anthropogenic Eco-degradation under Strategies Lags

Damages to ecosystems and the global environment, in general, have increasingly been become major concerns. Anthropogenic impacts on the environment are the prime contributors to our catastrophe bound environmental problems. The term anthropogenic designates an effect or object resulting from human activity. Anthropogenic environmental impacts are much more than just pollution accumulation, global warming or climate change. They also include changes to biophysical environments and ecosystems, depletion and destruction of natural resources, environmental degradation, acidification, massive species extinction, biodiversity loss, toxic soil and water, ecological crises and even ecological collapse. Altering the environment in the process of satisfying the needs of society can cause detrimental effects to the survival of human beings. Recent rapid technological advancement and economic growth have led to an unprecedented level of increment in environmental damages brought about by human deeds through industrial pollution and ecosystems degradation. Due to the geographical diffusion of harmful anthropogenic environmental impacts, unilateral response on the part of one nation or region is often ineffective. Major international environmental management initiatives, like the Paris Agreement, which focuses mainly on emissions reduction and permit trading can hardly be expected to provide an effective means to solve the problem.

Green technology development has been identified as one of the most effective instruments in solving our catastrophe bound environmental crisis. The positive externalities in green technology development are becoming significant and the negative externalities of environmental damages have long been observed. Moreover, human actions that lead to pollution creation, ecosystems degradation and green technology advancement are strategies having lag effects. Dynamic game theoretic analysis with durable strategies becomes one of the most realistic paradigms to study environmental problems.

This chapter presents a class of dynamic environmental games involving anthropogenic eco-degradation and green technology development under strategies lags. Section 1 gives an overview of pollution and ecosystem degradation.

D. W. K. Yeung and L. A. Petrosyan, *Durable-Strategies Dynamic Games*,
Theory and Decision Library C 50, https://doi.org/10.1007/978-3-030-92742-4_4

Section 2 develops a generic class of durable-strategies environmental games of eco-degradation. Section 3 presents an environmental game under durable strategies with explicit functional forms. Section 4 examines anthropogenic eco-degradation and green technology developments under strategy lags in a game theoretic framework. Section 5 contains an appendix and chapter notes are provided in Sect. 6.

1 Pollution and Ecosystem Degradation

Natural ecosystems are made up of abiotic factors and biotic factors. They contain abiotic resources such as air, water, soil, rocks and energy and interacting life forms of individual organisms such as plants, animals and micro-organisms that aggregate into populations which aggregate into ecological communities. Ecosystems are dynamic and changing over time. For an ecosystem to be stable and sustainable, it needs to be in equilibrium. Disruptions that cause a shift from one equilibrium to another can be reversible or non-reversible. In fact, our ability to exist continuously requires the coexistence of the biosphere and human civilization. We need a balanced environment in which the exploitation of resources, the direction of investments, the orientation of technological development and institutional change are all in harmony ecologically so that the current and future potentials to meet human needs and aspirations can be maintained.

As the human population and economic activities are growing rapidly with human beings increasingly dominate available resources, ecosystems begin to show increasing signs of stress. Loss of biodiversity, environmental degradation and conflict over resources among the dominant species are typical signs that a biological system is nearing a state of change, which could range from collapse of the dominant species, development of alternative biological communities, to the collapse of the entire system. Some changes can be reversed given adequate time for recovery. Some changes are irreversible and their impacts on our ecosystem can be permanent and potentially devastating.

Pollution is one of the most damaging elements to ecosystems and biodiversity. Two most widely spread forms of pollution are air pollution and water pollution. Air pollution is the release of pollutants like chemicals and particulates into the atmosphere. Common gaseous pollutants include carbon monoxide, sulfur dioxide, chlorofluorocarbons (CFCs) and nitrogen oxides produced by industry and motor vehicles. Photochemical ozone and smog are created as nitrogen oxides and hydrocarbons react to sunlight. Water pollution, by the discharge of commercial and industrial wastes (intentionally or unintentionally through spills in accidents) into surface waters. Discharges of untreated domestic sewage, chemical contaminants, chemical fertilizers, pesticide and human faeces into surface runoff flowing to surface waters constitute serious cases of water pollution. Dumping of toxic materials, radioactive water and plastics into oceans becomes a recent source of water pollution.

Pollution is not just the emission of harmful pollutants, it also damages the entire ecosystem by reducing its capability to regenerate. Ultimately, pollution affects the ability of ecosystems to provide ecosystem services, such as nutrient cycling, carbon cycling and clean water provision, on which the earth and human life are dependent. Millions of deaths are due to pollution and tens of millions of people's health conditions are affected every year.

Besides pollution, the degradation of ecosystems is also affected by many anthropogenic factors. Damaging anthropogenic impacts on the environment comes from six major categories of sources.

The first category of anthropogenic impacts is related to agriculture, fishery and irrigation. (i) The environmental impacts on agriculture involve a variety of factors ranging from the soil to water, the air, animal and soil diversity, plants and the food supply itself. Deforestation, irrigation problems, pollutants, soil degradation, genetic engineering and waste are also damages brought about by agriculture. (ii) The environmental impacts on fishery include the decrease or demise of the biomass of especially through overfishing, unsustainable fisheries and fisheries mismanagement. The disruption of the marine food chain and negative impact on other elements of the environment, for instance, the destruction of marine habitats such as coral reefs (see Oppenlander (2013)) are also often observed. (iii) Irrigation projects can have negative side effects that are often overlooked (see Thakkar (1999)). The massive depletion of freshwater severely damages surrounding ecosystems and contributes to the wipe-out of aquatic species (see Pearce (2006)).

The second category of anthropogenic impacts is related to the energy industry. (i) The environmental impact of coal combustion and coal mining is serious and diverse. Coal contains sulfur and other dangerous metals that emit into the air when coal is burned. (ii) The environmental impact of electricity generation is significant because of the heavy use of electricity in modern society. Although electricity is a relatively safe form of energy, the generation and transmission of electricity affects the environment. Almost all mechanisms for generating electricity release carbon dioxide and other greenhouse gases. (iii) The environmental impact of nuclear power stems from nuclear fuel cycle processes – like mining, processing, transporting and storing fuel and radioactive fuel waste. The release of radioisotopes poses a health danger to human populations, animals and plants. The environmental impacts of nuclear plant accidents persist for an extremely long time or even indefinitely. (iv) The environmental impact of petroleum is highly undesirable because its use and extraction are toxic to almost all forms of life.

The third category of anthropogenic impacts is related to manufactured products. The environmental impacts of manufactured goods like cleaning agents, nanotechnological products, paints, paper, plastics, pesticides, electronic products, cars, vehicles and buildings are substantial. Their high levels of consumption, resources utilized, pollutants emitted and waste generated cause significant damages to the ecosystems.

The fourth category of anthropogenic impacts is related to mining and transportation. (i) The environmental impact of mining can be devastating because some mining methods have significant negative environmental and public health effects. Soil erosion, formation of sinkholes, loss of biodiversity and contamination of soil and water by chemicals from mining processes create serious environmental damages. (ii) The environmental impact of transport is significant because it is a major user of energy and consumes most of the world's petroleum. With faster growth in the number of automobiles around the world emission of carbon dioxide in the transportation sector becomes one of the largest contributors to global warming. (iii) The environmental impact of aviation comes from noise, particulates and gases emitted by aircraft engines. The rapid growth of air travel in recent years contributes to an increase in total pollution attributable to aviation. (vi) The environmental impact of shipping comes mainly from greenhouse gas emissions and oil spills. Invasive species were brought into new areas through shipping, usually by attaching themselves to the ship's hull.

The fifth category of anthropogenic impacts is overconsumption and over-extraction. Overconsumption and over-extraction. lead to a situation where resource use has outpaced the sustainable capacity of the ecosystem. Prolonged over-extraction leads to environmental degradation and the eventual loss of resource bases and even irreversible damages to the ecosystem.

The sixth category of anthropogenic impacts is waste disposal, war and military activities. (i) Waste disposal contributes significantly to the degradation and contamination. The disposal of solid waste, electronic appliances, chemical waste, pharmaceuticals, radioactive materials, plastics, oil products and toxic waste is posting a danger to the environment. (ii) There are significant environmental impacts from war and military activities like nuclear bomb testing. Scorched earth tactics during or after war were not uncommon in traditional warfare throughout history. And modern technological warfare, with the use of missiles and chemical weapons, can cause even a far greater devastation effect.

Worthwhile to point out is that anthropogenic activities affecting pollution and/or the ecosystems cause durable effects. Deforestation by lumber firms would cause increases in carbon dioxide for a certain period of time. Toxic and chemical wastes from industrial production will continue to produce damaging elements that add to the pollution stock for a lengthy period. In addition, anthropogenic activities are carried out by many parties and externalities are prevalent. A dynamic environmental game under durable strategies will be presented in the next section.

2 Durable-Strategies Environmental Games of Eco-degradation

In addition to abatement efforts, the development of green technologies offers the most potent solution to alleviate environmental problems. In this section, we develop an analytical framework for formulating a general class of dynamic environmental games involving ecosystem degradation under durable interactive strategies. The game strategies include pollution generating outputs, ecosystems degrading outputs, clean outputs, pollution abatement efforts, ecosystems repairing efforts and green technologies R&D. Then, we characterize the game equilibrium solution.

2.1 Game Formulation

Consider a $n-$ player $T-$ stage dynamic environmental game in which player/region $i \in N$ produces three types of outputs. The first type of outputs $\overline{u}_k^{(\omega_1)i} \in \overline{U}^{(\omega_1)i} \subset R^{m^{(\omega_3)i}}$ emits pollutants over ω_1 stages. The second type of outputs $\overline{u}_k^{(\omega_2)i} \in \overline{U}^{(\omega_2)i} \subset R^{m^{(\omega_2)i}}$ causes damages to the ecosystems over ω_2 stages. The vector of pollution stocks is $x_k \in X \subset R^{n_1}$ and the dynamics of pollution accumulation is governed by the vector dynamical equation:

$$x_{k+1} = x_k + \sum_{j=1}^{n} m_k^{|k|j}(\overline{u}_k^{(\omega_1)j})$$
$$+ \sum_{j=1}^{n} \sum_{\tau=k-\omega_1+1}^{k-1} m_k^{|\tau|j}(\overline{u}_\tau^{(\omega_1)j}) - \sum_{j=1}^{n} b_k^j(u_k^{Ij}, x_k) - \lambda_k^x(x_k)$$
$$x_1 = x_1^0, \tag{2.1}$$

where $m_k^{|k|j}(\overline{u}_k^{(\omega_1)j})$ is the amount of pollutants created in stage k by player j's stage k polluting output $\overline{u}_k^{(\omega_1)j}$, $m_k^{|\tau|j}(\overline{u}_\tau^{(\omega_1)j})$ is the amount of pollutants created in stage k by player j's stage $\tau \in \{k-1, k-2, \cdots, k-\omega_1+1\}$ output, $b_k^j(u_k^{Ij}, x_k)$ is the amount of pollutants removed by u_k^{Ij} amounts of non-durable abatement effort from region j in stage k; $\lambda_k^x(x_k)$ is the natural decay of the pollutants; and $\overline{u}_\tau^{(\omega_1)j}$, for $\tau \in \{0, -1, -2, \cdots, -\omega_1+1\}$ and $j \in N$, is known with some or all of them being zero.

A vector of ecosystems $y_k \in Y \subset R^{n_2}$ which conditions generate benefits to the regions. The dynamics governing the evolution of the conditions of the ecosystems is

$$y_{k+1} = y_k + g_k(y_k) - \sum_{j=1}^{n} \sigma_k^{|k|j}(\overline{u}_k^{(\omega_2)j})$$
$$- \sum_{j=1}^{n} \sum_{\tau=k-\omega_2+1}^{k-1} \sigma_k^{|\tau|j}(\overline{u}_\tau^{(\omega_2)j}) + \sum_{j=1}^{n} \varepsilon_k^j(u_k^{IIj}, y_k)$$
$$- \vartheta_k(x_k) - \lambda_k^y(y_k),$$
$$y_1 = y_{1z}^0, \tag{2.2}$$

where $\sigma_k^{|\tau|j}(\overline{u}_\tau^{(\omega_2)j})$ are the damages of the ecosystems by output $\overline{u}_\tau^{(\omega_2)j}$ in stage $\tau \in \{k, k-1, \cdots, k-\omega_2+1\}$, u_k^{IIj} are the non-durable inputs in stage k for repairing the ecosystems, $\varepsilon_k^j(u_k^{IIj}, y_k)$ is the improvements by u_k^{IIj} to the ecosystems, $g_k(y_k)$

are the natural replenishments of the ecosystems and $\lambda_k^y(y_k)$ are the natural decay of ecosystems and $\vartheta_k(x_k)$ are the damages to ecosystems by pollution stocks x_k.

Again $\overline{u}_\tau^{(\omega_2)j}$, for $\tau \in \{0, -1, -2, \cdots, -\omega_1 + 1\}$ and $j \in N$, is known with some or all of them being zero. In the case of no human interventions in the ecosystems a natural equilibrium of the ecosystems is.

$$g_k(y_k) = \vartheta_k(x_k) + \lambda_k^y(y_k).$$

The third type of outputs is clean outputs which are produced by non-durable inputs and green technologies $z_k \in Z \subset R^{n_3}$ The dynamics governing the accumulation of green technologies is.

$$z_{k+1} = z_k + \sum_{j=1}^{n} a_k^{k|j}\left(\overline{u}_k^{(\omega_3)j}\right) + \sum_{j=1}^{n} \sum_{\tau-k-\omega_3+1}^{k-1} a_k^{|\tau j}\left(\overline{u}_\tau^{(\omega_3)j}\right) - \lambda_k^z(z_k), z_1 = z_1^0, \quad (2.3)$$

where $u_k^{(\omega_3)j} \in U^{(\omega_3)j} \subset R^{m_3^j}$ are the investments in the public green technologies by region j in stage k and it would take ω_3 stages to fully transmit the investments into the public technologies, $a_k^{|k|j}(u_k^{(\omega_2)j})$ are the additions to the public green technologies in stage k by player j's technology investment in stage k, the term $a_k^{|\tau|j}(\overline{u}_\tau^{(\omega_3)j})$ is the addition to the public green technologies in stage k by player j's technology investment in stage $\tau \in \{k-1, k-2, \cdots, k-\omega_3+1\}$ and $\lambda_k^z(z_k)$ are the obsolescence of the green technologies.

The values of $\overline{u}_\tau^{(\omega_3)j}$, for $\tau \in \{0, -1, -2, \cdots, -\omega_1 + 1\}$ and $j \in N$, are known with some or all of them being zero. The realized green technologies from the public technologies for region i are $\kappa_k^i z_k$. The region uses non-durable inputs $u_k^{IIIi} \in U^{IIIi} \subset R^{m^{IIIi}}$ together with its realized green technologies to produce clean outputs.

The payoff of region $i \in N$ is

$$\begin{aligned}
&\sum_{k=1}^{T} \Big(R_k^{(x)i}\left(\bar{u}_k^{(\varphi)i}\right) - \varphi_k^{(x)i}\left(\bar{u}_k^{(\varphi)i}\right) - c_k^{Ii}\left(u_k^{Ii}\right) \\
&\quad + R_k^{(y)i}\left(\bar{u}_k^{(\omega_2)i}\right) - \varphi_k^{(y)i}\left(\bar{u}_k^{(\omega_2)i}\right) - c_k^{IIi}\left(u_k^{IIi}\right) \\
&\quad + R_k^{(z)i}\left(q_k^{(z)i}\left(u_k^{III_i}, \kappa_k^i z_k\right)\right) - \varphi_k^{(z)i}\left(\bar{u}_k^{((a)i}\right) \\
&\quad - c_k^{IIL_k}\left(u_k^{III_i}\right) - h_k^i(x_k) + \chi_k^i(y_k)\Big)\delta_1^k \\
&\quad + \left(Q_{T+1}^{x(i)}(x_{T+1}) + Q_{T+1}^{y(i)}(y_{T+1}) + Q_{T+1}^{z(i)}\left(\kappa_{T+1}^i z_{T+1}\right) + \varpi_{T+1}^i\right)\delta_1^{T+1}. \quad (2.4)
\end{aligned}$$

In particular, $R_k^{(x)i}(\overline{u}_k^{(\omega_1)i})$ are the gross revenues from outputs creating pollution, $\varphi_k^{(x)i}(\overline{u}_k^{(\omega_1)i})$ are the costs of producing the output and the costs of pollution abatements are $c_k^{Ii}(u_k^{Ii})$. $R_k^{(y)i}(\overline{u}_k^{(\omega_2)i})$ are the gross revenues from outputs leading to damages in the ecosystems, $\varphi_k^{(y)i}(\overline{u}_k^{(\omega_2)i})$ are the costs of producing the outputs and the costs of improving the ecosystems are $c_k^{IIi}(u_k^{IIi})$. The term $(q_k^{(z)i}(u_k^{IIIi}, \kappa_k^i z_k))$ is the production functions of the clean outputs which use the region's realized green capitals $\kappa_k^i z_k$ and non-durable inputs $u_k^{III(1)i}$. The term $R_k^{(z)i}(q_k^{(z)i}(u_k^{IIIi}, \kappa_k^i z_k))$ is the gross revenues from clean outputs using green technologies and $c_k^{IIIi}(u_k^{IIIi})$ are the costs of non-durable inputs and the investment costs in public green technologies are $\varphi_k^{(z)i}(\overline{u}_k^{(\omega_3)i})$. The damages from pollutions are $h_k^i(x_k)$ and the environmental amenities from the ecosystems are $\chi_k^i(y_k)$. The terminal payoff of region i in stage $T+1$ is $\left(Q_{T+1}^{x(i)}(x_{T+1}) + Q_{T+1}^{y(i)}(y_{T+1}) + Q_{T+1}^{z(i)}(\kappa_{T+1}^i z_{T+1}) + \varpi_{T+1}^i\right)$.

2.2 Game Equilibrium

Invoking Theorem 3.1 in Chap. 2, we characterize the feedback equilibrium of the game (2.1–2.4) as follow.

Corollary 2.1 *Let* $(u_k^{I**}, u_k^{II**}, u_k^{III**}, \overline{u}_k^{(\omega_1)**}, \overline{u}_k^{(\omega_2)**}, \overline{u}_k^{(\omega_3)**})$ *be the set of feedback Nash equilibrium strategies and* $V^i(k, x, y, z; \underline{u}_{k-}^{(\omega_1)**}, \underline{u}_{k-}^{(\omega_2)**}, \underline{u}_{k-}^{(\omega_3)**})$ *be the feedback Nash equilibrium payoff of region i at stage k in the durable-strategies dynamic game (2.1–2.4). The functions* $V^i(k, x, y, z; \underline{u}_{k-}^{(\omega_1)**}, \underline{u}_{k-}^{(\omega_2)**}, \underline{u}_{k-}^{(\omega_3)**})$, *for* $k \in \{1, 2, \cdots, T+1\}$ *and* $i \in N$, *satisfy the following recursive equations:*

$$V^i(T+1, x, y, z; \underline{\overline{u}}_{(T+1)-}^{(\omega_1)**}, \underline{\overline{u}}_{(T+1)-}^{(\omega_2)**}, \underline{\overline{u}}_{(T+1)-}^{(\omega_3)**})$$

$$= \left(Q_{T+1}^{x(i)}(x_{T+1}) + Q_{T+1}^{y(i)}(y_{T+1}) + Q_{T+1}^{z(i)}(\kappa_{T+1}^i z_{T+1}) + \varpi_{T+1}^i \right)\delta_1^{T+1}, \tag{2.5}$$

$$V^i(k, x, y, z; \underline{\overline{u}}_{k-}^{(\omega_1)**}, \underline{\overline{u}}_{k-}^{(\omega_2)**}, \underline{\overline{u}}_{k-}^{(\omega_3)**})$$

$$= \max_{\substack{u_k^{Ii}, u_k^{IIi}, u_k^{IIIi}, \\ \overline{u}_k^{(\omega_1)i}, \overline{u}_k^{(\omega_2)i}, \overline{u}_k^{(\omega_3)i}}} \Bigg\{ \Bigg(R_k^{(x)i}(\overline{u}_k^{(\omega_1)i}) - \varphi_k^{(x)i}(\overline{u}_k^{(\omega_1)i})$$

$$- c_k^{Ii}\left(u_k^{Ii}\right) + R_k^{(y)i}\left(\overline{u}_k^{(\omega_2)i}\right) - \varphi_k^{(y)i}\left(\overline{u}_k^{(\omega_2)i}\right) - c_k^{IIi}\left(u_k^{IIi}\right)$$

$$+ R_k^{(z)i}\left(q_k^{(z)i}\left(u_k^{\mathrm{III}}, \kappa_k^i z\right)\right) - \varphi_k^{(z)i}\left(\overline{u}_k^{(\omega_3)i}\right) - c_k^{\mathrm{mi}}\left(u_k^{?\mathrm{i}}\right) - h_k^i(x) + \chi_k^i(y)\Bigg)\delta_1^k$$

$$+ V^i\left(k+1, x_{k+1}^{***}, y_{k+1}^{***}, z_{k+1}^{***}; \overline{u}_{(k+1)-}^{(\omega_1)}, \underline{u}_{(k+1)-}^{(\omega_2)*}, \underline{\vec{u}}_{(k+1)-}^{(\omega_3)***}\right)\Bigg\}, \tag{2.6}$$

where

$$x_{k+1}^{***} = x + m_k^{|k|i}(\overline{u}_k^{(\omega_1)i}) + \sum_{\substack{j \in N \\ j \neq i}} m_k^{|k|j}(\overline{u}_k^{(\omega_1)j**}) + \sum_{j=1}^{n} \sum_{\tau=k-\omega_1+1}^{k-1} m_k^{|\tau|j}(\overline{u}_\tau^{(\omega_1)j**})$$

$$-b_k^i(u_k^{Ii}, x) - \sum_{\substack{j \in N \\ j \neq i}} b_k^j(u_k^{Ij**}, x) - \lambda_k^x(x),$$

$$V^i(k, x, y, z; \overline{u}_{k-}^{(\omega_1)**}, \overline{u}_{k-}^{(\omega_2)**}, \overline{u}_{k-}^{(\omega_3)**})$$
$$= \max_{\substack{u_k^{Ii}, u_k^{IIi}, u_k^{IIIi}, \\ \overline{u}_k^{(\omega_1)i}, \overline{u}_k^{(\omega_2)i}, \overline{u}_k^{(\omega_3)i}}} \Big\{ \Big(R_k^{(x)i}(\overline{u}_k^{(\omega_1)i}) - \varphi_k^{(x)i}(\overline{u}_k^{(\omega_1)i})$$
$$- c_k^{Ii}\left(u_k^{Ii}\right) + R_k^{(y)i}\left(\overline{u}_k^{(\omega_2)i}\right) - \varphi_k^{(y)i}\left(\overline{u}_k^{(\omega_2)i}\right) - c_k^{IIi}\left(u_k^{IIi}\right)$$
$$+ R_k^{(z)i}\left(q_k^{(z)i}\left(u_k^{\mathrm{III}}, \kappa_k^i z\right)\right) - \varphi_k^{(z)i}\left(\overline{u}_k^{(\omega_3)i}\right) - c_k^{\mathrm{mi}}\left(u_k^{?i}\right) - h_k^i(x) + \chi_k^i(y)\Big)\delta_1^k$$
$$+ V^i\left(k+1, x_{k+1}^{***}, y_{k+1}^{***}, z_{k+1}^{***}; \overline{u}_{(k+1)-}^{(\omega_1)}, \underline{u}_{(k+1)-}^{(\omega_2)*}, \vec{\underline{u}}_{(k+1)-}^{(\omega_3)***}\right)\Big\},$$

Performing the indicated maximization operator in Corollary 2.1, we obtain the following game equilibrium conditions.

(i) Outputs creating pollution, that is $\overline{u}_k^{(\omega_1)i}$, will be produced up to the levels

$$\frac{\partial R_k^{(x)i}\left(\overline{u}_k^{(\omega_1)i}\right)}{\partial \overline{u}_k^{(\omega_1)i}} \delta_1^k = \frac{\partial \varphi_k^{(x)i}\left(\overline{u}_k^{(\omega_1)i}\right)}{\partial \overline{u}_k^{(\omega_1)i}} \delta_1^k - \frac{\partial V_{k+1}^i}{\partial x_{k+1}^{***}} \frac{\partial x_{k+1}^{***}}{\partial \overline{u}_k^{(\omega_1)i}} - \frac{\partial V_{k+1}^i}{\partial \overline{u}_k^{(\omega_1)i}}, \quad \text{for } i \in N, \tag{2.7}$$

where V_{k+1}^i is the short form of $V^i(k+1, x_{k+1}^{***}, y_{k+1}^{***}, z_{k+1}^{***}; \overline{u}_{(k+1)-}^{(\omega_1)***}, \overline{u}_{(k+1)-}^{(\omega_2)***}, \overline{u}_{(k+1)-}^{(\omega_3)***})$

In particular, $\frac{\partial R_k^{(x)i}\left(\overline{u}_k^{(\omega_1)i}\right)}{\partial \overline{u}_k^{(\omega_1)i}} \delta_1^k$ are the marginal gross revenues from outputs creating pollution, $\frac{\partial \varphi_k^{(x)i}\left(\overline{u}_k^{(\omega_1)i}\right)}{\partial \overline{u}_k^{(\omega_1)i}} \delta_1^k$ are the marginal costs of producing the polluting outputs, $-\frac{\partial V_{k+1}^i}{\partial x_{k+1}^{***}} \frac{\partial x_{k+1}^{***}}{\partial \overline{u}_k^{(\omega_1)i}}$ are the marginal effects of $u_k^{(\omega_1)i}$ on the region i's payoff through the increments in pollution in stage $k+1$ and $-\frac{\partial V_{k+1}^i}{\partial u_k^{(\omega_1)i}}$ are the marginal effects of $u_k^{(\omega_1)i}$ on the region i's payoff in pollution from stage $k+1$ to stage $k+\omega_1-1$. Thus, in a game equilibrium, outputs creating pollution will be produced up to the levels where the marginal gross revenues equal the sum of the marginal production costs and the marginal environmental effects from pollution.

(ii) Outputs damaging the ecosystems will be produced up to the levels

$$\frac{\partial R_k^{(y)i}\left(\overline{u}_k^{(\omega_2)i}\right)}{\partial \overline{u}_k^{(\omega_2)i}} \delta_1^k = \frac{\partial \varphi_k^{(y)i}\left(\overline{u}_k^{(\omega_2)i}\right)}{\partial \overline{u}_k^{(\omega_2)i}} \delta_1^k - \frac{\partial V_{k+1}^i}{\partial y_{k+1}^{***}} \frac{\partial y_{k+1}^{***}}{\partial \overline{u}_k^{(\omega_2)i}} - \frac{\partial V_{k+1}^i}{\partial \overline{u}_k^{(\omega_2)i}}, \quad \text{for } i \in N. \tag{2.8}$$

In particular, $\frac{\partial R_k^{(y)i}\left(\overline{u}_k^{(\omega_2)i}\right)}{\partial \overline{u}_k^{(\omega_2)i}}\delta_1^k$ are the marginal gross revenues from outputs damaging the ecosystems, $\frac{\partial \varphi_k^{(y)i}\left(\overline{u}_k^{(\omega_2)i}\right)}{\partial \overline{u}_k^{(\omega_2)i}}\delta_1^k\delta_1^k$ are the marginal costs of producing the ecosystems damaging outputs, $-\frac{\partial V_{k+1}^i}{\partial y_{k+1}^{***}}\frac{\partial y_{k+1}^{***}}{\partial \overline{u}_k^{(\omega_2)i}}$ are the marginal effects of $\overline{u}_k^{(\omega_2)i}$ on the region i's payoff through the damage of the ecosystems in stage $k+1$ and $-\frac{\partial V_{k+1}^i}{\partial \overline{u}_k^{(\omega_2)i}}$ are the marginal effects of $u_k^{(\omega_2)i}$ on the region i's payoff in the damage of the ecosystems in stage $k+1$ to stage $k+\omega_1-1$. Thus, in a game equilibrium, outputs damaging the ecosystem's pollution will be produced up to the levels where the marginal gross revenues equal the sum of the marginal production costs and the marginal environmental effects from damages to the ecosystems.

(iii) Investments for building up green technologies, that is $\overline{u}_k^{(\omega_3)i}$, will be employed up to the levels

$$\frac{\partial \varphi_k^{(z)i}\left(\overline{u}_k^{(\omega_3)i}\right)}{\partial \overline{u}_k^{(\omega_3)i}}\delta_1^k = \frac{\partial V_{k+1}^i}{\partial z_{k+1}^{***}}\frac{\partial z_{k+1}^{***}}{\partial \overline{u}_k^{(\omega_3)i}} + \frac{\partial V_{k+1}^i}{\partial \overline{u}_k^{(\omega_3)i}}, \quad \text{for } i \in N. \tag{2.9}$$

In particular, $\frac{\partial \varphi_k^{(z)i}\left(\overline{u}_k^{(\omega_3)i}\right)}{\partial \overline{u}_k^{(\omega_3)i}}\delta_1^k$ are the marginal costs of investments for building up green technologies, $\frac{\partial V_{k+1}^i}{\partial z_{k+1}^{***}}\frac{\partial z_{k+1}^{***}}{\partial \overline{u}_k^{(\omega_3)i}}$ are the marginal contributions to region i's payoff through the increments in green technologies in stage $k+1$ and $\frac{\partial V_{k+1}^i}{\partial \overline{u}_k^{(\omega_3)i}}$ are the marginal contributions to region i's payoff from the green technologies in stage $k+1$ to stage $k+\omega_1-1$.

(iv) Pollution abatement efforts are employed up to the levels where the marginal costs equal the marginal benefits, that is

$$\frac{\partial c_k^{Ii}\left(u_k^{Ii}\right)}{\partial u_k^{Ii}}\delta_1^k = \frac{\partial V_{k+1}^i}{\partial x_{k+1}^{***}}\frac{\partial x_{k+1}^{**+1}}{\partial u_k^{II}}, \quad \text{for } i \in N. \tag{2.10}$$

(v) Ecosystems improvement efforts are employed up to the levels where the marginal costs equal the marginal benefits, that is

$$\frac{\partial c_k^{IIi}\left(u_k^{IIi}\right)}{\partial u_k^{Ii}}\delta_1^k = \frac{\partial V_{k+1}^i}{\partial y_{k+1}^{***}}\frac{\partial y_{k+1}^{***}}{\partial u_k^{IIi}}, \quad \text{for } i \in N. \tag{2.11}$$

(vi) Non-durable inputs used to produce the clean outputs are employed up to the levels where the marginal revenues of the clean outputs equal the marginal costs, that is

$$\frac{\partial R_k^{(z)i}\left(q_k^{(z)i}(u_k^{III},\kappa_k^i z)\right)}{\partial q_k^{(z)i}(u_k^{IIIi},\kappa_k^i z)}\frac{\partial q_k^{(z)i}(u_k^{III},\kappa_k^i z)}{\partial u_k^{III}} = \frac{\partial c_k^{III}(u_k^{III})}{\partial u_k^{III}}, \text{ for } i \in N. \quad (2.12)$$

If the game equilibrium conditions in (2.7–2.12) satisfy the implicit function theorem, the game equilibrium strategies $(u_k^{I**}, u_k^{II**}, u_k^{III**}, \overline{u}_k^{(\omega_1)**}, \overline{u}_k^{(\omega_2)**}, \overline{u}_k^{(\omega_3)**})$, for $i \in N$, can be obtained as explicit functions of $(x, y, z; \overline{u}_{k-}^{(\omega_1)**}, \overline{u}_{k-}^{(\omega_2)**}, \overline{u}_{k-}^{(\omega_3)**})$. Substituting these functions into the stage k equations of (2.5), we can obtain $V^i(k, x, y, z; \overline{u}_{k-}^{(\omega_1)**}, \overline{u}_{k-}^{(\omega_2)**}, \overline{u}_{k-}^{(\omega_3)**})$, for $i \in N$ and $k \in \{1, 2, \cdots, T\}$. In particular, $V^i(k, x, y, z; \overline{u}_{k-}^{(\omega_1)**}, \overline{u}_{k-}^{(\omega_2)**}, \overline{u}_{k-}^{(\omega_3)**})$ reflects the payoff of region i which depends on the states (x, y, z) and the previously executed durable strategies $(\overline{u}_{k-}^{(\omega_1)**}, \overline{u}_{k-}^{(\omega_2)**}, \overline{u}_{k-}^{(\omega_3)**})$.

3 An Environmental Game under Durable Strategies

In this section, we present a representative dynamic environmental game involving anthropogenic eco-degradation under durable investments with explicit functional forms.

3.1 *Game Formulation*

Consider a $n-$ player $T-$ stage dynamic environmental game in which player/region $i \in N$ produces three types of outputs. The first type of output $\overline{u}_k^{(\omega_1)i} \in \overline{U}^{(\omega_3)i} \subset R^+$ emits pollutants over ω_1 stages. The second type of output $\overline{u}_k^{(\omega_2)i} \in \overline{U}^{(\omega_2)i} \subset R^+$ causes damages to the ecosystems over ω_2 stages. The level of pollution stocks is $x_k \in X \subset R$ and the dynamics of pollution accumulation is governed by the difference equation:

$$x_{k+1} = x_k + \sum_{j=1}^{n} m_k^{|k|j}\overline{u}_k^{(\omega_1)j} + \sum_{j=1}^{n}\sum_{\tau=k-\omega_1+1}^{k-1} m_k^{|\tau|j}\overline{u}_\tau^{(\omega_1)j} - \sum_{j=1}^{n} b_k^j u_k^{Ij}(x_k)^{1/2} - \lambda_k^x x_k$$

$$x_1 = x_1^0, \quad (3.1)$$

where $m_k^{|k|j}$, $m_k^{|\tau|j}$, b_k^j and $\lambda_k^x(x_k)$ are positive parameters; and $\overline{u}_\tau^{(\omega_1)j} = 0$, for $\tau \in \{0, -1, -2, \cdots, -\omega_1 + 1\}$ and $j \in N$.

The state of the ecosystem is reflected by $y_k \in Y \subset R$ which conditions generate benefits to the regions. Negative benefits are possible if the state of the ecosystem becomes bad enough. The dynamics governing the evolution of the condition of the ecosystem is

$$\begin{aligned} y_{k+1} &= y_k + g_k - \sum_{j=1}^{n} \sigma_k^{k|j} \overline{u}_k^{(\omega_2)j} - \sum_{j=1}^{n} \sum_{\tau=k-\omega_2+1}^{k-1} \sigma_k^{\tau|j} \overline{u}_\tau^{(\omega_2)j} \\ &+ \sum_{j=1}^{n} \varepsilon_k^j u_k^{IJj} - \vartheta_k x_k - \lambda_k^y y_k, \\ y_1 &= y_1^0, \end{aligned} \tag{3.2}$$

where $\sigma_k^{|\tau|j}$, ε_k^j, g_k, λ_k^y and ϑ_k are positive parameters.

The third type of output is a clean output which is produced by non-durable input $u_k^{IIIi} \in U^{IIIi} \subset R^+$ and green technology $z_k \in Z \subset R^{n_3}$. The dynamics governing the accumulation of green technology is

$$z_{k+1} = z_k + \sum_{j=1}^{n} a_k^{|k|j} \overline{u}_k^{(\omega_3)j} + \sum_{j=1}^{n} \sum_{\tau=k-\omega_3+1}^{k-1} a_k^{|\tau|j} \overline{u}_\tau^{(\omega_3)j} - \lambda_k^z z_k, \quad z_1 = z_1^0, \tag{3.3}$$

where $\overline{u}_k^{(\omega_3)j} \in \overline{U}^{(\omega_3)j} \subset R$ are the investments in the public green technology by region j in stage k and it would take ω_3 stages to fully transmit the investments into the public technology, $a_k^{|k|j}$, $a_k^{|\tau|j}$ and λ_k^z are positive parameters.

The realized green technologies from the public technologies for region i are $\kappa_k^i z_k$. The region uses a non-durable input $u_k^{IIIi} \in U^{IIIi} \subset R^{m^{III(1)i}}$ together with its realized green technologies to produce clean outputs.

The payoff of region $i \in N$ is

$$\begin{aligned} &\sum_{k=1}^{T} \Bigg(R_k^{(x)i} \overline{u}_k^{(\omega_1)i} - \varphi_k^{(x)i} \left(\overline{u}_k^{(\omega_1)i}\right)^2 - c_k^{Ii} \left(u_k^{Ii}\right)^2 + R_k^{(y)i} u_k^{(\omega_2)i} \\ &- \varphi_k^{(y)i} \left(\overline{u}_k^{(\omega_2)i}\right)^2 - c_k^{IIi} \left(u_k^{IIi}\right)^2 \\ &+ R_k^{(z)i} \left(u_k^{IIIi} \kappa_k^j z_k\right)^{1/2} - \varphi_k^{(z)i} \left(\overline{u}_k^{(\omega_3)i}\right)^2 - c_k^{IIIi} u_k^{IIIi} - h_k^i x_k + \chi_k^i y_k \Bigg) \delta^{k-1} \\ &+ \left(Q_{T+1}^{x(i)} x_{T+1} + Q_{T+1}^{y(i)} y_{T+1} + Q_{T+1}^{z(i)} \kappa_{T+1}^i z_{T+1} + \varpi_{T+1}^i \right) \delta^T, \end{aligned} \tag{3.4}$$

where $R_k^{(x)i}$, $R_k^{(y)i}$, $R_k^{(z)i}$, $\varphi_k^{(x)i}$, $\varphi_k^{(y)i}$, $\varphi_k^{(z)i}$, c_k^{Ii}, c_k^{IIi}, c_k^{IIIi}, h_k^i, χ_k^i, $Q_{T+1}^{y(i)}$, $Q_{T+1}^{z(i)}$ are positive parameters and $Q_{T+1}^{x(i)}$ is negative.

3.2 Game Equilibrium

Invoking Theorem 3.1 in Chap. 2, we characterize the feedback equilibrium of the game (3.1–3.4) as follow.

Corollary 3.1 *Let* $(\underline{u}_k^{I**}, \underline{u}_k^{II**}, \underline{u}_k^{III**}, \overline{\underline{u}}_k^{(\omega_1)**}, \overline{\underline{u}}_k^{(\omega_2)**}, \overline{\underline{u}}_k^{(\omega_3)**})$ *be the set of feedback Nash equilibrium strategies and* $V^i(k, x, y, z; \overline{\underline{u}}_{k-}^{(\omega_1)**}, \overline{\underline{u}}_{k-}^{(\omega_2)**}, \overline{\underline{u}}_{k-}^{(\omega_3)**})$ *be the feedback Nash equilibrium payoff of region i at stage k in the durable-strategies dynamic game* (3.1–3.4). *The functions* $V^i(k, x, y, z; \overline{\underline{u}}_{k-}^{(\omega_1)**}, \overline{\underline{u}}_{k-}^{(\omega_2)**}, \overline{\underline{u}}_{k-}^{(\omega_3)**})$, *for* $k \in \{1, 2, \cdots, T+1\}$ *and* $i \in N$, *satisfy the following recursive equations:*

$$V^i(T+1, x, y, z; \overline{\underline{u}}_{(T+1)-}^{(\omega_1)**}, \overline{\underline{u}}_{(T+1)-}^{(\omega_2)**}, \overline{\underline{u}}_{(T+1)-}^{(\omega_3)**})$$
$$= \left(Q_{T+1}^{x(i)} x + Q_{T+1}^{y(i)} y + Q_{T+1}^{z(i)} \kappa_{T+1}^{i} z + \varpi_{T+1}^{i}\right)\delta^T, \tag{3.5}$$

$$V^i(k, x, y, z; \overline{\underline{u}}_{k-}^{(\omega_1)**}, \overline{\underline{u}}_{k-}^{(\omega_2)**}, \overline{\underline{u}}_{k-}^{(\omega_3)**})$$
$$= \max_{\substack{u_k^{I(1)i}, u_k^{II(1)i}, u_k^{III(1)i}, \\ u_k^{(\omega_1)i}, u_k^{(\omega_2)i}, u_k^{(\omega_3)i}}}$$
$$\Bigg\{ \Bigg(R_k^{(x)i} \overline{u}_k^{(\omega_1)i} - \varphi_k^{(x)i} \left(\overline{u}_k^{(\omega_1)i}\right)^2 - c_k^{Ii} \left(u_k^{Ii}\right)^2 + R_k^{(y)i} \overline{u}_k^{(\omega_2)i} - \varphi_k^{(y)i} \left(\overline{u}_k^{(\omega_2)i}\right)^2$$
$$- c_k^{IIi} \left(u_k^{IIi}\right)^2 + R_k^{(z)i} \left(u_k^{IIIi} \kappa_k^j z\right)^{1/2} - \varphi_k^{(z)i} \left(\overline{u}_k^{(\omega_3)i}\right)^2 - c_k^{IIII_i} u_k^{IIIi} - h_k^i x + \chi_k^j y \Bigg)$$
$$\delta^{k-1} + V^i\left(k+1, x_{k+1}^{***}, y_{k+1}^{***}, z_{k+1}^{z+k}; , \overline{u}_{(k+1)-}^{(\omega_1)}, \underline{u}_{(k+1)-}^{(\omega_2)*k+*}, \underline{u}_{(k+1)-}^{(\omega_3)}\right)\Bigg\}, \tag{3.6}$$

where

$$x_{k+1}^{***} = x + m_k^{|k|i} \overline{u}_k^{(\omega_1)i} + \sum_{\substack{j\in N \\ j\neq i}} m_k^{|k|j} \overline{u}_k^{(\omega_1)j**}$$
$$+ \sum_{j=1}^{n} \sum_{\tau=k-\omega_1+1}^{k-1} m_k^{|\tau|j} \overline{u}_\tau^{(\omega_1)j**}$$
$$- b_k^i u_k^{Ii} (x)^{1/2} - \sum_{\substack{j\in N \\ j\neq i}} b_k^j u_k^{Ij**} (x)^{1/2} - \lambda_k^x x$$
$$y_{k+1}^{***} = y + g_k - \sigma_k^{|k|i} \overline{u}_k^{(\omega_2)i}$$
$$- \sum_{\substack{j \in N \\ j \neq i}} \sigma_k^{|k|j} \overline{u}_k^{(\omega_2)j**} - \sum_{j=1}^{n} \sum_{\tau=k-\omega_2+1}^{k-1} \sigma_k^{|\tau|j} \overline{u}_\tau^{(\omega_2)j**}$$

$$+\,\varepsilon_k^i u_k^{IIi} + \sum_{\substack{j\in N\\ j\neq i}} \varepsilon_k^j u_k^{IIj**} - \vartheta_k x - \lambda_k^y y$$

$$z_{k+1}^{***} = z + a_k^{|k|i}\bar{u}_k^{(\omega_3)i} + \sum_{\substack{j\in N\\ j\neq i}} a_k^{|k|j}\bar{u}_k^{(\omega_3)j**}$$

$$+\sum_{j=1}^{n}\sum_{\tau=k-\omega_3+1}^{k-1} a_k^{|\tau|j}\bar{u}_\tau^{(\omega_3)j**} - \lambda_k^z z$$

$$\underline{\bar{u}}_{(k+1)-}^{(\omega_1)***} = (\bar{u}_k^{(\omega_1)i}, \bar{u}_k^{(\omega_1)(\neq i)**}, \underline{\bar{u}}_{(k+1)-}^{(\omega_1)**} \cap \underline{\bar{u}}_{k-}^{(\omega_1)**}) \text{ and}$$

$$\underline{\bar{u}}_{(k+1)-}^{(\omega_3)***} = (\bar{u}_k^{(\omega_3)i}, \bar{u}_k^{(\omega_3)(\neq i)**}, \underline{\bar{u}}_{(k+1)-}^{(\omega_3)**} \cap \underline{\bar{u}}_{k-}^{(\omega_3)**})$$

Performing the indicated maximization operator in Corollary 2.1, we obtain the following game equilibrium conditions.

(i) The pollution creating output, that is $\bar{u}_k^{(\omega_1)i}$, will be produced up to the level where the marginal gross revenue equals the sum of the marginal production cost and the marginal environmental effects from pollution, that is

$$R_k^{(x)i}\delta^{k-1} = 2\varphi_k^{(x)i}\left(\bar{u}_k^{(\omega_2)i}\right)\delta_1^{k-1} - \frac{\partial V_{k+1}^i}{\partial x_{k+1}^{***}} m_k^{|k|i} - \frac{\partial V_{k+1}^i}{\partial \bar{u}_k^{(\omega_1)i}}, \text{ for } i \in N. \tag{3.7}$$

Condition (3.7) implies

$$\bar{u}_k^{(\omega_1)i} = \frac{1}{2\varphi_k^{(x)i}}\left[R_k^{(x)i} + \left(\frac{\partial V_{k+1}^i}{\partial x_{k+1}^{***}} m_k^{|k|i} + \frac{\partial V_{k+1}^i}{\partial \bar{u}_k^{(\omega_1)i}}\right)\delta^{1-k}\right].$$

(ii) The output that damages the ecosystem will be produced up to the level where the marginal gross revenue equals the sum of the marginal production cost and the marginal environmental effects from damages to the ecosystem, that is

$$R_k^{(y)i}\delta^{k-1} = 2\varphi_k^{(y)i}\left(\bar{u}_k^{(\omega_2)i}\right)\delta_1^{k-1} + \frac{\partial V_{k+1}^i}{\partial y_{k+1}^{y+8}}\sigma_k^{|k|i} - \frac{\partial V_{k+1}^i}{\partial \bar{u}_k^{(\omega_2)i}}, \text{ for } i \in N. \tag{3.8}$$

The condition in (3.8) implies

$$\bar{u}_k^{(\omega_2)i} = \frac{1}{2\varphi_k^{(y)i}}\left[R_k^{(y)i} + \left(\frac{\partial V_{k+1}^i}{\partial \bar{u}_k^{(\omega_1)i}} - \frac{\partial V_{k+1}^i}{\partial y_{k+1}^{***}}\sigma_k^{|k|i}\right)\delta^{1-k}\right].$$

(iii) Investments for building up green technologies, that is $\overline{u}_k^{(\omega_3)i}$, will be employed up to the level where the marginal cost of investment for building up the green technology equals the marginal contributions to region i's payoff by the green technology, that is

$$2\varphi_k^{(z)i}\overline{u}_k^{(\omega_3)i}\delta^{k-1} = \frac{\partial V_{k+1}^i}{\partial z_{k+1}^{***}}a_k^{|k|i} + \frac{\partial V_{k+1}^i}{\partial \overline{u}_k^{(\omega_3)i}}, \quad \text{for } i \in N. \tag{3.9}$$

The condition in (3.9) implies

$$\overline{u}_k^{(\omega_3)i} = \frac{1}{2\varphi_k^{(z)i}}\left(\frac{\partial V_{k+1}^i}{\partial z_{k+1}^{***}}a_k^{|k|i} + \frac{\partial V_{k+1}^i}{\partial \overline{u}_k^{(\omega_3)i}}\right)\delta^{1-k}$$

(iv) Pollution abatement effort is employed up to the level where the marginal cost of abatement effort equals the marginal benefit, that is

$$2c_k^{Ii}u_k^{Ii} = -\frac{\partial V_{k+1}^i}{\partial x_{k+1}^{i+1}}b_k^i(x)^{1/2}\delta^{1-k}, \quad \text{for } i \in N. \tag{3.10}$$

The condition in (3.10) implies

$$u_k^{Ii} = -\frac{1}{2c_k^{Ii}}\frac{\partial V_{k+1}^i}{\partial x_{k+1}^{***}}b_k^i(x)^{1/2}\delta^{1-k}.$$

(v) Ecosystem improvement effort is employed up to the level where the marginal improvement cost equal the marginal benefit, that is

$$2c_k^{IIi}u_k^{IIi} = \frac{\partial V_{k+1}^i}{\partial y_{k+1}^{***}}\varepsilon_k^i\delta^{1-k}, \quad \text{for } i \in N. \tag{3.11}$$

The condition in (3.11) implies

$$u_k^{IIi} = \frac{1}{2c_k^{IIi}}\frac{\partial V_{k+1}^i}{\partial y_{k+1}^{***}}\varepsilon_k^i\delta^{1-k}.$$

(vi) Non-durable input used to produce the clean output are employed up to the level where the marginal revenue from the clean output equals the marginal cost, that is

$$\frac{1}{2}\left(u_k^{IIIi}\right)^{-1/2}R_k^{(z)i}\left(\kappa_k^i z\right)^{1/2} = c_k^{IIII}, \text{ for } i \in N. \tag{3.12}$$

The condition in (3.12) implies

$$u_k^{IIIi} = \left(\frac{R_k^{(z)i}}{2c_k^{IIIi}}\right)^2 \kappa_k^i z$$

A feedback Nash equilibrium solution of the game (3.1–3.5) can be solved as:

Proposition 3.1 *System* (3.5–3.6) *admits a solution with the game equilibrium payoff of player i being*

$$V^i\left(k, x, y, z; \underline{\overline{u}}_{k-}^{((a_1)^*}, \underline{\overline{u}}_{k-}^{(\omega_2)^{*I*}}, \underline{\overline{u}}_{k-}^{(\omega_3)*}\right) = \left(A_k^i x + B_k^i y + C_k^i z + D_k^j\right)\delta^{k-1}, \quad (3.13)$$

where

$$A_k^i = -\frac{\left(\delta A_{k+1}^i b_k^i\right)^2}{4c_k^{Ii}} - h_k^i + \delta A_{k+1}^i \left(\sum_{j=1}^n \frac{\delta A_{k+1}^j \left(b_k^j\right)^2}{2c_k^{Ij}} + \left(1 - \lambda_k^x\right)\right) - B_{k+1}^i \vartheta_k,$$

$$B_k^i = \chi_k^i + \delta B_{k+1}^i \left(1 - \lambda_k^y\right),$$

$$C_k^i = \frac{\left(R_k^{(z)i}\right)^2 \kappa_k^i}{4\left(c_k^{IIIt}\right)} + \delta C_{k+1}^i \left(1 - \lambda_k^z\right),$$

D_k^i *is an expression that contains the previously executed controls* $(\underline{\overline{u}}_{k-}^{(\omega_1)**}, \underline{\overline{u}}_{k-}^{(\omega_2)**}, \underline{\overline{u}}_{k-}^{(\omega_3)**})$, *for* $i \in N$ *and* $k \in \{, 1, 2, \cdots, T\}$.

The game equilibrium strategies in stage $k \in \{1, 2, \cdots, T\}$ *can be obtained as:*

$$\overline{u}_k^{(\omega_1)t^{***}} = \frac{R_k^{(x)i} + \sum_{\tau=k}^{k+\omega_1-1} \delta^{\tau-k+1} A_{\tau+1}^i m_\tau^{|k|i}}{2\varphi_k^{(x)i}}, \quad \text{and } A_\tau^i = 0 \text{ for } \tau > T+1,$$

$$\overline{u}_{k,}^{(\omega_2)t^{***}} = \frac{R_k^{(y)i} - \sum_{\tau=k}^{k+\omega_2-1} \delta^{\tau-k+1} B_{\tau+1}^i \sigma_\tau^{|k|i}}{2\varphi_k^{(y)i}}, \quad \text{and } B_\tau^i = 0 \text{ for } \tau > T+1,$$

$$\overline{u}_k^{(\omega_3)i^{**}} = \frac{\sum_{\tau=k}^{k+\omega_3-1} \delta^{\tau-k+1} C_{\tau+1}^i a_\tau^{|k|i}}{2\varphi_k^{(z)i}}, \quad \text{and } C_\tau^i = 0 \text{ for } \tau > T+1,$$

$$u_k^{I***} = -\frac{\delta A_{k+1}^i b_k^i}{2c_k^{Ii}} x^{1/2}, \; u_k^{IIt***} = \frac{\delta B_{k+1}^i \varepsilon_k^i}{2c_k^{IIi}}, \; u_k^{III***} = \frac{\left(R_k^{(z)i}\right)^2 \kappa_k^i z}{4\left(c_k^{IIIi}\right)^2}; \quad \text{for } i \in N. \quad (3.14)$$

Proof See Appendix in Sect. 5. ■

The values of A_k^i, B_k^i and C_k^i, for $k \in \{,1,2,\cdots,T\}$, can be solved backwardly from stage T to stage 1. Substituting the strategies from (3.14) into the state dynamics (3.1–3.3), we obtain the game equilibrium evolutions of the state variables. In particular, the game equilibrium dynamics of pollution accumulation is

$$x_{k+1} = x_k + \sum_{j=1}^{n} m_k^{|k|j} \frac{R_k^{(x)j} + \sum_{\tau=k}^{k+\omega_1-1} \delta^{\tau-k+1} A_{\tau+1}^j m_\tau^{|k|j}}{2\varphi_k^{(x)i}}$$
$$+ \sum_{j=1}^{n} \sum_{t=k-\omega_1+1}^{k-1} m_k^{|t|j} \frac{R_t^{(x)j} + \sum_{\tau=t}^{t+\omega_1-1} \delta^{\tau-t+1} A_{\tau+1}^j m_\tau^{|t|j}}{2\varphi_t^{(x)i}}$$
$$+ \sum_{j=1}^{n} b_k^j \frac{\delta A_{k+1}^j b_k^j}{2c_k^{jj}} x_k - \lambda_k^x x_k, \quad x_1 = x_1^0. \tag{3.15}$$

To solve (3.15), we first start from the initial stage, that is stage 1, where strategies before stage 1 are zero. We can obtain x_2 using (3.15). With x_2 and (3.15) we can obtain x_3. Continuing with the process for stage 4 to stage $T+1$, we can obtain the game equilibrium path of the pollution stock $\{x_k\}_{k=1}^{T+1}$.

The dynamics governing the evolution of the conditions of the ecosystems in the game equilibrium is

$$y_{k+1} = y_k + g_k - \sum_{j=1}^{n} \sigma_k^{k_j j} \frac{R_k^{(y)j} - \sum_{\tau=k}^{k+\omega_2-1} \delta^{\tau-k+1} B_{\tau+1}^j \sigma_\tau^{k_k j}}{2\varphi_k^{(y)j}}$$
$$- \sum_{j=1}^{n} \sum_{t=k-\omega_2+1}^{k-1} \sigma_k^{ttj} \frac{R_t^{(y)j} - \sum_{\tau=t}^{t+\omega_2-1} \delta^{\tau-t+1} B_{\tau+1}^j \sigma_\tau^{|t|j}}{2\varphi_t^{(y)j}}$$
$$+ \sum_{j=1}^{n} \varepsilon_k^j \frac{\delta B_{k+1}^j \varepsilon_k^j}{2c_k^{IJj}} - \vartheta_k x_k - \lambda_k^y y_k,$$
$$y_1 = y_1^0 \text{ and } x_1 = x_1^0 \tag{3.16}$$

We first start from stage 1. We can obtain y_2 using (3.16) and x_1. With x_2, y_2 and (3.16) we can obtain y_3. Continuing with the process for stage 4 to stage $T+1$, we can obtain the game equilibrium path of the condition of the ecosystems $\{y_k\}_{k=1}^{T+1}$.

The game equilibrium dynamics governing the accumulation of green technology is

$$Z_{k+1} = z_k + \sum_{j=1}^{n} a_k^{|k|j} \frac{\sum_{\tau=k}^{k+\omega_3-1} \delta^{\tau-k+1} C_{\tau+1}^{j} a_\tau^{|k|j}}{2\varphi_k^{(z)j}}$$
$$+ \sum_{j=1}^{n} \sum_{t=k-\omega_3+1}^{k-1} a_k^{|\tau|j} \frac{\sum_{\tau=t}^{t+\omega_3-1} \delta^{\tau-t+1} C_{\tau+1}^{j} a_\tau^{|t|j}}{2\varphi_t^{(z)j}} - \lambda_k^z z_k,$$
$$z_1 = z_1^0. \tag{3.17}$$

Using (3.17), we can obtain the game equilibrium path of the public green technology $\{z_3\}_{k=1}^{T+1}$.

4 Anthropogenic Eco-degradation under Strategy Lags

Several new theoretical outcomes in dynamic environmental games can be obtained from the environmental games under durable strategies in the previous two sections.

(i) *Increase in Pollution even Production Stops*

The topic of pollution continues to increase even when production stops have been raised in recent years. The lagged effects of industrial production on pollution accumulation highlight the seriousness of the world's environmental problems. Corbett (2019) pointed out that even if we stop all new drilling and fracking immediately, the flood of toxic waste streams will continue to grow for decades. Rood (2014) stated that once we release the carbon dioxide stored in the fossil fuels we burn, it accumulates in and moves amongst the atmosphere, the oceans, the land and the plants and animals of the biosphere. The released carbon dioxide will remain in the atmosphere and once released the carbon dioxide is in our environment essentially forever. It does not go away, unless we, ourselves, remove it. This decades-long lag between cause and effect is due to the long time it takes to heat the ocean's huge mass. So even if carbon emissions stopped completely right now, as the oceans catch up with the atmosphere, the Earth's temperature would rise about another 1.1F. Scientists refer to this as committed warming. Rood (2017) further stressed that it is possible that even as emissions decrease, the carbon dioxide in the atmosphere will continue to increase. Pollution Solutions (2015) concluded that even if we were to cease all carbon emissions today, the temperature of the Earth would still continue to rise because there is a delayed period in which the carbon dioxide already released will continue to accumulate and move among the atmosphere and oceans of our planet.

The game equilibrium pollution accumulation dynamics in (3.15) in Sect. 3,

$$x_{k+1} = x_k + \sum\nolimits_{j=1}^{n} m_k^{|k|j} \frac{R_k^{(x)j} + \sum\nolimits_{\tau=k}^{k+\omega_1-1} \delta^{\tau-k+1} A_{\tau+1}^{j} m_\tau^{|k|j}}{2\varphi_k^{(x)i}}$$
$$+ \sum\nolimits_{j=1}^{n} \sum\nolimits_{\tau=k-\omega_1+1}^{k-1} m_k^{|\tau|j} \overline{u}_\tau^{(\omega_1)j**}$$
$$+ \sum_{j=1}^{n} b_k^j \frac{\delta A_{k+1}^j b_k^j}{2c_k^j} x_k - \lambda_k^x x_k, \quad x_1 = x_1^0,$$

is the first time that durable strategies are involved in dynamic games with pollution.

Given that pollution is a global problem, the number of regions could be very large. The lagged effects on pollution by industrial production, $\sum\nolimits_{j=1}^{n} \sum\nolimits_{\tau=k-\omega_1+1}^{k-1} m_k^{|\tau|j} \overline{u}_\tau^{(\omega_1)j**}$, could be enormous and capable of causing severe harm to the environment in the future. An important policy implication of lagged effects in pollutant emissions is the urgent call for cooperation to avert future calamity. Time lost in reaching an agreement could result in heavy costs from previous damaging actions that are still effective after the agreed measures have begun. In addition, the current main policy instrument of global pollution management in the Paris Agreement is reduction of carbon emissions. In the presence of strategies lags, carbon emissions control may not be an effective policy tool because the pollution stock would continue to grow even-though production and emissions are reduced to zero.

(ii) *Persistent Damage to the Ecosystems*

Human activities affect ecosystems in a wide variety of ways and whenever humans enter a habitat, they tend to reshape it for their own needs, destroying the resources that support other natural habitats living there long ago. In particular, there are human activities that affect the ecosystems but without emissions of pollutions. These include activities like lumbering, extraction of resources, urbanization, commercial activities leading to the destruction of land and genetic modification. As our population and economic activities continue to increase, deforestation becomes more and more wide spread. This has persistent devastating effects on the functioning of ecosystems. In addition, there are many types of over-exploitation of nature—in animal species, renewable resources, non-renewable resources, plants, marine species, soil and land. Johnson (2020) pointed out that all of the oceans of the world are fully exploited or over-exploited with most fisheries expected to collapse within the next 40 years if fishing practices are not changed. Soils are being depleted by human economic activities at a rapid rate, leading to desertification and loss of agricultural productivity. The destruction of biodiversity leads to an off balance of self-regeneration of the ecosystems. Ecosystems throughout the world are endangered. Damaged ecosystems occur when species within the system are lost, habitat is destroyed and/or the food web is affected. Because all species live in complex interdependent systems with interdependent relationships, the loss or change of any single species or abiotic factors has prolonged negative consequences on others within the ecosystem.

The game equilibrium dynamics governing the evolution of the conditions of the ecosystems in (3.16),

$$y_{k+1} = y_k + g_k - \sum_{j=1}^{n} \sigma_k^{|k|j} \frac{R_k^{(y)j} - \sum_{\tau=k}^{k+\omega_2-1} \delta^{\tau-k+1} B_{\tau+1}^j \sigma_\tau^{|k|j}}{2\varphi_k^{(y)j}}$$

$$-\sum_{j=1}^{n} \sum_{\tau=k-\omega_2+1}^{k-1} \sigma_k^{|\tau|j} \bar{u}_\tau^{(\omega_2)j**}$$

$$+\sum_{j=1}^{n} \varepsilon_k^j \frac{\delta B_{k+1}^j \varepsilon_k^j}{2c_k^{Ij}} - \vartheta_k x_k - \lambda_k^y y_k,$$

$$y_1 = y_1^0 \text{ and } x_1 = x_1^0,$$

is the first time that durable strategies are involved in dynamic games with ecosystem degradation.

Just like in the case of pollution, the number of regions could be very large. The lagged effects on ecosystems degradation by industrial production, $\sum_{j=1}^{n} \sum_{\tau=k-\omega_2+1}^{k-1} \sigma_k^{|\tau|j} u_\tau^{(\omega_2)j**}$, could be enormous and capable of causing severe harm to the environment in the future. Again, an important policy implication of lagged effects in anthropogenic ecosystem damages is an urgent call for actions to avert future calamity.

Moreover, the level of pollution also affects the regeneration of the ecosystems through the presence of $\vartheta_k x_k$ in (3.16). For example, pollution from mining contaminates rivers, poisons aquatic life and bio-accumulating in the food chain. Chemical pollutants, including pesticides and plastics, disrupt animal hormonal activity and reproduction, reducing biodiversity in water and on land. Air pollution causes global warming and climate change which affect virtually all ecosystems. Increased temperatures lead to changes in ocean currents and plant growth which affect food webs and relationships within ecosystems.

(iii) *Sluggishness in Green Technology Development*

Environmental technology or green technology involves the application of environmental science, green chemistry, environmental monitoring and electronic devices to monitor, model and conserve the natural environment and resources and to alleviate negative anthropogenic environmental impacts. It includes recycling, renewable energy (wind-power, solar-power, biomass, hydropower, biofuels), information technology, green transportation, green buildings, electric motors, green chemistry, lighting, grey-water and any other energy efficient appliances. More comprehensively, clean technology covers the segments in energy generation, energy storage, energy infrastructure, energy efficiency, transportation, water and wastewater, air and environment, materials manufacturing/industrial, agriculture, recycling and waste. Many green technologies have the features of public good capital and take a certain period of time for full transformation of investments into effective capitals.

The game equilibrium dynamics governing the accumulation of green technology in (3.17),

$$
\begin{aligned}
z_{k+1} &= z_k + \sum_{j=1}^{n} a_k^{|k|j} \frac{\sum_{\tau=k}^{k+\omega_3-1} \delta^{\tau-k+1} C_{\tau+1}^{j} a_\tau^{|k|j}}{2\varphi_k^{(z)j}} \\
&+ \sum_{j=1}^{n} \sum_{z=k-\omega_3+1}^{k-1} a_k^{|\tau|j} \bar{u}_\tau^{(\omega_3)j^{***}} - \lambda_k^z z_k, \quad z_1 = z_1^0,
\end{aligned}
$$

is the first time that durable strategies are involved in dynamic games with green technology.

Potential slow diffusion of knowledge leads to disincentive in public capital provision by individual regions. Regions whose gain declines from public green technology in later stages would lower their investment, although they may have cost advantages. In the case where the diffusion period is long while the game horizon is short, individual regions would tend to be less enthusiastic in investing in green technology because they cannot realize the full benefits of their investments.

5 Appendices

5.1 Appendix I: Proof of Proposition 3.1

Using (3.5), we have

$$
A_{T+1}^{i} = Q_{T+1}^{(i)}, \; B_{T+1}^{i} = Q_{T+1}^{r(i)}, \; C_{T+1}^{i} = Q_{T+1}^{(i)}, \; D_{T+1}^{i} = \overline{w}_{T+1}^{j}. \tag{5.1}
$$

Using the technique of backward induction, we consider first the last operational stage T. Using Proposition 3.1 and (3.7–3.12), we can express the game equilibrium strategies in stage T as:

$$
\begin{aligned}
\overline{u}_T^{(\omega_1)i**} &= \frac{R_T^{(x)i} + \delta A_{T+1}^{i} m_T^{|T|i}}{2\varphi_T^{(x)i}}, \\
\overline{u}_T^{(\omega_2)i**} &= \frac{R_T^{(y)i} - \delta B_{T+1}^{i} \sigma_T^{|T|i}}{2\varphi_T^{(y)i}}, \; \overline{u}_T^{(\omega_3)i**} = \frac{\delta C_{T+1}^{i} a_T^{|T|i}}{2\varphi_T^{(z)i}}, \\
u_T^{I_T^{**}} &= -\frac{\delta A_{T+1}^{i} b_T^{i}}{2c_T^{Ii}} x^{1/2},
\end{aligned}
$$

$$u_T^{II\tau**} = \frac{\delta B_{T+1}^i \varepsilon_T^i}{2c_T^{IIi}}, u_T^{III**} = \frac{\left(R_T^{(z)i}\right)^2 \kappa_T^i Z}{4\left(c_T^{IIIi}\right)^2}, \quad \text{for } i \in N. \tag{5.2}$$

Substituting the optimal strategies in (5.2) into the stage T equation of (3.6) we obtain

$$\begin{aligned}
&V^i\left(T, x, y, z; \overline{u}_{T-}^{(\omega_1)**}, \underline{u}_{T-}^{(\omega_2)**}, \overline{\underline{u}}_{T-}^{(\omega_3)**}\right) \\
&= (A_T x + B_T^i y + C_T z + D_T^i)\delta^{T-1} \\
&= \left(\frac{\left(R_T^{(x)i} - \delta A_{T+1}^i m_T^{|T|i}\right)\left(R_T^{(x)i} + \delta A_{T+1}^i m_T^{|T|i}\right)}{4\varphi_T^{(x)i}} - \frac{\left(\delta A_{T+1}^i b_T^i\right)^2}{4c_T^{Ii}} x\right. \\
&+ \frac{\left(R_T^{(y)i} + \delta B_{T+1}^i \sigma_T^{|T|i}\right)\left(R_T^{(y)i} - \delta B_{T+1}^i \sigma_T^{|T|i}\right)}{4\varphi_T^{(y)i}} - \frac{\left(\delta B_{T+1}^i \varepsilon_T^i\right)^2}{4c_T^{IIi}} \\
&\left.+ \frac{\left(R_T^{(z)i}\right)^2 \kappa_T^i Z}{4\left(c_T^{IIIi}\right)} - \frac{\left(\delta C_{T+1}^i a_T^{|T|i}\right)^2}{4\varphi_T^{(z)i}} - h_T^i x + \chi_T^j y\right) \delta^{T-1} \\
&+ \left[A_{T+1}^i \left(x + \sum_{j=1}^n m_T^{|T|j} \frac{R_T^{(x)j} + \delta A_{T+1}^j m_T^{|T|j}}{2\varphi_T^{(x)j}} + \sum_{j=1}^n \sum_{\tau=T-\omega_1+1}^{T-1} m_T^{|\tau|j} \overline{u}_\tau^{(\omega_1)j**}\right.\right. \\
&\left.+ \sum_{j=1}^n \frac{\delta A_{T+1}^j \left(b_T^j\right)^2}{2c_T^{Ij}} x - \lambda_T^x x\right) \\
&+ B_{T+1}^i \left(y + g_T - \sum_{j=1}^n \sigma_T^{|T|j} \frac{R_T^{(y)j} - \delta B_{T+1}^j \sigma_T^{|T|j}}{2\varphi_T^{(y)j}}\right. \\
&\left.- \sum_{j=1}^n \sum_{\tau=T-\omega_2+1}^{T-1} \sigma_T^{|\tau|j} \overline{u}_\tau^{(\omega_2)j**} + \sum_{j=1}^n \varepsilon_T^j \frac{\delta B_{T+1}^j \varepsilon_T^j}{2c_T^{IIj}} - \vartheta_T x - \lambda_T^y y\right) \\
&\left.+ C_{T+1}^i \left(z + \sum_{j=1}^n a_T^{|T|j} \frac{\delta C_{T+1}^j a_T^{|T|j}}{2\varphi_T^{(z)j}} + \sum_{j=1}^n \sum_{\tau=T-\omega_3+1}^{T-1} a_T^{|t|j} \overline{u}_\tau^{(\omega_3)j**} - \lambda_T^z z\right)\right] \\
&+ \varpi_{T+1}^i \delta^T, \quad \text{for } i \in N.
\end{aligned} \tag{5.3}$$

System (5.3) yields equations in which right-hand-side and left-hand-side are linear functions of

x, y and z, for $i \in N$. For (5.3) to hold, it is required that:

$$A_T^i x = -\frac{\left(\delta A_{T+1}^i b_T^i\right)^2}{4c_T^{Ii}} x - h_T^i x$$
$$+ \delta A_{T+1}^i \left(\sum_{j=1}^n \frac{\delta A_{T+1}^j \left(b_T^j\right)^2}{2c_T^{Ij}} + \left(1 - \lambda_T^x\right) \right) x - B_{T+1}^i \vartheta_T x,$$
$$B_T^i y = \chi_T^j y + \delta B_{T+1}\left(1 - \lambda_T^v\right) y,$$
$$C_T^i z = \frac{\left(R_T^{(z)i}\right)^2 \kappa_T^i z}{4\left(c_T^{III} i\right)} + \delta_{T+1}^i \left(1 - \lambda_T^z\right) z, \quad \text{for } i \in N. \tag{5.4}$$

From (5.4) we obtain:

$$A_T^i = -\frac{\left(\delta A_{T+1}^i b_T^i\right)^2}{4c_T^{Ii}} - h_T^i$$
$$+ \delta A_{T+1}^i \left(\sum_{j=1}^n \frac{\delta A_{T+1}^j \left(b_T^j\right)^2}{2c_T^{Ij}} + \left(1 - \lambda_T^x\right) \right) - B_{T+1}^i \vartheta_T,$$
$$B_T^i = \chi_T^j + \delta_{T+1}^i \left(1 - \lambda_T^y\right),$$
$$C_T^i = \frac{\left(R_T^{(z)i}\right)^2 \kappa_T^i}{4\left(c_T^{III}\right)} + \delta C_{T+1}^i \left(1 - \chi_T^z\right), \quad \text{for } i \in N. \tag{5.5}$$

In addition,

$$D_T^i = \left(\frac{\left(R_T^{(x)i} - \delta A_{T+1}^i m_T^{|T|i}\right)\left(R_T^{(x)i} + \delta A_{T+1}^i m_T^{|T|i}\right)}{4\varphi_T^{(x)i}} \right.$$
$$\left. + \frac{\left(R_T^{(y)i} + \delta B_{T+1}^i \sigma_T^{|T|i}\right)\left(R_T^{(y)i} - \delta B_{T+1}^i \sigma_T^{|T|i}\right)}{4\varphi_T^{(y)i}} - \frac{\left(\delta B_{T+1}^i \varepsilon_T^i\right)^2}{4c_T^{IIi}} - \frac{\left(\delta C_{T+1}^i a_T^{|T|i}\right)^2}{4\varphi_T^{(z)i}} \right)$$
$$+ \left[A_{T+1}^i \left(\sum_{j=1}^n m_T^{|T|j} \frac{R_T^{(x)j} + \delta A_{T+1}^j m_T^{|T|j}}{2\varphi_T^{(x)j}} + \sum_{j=1}^n \sum_{\tau=T-\omega_1+1}^{T-1} m_T^{|\tau|j} \overline{u}_\tau^{(\omega_1)} j^{**} \right) \right.$$
$$+ B_{T+1}^i \left(g_T^i + \sum_{j=1}^n \varepsilon_T^j \frac{\delta B_{T+1}^j \varepsilon_T^j}{2c_T^{IIj}} - \sum_{j=1}^n \sigma_T^{|T|j} \frac{R_T^{(y)j} - \delta B_{T+1}^j \sigma_T^{|T|j}}{2\varphi_T^{(y)j}} \right.$$
$$\left. - \sum_{j=1}^n \sum_{\tau=T-\omega_2+1}^{T-1} \sigma_T^{|\tau|j} \overline{u}_\tau^{(\omega_2) j^{**}} \right)$$

$$+ C^i_{T+1}\left(\sum_{j=1}^{n} a_T^{|T|j}\frac{\delta C^j_{T+1}a_T^{|T|j}}{2\varphi_T^{(z)j}} + \sum_{j=1}^{n}\sum_{\tau=T-\omega_3+1}^{T-1} a_T^{|\tau|j}\overline{u}_\tau^{(\omega_3)j**}\right) + \varpi^i_{T+1}\Bigg]\delta,$$

for $i \in N$. (5.6)

which is a function of previously executed controls $(\underline{\overline{u}}_{T-}^{(\omega_1)**}, \underline{\overline{u}}_{T-}^{(\omega_2)**}, \underline{\overline{u}}_{T-}^{(\omega_3)**})$ Then we move to stage $T-1$.

Using $V^i\left(T, x, y, z; \underline{\overline{u}}_{T-}^{(\omega_1)=\overline{u}}, \underline{\overline{u}}_{T-}^{(\omega_2)**}, \underline{\overline{u}}_{T-}^{(\omega_3)=*}\right) = (A_T x + B^i_T y + C_T z + D^i_T)\delta^{T-1}$ derived in (5.5–5.6) and the stage $T-1$ equation in (3.6), the game equilibrium strategies in stage $T-1$ can be expressed as:

$$\begin{aligned}
\overline{u}_{T-1}^{(\omega_1)i**} &= \frac{R^{(x)i}_{T-1} + \delta A^i_T m_{T-1}^{|T-1|i} + \delta^2 A^i_{T+1} m_T^{|T-1|i}}{2\varphi^{(x)i}_{T-1}},\\
\overline{u}_{T-1}^{(\omega_2)i**} &= \frac{R^{(y)i}_{T-1} - \delta B^i_T \sigma_{T-1}^{|T-1|i} - \delta^2 B^i_{T+1}\sigma_T^{|T-1|i}}{2\varphi^{(y)i}_{T-1}},\\
\overline{u}_{T-1}^{(\omega_3)i**} &= \frac{\delta C^i_T a_{T-1}^{T-1i} + \delta^2 C^i_{T+1} a_T^{T-1i}}{2\varphi^{(z)i}_{T-1}},\\
u_{T-1}^{I***} &= -\frac{\delta A^i_T b^i_{T-1}}{2c^{Ii}_{T-1}}x^{1/2},\\
u_{T-1}^{II_i*} &= \frac{\delta B^i_T \varepsilon^i_{T-1}}{2c^{III}_{T-1}},\\
u_{T-1}^{III^{i**}} &= \frac{\left(R^{(z)i}_{T-1}\right)^2 \kappa^i_{T-1} z}{4\left(c^{IIIi}_{T-1}\right)^2}, \quad \text{for } i \in N.
\end{aligned} \tag{5.7}$$

Substituting the optimal strategies in (5.7) into the stage $T-1$ equation of (3.6) we obtain

$$V^i\left(T-1, x, y, z; , \underline{\overline{u}}_{(T-1)-}^{(\omega_1)**}, \underline{\overline{u}}_{(T-1)-}^{(\omega_2)**}, \underline{\overline{u}}_{(T-1)-}^{(\omega_3)**}\right)$$

$$= \left(A^i_{T-1}x + B^i_{T-1}y + C^i_{T-1}z + D^j_{T-1}\right)\delta^{T-2}$$

$$= \Bigg(\frac{\left(R^{(x)i}_{T-1} - \delta A^i_T m_{T-1}^{|T-1|i} - \delta^2 A^i_{T+1} m_T^{|T-1|i}\right)\left(R^{(x)i}_{T-1} + \delta A^i_T m_{T-1}^{|T-1|i} + \delta^2 A^i_{T+1} m_T^{|T-1|i}\right)}{4\varphi^{(x)i}_{T-1}}$$

$$- \frac{\left(\delta A^i_T b^i_{T-1}\right)^2}{4c^{Ii}_{T-1}}x$$

$$+ \frac{\left(R^{(y)i}_T + \delta B^i_T \sigma_{T-1}^{|T-1|i} + \delta^2 B^i_{T+1}\sigma_T^{|T-1|i}\right)\left(R^{(y)i}_{T-1} - \delta B^i_T \sigma_{T-1}^{|T-1|i} - \delta^2 B^i_{T+1}\sigma_T^{|T-1|i}\right)}{4\phi^{(y)i}_{T-1}}$$

$$-\frac{\left(\delta B_T^i\varepsilon_{T-1}^i\right)^2}{4c_{T-1}^{IIi}}$$

$$+\frac{\left(R_{T-1}^{(z)i}\right)^2\kappa_{T-1}^i z}{4\left(c_{T-1}^{IIIi}\right)}-\frac{\left(\delta C_T^i a_{T-1}^{|T-1|i}+\delta^2 C_{T+1}^i a_T^{|T-1|i}\right)^2}{4\varphi_{T-1}^{(z)i}}-h_{T-1}^i x+\chi_{T-1}^j y\Bigg)\delta^{T-2}$$

$$+\Bigg[A_T^i\Bigg(x+\sum_{j=1}^n m_{T-1}^{|T-1|j}\frac{R_{T-1}^{(x)j}+\delta A_T^j m_{T-1}^{|T-1|j}+\delta^2 A_{T+1}^j m_T^{|T-1|j}}{2\varphi_{T-1}^{(x)j}}$$

$$+\sum_{j=1}^n\sum_{\tau=T-1-\omega_1+1}^{T-2} m_{T-1}^{|\tau|j}\overline{u}_\tau^{(\omega_1)j**}+\sum_{j=1}^n\frac{\delta A_T^j\left(b_{T-1}^j\right)^2}{2c_{T-1}^{Ij}}x-\lambda_{T-1}^x x\Bigg)$$

$$+B_T^i\Bigg(y+g_{T-1}-\sum_{j=1}^n\sigma_{T-1}^{|T-1|j}\frac{R_{T-1}^{(y)j}-\delta B_T^j\sigma_{T-1}^{|T-1|j}-\delta^2 B_{T+1}^j\sigma_T^{|T-1|j}}{2\varphi_{T-1}^{(y)j}}$$

$$-\sum_{j=1}^n\sum_{\tau=T-1-\omega_2+1}^{T-2}\sigma_{T-1}^{|\tau|j}\overline{u}_\tau^{(\omega_2)j**}+\sum_{j=1}^n\varepsilon_{T-1}^j\frac{\delta B_T^j\varepsilon_{T-1}^j}{2c_{T-1}^{IIj}}-\vartheta_{T-1}x-\lambda_{T-1}^y y\Bigg)$$

$$+C_T^i\Bigg(z+\sum_{j=1}^n a_{T-1}^{|T-1|j}\frac{\delta C_T^j a_{T-1}^{|T-1|j}+\delta^2 C_{T+1}^j a_T^{|T-1|j}}{2\varphi_{T-1}^{(z)j}}$$

$$+\sum_{j=1}^n\sum_{\tau=T-1-\omega_3+1}^{T-2} a_{T-1}^{|\tau|j}\overline{u}_\tau^{(\omega_3)j**}-\lambda_{T-1}^z z\Bigg)+D_T^{i***}\Bigg]\delta^{T-1},$$

for $i\in N$, (5.8)

where D_{T+1}^{i***} is D_T^i in (5.6) with

$$\overline{u}_{T-1}^{(\omega))j**}=\frac{R_{T-1}^{(x)j}+\delta A_T^j m_{T-1}^{|T-1|j}+\delta^2 A_{T+1}^j m_T^{|T-1|j}}{2\varphi_{T-1}^{(x)j}},$$

$$\overline{u}_{T-1}^{(\omega_2)j**}=\frac{R_{T-1}^{(y)j}-\delta B_T^j\sigma_{T-1}^{|T-1|j}-\delta^2 B_{T+1}^j\sigma_T^{|T-1|j}}{2\varphi_{T-1}^{(y)j}},$$

$$\overline{u}_{T-1}^{(\omega_3)j**}=\frac{\delta C_T^j a_{T-1}^{|T-1|j}+\delta^2 C_{T+1}^j a_T^{|T-1|j}}{2\varphi_{T-1}^{(z)j}}$$

System (5.8) yields equations in which right-hand-side and left-hand-side are linear functions of
x, y and z, for $i\in N$. For (5.8) to hold, it is required that:

$$A_{T-1}^{i} = -\frac{\left(\delta A_T^i b_{T-1}^i\right)^2}{4c_{T-1}^{I_i}} - h_{T-1}^i + \delta A_T^i \left(\sum\nolimits_{j=1}^{n} \frac{\delta A_T^j \left(b_{T-1}^j\right)^2}{2c_{T-1}^{IJ}} + \left(1 - \lambda_{T-1}^x\right) \right) - B_T^i \vartheta_{T-1},$$

$$B_{T-1}^i = \chi_{T-1}^j + \delta B_T^i \left(1 - \lambda_{T-1}^v\right),$$

$$C_{T-1}^i = \frac{\left(R_{T-1}^{(z)i}\right)^2 \kappa_{T-1}^i}{4\left(c_{T-1}^{IIIi}\right)} + \delta C_T^i \left(1 - \lambda_{T-1}^z\right), \quad \text{for } i \in N. \tag{5.9}$$

In addition,

$$D_{T-1}^i = \left(\frac{\left(R_{T-1}^{(x)i} - \delta A_T^i m_{T-1}^{|T-1|i} - \delta^2 A_{T+1}^i m_T^{|T-1|i|}\right)\left(R_{T-1}^{(x)i} + \delta A_T^i m_{T-1}^{|T-1|i} + \delta^2 A_{T+1}^i m_T^{|T-1|i}\right)}{4\varphi_{T-1}^{(x)i}} \right.$$
$$+ \frac{\left(R_T^{(y)i} + \delta B_T^i \sigma_{T-1}^{|T-1|i} + \delta^2 B_{T+1}^i \sigma_T^{|T-1|i}\right)\left(R_{T-1}^{(y)i} - \delta B_T^i \sigma_{T-1}^{|T-1|i} - \delta^2 B_{T+1}^i \sigma_T^{|T-1|i}\right)}{4\varphi_{T-1}^{(y)i}}$$
$$\left. - \frac{\left(\delta B_T^i \varepsilon_{T-1}^i\right)^2}{4c_{T-1}^{IIi}} - \frac{\left(\delta C_T^i a_{T-1}^{|T-1|i} + \delta^2 C_{T+1}^i a_T^{|T-1|i}\right)^2}{4\varphi_{T-1}^{(z)i}} \right)$$
$$+ \left[A_T^i \left(\sum_{j=1}^{n} m_{T-1}^{|T-1|j} \frac{R_{T-1}^{(x)j} + \delta A_T^j m_{T-1}^{|T-1|j} + \delta^2 A_{T+1}^j m_T^{|T-1|j}}{2\varphi_{T-1}^{(x)j}} + \sum_{j=1}^{n} \sum_{\tau=T-\omega_1}^{T-2} m_{T-1}^{|\tau|j} \overline{u}_\tau^{(\omega_1)j**} \right) \right.$$
$$+ B_T^i \left(g_{T-1}^i + \sum_{j=1}^{n} \varepsilon_{T-1}^j \frac{\delta B_T^j \varepsilon_{T-1}^j}{2c_{T-1}^{IIj}} - \sum_{j=1}^{n} \sigma_{T-1}^{|T-1|j} \frac{R_{T-1}^{(y)j} - \delta B_T^j \sigma_{T-1}^{|T-1|j} - \delta^2 B_{T+1}^j \sigma_T^{|T-1|j}}{2\varphi_{T-1}^{(y)j}} \right.$$
$$\left. - \sum_{j=1}^{n} \sum_{\tau=T-\omega_2}^{T-2} \sigma_{T-1}^{|\tau|j} u_\tau^{(\omega_2)j**} \right)$$
$$\left. + C_T^i \left(\sum_{j=1}^{n} a_{T-1}^{|T-1|j} \frac{\delta C_T^j a_{T-1}^{|T-1|j} + \delta^2 C_{T+1}^j a_T^{|T-1|j}}{2\varphi_{T-1}^{(z)j}} + \sum_{j=1}^{n} \sum_{\tau=T-\omega_3}^{T-2} a_{T-1}^{|\tau|j} u_\tau^{(\omega_3)j**} \right) + D_T^{i***} \right] \delta,$$

$$\text{for } i \in N, \tag{5.10}$$

which is a function of previously executed controls $(\overline{\underline{u}}_{(T-1)-}^{(\omega_1)**}, \overline{\underline{u}}_{(T-1)-}^{(\omega_2)**}, \overline{\underline{u}}_{(T-1)-}^{(\omega_3)**})$. Then, we move to stage $T-2$.

Using $V^i\left(T-1, x, y, z; , \underline{u}_{(T-1)-}^{(\omega_1)*}, \overline{u}_{(T-1)-}^{(\omega_2)**}, \overline{u}_{(T-1)-}^{(\omega_3)**}\right) = \left(A_{T-1}^i x + B_{T-1}^i y + C_{T-1}^i z + D_{T-1}^j\right)\delta^{T-2}$ derived in (5.9–5.10) and the stage $T-2$ equation in (3.6), the game equilibrium strategies in stage $T-2$ can be expressed as:

$$\overline{u}_{T-2}^{(\omega_1)i**} = \frac{R_{T-2}^{(x)i} + \delta A_{T-1}^{i} m_{T-2}^{|T-2|i} + \delta^2 A_{T}^{i} m_{T-1}^{|T-2|i} + \delta^3 A_{T+1}^{i} m_{T}^{|T-2|i}}{2\varphi_{T-2}^{(x)i}},$$

$$\overline{u}_{T-2}^{(\omega_2)i**} = \frac{R_{T-2}^{(y)i} - \delta B_{T-1}^{i} \sigma_{T-2}^{|T-2|} - \delta^2 B_{T}^{i} \sigma_{T-1}^{|T-2|i} - \delta^3 B_{T+1}^{i} \sigma_{T}^{|T-2|i}}{2\varphi_{T-2}^{(x)i}},$$

$$\overline{u}_{T-2}^{(\omega_3)i**} = \frac{\delta C_{T-1}^{i} a_{T-2}^{|T-2|i} + \delta^2 C_{T}^{i} a_{T-1}^{|T-2|i} + \delta^3 C_{T+1}^{i} a_{T}^{|T-2|i}}{2\varphi_{T-2}^{(z)i}},$$

$$u_{T-2}^{Ii**} = -\frac{\delta A_{T-1}^{i} b_{T-2}^{i}}{2c_{T-2}^{Ii}} x^{1/2},$$

$$u_{T-2}^{IIi**} = \frac{\delta B_{T-1}^{i} \varepsilon_{T-2}^{i}}{2c_{T-2}^{IIi}}, u_{T-2}^{IIIi**} = \frac{\left(R_{T-2}^{(z)i}\right)^2 \kappa_{T-2}^{i} Z}{4\left(c_{T-2}^{IIIi}\right)^2}, \quad \text{for } i \in N. \tag{5.11}$$

Substituting the optimal strategies in (5.11) into the stage $T-2$ equation of (3.6) we obtain a system of equations in which right-hand-side and left-hand-side are linear functions of x, y and z, for $i \in N$. For (5.11) to hold, it requires that:

$$A_{I-2} = -\frac{\left(\delta A_{T-1}^{i} b_{T-2}^{i}\right)^2}{4c_{T-2}^{Ii}} - h_{T-2}^{j}$$
$$+ \delta A_{T-1}^{i} \left(\sum_{j=1}^{n} \frac{\delta A_{T-1}^{j} \left(b_{T-2}^{j}\right)^2}{2c_{T-2}^{Ij}} + \left(1 - \lambda_{T-2}^{x}\right) \right) - B_{T-1}^{i} \vartheta_{T-2},$$
$$B_{T-2}^{i} = \pi_{T-2}^{i} + \delta B_{T-1}^{i} \left(1 - \lambda_{T-2}^{y}\right),$$
$$C_{T-2}^{i} = \frac{\left(R_{T-2}^{(w)i}\right)^2 \kappa_{T-2}^{i}}{4\left(c_{T-2}^{(\omega)i}\right)} + \delta C_{T-1}^{i} \left(1 - \lambda_{T-2}^{z}\right), \quad \text{for } i \in N. \tag{5.12}$$

In addition, D_{T-2}^{i} is an expression that contains the previously executed controls $(\overline{u}_{(T-2)-}^{(\omega_1)**}, \overline{u}_{(T-2)-}^{(\omega_2)**}, \overline{u}_{(T-2)-}^{(\omega_3)**})$.

Following the above analysis for stage, the game equilibrium strategies in stage $k \in \{T, T-1, \cdots, 1\}$ can be expressed as:

$$\overline{u}_{k}^{(\omega_1)i***} = \frac{R_{k}^{(x)i} + \sum_{\tau=k}^{k+\alpha_1-1} \delta^{\tau-k+1} A_{\tau+1}^{i} m_{\tau}^{|k|i}}{2\varphi_{k}^{(x)i}}, \quad \text{and} \quad A_{\tau}^{i} = 0 \quad \text{for} \quad \tau > T+1$$

$$\overline{u}_{k}^{(\omega_2)i***} = \frac{R_{k}^{(y)i} - \sum_{\tau=k}^{k+\omega_2-1} \delta^{\tau-k+1} B_{\tau+1}^{i} \sigma_{\tau}^{|k|i}}{2\varphi_{k}^{(y)i}}, \quad \text{and} \quad B_{\tau}^{i} = 0 \quad \text{for} \quad \tau > T+1,$$

$$\overline{u}_{k}^{(\omega_2)i***} = \frac{R_{k}^{(y)i} - \sum_{\tau=k}^{k+\omega_2-1} \delta^{\tau-k+1} B_{\tau+1}^{i} \sigma_{\tau}^{|k|i}}{2\varphi_{k}^{(y)i}}, \quad \text{and} \quad C_{\tau}^{i} = 0 \quad \text{for} \quad \tau > T+1,$$

$$u_k^{Ii**} = -\frac{\delta A_{k+1}^i b_k^i}{2c_k^{Ii}} x^{1/2},\ u_k^{IIi**} = \frac{\delta B_{k+1}^i \varepsilon_k^i}{2c_k^{IIi}},\ u_k^{IIIi**} = \frac{\left(R_k^{(z)i}\right)^2 \kappa_k^i z}{4\left(c_k^{III_i}\right)^2};$$

for $i \in N$. (5.13)

Substituting the optimal strategies in (5.13) into the stage k equation of (3.6) we obtain

$$V^i\left(k, x, y, z; \overline{u}_{k-}^{(\omega_1)^{**}}, \overline{u}_{k-}^{(\omega_2)^{**}}, \underline{u}_{k-}^{(\omega_3)^{**}}\right) = \left(A_k^i x + B_k^i y + C_k^i z + D_k^j\right)\delta^{k-1}$$

where

$$A_k^i = -\frac{\left(\delta A_{k+1}^i b_k^i\right)^2}{4c_k^{Ii}} - h_k^i + \delta A_{k+1}^i \left(\sum_{j=1}^n \frac{\delta A_{k+1}^j \left(b_k^j\right)^2}{2\phi_k^j} + \left(1 - \lambda_k^x\right)\right) - B_{k+1}^i \vartheta_k,$$

$$B_k^i = \chi_k^j + \delta B_{k+1}^i \left(1 - \lambda_k^y\right),$$

$$C_k^i = \frac{\left(R_k^{(z)i}\right)^2 \kappa_k^i}{4\left(c_k^{IIIi}\right)} + \delta C_{k+1}^i \left(1 - \lambda_k^z\right),$$

and D_k^i is an expression which contains the previously executed controls

$$\left(\underline{\overline{u}}_{k-}^{(\omega_1)**}, \underline{\overline{u}}_{k-}^{(\omega_2)**}, \underline{\overline{u}}_{k-}^{(\omega_3)**}\right) \text{ for } i \in N. \quad (5.14)$$

In particular, the term involving $\underline{\overline{u}}_{k-}^{(\omega_1)**}$ in D_k^i is

$$\delta A_{k+1}^i \sum_{j=1}^n \sum_{\tau=k-\omega_1+1}^{k-1} m_k^{|\tau|j} \overline{u}_\tau^{(\omega_1)j**} + \delta^2 A_{k+2}^i \sum_{j=1}^n \sum_{\tau=k-\sigma_1+2}^{k-1} m_{k+1}^{|\tau|j} \overline{u}_\tau^{(\omega_1)j**}$$

$$+ \delta^3 A_{k+3}^i \sum_{j=1}^n \sum_{r=k-\omega_1+3}^{k-1} m_{k+2}^{|\tau|j} \overline{u}_\tau^{(\omega_1)j**}$$

$$+ \delta^4 A_{k+4}^i \sum_{j=1}^n \sum_{\tau=k-\omega_1+4}^{k-1} m_{k+3}^{|\tau|j} \overline{u}_\tau^{(\omega_1)j**} + \cdots$$

$$+ \delta^{\omega_1-1} A_{k+\omega_1-1}^i \sum_{j=1}^n \sum_{\tau=k-1}^{k-1} m_{k+\omega_2-2}^{|\tau|j} \overline{u}_\tau^{(\omega_1)j**} \quad (5.15)$$

The term involving $\underline{\overline{u}}_{k-}^{(\omega_2)**}$ in D_k^i is

$$- \left(\delta B^i_{k+1} \sum_{j=1}^{n} \sum_{\tau=k-\omega_2+1}^{k-1} \sigma_k^{|\tau|j} \overline{u}_\tau^{(\omega_2)j^{**}} + \delta^2 B^i_{k+2} \sum_{j=1}^{n} \sum_{\tau=k-\omega_2+2}^{k-1} \sigma_{k+1}^{|\tau|j} \overline{u}_\tau^{(\omega_2)j^{**}} \right.$$
$$+ \delta^3 B^i_{k+3} \sum_{j=1}^{n} \sum_{\tau=k-\omega_2+3}^{k-1} \sigma_{k+2}^{|\tau|j} \overline{u}_\tau^{(\omega_2)j^{**}}$$
$$+ \delta^4 B^i_{k+4} \sum_{j=1}^{n} \sum_{\tau=k-\omega_2+4}^{k-1} \sigma_{k+3}^{|\tau|j} \overline{u}_\tau^{(\omega_2)j^{***}} + \cdots$$
$$\left. + \delta^{\omega_2-1} B^i_{k+\omega_2-1} \sum_{j=1}^{n} \sum_{\tau=k-1}^{k-1} \sigma_{k+\omega_2-2}^{|\tau|j} \overline{u}_\tau^{(\omega_3)j^{**}} \right) \tag{5.16}$$

The term involving $\overline{u}_{k-}^{(\omega_3)**}$ in D_k^i is

$$\delta C^i_{k+1} \sum_{j=1}^{n} \sum_{\tau=k-\sigma_3+1}^{k-1} a_k^{|\tau|j} \overline{u}_\tau^{(\omega_3)j^{**}} + \delta^2 C^i_{k+2} \sum_{j=1}^{n} \sum_{r=k-\sigma_3+2}^{k-1} a_{k+1}^{|\tau|j} \overline{u}_\tau^{(\omega_3)j'^{**}}$$
$$+ \delta^3 C^i_{k+3} \sum_{j=1}^{n} \sum_{\tau=k-\omega_3+3}^{k-1} a_{k+2}^{|\tau|j} \overline{u}_\tau^{(\omega_3)j^{**}}$$
$$+ \delta^4 C^i_{k+4} \sum_{j=1}^{n} \sum_{\tau=k-\omega_3+4}^{k-1} a_{k+3}^{|\tau|j} \overline{u}_\tau^{(\omega_3)j^{**}} + \cdots$$
$$+ \delta^{\omega_3-1} C^i_{k+\omega_3-1} \sum_{j=1}^{n} \sum_{\tau=k-1}^{k-1} a_{k+\omega_3-2}^{|\tau|j} \overline{u}_\tau^{(\omega_3)j^{**}}. \tag{5.17}$$

6 Chapter Notes

Environmental games have been one of the most studied areas in dynamic games. This Chapter presents a class of dynamic environmental games that lead to anthropogenic degradation under strategies lags. The games take into consideration pollution accumulation, ecosystems degradation and green technology investments. Early studies of dynamic transboundary games can be found in Plourde and Yeung (1989), Kaitala et al. (1991),Kaitala et al. (1992a),Kaitala et al. (1991b) and, Long (1992), Yeung (1992), Hoel (1992, 1993) and Dockner and Long (1993), Martin et al. (1993) and Yeung and Cheung (1994). Subsequent refinement of dynamic environmental games includes those of Haurie (1995 and 2005), Dockner and Nishimura (1999), Jørgensen and Yeung (1999), Feenstra et al. (2001, 2002). Fredji et al. (2004),Fredji et al. (2006), Haurie (2005), Krawczyk (2005), Kossioris et al. (2008), Tidball and

Zaccour (2005 and Tidball and Zaccour (2009), Jørgensen (2010), Li and Guo (2019) and Tilman et al. (2020). Collections of articles on dynamic environmental games can be found in the surveys of Long (2010) and Jørgensen et al. (2010).

Studies on using green technology to alleviate environmental damages had been presented in Billatos and Basaly (1997), Ane (2000), Kazi and Makhija (2013), Eis et al. (2016), Yeung and Petrosyan (2016b) and Yeung et al. (2017).

Finally, the terminal payoffs of the regions can be extended to include previously executed strategies that have effects after stage T. In particular, the terminal payoff of region i in stage $T+1$ can be expressed as

$$Q^{x(i)}_{T+1}(x_{T+1}) + Q^{y(i)}_{T+1}(y_{T+1}) + Q^{z(i)}_{T+1}\left(\kappa^i_{T+1} z_{T+1}\right) + q^{x(i)}_{T+1}\left(\overline{u}^{(\omega_1)i}_{(T+1)-}\right)$$
$$+ q^{y(i)}_{T+1}\left(\overline{u}^{(\omega_2)i}_{(T+1)-}\right) + q^{z(i)}_{T+1}\left(\overline{u}^{(\omega_3)i}_{(T+1)-}\right) + \overline{w}^i_{T+1}\Big), \text{ for } i \in N.$$

Problem 2 in Sect. 7 deals with issue.

7 Problems

1. (Warmup exercise) Consider a 2-player 5-stage dynamic environmental game in which region $i \in \{1, 2\}$ produces two types of outputs. The first type of output $\overline{u}^{(2)i}_k \in \overline{U}^{(2)i} \subset R$ emits pollutants over 2 stages. The second type of output is a clean output $u^{Ii}_k \in U^{Ii} \subset R$ which uses green technologies.

The pollution stock is $x_k \in X \subset R$ and the dynamics of pollution accumulation is governed by the dynamical equation:

$$x_{k+1} = x_k + \overline{u}^{(2)1}_k + 1.5\overline{u}^{(2)2}_k + \sum_{j=1}^{2} 0.5\overline{u}^{(2)j}_{k-1} - \sum_{j=1}^{3} 0.1 u^{\text{IIj}}_k (x_k)^{1/2} - 0.3x_k,\ x_1 = 15,$$

where u^{IIi}_1 is the pollution abatement effort of region i.

The net economic gain from the clean output in region 1 is $2u^{I1}_k - 0.5(u^{I1}_k)^2$. The net economic gain from the clean output in region 2 is $3u^{I2}_k - (u^{I2}_k)^2$.

The cost of pollution abatement is $2(u^{IIj}_k)^2$ and the damage from pollutions is $-0.2x_k$.

The payoff of region 1 is

$$\sum_{k=1}^{5}\left(4\overline{u}^{(2)1}_k - \left(\overline{u}^{(2)1}_k\right)^2 + 2u^{I1}_k - 0.5\left(u^{II}_k\right)^2 - 2\left(u^{II1}_k\right)^2 - 0.2x_k\right)(0.9)^{k-1}$$
$$+\left(-2x_6 + 150\right)(0.9)^5$$

The payoff of region 2 is

$$\sum_{k=1}^{5}\left(3\overline{u}_k^{(2)1}-\left(\overline{u}_k^{(2)1}\right)^2+3u_k^{I1}-\left(u_k^{I1}\right)^2-2\left(u_k^{II1}\right)^2-0.2x_k\right)$$
$$(0.9)^{k-1}+(-2_6+100)(0.9)^5.$$

Characterize a feedback Nash equilibrium solution with (i) the game equilibrium strategies and (ii) the game equilibrium value of the regions.

2. Consider a 3-player 8-stage dynamic environmental game in which region $i \in \{1, 2, 3\}$ produces three types of outputs. The first type of output $\overline{u}_k^{(\omega_1)i} \in \overline{U}^{(\omega_1)i} \subset R$ emits pollutants over 2 stages. The second type of output $\overline{u}_k^{(\omega_2)i} \in \overline{U}^{(\omega_2)i} \subset R$ causes damages to the ecosystems over 4 stages. The third type of output is a clean output that uses green technologies.

The pollution stock is $x_k \in X \subset R$ and the dynamics of pollution accumulation is governed by the dynamical equation:

$$x_{k+1} = x_k + \sum_{j=1}^{3}\overline{u}_k^{(\omega_1)j} + \sum_{j=1}^{3}\sum_{\tau=k-1}^{k-1} 0.5\overline{u}_\tau^{(\omega_1)j} - \sum_{j=1}^{3} 0.1u_k^{Ij}(x_k)^{1/2} - 0.2x_k, \quad x_1 = 10$$

where u_1^{Ij} is the pollution abatement effort of region j.

The dynamics governing the evolution of the condition of the ecosystems $y_k \in Y \subset R$ is $y_{k+1} = y_k + 80 - \sum_{j=1}^{3} 1.5\overline{u}_k^{(\omega_2)j} - \sum_{j=1}^{3}\sum_{\tau=k-3}^{k-1} \overline{u}_\tau^{(\omega_2)j} + \sum_{j=1}^{3} 0.5u_k^{IIj} - x_k - 0.05y_k$,

$y_1 = 8000$,

where u_1^{IIj} is the ecosystem improvement effort of region j.

There is a public green technology $z_k \in Z \subset R$ that generates economic gains to the regions. The dynamics governing the built up of green technology is

$$z_{k+1} = z_k + \sum_{j=1}^{3}\overline{u}_k^{(\omega_3)j} + \sum_{j=1}^{3}\sum_{\tau=k-2}^{k-1} 0.5\overline{u}_\tau^{(\omega_3)j} - 0.1z_k, \quad z_1 = 50$$

where $\overline{u}_k^{(\omega_3)j}$ is the investment in green technology by region $j \in \{1, 2, 3\}$ in stage k.

The economic gain from the green technology for region 1 is $\Omega_k^1 z_k = 0.8z_k$, for region 2 is $\Omega_k^2 z_k = z_k$ and for region 3 is $\Omega_k^3 z_k = 1.2z_k$. The cost of investment in green technology is $2(\overline{u}_k^{(\omega_3)j})^2$ for region j.

The cost of pollution abatement is $2(u_k^{Ij})^2$, the cost of improving the ecosystems is $(u_k^{IIi})^2$. The damage from pollutions is x_k and the environmental amenity from the ecosystems is $10y_k$.

The payoff of region $i \in N$ is

$$
\sum_{k=1}^{8} \Big(8\bar{u}_k^{(\omega_1)i} - 2\big(\bar{u}_k^{(\omega_1)i}\big)^2 - 2\big(u_k^{Ij}\big)^2 + 5\bar{u}_k^{(\omega_2)i} \\
- \big(\bar{u}_k^{(\omega_2)i}\big)^2 - \big(u_k^{IIi}\big)^2 + \Omega_k^j z_k - 2\big(\bar{u}_1^{(\omega_3)i}\big)^2 \\
- x_k + 10 y_k \Big)\delta^{k-1} + (-2x_{T+1} + 20 y_{T+1} + z_{T+1} + 750)\delta^8 .
$$

Characterize a feedback Nash equilibrium solution mwith (i) the game equilibrium strategies and (ii) the game equilibrium value of the regions.

3. Consider the game (3.1–3.4) in Sect. 3. The terminal payoff of region i in stage $T+1$ is replaced by

$$
\Big(Q_{T+1}^{x(i)} x_{T+1} + Q_{T+1}^{y(i)} y_{T+1} + Q_{T+1}^{z(i)} \kappa_{T+1}^i z_{T+1} \\
+ q_{T+1}^{x(i)} \overline{u}_{(T+1)-}^{(\omega_1)i} + q_{T+1}^{y(i)} \overline{u}_{(T+1)-}^{(\omega_2)i} + q_{T+1}^{z(i)} \overline{u}_{(T+1)-}^{(\omega_3)i} + \varpi_{T+1}^i \Big),
$$

Characterize a feedback Nash equilibrium solution with (i) the game equilibrium strategies and (ii) the game equilibrium payoffs of the regions.

References

Ane RP (2000) Stimulating innovation in green technology. Policy Alternatives and Opportunities, American Behav Scientist 44(2):188–212

Billatos S, Basaly N (1997) In: Green technology and design for the environment. CRC Press, pp 312. ISBN 9781560324607

Corbett J (2019) Toxic waste will continue to grow for decades even if all U.S. drilling and fracking halts today, New Report Says. EcoWatch, June 19, 2019. https://www.ecowatch.com/epas-mis management-of-toxic-fossil-fuel-waste-2638917390.html?rebelltitem=1#rebelltitem1

Dockner E, Long NV (1993) International pollution control: cooperative versus noncooperative strategies. J Environ Econ Manag 24:13–29

Dockner E, Nishimura K (1999) Transboundary pollution in a dynamic game model. Jpn Econ Rev 50(4):443–456

Eis J, Bishop R, Gradwell P (2016) Galvanising low-carbon innovation. A new climate economy working paper for seizing the global opportunity: partnerships for better growth and a better climate. London, New Climate Economy, pp 1–28

Feenstra T, Kort P, de Zeeuw A (2001) Environmental policy instruments in an international duopoly with feedback investment strategies. J Econ Dyn Control 25:1665–1687

Feenstra T, Kort P, de Zeeuw A (2002) International competition and investment in abatement: taxes versus standards. In: Marsiliani L, Rauscher M, Withagen C (eds) Environmental economics and the international economy. Kluwer, Dordrecht, pp 89–98

Fredj K, Martín-Herrán G, Zaccour G (2004) Slowing deforestation pace through subsidies: a differential game. Automatica 40:301–309. https://doi.org/10.1016/j.automatica.2003.10.020

Fredj K, Martìn-Herràn G, Zaccour G (2006) Incentive mechanisms to enforce sustainable forest exploitation. Environ Model Assessment 11:145–156

Haurie A (1995) Environmental coordination in dynamic oligopolistic markets. Group Decis Negot 4(1):39–57

Haurie A (2005) A two-timescale stochastic game framework for climate change policy assessment. In: Haurie A, Zaccour G (eds), Dynamic games: Theory and applications. GERAD twenty-fifth anniversary series, New York, Springer, pp 193–211

Hoel M (1992) Emission taxes in a dynamic international game of CO2 emissions. In: Pethig R (ed) Conflict and cooperation in managing environmental resources. Springer, Berlin, pp 39–70

Hoel M (1993) Intertemporal properties of an international carbon tax. Resour Energy Econ 15(1):51–70

Johnson S (2020) "What damages an ecosystem?" sciencing.com. https://sciencing.com/damages-ecosystem-8355512.html. 31 March 2020

Jørgensen S (2010) A dynamic game of waste management. J Econ Dyn Control 34:258–265. https://doi.org/10.1016/j.jedc.2009.09.005

Jørgensen S, Yeung DWK (1999) Inter- and intragenerational renewable resource extraction. Ann Oper Res 88:275–289

Kaitala V, Pohjola M, Tahvonen O (1992) An economic analysis of transboundary air pollution between Finland and the Soviet Union. Scandinavian J Econ 94:409–424

Kaitala V, Pohjola M, Tahvonen O (1991) Transboundary air pollution between Finland and the U.S.S.R.: A dynamic acid rain game. In: Hamalainen R, Ehtamo H (eds) Dynamic games in the economics and management of pollution, Dynamic games in economic analysis. Berlin, Springer, pp 183–192

Kaitala V, Pohjola M, Tahvonen O (1992a) Transboundary air pollution and soil acidification: A dynamic analysis of an acid rain game between Finland and the U.S.S.R. Environ Resource Econ 2(2):161–181

Kazi AG, Makhija A (2013) Green technology and protection of environment. Published on May 28, 2013. Published in: Technology, Business

Kossioris G, Plexousakis M, Xepapadeas A, de Zeeuw A, Mäler K-G (2008) Feedback Nash equilibria for non-linear differential games in pollution control. J Econ Dyn Control 32(4):1312–1331

Krawczyk JB (2005) Coupled constraint Nash equilibria in environmental games. Resour Energy Econ 27(2):157–181

Li H, Guo G (2019) Dynamic decision of transboundary basin pollution under emission permits and pollution abatement. Physica A: Statist Mech its Appl 532(15):121869

Long NV (1992) Pollution control: a differential game approach. Annals of Operational Res 37:283–296

Long NV (2010) A survey of dynamic games in economics, World Scientific

Martin WE, Patrick RH, Tolwinski B (1993) A dynamic game of a transboundary pollutant with asymmetric players. J Environ Econ Manag 24:1–12

Oppenlander R (2013) Food choice and sustainability. minneapolis, MN: Langdon Street Press, pp 120–123. ISBN 978-1-62652-435-4

Pearce R (2006) When the rivers run dry: water—the defining crisis of the twenty-first century, Beacon Press, ISBN 0807085731

Plourde C, Yeung DWK (1989) Harvesting of a transboundary replenishable fish stock: a non cooperative game solution. Mar Resour Econ 6:57–70

Rood RB (2014) What would happen to the climate if we stopped emitting greenhouse gases today? The Conversation, December 11, 2014. https://www.iflscience.com/environment/what-would-happen-climate-if-we-stopped-emitting-greenhouse-gases-today/

Rood RB (2017) If we stopped emitting greenhouse gases right now, would we stop climate change? the conversation, July 5, 2017, theconversation.com/if-we-stopped-emitting-greenhouse-gases-right-now-would-we-stop-climate-change-78882

Thakkar H (1999) Assessment of irrigation in India. World Commission on Dams

Tidball M, Zaccour G (2005) An environmental game with coupling constraints. Environ Model Assess 10:153–158

Tidball M, Zaccour G (2009) A differential environmental game with coupling constraints. Optimal Control Appl Methods 30:197–2007

Tilman AR, Plotkin JB, Akçay E (2020) Evolutionary games with environmental feedbacks. Nat Commun 11:915. https://doi.org/10.1038/s41467-020-14531-6

Yeung DWK (1992) A Differential game of industrial pollution management. Ann Oper Res 37:297–311

Yeung DWK, Petrosyan LA (2016) Subgame consistent cooperation: a comprehensive treatise. Springer

Yeung DWK, Cheung MT (1994) Capital accumulation subject to pollution control: a differential game with a feedback nash equilibrium. In: Basar T, Haurie A (eds) Advances in dynamic games and applications. Annals of the international society of dynamic games. vol 1. Boston, Birkhäuser, pp 289–300

Chapter 5
Durable-Strategies Cooperative Dynamic Games

It is well known that non-cooperative behaviours among participants would, in general, lead to an outcome that is not Pareto optimal. Worse still, highly undesirable outcomes (like the prisoner's dilemma) and even devastating results (like the tragedy of the commons) could appear when the involved participants only care about their individual self-seeking optimization in a non-cooperative situation. In a dynamic world, non-cooperative behaviours with durable effects guided by short-sighted individual rationality could be a source for a series of disastrous consequences in the future. The phenomenon of the 'inter-temporal tragedy of temporal individual rationality' is emerging in our era of rapid globalization. Cooperation suggests the best possibility of obtaining socially optimal and group efficient solutions to decision problems involving strategic actions. In addition, durable strategies affecting both the decision-makers' payoffs and the dynamics of the state variables are prevalent in real-life situations. This chapter presents a generic class of cooperative dynamic games with multiple durable strategies of different lags affecting both the players' payoffs and the state dynamics. Measurements of efficiency loss by self-seeking optimization are examined in Sect. 4. A mathematical appendix is given in Sect. 5 and chapter notes are provided in Sect. 6. Problem sets are supplied in Sect. 7.

1 Cooperative Dynamic Games with Durable Strategies

In this section, we develop a $T-$ stage, $n-$ player nonzero-sum discrete-time cooperative dynamic game with non-durable and durable strategies affecting the players' payoffs and the state dynamics. We use $u_k^i \in U^i \subset R^{m^i}$ to denote the set of non-durable control strategies of player i. We use $\overline{u}_k^i = (\overline{u}_k^{(2)i}, \overline{u}_k^{(3)i}, \cdots, \overline{u}_k^{(\omega_i)i})$ to denote the set of durable strategies of player i, where $\overline{u}_k^{(\zeta)i} \in \overline{U}^{(\zeta)i} \subset R^{m_{(\zeta)^i}}$ for $\zeta \in \{2, 3, \cdots, \omega_i\}$. In particular, $\overline{u}_k^{(2)i}$ are non-durable strategies that have effects in stages k and $k+1$. The strategies $\overline{u}_k^{(3)i}$ are durable strategies that have effects within

D. W. K. Yeung and L. A. Petrosyan, *Durable-Strategies Dynamic Games*,
Theory and Decision Library C 50, https://doi.org/10.1007/978-3-030-92742-4_5

stage k to stage $k+2$. The strategies $\overline{u}_k^{(\omega)i}$ are durable strategies that have effects within stages $k, k+1, \cdots, k+\omega-1$. The state at stage k is $x_k \in X \subset R^m$ and the state space is common for all players. The single-stage payoff of player i in stage k is

$$g_k^i(x_k, \underline{u}_k, \overline{u}_k; \overline{u}_{k-}), \text{ for } k \in \{1, 2, \cdots, T\} \text{ and } i \in \{1, 2, \cdots, n\} \equiv N,$$

where $\underline{u}_k = (u_k^1, u_k^2, \cdots, u_k^n)$ is the set of durable strategies of all the n players, $\overline{u}_k = (\overline{u}_k^1, \overline{u}_k^2, \cdots, \overline{u}_k^n)$ is the set of durable strategies of all the n players and $\overline{u}_{k-} = (\overline{u}_{k-}^1, \overline{u}_{k-}^2, \cdots, \overline{u}_{k-}^n)$ is the set of strategies which are executed before stage k by all players but still in effect in stage k.

The payoff of player i is:

$$\sum_{k=1}^{T} g_k^i(x_k, \underline{u}_k, \overline{u}_k; \overline{u}_{k-})\delta_1^k + q_{T+1}^i(x_{T+1}; \overline{u}_{(T+1)-})\delta_1^{T+1}, \tag{1.1}$$

where $q_{T+1}^i(x_{T+1}; \overline{u}_{(T+1)-})$ is the terminal payoff of player i.

The state dynamics is characterized by a vector of difference equations:

$$x_{k+1} = f_k(x_k, \underline{u}_k, \overline{u}_k; \overline{u}_{k-}), \quad x_1 = x_1^0, \text{ for } k \in \{1, 2, \cdots, T\}, \tag{1.2}$$

where $\overline{u}_{1-}$ are given in stage 1 and some or all of previously executed strategies can be zero.

The controls executed before the start of the operation in stage 1, that is $\underline{u}_{1-}$, are known and some or all of them can be zeros. The $g_k^i(x_k, \underline{u}_k, \overline{u}_k; \overline{u}_{k-})$, $f_k^i(x_k, \underline{u}_k, \overline{u}_k; \overline{u}_{k-})$ and $q_{T+1}^i(x_{T+1}; \overline{u}_{(T+1)-})$ are continuously differentiable functions.

The information set of every player includes the knowledge in.

(i) all the possible moves by himself and other players, that is u_k^i and $\overline{u}_k^i$, for $k \in \{1, 2, \cdots, T\}$ and $i \in N$;

(ii) the set of controls which are executed before stage k by all players but still in effect in stage k, that is $\overline{u}_{k-} = (\overline{u}_{k-}^1, \overline{u}_{k-}^2, \cdots, \overline{u}_{k-}^n)$, for $k \in \{1, 2, \cdots, T\}$;

(iii) the payoff functions of all players, that is $\sum_{k=1}^{T} g_k^i(x_k, \underline{u}_k, \overline{u}_k; \overline{u}_{k-})\delta_1^k$ $+q_{T+1}^i(x_{T+1}; \overline{u}_{(T+1)-})\delta_1^{T+1}$, for $i \in N$; and

(iv) the state dynamics $x_{k+1} = f_k(x_k, \underline{u}_k, \overline{u}_k; \overline{u}_{k-})$ and the values of present and past states $(x_k, x_{k-1}, \cdots, x_1)$.

To exploit the potential gains from cooperation, the players agree to act cooperatively and distribute the payoffs among themselves according to an agreed-upon gain sharing optimality principle.

1.1 Efficiency and Group Optimality Under Cooperation

Consider the case when the players agree to cooperate and distribute the payoff among themselves according to an optimality principle. Two crucial properties that a cooperative scheme has to satisfy are group optimality and individual rationality. Group optimality ensures that the joint payoff of all the players under cooperation is maximized. Group optimality is a Pareto optimal condition under which yields the maximized efficiency that can be achieved. Failure to fulfil group optimality leads to the condition where the participants prefer to deviate from the agreed-upon solution plan in order to extract the unexploited gains. To achieve group optimality, the players will maximize their joint payoff by solving the dynamic optimization problem which maximizes

$$\sum_{j=1}^{n}\sum_{k=1}^{T} g_k^j(x_k, \underline{u}_k, \overline{u}_k; \overline{u}_{k-})\delta_1^k + \sum_{j=1}^{n} q_{T+1}^j(x_{T+1}; \overline{u}_{(T+1)-})\delta_1^{T+1}, \tag{1.3}$$

subject to (1.2).

An optimal solution to the joint maximization problem (1.2–1.3) can be characterized by the theorem below.

Theorem 1.1 *Let $W(k, x; \overline{u}_{k-})$ be the maximal value of the joint payoff*

$$\sum_{j=1}^{n}\sum_{t=k}^{T} g_t^j(x_t, \underline{u}_t, \overline{u}_t; \overline{u}_{t-})\delta_1^t + \sum_{j=1}^{n} q_{T+1}^j(x_{T+1}; \overline{u}_{(T+1)-})\delta_1^{T+1},$$

for the joint payoff maximization problem (1.2–1.3) *starting at stage k with state $x_k = x$ and previously executed controls $\overline{u}_{k-}$, then the function $W(k, x; \overline{u}_{k-})$ satisfies the following system of recursive equations:*

$$W(T+1, x; \overline{u}_{(T+1)-}) = \sum_{j=1}^{n} q_{T+1}^j(x; \overline{u}_{(T+1)-})\, \delta_1^{T+1}, \tag{1.4}$$

$$\begin{aligned} &W(k, x; \overline{u}_{k-}) \\ &= \max_{\underline{u}_k, \overline{u}_k} \left\{ \sum_{j=1}^{n} g_k^j(x, \underline{u}_k, \overline{u}_k; \overline{u}_{k-})\delta_1^k + W[k+1, f_k(x, \underline{u}_k, \overline{u}_k; \overline{u}_{k-}); \overline{u}_{(k+1)-}] \right\} \\ &= \max_{\underline{u}_k, \overline{u}_k} \left\{ \sum_{j=1}^{n} g_k^j(x, \underline{u}_k, \overline{u}_k; \overline{u}_{k-})\delta_1^k + W[k+1, f_k(x, \underline{u}_k, \overline{u}_k; \overline{u}_{k-}); \overline{u}_k, \overline{u}_{(k+1)-} \cap \overline{u}_{k-}] \right\}, \end{aligned}$$

$$\text{for } k \in \{1, 2, \cdots, T\}. \tag{1.5}$$

Proof The conditions in (1.4–1.5) satisfy the optimal conditions of the dynamic optimization technique with durable controls in Theorem 1.1 of Chap. 2 and hence an optimal solution to the control problem results. ■

We use $\{\underline{u}_k^*, \overline{u}_k^*\}$ for $k \in \{1, 2, \cdots, T\}$ to denote the optimal control strategies derived from Theorem 1.1. Substituting these optimal controls into the state dynamics (1.2), one can obtain the dynamics of the optimal cooperative trajectory as:

$$x_{k+1} = f_k(x_k, \underline{u}_k^*, \overline{u}_k^*; \overline{u}_{k-}^*), \quad \text{for } k \in \{1, 2, \cdots, T\} \text{ and } x_1 = x_1^0. \tag{1.6}$$

We use $\left\{x_k^*\right\}_{k=1}^{T+1}$ to denote the solution to (1.6) which yields the optimal cooperative state trajectory. The Pareto group optimal joint payoff of the players over the cooperative duration from stage 1 to stage T can be expressed as

$$W(1, x_1^0; \overline{u}_{1-}) = \sum_{j=1}^{n} \sum_{t=k}^{T} g_t^j(x_t^*, \underline{u}_t^*, \overline{u}_t^*; \overline{u}_{t-}^*)\, \delta_1^t + \sum_{j=1}^{n} q_{T+1}^j(x_{T+1}^*; \overline{u}_{(T+1)-}^*)\, \delta_1^{T+1}. \tag{1.7}$$

Obtaining $W(1, x_1^0; \overline{u}_{1-})$ guarantees the maximal joint payoff can be distributed to the players. Let

$$\xi(1, x_1^0; \overline{u}_{1-}) = [\xi^1(1, x_1^0; \overline{u}_{1-}), \xi^2(1, x_1^0; \overline{u}_{1-}), \cdots, \xi^n(1, x_1^0; \overline{u}_{1-})], \tag{1.8}$$

denote the agreed-upon distribution of cooperative payoffs among the players. In particular, player i would receive $\xi^i(1, x_1^0; \underline{u}_{1-}^*)$, for $i \in N$.

To satisfy group optimality in the cooperative scheme, one of the conditions is that the imputation vector $\xi(1, x_1^0; \overline{u}_{1-})$ in the outset of the game has to satisfy

$$W(1, x_1^0; \overline{u}_{1-}) = \sum_{j=1}^{n} \xi^j(1, x_1^0; \overline{u}_{1-}). \tag{1.9}$$

This condition guarantees the maximal joint payoff is distributed to the players.

1.2 Individual Rationality

Individual rationality is required to hold so that the payoff allocated to any player under cooperation will be no less than his non-cooperative payoff. Failure to guarantee individual rationality leads to the condition where the concerned participants would deviate from the agreed-upon solution plan and play non-cooperatively. For individual rationality to be satisfied, the payoffs received by the players under cooperation have to be no less than their non-cooperative payoffs along the cooperative state trajectory. The non-cooperative payoffs of the players in a Nash equilibrium of the dynamic game (1.1–1.2) are characterized in Theorem 3.1 in Chap. 2. Invoking this Theorem, the payoff of player i over stages 1 to stages $T+1$ in a non-cooperative equilibrium is

$$V^i(1, x_1^0; \underline{\overline{u}}_{1-}) = \sum_{t=k}^{T} g_t^j(x_t^{**}, \underline{u}_t^{**}, \underline{\overline{u}}_t^{**}; \underline{\overline{u}}_{t-}^{**})\delta_1^t + \sum_{j=1}^{n} q_{T+1}^j(x_{T+1}^{**}; \underline{\overline{u}}_{(T+1)-}^{**})\delta_1^{T+1}, \quad \text{for } i \in N, \tag{1.10}$$

where $\underline{u}_t^{**}$ is the game equilibrium strategies and x_t^{**} is the game equilibrium trajectory of the state variables.

For individual rationality to be maintained, one of the conditions is that the payoff of player i under cooperation must be no less than his non-cooperative payoff in the outset of the game, that is

$$\xi^i(1, x_1^0; \underline{\overline{u}}_{1-}) \geq V^i(1, x_1^0; \underline{\overline{u}}_{1-}), \quad \text{for } i \in N. \tag{1.11}$$

If condition (1.11) is not satisfied, no agreement can be reached in stage 1.

2 Dynamically Consistent Solutions and Payment Mechanism

Though group optimality and individual rationality in the outset of the game constitute two crucial properties for cooperation, their fulfilment does not necessarily guarantee these conditions that group optimality and individual rationality will be satisfied throughout the cooperative duration. Moreover, even-though group optimality and individual rationality are guaranteed to be satisfied throughout the cooperative duration, their fulfilment does not necessarily guarantee a dynamically stable solution in cooperation because there is no guarantee that the agreed-upon optimality principle is fulfilled throughout the cooperative duration.

The question of dynamic stability in differential games has been explored rigorously in the past four decades. Haurie (1976) discussed the problem of instability in extending the Nash bargaining solution to differential games. Petrosyan (1977) formalized mathematically the notion of dynamic stability in solutions of differential games. Petrosyan and Danilov (1979 and 1985) introduced the notion of "imputation distribution procedure" for cooperative solutions. To ensure stability in dynamic cooperation over time, a stringent condition is required: the specific agreed-upon optimality principle must be maintained at any instant of time throughout the game along the optimal state trajectory. This condition is the notion of dynamical/subgame consistency. Crucial to the derivation of a subgame consistent cooperative solution is the formulation of a payment distribution mechanism that would lead to the realization of the solution. Yeung and Petrosyan (2004 and 2010) developed generalized method for the derivation of analytically tractable subgame consistent solutions in stochastic differential games and stochastic dynamic games. This has made possible the rigorous study of subgame consistent solutions in continuous-time dynamic cooperation. Applications and exegeses of subgame consistent solutions in differential games and stochastic differential games can be found in Yeung (2005, 2006 and 2010) and Yeung and Petrosyan (2006a, 2007a, 2007b, 2007c, 2008, 2013 and 2014a).

2.1 Dynamical Consistency

For dynamical/subgame consistency to be satisfied, the players' agreed-upon optimality principle will be effective along the cooperative state trajectory x_k^* contingent upon $\underline{u}_{k-}^*$. Hence, the agreed-upon imputation $\xi(1, x_1^0; \underline{u}_{1-})$ in (1.8) has to be maintained at all stages $k \in \{1, 2, \cdots, T\}$ along the cooperative trajectory $\{x_k^*\}_{k=1}^{T+1}$. Therefore, the cooperative payoff given to player i has to satisfy the following condition.

We use

$$\xi(k, x_k^*; \underline{u}_{k-}^*) = [\xi^1(k, x_k^*; \underline{u}_{k-}^*), \xi^2(k, x_k^*; \underline{u}_{k-}^*), \cdots,$$
$$\xi^n(k, x_k^*; \underline{u}_{k-}^*)], \text{ for } k \in \{1, 2, \cdots, T\}, \tag{2.1}$$

to denote the agreed distribution of cooperative payoffs among the players along the cooperative trajectory $\{x_k^*\}_{k=1}^{T}$ given the previously executed controls $\underline{u}_{k-}^*$.

To satisfy group optimality along the cooperative trajectory $\{x_k^*\}_{k=1}^{T+1}$, it is required that

$$W(k, x^*; \underline{u}_{k-}^*) = \sum_{j=1}^{n} \xi^j(k, x_k^*; \underline{u}_{k-}^*), \quad \text{for } k \in \{1, 2, \cdots, T\}. \tag{2.2}$$

The condition in (2.2) guarantees the maximal joint payoff is distributed to the players throughout the cooperation duration.

For individual rationality to be maintained throughout the game, the payoff that player i receives under cooperation along the cooperative trajectory $\{x_k^*\}_{k=1}^{T+1}$ must be greater than or equal to his non-cooperative payoff, therefore the chosen imputation must satisfy the condition:

$$\xi^i(k, x_k^*; \underline{u}_{k-}^*) \geq V^i(k, x_k^*; \underline{u}_{k-}^{**}), \tag{2.3}$$

for $i \in N$ and $k \in \{1, 2, \cdots, T\}$.

The condition in (2.3) guarantees the fulfilment of individual rationality throughout the cooperation duration so that the payoff allocated to any player under cooperation will be no less than his non-cooperative payoff. Failure to guarantee individual rationality at any stage would lead to the condition where the concerned participants would deviate from the agreed-upon solution plan and play non-cooperatively.

2.2 Imputation Distribution Procedure

Crucial to the analysis is the derivation of an Imputation Distribution Procedure (IDP) leading to the realization of the agreed imputations in (2.1). To do this, we follow Yeung and Petrosyan (2010, 2016a and 2019) and use $\beta_k^i(x_k^*; \underline{\overline{u}}_{k-}^*)$ to denote the payment that player i receives in stage k under the cooperative agreement along the cooperative trajectory $\{x_k^*\}_{k=1}^T$ with durable strategies executive before but still in effect being $\underline{\overline{u}}_{k-}^*$. The payment scheme involving $\beta_k^i(x_k^*; \underline{\overline{u}}_{k-}^*)$ constitutes an IDP in the sense that the payoff to player i over the stages from k to $T+1$ satisfies the condition:

$$\xi^i(k, x_k^*; \underline{\overline{u}}_{k-}^*) = \beta_k^i(x_k^*; \underline{\overline{u}}_{k-}^*)\delta_1^k + \left\{ \sum_{\zeta=k+1}^{T} \beta_\zeta^i(x_\zeta^*; \underline{\overline{u}}_{\zeta-}^*)\,\delta_1^\zeta + q_{T+1}^i(x_{T+1}^*; \underline{\overline{u}}_{(T+1)-}^*)\,\delta_1^{T+1} \right\}, \tag{2.4}$$

for $i \in N$ and $k \in \{1, 2, \cdots, T\}$.

A theorem for the derivation of $\beta_k^i(x_k^*; \underline{\overline{u}}_{k-}^*)$, for $k \in \{1, 2, \cdots, T\}$ and $i \in N$, that satisfies (2.4) is provided below.

Theorem 2.1 *The agreed-upon imputation $\xi(k, x_k^*; \underline{\overline{u}}_{k-}^*)$, for $k \in \{1, 2, \cdots, T\}$ along the cooperative trajectory $\{x_k^*\}_{k=1}^T$, can be realized by a payment.*

$$\beta_k^i(x_k^*; \underline{\overline{u}}_{k-}^*) = (\delta_1^k)^{-1}[\xi^i(k, x_k^*; \underline{\overline{u}}_{k-}^*) - \xi^i\left(k+1, f_k(x_k^*, \underline{u}_k^*, \overline{\underline{u}}_k^*; \underline{\overline{u}}_{k-}^*); \underline{\overline{u}}_{(k+1)-}^*\right)] \tag{2.5}$$

given to player $i \in N$ at stage $k \in \{1, 2, \cdots, T\}$.

Proof Using (2.4) one can obtain.

$$\xi^i(k+1, x_{k+1}^*; \underline{\overline{u}}_{k-}^*) = B_{k+1}^i(x_{k+1}^*; \underline{\overline{u}}_{k-}^*)\delta_1^{k+1} + \left\{ \sum_{\zeta=k+2}^{T} B_\zeta^i(x_\zeta^*; \underline{\overline{u}}_{\zeta-}^*)\delta_1^\zeta + q_{T+1}^i(x_{T+1}^*; \underline{\overline{u}}_{(T+1)-}^*)\delta_1^{T+1} \right\}, \tag{2.6}$$

Upon substituting (2.6) into (2.4) yields

$$\xi^i(k, x_k^*; \underline{\overline{u}}_{k-}^*) = \beta_k^i(x_k^*; \underline{\overline{u}}_{k-}^*)\delta_1^k + \xi^i(k+1, x_{k+1}^*; \underline{\overline{u}}_{(k+1)-}^*),$$

which can be expressed as

$$\xi^i(k, x_k^*; \underline{\overline{u}}_{k-}^*) = \beta_k^i(x_k^*; \underline{\overline{u}}_{k-}^*)\delta_1^k + \xi^i\left(k+1, f_k(x_k^*, \underline{u}_k^*, \underline{\overline{u}}_k^*; \underline{\overline{u}}_{k-}^*); \underline{u}_{(k+1)-}^*\right). \quad (2.7)$$

From (2.7) one can obtain Theorem 2.1. ■

The payment scheme in Theorem 2.1 gives rise to the realization of the imputation guided by the agreed-upon optimality principle and constitutes a dynamically consistent payment scheme. More specifically, the payment of $\beta_k^i(x_k^*; \underline{\overline{u}}_{k-}^*)$ allotted to player $i \in N$ in stage $k \in \{1, 2, \cdots, T\}$ will establish a cooperative plan that matches with the agreed-upon imputation to every player along the cooperative path. It is worth noting that formula (2.5) in Theorem 2.1 is a new formulation in that the term $\xi^i\left(k+1, f_k(x_k^*, \underline{u}_k^*, \underline{\overline{u}}_k^*; \underline{\overline{u}}_{k-}^*); \underline{\overline{u}}_{(k+1)-}^*\right)$ appears in Yeung and Petrosyan (2019) instead of $\xi^i\left(k+1, f_k(x_k^*, \underline{u}_k^*)\right)$ as in Yeung and Petrosyan (2016).

Finally, under cooperation, all players would use the cooperative strategies and the payoff that player i will directly receive at stage k along the cooperative trajectory $\{x_k^*\}_{k=1}^{T+1}$ with previously executed durable strategies $\underline{u}_{k-}^*$ becomes $g_k^i(x_k^*, \underline{u}_k^*, \underline{\overline{u}}_k^*; \underline{\overline{u}}_{k-}^*)$. However, according to the agreed-upon imputation, player i will receive $\beta_k^i(x_k^*; \underline{\overline{u}}_{k-}^*)$ at stage k. Therefore, a side-payment

$$\pi_k^i(x_k^*; \underline{\overline{u}}_{k-}^*) = \beta_k^i(x_k^*; \underline{\overline{u}}_{k-}^*) - g_k^i(x_k^*, \underline{u}_k^*, \underline{\overline{u}}_k^*; \underline{\overline{u}}_{k-}^*), \quad (2.8)$$

for $k \in \{1, 2, \cdots, T\}$,
has to be given to player $i \in N$ to yield the cooperative imputation $\xi^i(k, x_k^*; \underline{\overline{u}}_{k-}^*)$.

For illustration sake, we consider some examples of gain sharing optimality principles.

Case I: Consider the case of an optimality principle which divides the excess of the total cooperative payoff over the sum of individual non-cooperative payoffs equally. According to this optimality principle, the imputation to player i is

$$\xi^i(k, x_k^*; \underline{\overline{u}}_{k-}^*) = V^i(k, x_k^*; \underline{\overline{u}}_{k-}^*) + \frac{1}{n}\left(W(k, x_k^*; \underline{\overline{u}}_{k-}^*) - \sum_{j=1}^{n} V^j(k, x_k^*; \underline{\overline{u}}_{k-}^*)\right), \quad (2.9)$$

for $i \in N$ at stage $k \in \{1, 2, \cdots, T\}$.

Applying Theorem 2.1 , a payment

$$\begin{aligned}\beta_k^i(x_k^*; \underline{\overline{u}}_{k-}^*) = (\delta_1^k)^{-1}\big\{&V^i(k, x_k^*; \underline{\overline{u}}_{k-}^*)\\ &+ \frac{1}{n}\left(W(k, x_k^*; \underline{\overline{u}}_{k-}^*) - \sum_{j=1}^{n} V^j(k, x_k^*; \underline{\overline{u}}_{k-}^*)\right)\\ &- \big[V^i(k+1, f_k(x_k^*, \underline{u}_k^*, \underline{\overline{u}}_k^*; \underline{\overline{u}}_{k-}^*); \underline{\overline{u}}_{(k+1)-}^*)\end{aligned}$$

$$+\frac{1}{n}\big(W(k+1,\, f_k(x_k^*, \underline{u}_k^*, \overline{u}_k^*; \overline{u}_{k-}^*); \overline{u}_{(k+1)-}^*)$$

$$-\sum_{j=1}^{n} V^j(k+1,\, f_k(x_k^*, \underline{u}_k^*, \overline{u}_k^*; \overline{u}_{k-}^*); \overline{u}_{(k+1)-}^*);\ \overline{u}_{(k+1)-}^*)\big)\big]\big\} \quad (2.10)$$

will be given to player $i \in N$ at stage $k \in \{1, 2, \cdots, T\}$.

Case II: Consider the case of an optimality principle that shares the cooperative payoff proportional to players' non-cooperative payoffs. According to this optimality principle, the imputation to player i is

$$\xi^i(k, x_k^*; \overline{u}_{k-}^*) = \frac{V^i(k, x_k^*; \overline{u}_{k-}^*)}{\sum_{j=1}^{n} V^j(k, x_k^*; \overline{u}_{k-}^*)} W(k, x_k^*; \overline{u}_{k-}^*), \quad (2.11)$$

for $i \in N$ at stage $k \in \{1, 2, \cdots, T\}$.

Applying Theorem 2.1, a payment

$$\beta_k^i(x_k^*; \overline{u}_{k-}^*) = (\delta_1^k)^{-1}\Bigg[\frac{V^i(k, x_k^*; \overline{u}_{k-}^*)}{\sum_{j=1}^{n} V^j(k, x_k^*; \overline{u}_{k-}^*)} W(k, x_k^*; \overline{u}_{k-}^*)$$

$$-\frac{V^i(k+1,\, f_k(x_k^*, \underline{u}_k^*, \overline{u}_k^*; \overline{u}_{k-}^*); \overline{u}_{(k+1)-}^*); \overline{u}_{(k+1)-}^*)}{\sum_{j=1}^{n} V^j(k+1,\, f_k(x_k^*, \underline{u}_k^*, \overline{u}_k^*; \overline{u}_{k-}^*); \overline{u}_{(k+1)-}^*); \overline{u}_{(k+1)-}^*)}$$

$$\times\, W(k+1,\, f_k(x_k^*, \underline{u}_k^*, \overline{u}_k^*; \overline{u}_{k-}^*); \overline{u}_{(k+1)-}^*)\Bigg] \quad (2.12)$$

will be given to player $i \in N$ at stage $k \in \{1, 2, \cdots, T\}$.

3 An Illustration in Cooperative Public Capital Provision

Consider the oligopoly with n firms in Sect. 4 of Chap. 2. The planning horizon of these firms involves T stages. The firms use knowledge-based public capital $x_k \in X \subset R^+$ to produce its output. Capital investment takes ω stages to complete its built up into the public capital stock. The accumulation process of the public capital stock is governed by the dynamics

$$x_{k+1} = x_k + \sum_{j=1}^{n} a_k^{|k|j} \overline{u}_k^{(\omega)j} + \sum_{j=1}^{n} \sum_{\tau=k-\omega+1}^{k-1} a_k^{|\tau|j} \overline{u}_\tau^{(\omega)j} - \lambda_k x_k, \quad x_1 = x_1^0, \quad (3.1)$$

where $\bar{u}_k^{(\omega)j}$ is the investment in the knowledge-based public capital by firm j in stage k, $a_k^{|k|j} \overline{u}_k^{(\omega)j}$ is the addition to the public capital in stage k by firm j's capital building investment $\overline{u}_k^{(\omega)j}$ in stage k, $a_k^{|\tau|j} \overline{u}_\tau^{(\omega)j}$ is the addition to the public capital in stage k by capital building investment $\overline{u}_{\tau}^{(\omega)j}$ in stage $\tau \in \{k-1, k-2, \cdots, k-\omega+1\}$ and λ_k is the rate of obsolescence of the public capital.

Firm $i \in N$ uses the knowledge-based public capital together with non-durable productive inputs u_k^i to produce outputs. The revenue from the output produced is $R_k^i (u_k^i)^{\frac{1}{2}} (x_k)^{\frac{1}{2}}$. The non-durable input cost is $c_k^i u_k^i$ and the cost of capital investment is $\varphi_k^i (\overline{u}_k^{(\omega)i})^2$.

The payoff of the firm i is

$$\sum_{k=1}^{T} \left([R_k^i (u_k^i)^{1/2} (x_k)^{1/2} - c_k^i u_k^i - \varphi_k^i (\overline{u}_k^{(\omega)i})^2] \right) \delta^{k-1}$$
$$+ \left(Q_{T+1}^i x_{T+1} + \sum_{\tau=T+1-\omega+1}^{T} v_{T+1}^{|\tau|i} \overline{u}_{\tau}^{(\omega)i} + \varpi_{T+1}^i \right) \delta^T, \tag{3.2}$$

where $\left(Q_{T+1}^i x_{T+1} + \sum_{\tau=T+1-\omega+1}^{T} v_{T+1}^{|\tau|i} \overline{u}_{\tau}^{(\omega)i} + \varpi_{T+1}^i \right)$ is the salvage value of firm i in stage $T+1$ and $\delta = (1+r)^{-1}$ is the discount factor.

In Sect. 4 of Chap. 2, the non-cooperative game equilibrium payoff of firm i is

$$V^i(k, x; \underline{u}_{k-}^{(\omega)**}) = (A_k^i x + C_k^i)\delta^{k-1}, \tag{3.3}$$

where $A_{T+1}^i = Q_{T+1}^i$ and $C_{T+1}^i = \sum_{\tau=T+1-\omega+1}^{T} v_{T+1}^{|\tau|i} \overline{u}_{\tau}^{(\omega)i} + \varpi_{T+1}^i$, for $i \in N$;

$$A_k^i = \frac{(P_k^i)^2}{4c_k^i} + \delta A_{k+1}^i (1 - \lambda_k), \quad \text{for } k \in \{1, 2, \cdots, T\} \text{ and } i \in N, \tag{3.4}$$

and C_k^i is an expression that contains the previously executed controls $\underline{\overline{u}}_{k-}^{(\omega)**}$. The game equilibrium strategies in stage k are

$$u_k^{i**} = \frac{(R_k^i)^2 x}{4(c_k^i)^2}, k \in \{1, 2, \cdots, T\} \text{ and}$$

$$\overline{u}_k^{(\omega)i**} = \frac{\sum_{\tau=k}^{k+\omega-1} \delta^{\tau-k+1} A_{\tau+1}^i a_{\tau}^{|k|i} + \delta^{(T+1-k)} v_{T+1}^{|k|i}}{2\phi_k^i},$$

for $k \in \{T-\omega+2, T-\omega+3, \cdots, T\}$ and $A_\tau = 0$ for $\tau > T+1$,

$$\overline{u}_k^{(\omega)i**} = \frac{\sum_{\tau=k}^{k+\omega-1} \delta^{\tau-k+1} A_{\tau+1}^i a_{\tau}^{|ki|}}{2\phi_k^i}, \quad k \in \{1, 2, \cdots, T-\omega+1\}. \tag{3.5}$$

and the game equilibrium public capital accumulation path can be obtained as:

$$x_{k+1} = x_k + \sum_{j=1}^{n} a_k^{|k|j} \overline{u}_k^{(\omega)j^{**}} + \sum_{j=1}^{n} \sum_{t=k-\omega+1}^{k-1} a_k^{|t|j} \overline{u}_t^{(\omega)j^{**}} - \lambda_k x_k, \quad x_1 = x_1^0, \tag{3.6}$$

where $\overline{u}_k^{(\omega)j^{**}}$ and $\overline{u}_t^{(\omega)j^{**}}$ are obtained from (3.5).

We use $\{x_k^{**}\}_{k=1}^{T+1}$ to denote the solution to (3.6).

Given the positive externality in public capital, the firms agree to enhance their payoffs through cooperation. They also agree to share the total cooperative payoff proportional to their non-cooperative payoffs. To secure group optimality the participating firms seek to maximize their joint payoff by solving the following dynamic optimization problem:

$$\max_{\substack{u_k^i, \overline{u}_k^{(\omega)i}, i \in N \\ k \in \{1,2,\cdots,T\}}} \Bigg\{ \sum_{i=1}^{n} \Bigg[\sum_{k=1}^{T} \Big([R_k^i (u_k^i)^{1/2} (x_k)^{1/2} - c_k^i u_k^i - \phi_k^i (\overline{u}_k^{(\omega)i})^2] \Big) \delta^{k-1}$$

$$+ \left(Q_{T+1}^i x_{T+1} + \sum_{\tau=T+1-\omega+1}^{T} v_{T+1}^{|\tau|i} \overline{u}_\tau^{(\omega)i} + \varpi_{T+1}^i \right) \delta^T \Bigg] \Bigg\}, \tag{3.7}$$

subject to the public capital accumulation dynamics (3.1).

Invoking Theorem 1.1, a set of optimal cooperative control strategies for dynamic optimization problem (3.7) can be obtained by solving the following system of recursive equations:

$$W(T+1, x; \underline{\overline{u}}_{(T+1)-}^{(\omega)}) = \sum_{i=1}^{n} \left(Q_{T+1}^i x + \sum_{\tau=T+1-\omega+1}^{T} v_{T+1}^{|\tau|i} \overline{u}_\tau^{(\omega)i} + \varpi_{T+1}^i \right) \delta^T, \tag{3.8}$$

$$W(k, x; \underline{\overline{u}}_{k-}^{(\omega)}) = \max_{u_k^i, \overline{u}_k^{(\omega)i}, i \in N} \Bigg\{ \sum_{i=1}^{n} \Big([R_k^i (u_k^i)^{1/2} (x)^{1/2} - c_k^i u_k^i - \phi_k^i (\overline{u}_k^{(\omega)i})^2] \Big) \delta^{k-1}$$

$$+ W\left(k+1, x + \sum_{j=1}^{n} a_k^{|k|j} \overline{u}_k^{(\omega)j} + \sum_{j=1}^{n} \sum_{\tau=k-\omega+1}^{k-1} a_k^{|\tau|j} \overline{u}_\tau^{(\omega)j} - \lambda_k x; \underline{\overline{u}}_{(k+1)-}^{(\omega)} \right),$$

for $k \in \{1, 2, \cdots, T\}$. (3.9)

Performing the indicated maximization in (3.9) and solving the system (3.8–3.9) one can obtain the maximized joint payoff under cooperation $W(k, x; \underline{\overline{u}}_{k-}^{(\omega)})$, which reflects the maximized joint payoffs of the firm covering stage $k \in \{1, 2, \cdots, T\}$ to stage $T+1$, as follows.

Proposition 3.1 *System* (3.8–3.9) *admits a solution with the optimal cooperative joint payoff of the firms being.*

$$W(k, x; \underline{\overline{u}}_{k-}^{(\omega)}) = (A_k x + C_k) \delta^{k-1}, \tag{3.10}$$

where $A_{T+1} = \sum_{i=1}^{n} Q^i_{T+1}$ *and* $C_{T+1} = \sum_{i=1}^{n} \left(\sum_{\tau=T+1-\omega+1}^{T} v^{|\tau|i}_{T+1} \overline{u}^{(\omega)i}_{\tau} + \varpi^i_{T+1} \right)$,

$$A_k = \sum_{i=1}^{n} \frac{(R^i_k)^2}{4c^i_k} + \delta A_{k+1}(1-\lambda_k), \quad \text{for } k \in \{1, 2, \cdots, T\}, \tag{3.11}$$

and C_k *is an expression that contains the previously executed controls* $\underline{\overline{u}}^{(\omega)}_{k-}$. *The terms involving* $\underline{\overline{u}}^{(\omega)}_{k-}$ *in the* C_k *is*

$$\delta A_{k+1} \sum_{j=1}^{n} \sum_{\tau=k-\omega+1}^{k-1} a^{|\tau|j}_{k} u^{(\omega)j}_{\tau} + \delta^2 A_{k+2} \sum_{j=1}^{n} \sum_{\tau=k-\omega+2}^{k-1} a^{|\tau|j}_{k+1} \overline{u}^{(\omega)j}_{\tau}$$

$$+ \delta^3 A_{k+3} \sum_{j=1}^{n} \sum_{\tau=k-\omega+3}^{k-1} a^{|\tau|j}_{k+2} u^{(\omega)j}_{\tau} + \cdots$$

$$\cdots + \delta^{\omega-1} A_{k+\omega-1} \sum_{j=1}^{n} \sum_{\tau=k-1}^{k-1} a^{|\tau|j}_{k+\omega-2} \overline{u}^{(\omega)j}_{\tau}$$

$$+ \delta^{(T+1-k)} \sum_{j=1}^{n} \sum_{\tau=T-\omega+2}^{k-1} v^{|\tau|j}_{T+1} \overline{u}^{(\omega)j}_{\tau},$$

for $k \in \{T-\omega+2, T-\omega+3, \cdots, T\}$ *and* $A_k = 0$ *for* $k > T+1$;

$$\delta A_{k+1} \sum_{j=1}^{n} \sum_{\tau=k-\omega+1}^{k-1} a^{|\tau|j}_{k} u^{(\omega)j}_{\tau} + \delta^2 A_{k+2} \sum_{j=1}^{n} \sum_{\tau=k-\omega+2}^{k-1} a^{|\tau|j}_{k+1} \overline{u}^{(\omega)j}_{\tau}$$

$$+ \delta^3 A_{k+3} \sum_{j=1}^{n} \sum_{\tau=k-\omega+3}^{k-1} a^{|\tau|j}_{k+2} u^{(\omega)j}_{\tau} + \cdots + \delta^4 A_{k+4} \sum_{j=1}^{n} \sum_{\tau=k-\omega+4}^{k-1} a^{|\tau|j}_{k+3} \overline{u}^{(\omega)j}_{\tau}$$

$$+ \cdots + \delta^{\omega-1} A_{k+\omega-1} \sum_{j=1}^{n} \sum_{\tau=k-1}^{k-1} a^{|\tau|j}_{k+\omega-2} \overline{u}^{(\omega)j}_{\tau}, \tag{3.12}$$

for $k \in \{1, 2, \cdots, T-\omega+1\}$.

Proof See Appendix. ■

Using Proposition 3.1, one can obtain the optimal cooperative in stage k as (see derivation details in Appendix):

$$u^{i*}_k = \frac{(R^i_k)^2 x}{4(c^i_k)^2}, \text{ for } k \in \{1, 2, \cdots, T\}; \text{ and}$$

$$\overline{u}^{(\omega)i*}_k = \frac{\sum_{\tau=k}^{k+\omega-1} \delta^{\tau-k+1} A_{\tau+1} a^{|k|i}_{\tau} + \delta^{(T+1-k)} v^{|k|i}_{T+1}}{2\phi^i_k},$$

for $k \in \{T-\omega+2, T-\omega+3, \cdots, T\}$ and $A_\tau = 0$ for $\tau > T+1$,

$$\overline{u}_k^{(\omega)i^*} = \frac{\sum_{\tau=k}^{k+\omega-1} \delta^{\tau-k+1} A_{\tau+1} a_\tau^{|k|i}}{2\phi_k^i}, \quad \text{for } k \in \{1, 2, \cdots, T-\omega+1\}. \tag{3.13}$$

Substituting the optimal cooperative level of public capital investment $\underline{u}_{k-}^{(\omega)}$ into (3.1) yields the optimal cooperative public capital accumulation path as:

$$x_{k+1} = x_k + \sum_{j=1}^{n} a_k^{|k|j} \overline{u}_k^{(\omega)j^*} + \sum_{j=1}^{n} \sum_{t=k-\omega+1}^{k-1} a_k^{|t|j} \overline{u}_t^{(\omega)j^*} - \lambda_k x_k, \quad x_1 = x_1^0, \tag{3.14}$$

where $\overline{u}_k^{(\omega)j^*}$ and $\overline{u}_t^{(\omega)j^*}$ are given in (3.13).

Equation (3.14) is a first order linear difference equation that can be solved by standard techniques. We use $\{x_k^*\}_{k=1}^{T+1}$ to denote the solution of the optimal public capital accumulation path in (3.14).

Given the agreed-upon sharing of the total cooperative payoff proportional to the firms' non-cooperative payoffs, the cooperative payoff of firm i along the cooperative trajectory $\{x_k^*\}_{k=1}^{T+1}$ has to fulfil:

$$\xi^i(k, x_k^*; \underline{u}_{k-}^{(\omega)^*}) = \frac{V^i(k, x_k^*; \underline{u}_{k-}^{(\omega)^*})}{\sum_{j=1}^{n} V^j(k, x_k^*; \underline{u}_{k-}^{(\omega)^*})} W(k, x_k^*; \underline{u}_{k-}^{(\omega)^*}),$$
$$\text{for } i \in N\ k \in \{1, 2, \cdots, T\}. \tag{3.15}$$

Invoking Theorem 2.1, to maintain the condition in (3.15), a payment equalling

$$\begin{aligned}
\beta_k^i(x_k^*; \underline{u}_{k-}^{(\omega)^*}) = \delta^{1-k} \Bigg[& \frac{V^i(k, x_k^*; \underline{u}_{k-}^{(\omega)^*})}{\sum_{j=1}^{n} V^j(k, x_k^*; \underline{u}_{k-}^{(\omega)^*})} W(k, x_k^*; \underline{u}_{k-}^{(\omega)^*}) \\
& - \frac{V^i(k+1, x_{k+1}^*; \underline{u}_{(k+1)-}^{(\omega)^*})}{\sum_{j=1}^{n} V^j(k+1, x_{k+1}^*; \underline{u}_{(k+1)-}^{(\omega)^*})} W(k+1, x_{k+1}^*; \underline{u}_{k-}^{(\omega)^*}); \underline{u}_{(k+1)-}^{(\omega)^*}) \Bigg] \\
= & \frac{(A_k^i x_k^* + C_k^i)}{\sum_{j=1}^{n} (A_k^j x_k^* + C_k^j)} (A_k x_k^* + C_k^i) \\
& - \frac{(A_{k+1}^i x_{k+1}^* + C_{k+1}^i)}{\sum_{j=1}^{n} (A_{k+1}^j x_{k+1}^* + C_{k+1}^j)} (A_{k+1} x_{k+1}^* + C_{k+1}) \delta
\end{aligned} \tag{3.16}$$

will be given to firm $i \in N$ at stage $k \in \{1, 2, \cdots, T\}$.

Under cooperation, all firms would use the cooperative strategies and the payoff that firm i will directly receive at stage k along the cooperative trajectory $\{x_k^*\}_{k=1}^{T+1}$ becomes

$$[R_k^i(u_k^{i^*})^{1/2}(x_k^*)^{1/2} - c_k^i u_k^{i^*} - \phi_k^i(\overline{u}_k^{(\omega)i^*})^2].$$

However, according to the agreed-upon imputation, firm i will receive $\beta_k^i(x_k^*; \underline{\overline{u}}_{k-}^{(\omega)^*})$ at stage k. Therefore, a side-payment

$$\pi_k^i(x_k^*; \underline{\overline{u}}_{k-}^{(\omega)^*}) = \beta_k^i(x_k^*; \underline{\overline{u}}_{k-}^{(\omega)^*}) - [R_k^i(u_k^{i^*})^{1/2}(x_k^*)^{1/2} - c_k^i u_k^{i^*} - \phi_k^i(\overline{u}_k^{(\omega)i^*})^2], \quad (3.17)$$

for $k \in \{1, 2, \cdots, T\}$,
has to be given to firm $i \in N$ to yield the cooperative imputation

$$\xi^i(k, x_k^*; \underline{\overline{u}}_{k-}^{(\omega)^*}) = \frac{V^i(k, x_k^*; \underline{\overline{u}}_{k-}^{(\omega)^*})}{\sum_{j=1}^n V^j(k, x_k^*; \underline{\overline{u}}_{k-}^{(\omega)^*})} W(k, x_k^*; \underline{\overline{u}}_{k-}^{(\omega)^*}).$$

Finally, comparing A_k^i in (3.4) with A_k in (3.11) shows that $A_k > A_k^i$, for $i \in N$ and $k \in \{1, 2, \cdots, T\}$. Hence, the optimal cooperative investment in public capital by firm i is larger than the non-cooperative investment in public capital by firm i, that is

$$\overline{u}_k^{(\omega)i^*} = \frac{\sum_{\tau=k}^{k+\omega-1} \delta^{\tau-k+1} A_{\tau+1} a_\tau^{|k|i} + \delta^{(T+1-k)} v_{T+1}^{|k|i}}{2\phi_k^i} > \overline{u}_k^{(\omega)i^{**}}$$
$$= \frac{\sum_{\tau=k}^{k+\omega-1} \delta^{\tau-k+1} A_{\tau+1}^i a_\tau^{|k|i} + \delta^{(T+1-k)} v_{T+1}^{|k|i}}{2\phi_k^i},$$

for $k \in \{T - \omega + 2, T - \omega + 3, \cdots, T\}$ and

$$\overline{u}_k^{(\omega)i^*} = \frac{\sum_{\tau=k}^{k+\omega-1} \delta^{\tau-k+1} A_{\tau+1} a_\tau^{|k|i}}{2\phi_k^i} > \overline{u}_k^{(\omega)i^{**}}$$
$$= \frac{\sum_{\tau=k}^{k+\omega-1} \delta^{\tau-k+1} A_{\tau+1}^i a_\tau^{|k|i}}{2\phi_k^i},$$

for $k \in \{1, 2, \cdots, T - \omega + 1\}$.

With investments in public capital by firms under cooperation being larger than those under non-cooperation, the public capital stock under cooperation is larger than that under non-cooperation, that is $x_k^* > x_k^{**}$, for $k \in \{2, 3, \cdots, T\}$. With the internalization of the positive externalities of public capital and the individually rational payoff sharing scheme $\xi^i(k, x_k^*; \underline{\overline{u}}_{k-}^{(\omega)^*})$ in (3.15), the firm's cooperative payoff is larger than its non-cooperative counterpart.

Worth noting is that in the case of diffusion lags in the capital accumulation process, the potential positive externalities of public capital investment in future stages can be substantial. Long term cooperation is preferable for the full realization of the transformation of investment into public capital.

4 Efficiency Loss and Self-seeking Optimization

Decision problems involving strategic interactions are usually handled by self-seeking optimization in the absence of an agreement. Cooperative optimization represents the best way of obtaining Pareto optimal solutions to these problems with maximum efficiency. Recently, a game theory measure that evaluates how the efficiency of a system degrades due to self-seeking optimization of players had been proposed by Koutsoupias and Papadimitriou (2009), namely the Price of Anarchy (PoA). It is a general notion originated in transportation studies that can be extended to diverse systems and notions of efficiency. The measurement of inefficiency of non-cooperative equilibrium had also been studied in Dubey (1986). The essential idea of PoA is to see how poor an outcome one can get in a competitive situation (congestion game, routing internet packets, machine scheduling, etc.) when the players act in their own interests (fulfilling individuality) versus the Pareto optimum that can be obtained for the same situation assuming that the players can cooperate (fulfilling optimality collectively).

The formal definition of the Price of Anarchy (PoA) is the ratio of the most efficient (Pareto optimal) sum of players' payoffs to the sum of the players' payoffs in a non-cooperative equilibrium. In this chapter, the most efficient sum is the group optimal joint payoff $W(k, x_k^*; \underline{\overline{u}}_{k-}^{(\omega)^*})$. We choose the sum of the players' payoffs under a feedback Nash equilibrium, that is $\sum_{j=1}^{n} V^j(k, x_k^*; \underline{\overline{u}}_{k-}^{(\omega)^*})$, as the non-cooperative equilibrium. A PoA measure can then be obtained as

$$PoA(k, x_k^*; \underline{\overline{u}}_{k-}^{(\omega)^*}) = \frac{W(k, x_k^*; \underline{\overline{u}}_{k-}^{(\omega)^*})}{\sum_{j=1}^{n} V^j(k, x_k^*; \underline{\overline{u}}_{k-}^{(\omega)^*})}. \tag{4.1}$$

Three related measures can be derived.

(i) *Percentage of Efficiency Loss*

$$PEL(k, x_k^*; \underline{\overline{u}}_{k-}^{(\omega)^*}) = \frac{W(k, x_k^*; \underline{\overline{u}}_{k-}^{(\omega)^*}) - \sum_{j=1}^{n} V^j(k, x_k^*; \underline{\overline{u}}_{k-}^{(\omega)^*})}{W(k, x_k^*; \underline{\overline{u}}_{k-}^{(\omega)^*})}$$
$$= 1 - (PoA)^{-1}.$$

The Percentage of Efficiency Loss indicates the potential percentage loss of efficiency brought about by self-seeking optimization along the cooperative trajectory x_k^*. It measures the loss of efficiency if the players switch from cooperation to non-cooperative behaviours in stage $k \in \{1, 2, \cdots, T\}$. ■

(ii) *Percentage of Efficiency Gain*

$$PEG(k, x_k^*; \overline{\underline{u}}_{k-}^{(\omega)^*}) = \frac{W(k, x_k^*; \overline{\underline{u}}_{k-}^{(\omega)^*}) - \sum_{j=1}^n V^j(k, x_k^*; \overline{\underline{u}}_{k-}^{(\omega)^*})}{\sum_{j=1}^n V^j(k, x_k^*; \overline{\underline{u}}_{k-}^{(\omega)^*})}$$
$$= PoA - 1. \qquad (4.2)$$

The Percentage of Efficiency Gain indicates the potential percentage gain brought about by Pareto optimization if the players maintain cooperation at any stage $k \in \{1, 2, \cdots, T\}$ along the cooperative trajectory x_k^*. ■

(iii) *The Cost of Delay*

Consider the case when the players defer the starting of their cooperative scheme from stage 1 to stage $\tau + 1$. The loss in payoffs due to the delay in cooperation is measured by the Cost of Delay:

$$COD(1, \tau + 1) = W(1, x_1^*; \overline{\underline{u}}_{1-}^{(\omega)^*}) - W(\tau, x_\tau^*; \overline{\underline{u}}_{\tau-}^{(\omega)^*})$$
$$- [\sum_{j=1}^n V^j(1, x_1^{**}; \overline{\underline{u}}_{1-}^{(\omega)^{**}}) - \sum_{j=1}^n V^j(\tau, x_\tau^{**}; \overline{\underline{u}}_{\tau-}^{(\omega)^{**}})]$$
$$+ W(\tau + 1, x_{\tau+1}^*; \overline{\underline{u}}_{(\tau+1)-}^{(\omega)^*}) - W(\tau + 1, x_{\tau+1}^{**}; \overline{\underline{u}}_{(\tau+1)-}^{(\omega)^{**}}), \qquad (4.3)$$

where $W(1, x_1^*; \overline{\underline{u}}_{1-}^{(\omega)^*}) - W(\tau, x_\tau^*; \overline{\underline{u}}_{\tau-}^{(\omega)^*})$ is the optimal joint payoff from stage 1 to stage τ, $[\sum_{j=1}^n V^j(1, x_1^{**}; \overline{\underline{u}}_{1-}^{(\omega)^{**}}) - \sum_{j=1}^n V^j(\tau, x_\tau^{**}; \overline{\underline{u}}_{\tau-}^{(\omega)^{**}})]$ is the sum of the players' payoff from stage 1 to stage τ under a feedback Nash equilibrium,

$W(\tau + 1, x_{\tau+1}^*; \underline{u}_{(\tau+1)-}^{(\omega)^*})$ is the optimal cooperative joint payoff in stage $\tau + 1$ given that the vector of states is $x_{\tau+1}^*$ and the previously executed (but still effective) strategies are $\underline{u}_{(\tau+1)-}^{(\omega)^*}$ and $W(\tau + 1, x_{\tau+1}^{**}; \overline{\underline{u}}_{(\tau+1)-}^{(\omega)^{**}})$ is the optimal cooperative joint payoff in stage $\tau+1$ given that the vector of states is $x_{\tau+1}^{**}$ and the previously executed (but still effective) strategies are $\overline{\underline{u}}_{(\tau+1)-}^{(\omega)^{**}}$.

In particular, the term

$$W(1, x_1^*; \overline{\underline{u}}_{1-}^{(\omega)^*}) - W(\tau, x_\tau^*; \overline{\underline{u}}_{\tau-}^{(\omega)^*})$$
$$- \left[\sum_{j=1}^n V^j(1, x_1^{**}; \overline{\underline{u}}_{1-}^{(\omega)^{**}}) - \sum_{j=1}^n V^j(\tau, x_\tau^{**}; \overline{\underline{u}}_{\tau-}^{(\omega)^{**}})\right]$$

yields the loss in payoff from stage 1 to stage τ for choosing a feedback equilibrium outcome rather than a Pareto optimal solution.

The term $W\left(\tau + 1, x_{\tau+1}^*; \overline{\underline{u}}_{(\tau+1)-}^{(\omega)^*}\right) - W\left(\tau + 1, x_{\tau+1}^{**}; \overline{\underline{u}}_{(\tau+1)-}^{(\omega)^{**}}\right)$ reflects the loss in the cooperative joint payoff from stage $\tau + 1$ to the end of the game if cooperation starts in stage $\tau + 1$ rather than stage 1.

Hence, $COD(1, \tau + 1)$ gives the loss in payoffs due to the delay in cooperation. Schematically, $COD(1, \tau + 1)$ can be expressed as.

$$Cost\,of\,Delay = Optimal\,Cooperative\,Payoff\,in\,stage\,1\,to\,stage\,\tau$$
$$- Sum\,of\,non-cooperative$$
$$Payoffs\,in\,stage\,1\,to\,stage\,\tau stage\,\tau$$
$$+ Optimal\,Cooperative\,Payoff\,in\,stage\,\tau+1\,to\,stage$$
$$T+1\,with\,x^*_{\tau+1}\,and\,\underline{\bar{u}}^*_{(\tau+1)}$$
$$- Optimal\,Cooperative\,Payoff\,in\,stage\,\tau+1\,to\,stage$$
$$T+1\,with\,x^{**}_{\tau+1}\,and\,\underline{\bar{u}}^{**}_{(\tau+1)-}.$$

■

In a dynamic durable strategies setting, worth-noting is that the delay in cooperation can cause very significant loss because of (i) the state $x^{**}_{\tau+1}$ along the game equilibrium path can reach a potentially devastating condition and (ii) the previously executed strategies $\underline{\bar{u}}^{(\omega)^{**}}_{\tau+1}$ under non-cooperation from stage 1 to stage τ can generate a heavy burden to the future payoffs and state dynamics.

Some alternative measures can also be derived.

An alternative measure of PoA measure can then be obtained by choosing the sum of non-cooperative payoffs along the game equilibrium path x^{**}_k, that is

$$PoA = \frac{W(k, x^*_k; \underline{\bar{u}}^{(\omega)^*}_{k-})}{\sum_{j=1}^n V^j(k, x^{**}_k; \underline{\bar{u}}^{(\omega)^{**}}_{k-})}, \tag{4.4}$$

if and the sum of non-cooperative payoffs along the game equilibrium path x^{**}_k is used instead of the sum of non-cooperative payoffs along the optimal cooperative path x^*_k.

A variant of the Percentage of Efficiency Loss—$PEL[k, (x^*_k, x^{**}_k), W]$—can be obtained as

$$PEL[k, (x^*_k, x^{**}_k), W] = PEL[k, (x^*_k, x^{**}_k), W]$$
$$= \frac{W(k, x^*_k; \underline{\bar{u}}^{(\omega)^*}_{k-}) - \sum_{j=1}^{n} V^j(k, x^{**}_k; \underline{\bar{u}}^{(\omega)^{**}}_{k-})}{W(k, x^*_k; \underline{\bar{u}}^{(\omega)^*}_{k-})}. \tag{4.5}$$

The $PEL[k, (x^*_k, x^{**}_k), W]$ measures the percentage loss in efficiency along the game equilibrium path x^{**}_k if the players continue to act non-cooperatively throughout the game relative to the optimal cooperative situation along the optimal trajectory x^*_k.

Similarly, a variant of the Percentage of Efficiency Gain-$PEL[k, (x^*_k, x^{**}_k), V]$—can be obtained as

$$PEL[k, (x^*_k, x^{**}_k), V] = \frac{W(k, x^*_k; \underline{\bar{u}}^{(\omega)*}_{k-}) - \sum_{j=1}^n V^j(k, x^{**}_k; \underline{\bar{u}}^{(\omega)^{**}}_{k-})}{\sum_{j=1}^n V^j(k, x^{**}_k; \underline{\bar{u}}^{(\omega)^{**}}_{k-})}. \tag{4.6}$$

The $PEL[k, (x_k^*, x_k^{**}), V] =$ measures the percentage gain in efficiency using the optimal joint payoff along the optimal trajectory x_k^* and the sum of non-cooperative payoffs along the game equilibrium path x_k^{**}.

Finally, if optimality principle which shares the excess of the total cooperative payoff over the sum of individual non-cooperative payoffs proportional to the players' non-cooperative payoffs is agreed upon, then we have the cooperative payoff to player i as

$$\begin{aligned}\xi^i(k, x_k^*; \underline{\overline{u}}_{k-}^{(\omega)^*}) &= \frac{V^i(k, x_k^*; \underline{\overline{u}}_{k-}^{(\omega)^*})}{\sum_{j=1}^n V^j(k, x_k^*; \underline{\overline{u}}_{k-}^{(\omega)^*})} W(k, x_k^*; \underline{\overline{u}}_{k-}^{(\omega)^*}) \\ &= PoA(k, x_k^*; \underline{\overline{u}}_{k-}^{(\omega)^*}) V^i(k, x_k^*; \underline{\overline{u}}_{k-}^{(\omega)^*}), \end{aligned} \tag{4.7}$$

for $i \in N$ at stage $k \in \{1, 2, \cdots, T\}$.

Under this gain sharing optimality principle, each player's cooperative payoff equals his non-cooperative payoff multiplied by the Price of Anarchy.

5 Appendices

5.1 Appendix I: Proof of Proposition 3.1

Using (3.8) and Proposition 3.1, we have

$$A_{T+1} = \sum_{i=1}^n Q_{T+1}^i \text{ and } C_{T+1} = \sum_{i=1}^n \left(\sum_{\tau=T+1-\omega+1}^T v_{T+1}^{|\tau|i} \overline{u}_\tau^{(\omega)i} + \varpi_{T+1}^i \right). \tag{5.1}$$

Using Proposition 3.1 and (3.9), we can express the optimal cooperative strategies in stage T as:

$$u_T^i = \frac{(R_T^i)^2 x}{4(c_T^i)^2} \text{ and } \overline{u}_T^{(\omega)i} = \frac{\delta(A_{T+1} a_T^{|T|i} + v_{T+1}^{|T|i})}{2\phi_T^i}, \text{ for } i \in N. \tag{5.2}$$

Substituting the optimal cooperative strategies in (5.2) into the stage T equations of (3.9) we can obtain

$$\begin{aligned}(A_T x + C_T)\delta^{k-1} &= \sum_{i=1}^n \left(\frac{(R_T^i)^2 x}{4c_T^i} - \frac{(\delta(A_{T+1} a_T^{|T|i} + v_{T+1}^{|T|i}))^2}{4\phi_T^i} \right) \delta^{T-1} \\ &+ \left[A_{T+1} \left(x + \sum_{j=1}^n a_T^{|T|j} \frac{\delta(A_{T+1} a_T^{|T|j} + v_{T+1}^{|T|j})}{2\phi_T^j} \right. \right.\end{aligned}$$

$$+\sum_{j=1}^{n}\sum_{\tau=T-\omega+1}^{T-1} a_T^{|\tau|j}\overline{u}_\tau^{(\omega)j} - \lambda_T x\Bigg)$$

$$+\sum_{i=1}^{n}\left(v_{T+1}^{|T|i}\frac{\delta(A_{T+1}a_T^{|T)|} + v_{T+1}^{|T|i})}{2\phi_T^i}\right.$$

$$+\sum_{\tau=T+1-\omega+1}^{T-1} v_{T+1}^{|\tau|i}\overline{u}_\tau^{(\omega)i} + \varpi_{T+1}^i)]\delta^T. \tag{5.3}$$

The right-hand-side and the left-hand-side of the equations in (5.3) are linear functions of x. For (5.3) to hold, it is required:

$$A_T = \sum_{i=1}^{n}\frac{(R_T^i)^2}{4c_T^i} + \delta A_{T+1}(1-\lambda_T). \tag{5.4}$$

In addition,

$$C_T = -\sum_{i=1}^{n}\left(-\frac{(\delta(A_{T+1}a_T^{|T|i} + v_{T+1}^{|T|i}))^2}{4\phi_T^i}\right)$$
$$+\left[A_{T+1}\left(\sum_{j=1}^{n} a_T^{|T|j}\frac{\delta(A_{T+1}a_T^{|T|j} + v_{T+1}^{|T|j})}{2\phi_T^j} + \sum_{j=1}^{n}\sum_{\tau=T-\omega+1}^{T-1} a_T^{|\tau|j}\overline{u}_\tau^{(\omega)j}\right)\right.$$
$$\left.+\sum_{i=1}^{n}\left(v_{T+1}^{|T|i}\frac{\delta(A_{T+1}a_T^{|T)|} + v_{T+1}^{|T|i})}{2\phi_T^i} + \sum_{\tau=T+1-\omega+1}^{T-1} v_{T+1}^{|\tau|i}\overline{u}_\tau^{(\omega)i} + \varpi_{T+1}^i\right)\right]\delta, \tag{5.5}$$

which is a function of previously executed controls $\underline{\overline{u}}_{T-}^{(\omega)}$.

Then we move to stage $T-1$.

Using $(A_T x + C_T)\delta^{T-1}$ in (5.3–5.5) and the stage $T-1$ equation in (3.9), the optimal cooperative strategies in stage $T-1$ can be obtained as:

$$u_{T-1}^i = \frac{(R_{T-1}^i)^2 x}{4(c_{T-1}^i)^2} \text{ and } \overline{u}_{T-1}^{(\omega)i} = \frac{\delta A_T a_{T-1}^{|T-1|i} + \delta^2(A_{T+1}a_T^{|T-1|i} + v_{T+1}^{|T-1|i})}{2\phi_{T-1}^i},$$
$$\text{for } i \in N. \tag{5.6}$$

Invoking Proposition 3.1 and substituting the optimal cooperative strategies in (5.6) into the stage $T-1$ equation of (3.9) we obtain

$$(A_{T-1}x + C_{T-1})\delta^{T-2} = \sum_{i=1}^{n}\left(\frac{(R_{T-1}^i)^2 x}{4c_{T-1}^i}\right.$$

$$- \frac{(\delta A_T a_{T-1}^{|T-1|i} + \delta^2 (A_{T+1} a_T^{|T-1|i} + v_{T+1}^{|T-1|i}))^2}{4\phi_{T-1}^i} \Bigg)$$

$$+ \left[\delta A_T \left(x + \sum_{j=1}^{n} a_{T-1}^{|T-1|j} \frac{\delta A_T a_{T-1}^{|T-1|j} + \delta^2 (A_{T+1} a_T^{|T-1|j} + v_{T+1}^{|T-1|j})}{2\phi_{T-1}^j} \right.\right.$$

$$\left. + \sum_{j=1}^{n} \sum_{\tau=T-\omega}^{T-2} a_{T-1}^{|\tau|j} \overline{u}_\tau^{(\omega)j} - \lambda_{T-1} x \right)$$

$$- \delta \sum_{i=1}^{n} \frac{(\delta (A_{T+1} a_T^{|T|i} + v_{T+1}^{|T|i}))^2}{4\phi_T^i}$$

$$+ \delta^2 A_{T+1} \left(\sum_{j=1}^{n} a_T^{|T|j} \frac{\delta (A_{T+1} a_T^{|T|j} + v_{T+1}^{|T|j})}{2\phi_T^j} \right.$$

$$+ \sum_{j=1}^{n} a_T^{|T-1|j} \frac{\delta A_T a_{T-1}^{|T-1|j} + \delta^2 (A_{T+1} a_T^{|T-1|j} + v_{T+1}^{|T-1|j})}{2\phi_{T-1}^j}$$

$$\left. + \sum_{j=1}^{n} \sum_{\tau=T-\omega+1}^{T-2} a_T^{|\tau|j} \overline{u}_\tau^{(\omega)j} \right)$$

$$+ \delta^2 \sum_{i=1}^{n} \left(v_{T+1}^{|T|i} \frac{\delta (A_{T+1} a_T^{|T|} + v_{T+1}^{|T|i})}{2\phi_T^i} \right.$$

$$+ v_{T+1}^{|T-1|i} \frac{\delta A_T a_{T-1}^{|T-1|i} + \delta^2 (A_{T+1} a_T^{|T-1|i} + v_{T+1}^{|T-1|i})}{2\phi_{T-1}^i}$$

$$\left.\left. + \sum_{\tau=T+1-\omega+1}^{T-2} v_{T+1}^{|\tau|i} \overline{u}_\tau^{(\omega)i} + \varpi_{T+1}^i \right) \right]. \tag{5.7}$$

The right-hand-side and the left-hand-side of the equations in (5.7) are linear functions of x. For (5.7) to hold, it is required:

$$A_{T-1} = \sum_{i=1}^{n} \frac{(R_{T-1}^i)^2}{4c_{T-1}^i} + \delta A_T (1 - \lambda_{T-1}). \tag{5.8}$$

In addition,

$$C_{T-1} = - \sum_{i=1}^{n} \frac{(\delta A_T a_{T-1}^{|T-1|i} + \delta^2 (A_{T+1} a_T^{|T-1|i} + v_{T+1}^{|T-1|i}))^2}{4\phi_{T-1}^i}$$

$$+\Bigg[\delta A_T\Bigg(\sum_{j=1}^{n} a_{T-1}^{|T-1|j}\,\frac{\delta A_T a_{T-1}^{|T-1|j}+\delta^2(A_{T+1}a_T^{|T-1|j}+v_{T+1}^{|T-1|j})}{2\phi_{T-1}^j}$$

$$+\sum_{j=1}^{n}\sum_{\tau=T-\omega}^{T-2} a_{T-1}^{|\tau|j}\overline{u}_\tau^{(\omega)j}\Bigg)-\delta\sum_{i=1}^{n}\frac{(\delta(A_{T+1}a_T^{|T|i}+v_{T+1}^{|T|i}))^2}{4\phi_T^i}$$

$$+\delta^2 A_{T+1}\Bigg(\sum_{j=1}^{n} a_T^{|T|j}\,\frac{\delta(A_{T+1}a_T^{|T|j}+v_{T+1}^{|T|j})}{2\phi_T^j}$$

$$+\sum_{j=1}^{n} a_T^{|T-1|j}\,\frac{\delta A_T a_{T-1}^{|T-1|j}+\delta^2(A_{T+1}a_T^{|T-1|j}+v_{T+1}^{|T-1|j})}{2\phi_{T-1}^j}$$

$$+\sum_{j=1}^{n}\sum_{\tau=T-\omega+1}^{T-2} a_T^{|\tau|j}\overline{u}_\tau^{(\omega)j}\Bigg)$$

$$+\delta^2\sum_{i=1}^{n}\Bigg(v_{T+1}^{|T|i}\,\frac{\delta(A_{T+1}a_T^{|T)|}+v_{T+1}^{|T|i})}{2\phi_T^i}$$

$$+v_{T+1}^{|T-1|i}\,\frac{\delta A_T a_{T-1}^{|T-1|i}+\delta^2(A_{T+1}a_T^{|T-1|i}+v_{T+1}^{|T-1|i})}{2\phi_{T-1}^i}$$

$$+\sum_{\tau=T+1-\omega+1}^{T-2} v_{T+1}^{|\tau|i}\overline{u}_\tau^{(\omega)i}+\overline{\varpi}_{T+1}^i\Bigg)\Bigg], \tag{5.9}$$

which is a function of previously executed controls $\underline{\overline{u}}_{(T-1)-}^{(\omega)}$.

Then we move to stage $T-2$.

Using $(A_{T-1}x+C_{T-1})\delta^{T-2}$ in (5.7–5.9) and the stage $T-2$ equation in (3.9), the optimal cooperative strategies in stage $T-2$ can be obtained as:

$$u_{T-2}^i=\frac{(R_{T-2}^i)^2 x}{4(c_{T-2}^i)^2}\ \text{and}\ \overline{u}_{T-2}^{(\omega)i}$$

$$=\frac{\delta A_{T-1}a_{T-2}^{|T-2|i}+\delta^2 A_T a_{T-1}^{|T-2|i}+\delta^3(A_{T+1}a_T^{|T-2|i}+v_{T+1}^{|T-2|i})}{2\phi_{T-2}^i},$$

$$\text{for } i\in N. \tag{5.10}$$

Substituting the optimal cooperative strategies in (5.10) into the stage $T-2$ equations of (4.4) we obtain a system of equations which right-hand-side and left-hand-side are linear functions of x. For the equation to hold, it is required:

$$A_{T-2} = \sum\nolimits_{i=1}^{n} \frac{(R_{T-2}^i)^2}{4c_{T-2}^i} + \delta A_{T-1}(1 - \lambda_{T-2}). \tag{5.11}$$

In addition, C_{T-2} is an expression that contains all the terms on the right-hand-side of the equations which do not involve x including the previously executed controls $\underline{\overline{u}}_{(T-2)-}^{(\omega)}$.

Following the above analysis for stage $k \in \{1, 2, \cdots, T\}$ the optimal cooperative strategies in stage k can be expressed as:

$$u_k^i = \frac{(R_k^i)^2 x}{4(c_k^i)^2}, \quad \text{for } k \in \{1, 2, \cdots, T\}; \text{ and}$$

$$\overline{u}_k^{(\omega)i} = \frac{\sum_{\tau=k}^{k+\omega-1} \delta^{\tau-k+1} A_{\tau+1} a_\tau^{|k|i} + \delta^{(T+1-k)} v_{T+1}^{|k|i}}{2\phi_k^i},$$

For $k \in \{T - \omega + 2, T - \omega + 3, \cdots, T\}$ and $A_\tau = 0$ for $\tau > T + 1$,

$$\overline{u}_k^{(\omega)i} = \frac{\sum_{\tau=k}^{k+\omega-1} \delta^{\tau-k+1} A_{\tau+1} a_\tau^{|k|i}}{2\phi_k^i}, \quad \text{for } k \in \{1, 2, \cdots, T - \omega + 1\}. \tag{5.12}$$

Substituting the optimal cooperative strategies in (5.12) into the stage k equation of (4.4) we can obtain the optimal cooperative joint payoff as

$$(A_k x + C_k)\delta^{k-1}, \quad \text{for } i \in N, \tag{5.13}$$

where

$$A_k = \sum_{i=1}^{n} \frac{(R_k^i)^2}{4c_k^i} + \delta A_{k+1}(1 - \lambda_k), \tag{5.14}$$

and C_k is an expression that contains the previously executed controls $\underline{\overline{u}}_{k-}^{(\omega)}$.

With $A_{T+1} = Q_{T+1}$, one can use (5.14) to obtain the values of A_k, for $k \in \{T, T-1, \cdots, 1\}$. Invoking (5.12), the optimal cooperative levels of capital investment $\underline{\overline{u}}_k^{(\omega)}$, for $k \in \{T, T-1, \cdots, 1\}$, can be obtained. Substituting the optimal cooperative level of capital investment $\underline{\overline{u}}_k^{(\omega)}$ into (3.1) yields the optimal cooperative capital accumulation path as:

$$x_{k+1} = x_k + \sum_{j=1}^{n} a_k^{|k|j} \overline{u}_k^{(\omega)j} + \sum_{j=1}^{n} \sum_{t=k-\omega+1}^{k-1} a_k^{|t|j} \overline{u}_t^{(\omega)j} - \lambda_k x_k, \quad x_1 = x_1^0, \tag{5.15}$$

where $\overline{u}_k^{(\omega)j}$ and $\overline{u}_t^{(\omega)j}$ are given in (5.12).

Moreover, the terms involving $\underline{\overline{u}}_{k-}^{(\omega)}$ in the C_k can be obtained as

$$\delta A_{k+1}\sum_{j=1}^{n}\sum_{\tau=k-\omega+1}^{k-1} a_k^{|\tau|j} u_\tau^{(\omega)j} + \delta^2 A_{k+2}\sum_{j=1}^{n}\sum_{\tau=k-\omega+2}^{k-1} a_{k+1}^{|\tau|j}\overline{u}_\tau^{(\omega)j}$$

$$+\,\delta^3 A_{k+3}\sum_{j=1}^{n}\sum_{\tau=k-\omega+3}^{k-1} a_{k+2}^{|\tau|j} u_\tau^{(\omega)j} + \cdots \cdots + \delta^{\omega-1} A_{k+\omega-1}\sum_{j=1}^{n}\sum_{\tau=k-1}^{k-1} a_{k+\omega-2}^{|\tau|j}\overline{u}_\tau^{(\omega)j}$$

$$+\,\delta^{(T+1-k)}\sum_{j=1}^{n}\sum_{\tau=T-\omega+2}^{k-1} v_{T+1}^{|\tau|j}\overline{u}_\tau^{(\omega)j},$$

for $k \in \{T-\omega+2, T-\omega+3, \cdots, T\}$ and $A_k = 0$ for $k > T+1$;

$$\delta A_{k+1}\sum_{j=1}^{n}\sum_{\tau=k-\omega+1}^{k-1} a_k^{|\tau|j} u_\tau^{(\omega)j} + \delta^2 A_{k+2}\sum_{j=1}^{n}\sum_{\tau=k-\omega+2}^{k-1} a_{k+1}^{|\tau|j}\overline{u}_\tau^{(\omega)j}$$

$$+\,\delta^3 A_{k+3}\sum_{j=1}^{n}\sum_{\tau=k-\omega+3}^{k-1} a_{k+2}^{|\tau|j} u_\tau^{(\omega)j} + \cdots + \delta^4 A_{k+4}\sum_{j=1}^{n}\sum_{\tau=k-\omega+4}^{k-1} a_{k+3}^{|\tau|j}\overline{u}_\tau^{(\omega)j}$$

$$+\cdots+\delta^{\omega-1} A_{k+\omega-1}\sum_{j=1}^{n}\sum_{\tau=k-1}^{k-1} a_{k+\omega-2}^{|\tau|j}\overline{u}_\tau^{(\omega)j}, \tag{5.16}$$

for $k \in \{1, 2, \cdots, T-\omega+1\}$. ■

6 Chapter Notes

This chapter presents a generic class of cooperative dynamic games (originated in Petrosyan and Yeung (2020a)) with durable controls in which the lagged effects are allowed to affect both the player's payoffs and the state dynamics. A subgame consistent solution for cooperative dynamic games with durable controls is derived. It expands the Yeung and Petrosyan (2019) game to a comprehensive theory of durable control games with the new features of (i) multiple durable controls with lags of different durations and (ii) durable controls affecting both the players' payoffs and state dynamics. The analysis widens the application of dynamic games to many problems where durable controls appear in the players' payoffs and the state dynamics.

Subgame/dynamic consistency in cooperative games was developed by Yeung and Petrosyan (2004 and 2010). It plays an important role in obtaining an agreeable cooperative scheme. The analysis on subgame consistent solution was extended to randomly furcating stochastic differential games in which both the state dynamics and future payoffs are stochastic in Petrosyan and Yeung (2007). Applications of subgame consistent solutions in randomly furcating stochastic differential games are found in Petrosyan and Yeung (2006), Yeung (2008) and Yeung and Petrosyan (2012b). Yeung (2011) analyzed subgame consistent solutions in differential games with asynchronous players' horizons. In developing the analysis for games with nontransferable payoffs Yeung (2004) derived the nontransferable individual payoff functions under cooperation in stochastic differential games with nontransferable payoffs. Yeung and Petrosyan (2005) and Yeung et al. (2007) analyzed subgame consistent solutions in cooperative stochastic differential games with nontransferable payoffs. For discrete-time analyses, Yeung and Petrosyan (2010) developed a generalized method for the derivation of analytically tractable subgame consistent solutions in stochastic dynamic games. This has made possible the rigorous study of subgame consistent solutions in discrete-time dynamic cooperation. Yeung (2014) analyzed subgame consistent solutions in a cooperative dynamic game of environmental management with the possibility of switching the choice of control. To accommodate the possibility of uncertain game duration Yeung and Petrosyan (2011) developed subgame consistent solution mechanisms for cooperative dynamic games with random horizons. The analysis was extended to the case where the state dynamics and the game horizon are stochastic in Yeung and Petrosyan (2012b). The notion of subgame consistency and solution mechanisms for randomly furcating cooperative stochastic dynamic games were developed by Yeung and Petrosyan (2013b). Applications of subgame consistent cooperative solution in randomly furcating stochastic dynamic games in collaborative provision of public goods was given in Yeung and Petrosyan (2014b). Yeung and Petrosyan (2014c) developed subgame consistent cooperative solution mechanisms for randomly furcating stochastic dynamic games with uncertain horizons. Subgame consistent studies involving the Shapley value can be found in Petrosyan and Yeung (2020b) and Yeung and Petrosyan (2018). The books by Yeung and Petrosyan (2012a and 2016a) provides detailed collections of subgame consistent cooperative differential games and dynamic games. Analyses on Price of Anarchy in dynamic games are provided by Parilina et al. (2017), Zhu and Başar (2010) and Başar and Zhu (2011).

7 Problems

1. (Warmup exercise) Consider a 2-firm oligopoly. The planning horizon of these firms involves five stages. The firms use knowledge-based public capital $x_k \in X \subset R^+$ to produce its output. Public capital investment takes two stages to be built up into the capital stock. The accumulation process of the public capital stock is governed by the dynamics

$$x_{k+1} = x_k + 2\overline{u}_k^{(2)1} + 2\overline{u}_k^{(2)2} + \sum_{j=1}^{2} \sum_{\tau=k-1}^{k-1} 0.5\overline{u}_\tau^{(2)j} - 0.1x_k, \quad x_1 = 80,$$

where $\overline{u}_k^{(2)j}$ is the investment in the knowledge-based capital by firm j in stage k.

Using the public capital, firm 1 yields a net revenue $4x_k - 2(\overline{u}_k^{(2)1})^2$ and firm 2 yields a net revenue $3x_k - 2(\overline{u}_k^{(2)2})^2$.

The payoff of firm 1 is $\sum_{k=1}^{5}\left([4x_k - 2(\overline{u}_k^{(2)1})^2]\right)(0.9)^{k-1} + (3x_6 + 95)(0.9)^5$.

The payoff of firm 2 is $\sum_{k=1}^{5}\left([3x_k - (\overline{u}_k^{(2)2})^2]\right)(0.9)^{k-1} + (2x_6 + 110)(0.9)^5$.

The firms agree to cooperate and maximize their joint payoff and share the total cooperative payoff proportional to their non-cooperative payoffs.

Derive (i) the optimal cooperative investment strategies and (ii) the optimal cooperative joint payoff.

2. Consider an oligopoly with n firms. The planning horizon of these firms involves T stages. The firms use knowledge-based public capital $x_k \in X \subset R^+$ to produce its output. Public capital investment takes time to be built up into the capital stock. The accumulation process of the capital stock is governed by the dynamics

$$x_{k+1} = x_k + \sum_{j=1}^{n} a_k^{|k|j}\overline{u}_k^{(\omega)j} + \sum_{j=1}^{n}\sum_{\tau=k-\omega+1}^{k-1} a_k^{|\tau|j}\overline{u}_\tau^{(\omega)j} - \lambda_k x_k, \quad x_1 = x_1^0,$$

where $\overline{u}_k^{(\omega)j}$ is the investment in the knowledge-based capital by firm j in stage k.

Firm i's revenue of the firm is related to the capital stock and yields a revenue $P_k^i x_k$ and the cost of capital investment is $\vartheta_k^i(\overline{u}_k^{(\omega)i})^2$. The payoff of the firm i is

$$\sum_{k=1}^{T}\left([P_k^i x_k - \vartheta_k^i(\overline{u}_k^{(\omega)i})^2]\right)\delta^{k-1} + \left(Q_{T+1}^i x_{T+1} + \varpi_{T+1}^i\right)\delta^T,$$

where $\left(Q_{T+1}^i x_{T+1} + \varpi_{T+1}^i\right)$ is the salvage value of firm i in stage $T+1$.

The firms agree to cooperate and maximize their joint payoff

$$\sum_{i=1}^{n}\sum_{k=1}^{T}\left([P_k^i x_k - \vartheta_k^i(\overline{u}_k^{(\omega)i})^2]\right)\delta^{k-1} + \sum_{i=1}^{n}\left(Q_{T+1}^i x_{T+1} + \varpi_{T+1}^i\right)\delta^T,$$

and share the total cooperative payoff proportional to their non-cooperative payoffs.

Derive (i) the optimal cooperative investment strategies, (ii) the optimal cooperative joint payoff and (iii) the cooperative capital accumulation trajectory.

3. Obtain the Price of Anarchy $PoA(k, x_k^*; \underline{\overline{u}}_{k-}^{(\omega)*}) = \frac{W(k, x_k^*; \underline{\overline{u}}_{k-}^{(\omega)*})}{\sum_{j=1}^{n} V^j(k, x_k^*; \underline{\overline{u}}_{k-}^{(\omega)*})}$.

References

Başar T, Zhu Q (2011) Prices of anarchy, information, and cooperation in differential games. Dyn Games Appl 1:50–73. https://doi.org/10.1007/s13235-010-0002-3

Dubey P (1986) Inefficiency of Nash equilibria. Math Oper Res 11(1):1–8

Haurie A (1976) A note on nonzero-sum differential games with bargaining solutions. J Optim Theory Appl 18:31–39

Koutsoupias E, Papadimitriou C (2009) Worst-case Equilibria. Comput Sci Rev 3(2):65–69. Archived from the original on 2016–03–13. Retrieved 2010–09–12

Parilina E, Sedakov A, Zaccour G (2017) Price of anarchy in a linear-state stochastic dynamic game. Europ J Operat Res Elsevier 258(2):790–800

Petrosyan LA (1977) Stability of solutions of nonzero sum differential games with many participants. Viestnik (Transactions) of Leningrad University, 19, pp 6–52

Petrosyan LA, Danilov NN (1979) Stability of the solutions in nonantagonistic differential games with transferable payoffs. (Russian) Vestnik Leningrad. Univ. Mat. Mekh. Astronom., vyp. 1, vol 134. pp 52–59

Petrosyan LA, Danilov NN (1985) Cooperative differential games and their applications. Tomsk Gos Univ, Tomsk. pp 276

Petrosyan LA, Yeung DWK (2006) Dynamically stable solutions in randomly-furcating differential games, transactions of the Steklov institute of mathematics, 253. Supplement 1:S208–S220

Petrosyan LA, Yeung DWK (2007) Subgame-consistent cooperative solutions in randomly-furcating stochastic differential games. Int J Mathem Comput Model (Special Issue on Lyapunov's Methods in Stability and Control). 45:1294–1307

Petrosyan LA, Yeung DWK (2020) Cooperative dynamic games with durable controls: theory and application. Dynamic Games Appl 10:872–896. https://doi.org/10.1007/s13235-019-00336-w

Petrosyan LA, Yeung DWK (2020b) Construction of dynamically stable solutions in differential network games. In: Tarasyev A, Maksimov V, Filippova T (eds) Stability, control and differential games (SCGD2019), Lecture Notes in Control and Information Sciences, Springer, pp 51–61

Yeung DWK (2004) Nontransferable individual payoff functions under stochastic dynamic cooperation. Int Game Theory Rev 6:281–289

Yeung DWK (2005) Subgame consistent dormant-firm cartel. In: Haurie A, Zaccour G (eds) Dynamic games and applications. Springer, Berlin, pp 255–271

Yeung DWK (2006) Solution mechanisms for cooperative stochastic differential games. Int Game Theory Rev 8(2):309–326

Yeung DWK (2008) Dynamically consistent solution for a pollution management game in collaborative abatement with uncertain future payoffs. Special Issue on Frontiers in Game Theory: In Honour of Nash JF (ed) International Game Theory Review 10(4):517–538

Yeung DWK (2010) Subgame consistent shapley value imputation for cost-saving joint ventures. Mathemat Game Theory Appl 2(3):137–149

Yeung DWK (2011) Dynamically consistent cooperative solutions in differential games with asynchronous players' horizons. Ann Internat Soc Dynam Games 11:375–395

Yeung DWK (2014) Dynamically consistent collaborative environmental management with production technique choices. Ann Oper Res 220(1):181–204. https://doi.org/10.1007/s10479-011-0844-0

Yeung DWK, Petrosyan LA (2004) Subgame consistent cooperative solution in stochastic differential games. J Optim Theory Appl 120(3):651–666

Yeung DWK, Petrosyan LA (2005) Subgame consistent solution of a cooperative stochastic differential game with nontransferable payoffs. J Optim Theory Appl 124(3):701–724

Yeung DWK, Petrosyan LA (2006) Dynamically stable corporate joint ventures. Automatica 42:365–370

Yeung DWK, Petrosyan LA (2007a) Managing catastrophe-bound industrial pollution with game-theoretic algorithm: the St Petersburg initiative, contributions to game theory and management, Petrosyan LA, NA Zenkevich (eds) St Petersburg State University, pp 524–538

Yeung DWK, Petrosyan LA (2007) The crux of dynamic economic cooperation: subgame consistency and equilibrating transitory compensation. Game Theory Appl 11:207–221

Yeung DWK, Petrosyan LA (2007) The tenet of transitory compensation in dynamically stable cooperation. Int J Tomograph Statist 7:60–65

Yeung DWK, Petrosyan LA (2008) A cooperative stochastic differential game of transboundary industrial pollution. Automatica 44(6):1532–1544

Yeung DWK, Petrosyan LA (2010) Subgame consistent solutions for cooperative stochastic dynamic games. J Optim Theory Appl 145:579–596

Yeung DWK, Petrosyan LA (2011) Subgame consistent cooperative solution of dynamic games with random horizon. J Optim Theory Appl 150:78–97

Yeung DWK, Petrosyan LA (2012a) Subgame consistent economic optimization: an advanced cooperative dynamic game analysis. Boston, Birkhäuser, pp 395. ISBN 978-0-8176-8261-3

Yeung DWK, Petrosyan LA (2012b) Subgame consistent solution for cooperative stochastic dynamic games with random horizon. Int Game Theory Rev 14(2). https://doi.org/10.1142/S0219198912500120

Yeung DWK, Petrosyan LA (2013) Subgame-consistent cooperative solutions in randomly furcating stochastic dynamic games. Math Comput Model 57(3–4):976–991

Yeung DWK, Petrosyan LA (2014) Subgame consistent cooperative solution of stochastic dynamic game of public goods provision. Contrib Game Theory Manage 7:404–414

Yeung DWK, Petrosyan LA (2014) Subgame consistent cooperative provision of public goods under accumulation and payoff uncertainties. In: Haunschmied J, Veliov VM, Wrzaczek S (eds) Dynamic games in economics. Springer, Berlin Heidelberg, pp 289–315

Yeung DWK, Petrosyan LA (2014c) Subgame consistent cooperative solutions for randomly furcating stochastic dynamic games with uncertain horizon. Int Game Theory Rev 16(2):1440012.01–1440012.29.

Yeung DWK, Petrosyan LA (2016a) A cooperative dynamic environmental game of subgame consistent clean technology development. Int Game Theory Rev 18(2):164008.01–164008.23

Yeung DWK, Petrosyan LA (2018) Nontransferable utility cooperative dynamic games. In: Basar T, Zaccour G (eds) Handbook of dynamic game theory, vol 1. Springer, pp 633–670

Yeung DWK, Petrosyan LA (2019) Cooperative dynamic games with control lags. Dyn Games Appl 9(2):550–567. https://doi.org/10.1007/s13235-018-0266-6

Yeung DWK, Petrosyan LA, Yeung PM (2007) Subgame consistent solutions for a class of cooperative stochastic differential games with nontransferable payoffs. Ann Internat Soc Dynam Games 9:153–170

Zhu Q, Başar T (2010) Price of anarchy and price of information in N-person linear-quadratic differential games. In: Proc American control conf (ACC), Baltimore, Maryland, June 2010

Chapter 6
Eco-Degradation Management Under Durable Strategies: Efficiency Maximization and Sustainable Imputation

Anthropogenic eco-degradation has become one of the world's most complex and urgent problems to be dealt with by scientists, policy-makers and economists alike. Self-seeking maximization by nations and regions portrays the situation of industrial civilization on the verge of suicide, destroying its environmental conditions of existence with people being held as prisoners on a runaway catastrophe-bound train. Unilateral response of a single nation or region deems to be ineffective; only through cooperation can a viable solution be made. This chapter presents a general class of cooperative dynamic environmental games involving anthropogenic eco-degradation and green technology development under durable strategies. Section 1 provides an overview of collaboration in ecosystem degradation management. Section 2 develops a generic class of durable-strategies cooperative environmental games of eco-degradation management. Section 3 derives a dynamically stable imputation distribution procedure leading to the realization of the agreed-upon payoff imputations so that no region will have inventive to deviate from the cooperation scheme within the cooperation duration. Section 4 presents a cooperative environmental game under durable strategies with explicit functional forms. Section 5 examines efficiency gain and sustainable imputation under the cooperation scheme. Section 6 contains an appendix and chapter notes are provided in Sect. 7.

1 Collaboration in Ecosystem Degradation Management

After several decades of rapid technological advancement and economic growth, alarming levels of pollutions and environmental degradation are emerging all over the world. Due to the geographical diffusion of pollutants and cross-national effects of ecosystem degradation, unilateral response of one nation or region is often ineffective. Self-seeking optimization by individual nations/regions (as shown in Chap. 4) can produce devastating outcomes. Reports are portraying the situation as an advanced

D. W. K. Yeung and L. A. Petrosyan, *Durable-Strategies Dynamic Games*,
Theory and Decision Library C 50, https://doi.org/10.1007/978-3-030-92742-4_6

civilization on the verge of suicide, destroying its environmental conditions of existence with people being held as prisoners on a runaway catastrophe-bound train. Although global cooperation in environmental control holds out the best promise of effective action, success has been limited. This is the result of many hurdles, ranging from agreeable individual commitments, cost-sharing and disparities in future developments under cooperation to expected effectiveness of the policy instruments. It is hard to be convinced that multinational joint initiatives like the Paris Agreement can offer a long-term solution because there is no guarantee that participants will always be better off within the entire cooperation duration. To ensure stability in dynamic environmental cooperation over time, a stringent condition is required: the specific agreed-upon optimality principle must be maintained at any instant of time throughout the game along the optimal state trajectory. It is precisely due to the lack of this kind of foresight that current cooperative schemes would fail to provide an effective mean to avert disaster. This is a 'classic' game-theoretic problem.

Moreover, existing international environmental cooperation arrangements focus mainly on emissions reduction, permit trading and regulations. The focus on pollution emissions only leaves out the needs of two important elements in alleviating anthropogenic eco-degradation. The first is the improvement (and the reduction in damages) of the ecosystems which impacts are much more than just pollution accumulation, global warming or climate change. Altering the ecosystems in the processes of satisfying the needs of society can cause detrimental effects to the survival of human beings. The second is the development of environmentally clean technology into efficient and affordable means of production plays a key role in effectively solving the continual worsening global industrial pollution problem and meeting the industrial growth needs.

Although cooperation in environmental control holds out the best promise of effective actions, a number of crucial issues remain unresolved in the cooperation scheme. These crucial issues include:

(i) The inadequacy of the policy instruments of pollutant emissions in resolving the current environmental problem.
(ii) The lack of guarantee that the participating nations would not be worse off (compared to their non-cooperative well-being) throughout the cooperation duration.
(iii) The economic viability of the policy for all participants after joining the cooperation scheme is not ensured.
(iv) The original positions those nations are entitled to hold in the initial stage may not be maintained in all the subsequent stages of the cooperation scheme.
(v) The cooperation scheme does not yield Pareto optimal efficiency.
(vi) Cooperative development of clean technology as a long-term effective remedy has not been seriously deliberated.

An additional problem is that game strategies like pollution generating outputs, ecosystems degrading outputs and green technologies R&D often are durable strategies.

In the sequel, we present a generic class of cooperative dynamic environmental games with durable strategies which include:

(i) pollution accumulation processes which are affected by polluting outputs with durable effects and abatement efforts,
(ii) ecosystem degradation dynamics which are affected by ecosystems damaging activities with durable effects and remedying efforts,
(iii) public green technology development dynamics which are affected by technology investments which take time to diffuse and realized,
(iv) outputs which emit pollution,
(v) ecosystems degrading activities which damage ecosystems,
(vi) clean outputs produced with green technology,
(vii) investments in pollution abatements, in ecosystem regeneration efforts and in green technology,
(viii) a cooperative scheme with group optimality and individual rationality,
(ix) a subgame consistent cooperative solution and
(x) a sustainable imputation distribution procedure that leads to the realization of the subgame consistent solution.

2 Durable-Strategies Cooperative Environmental Games

In this section, we first develop an analytical framework of a general class of cooperative dynamic environmental games involving ecosystem degradation under durable strategies. The game strategies include pollution generating outputs, ecosystems degrading outputs, clean outputs, pollution abatement efforts, ecosystems repairing efforts and green technologies R&D. Then, we study the group optimal solution, the gain sharing optimality principle and the requirement of individual rationality.

2.1 Game Formulation

Consider a cooperative counterpart of the game-theoretic framework in Chap. 4. Again, the basic framework involves a $n-$ region $T-$ stage dynamic environmental game in which region $i \in N$ produces three types of outputs. The first type of outputs $\overline{u}_k^{(\omega_1)i} \in \overline{U}^{(\omega_1)i} \subset R^{m^{(\omega_1)i}}$ emits pollutants over ω_1 stages. The second type of outputs $\overline{u}_k^{(\omega_2)i} \in \overline{U}^{(\omega_2)i} \subset R^{m^{(\omega_2)i}}$ causes damages to the ecosystems over ω_2 stages. The third type of outputs are clean outputs which use green technologies.

The vector of pollution stocks is $x_k \in X \subset R^{n_1}$ and the dynamics of pollution accumulation is governed by the vector dynamical equation:

$$x_{k+1} = x_k + \sum_{j=1}^{n} m_k^{|k|j}(\overline{u}_k^{(\omega_1)j}) + \sum_{j=1}^{n} \sum_{\tau=k-\omega_1+1}^{k-1} m_k^{|\tau|j}(\overline{u}_\tau^{(\omega_1)j})$$

$$-\sum_{j=1}^{n} b_k^j(u_k^{Ij}, x_k) - \lambda_k(x_k), \quad x_1 = x_1^0, \tag{2.1}$$

where $m_k^{|k|j}(\overline{u}_k^{(\omega_1)j})$ is the amount of pollutants created in stage k by region j's stage k polluting output $\overline{u}_k^{(\omega_1)j}$, $m_k^{|\tau|j}(\overline{u}_\tau^{(\omega_1)j})$ is the amount of pollutants created in stage k by region j's stage $\tau \in \{k-1, k-2, \cdots, k-\omega_1+1\}$ output, $b_k^j(u_k^{Ij}, x_k)$ is the amount of pollutants removed by u_k^{Ij} amounts of abatement effort from region j; $\lambda_k^x(x_k)$ is the natural of decay of the pollutants; and $\overline{u}_\tau^{(\omega_1)j}$, for $\tau \in \{0, -1, -2, \cdots, -\omega_1+1\}$ and $j \in N$, is known with some or all of them being zero.

A vector of ecosystems $y_k \in Y \subset R^{n_2}$ which conditions generate benefits to the regions. The dynamics governing the evolution of the conditions of the ecosystems is

$$\begin{aligned} y_{k+1} = & y_k + g_k(y_k) - \sum_{j=1}^{n} \sigma_k^{|k|j}(\overline{u}_k^{(\omega_2)j}) - \sum_{j=1}^{n} \sum_{\tau=k-\omega_2+1}^{k-1} \sigma_k^{|\tau|j}(\overline{u}_\tau^{(\omega_2)j}) \\ & + \sum_{j=1}^{n} \varepsilon_k^j(u_k^{IIj}) - \vartheta_k(x_k) - \lambda_k^y(y_k), \quad y_1 = y_1^0 \end{aligned} \tag{2.2}$$

where $\sigma_k^{|\tau|j}(\overline{u}_\tau^{(\omega_2)j})$ are the damages of the ecosystems by output $\overline{u}_\tau^{(\omega_2)j}$ in stage $\tau \in \{k, k-1, \cdots, k-\omega_2+1\}$, u_k^{IIj} are the inputs in stage k for repairing the ecosystems, $\varepsilon_k^j(u_k^{IIj})$ is the improvements by u_k^{IIj} to the ecosystems, $g_k(y_k)$ are the natural replenishments of the ecosystems and $\lambda_k^y(y_k)$ are the natural decay of ecosystems and $\vartheta_k(x_k)$ are the damages to ecosystems by pollution stocks x_k.

In the case of no human intervention in the ecosystems a natural equilibrium of the ecosystems is

$$g_k(y) = \vartheta_k(x_k) + \lambda_k^y(y_k).$$

There is a vector of public green technologies $z_k \in Z \subset R^{n_3}$ which are used to produce clean outputs. The dynamics governing the accumulation of green technologies is

$$z_{k+1} = z_k + \sum_{j=1}^{n} a_k^{|k|j}(\overline{u}_k^{(\omega_3)j}) + \sum_{j=1}^{n} \sum_{\tau=k-\omega_3+1}^{k-1} a_k^{|\tau|j}(\overline{u}_\tau^{(\omega_3)j}) - \lambda_k^z(z_k), \quad z_1 = z_1^0, \tag{2.3}$$

where $\overline{u}_k^{(\omega_3)j} \in \overline{U}^{(\omega_3)j} \subset R^{m^{(\omega_3)i}}$ are the investments in the public green technologies by region j in stage k, it would take ω_3 stages to fully transmit the investments into the public technologies, $a_k^{|k|j}(\overline{u}_k^{(\omega_3)j})$ are the additions to the public green technologies in stage k by region j's capital building investment in stage k, the term $a_k^{|\tau|j}(\overline{u}_\tau^{(\omega_3)j})$

the public green technologies in stage k by region j's capital building investment in stage $\tau \in \{k-1, k-2, \cdots, k-\omega_3+1\}$ and $\lambda_k^z(z_k)$ are the obsolescence of the green technologies.

The realized green technologies from the public technologies for region i are $\kappa_k^i z_k$. The region uses non-durable inputs $u_k^{IIIj} \subset R^{m_3^i}$ together with its green technologies to produce clean outputs.

The sum of net gains from the three types of outputs produced in region i is

$$R_k^{(x)i}(\overline{u}_k^{(\omega_1)i}) - \varphi_k^{(x)i}(\overline{u}_k^{(\omega_1)i}) + R_k^{(y)i}(\overline{u}_k^{(\omega_2)i}) - \varphi_k^{(y)i}(\overline{u}_k^{(\omega_2)i})$$
$$+ R_k^{(z)i}(q_k^{(z)i}(u_k^{IIIi}, \kappa_k^i z_k)) - c_k^{IIIi}(u_k^{IIIi}) - \varphi_k^{(z)i}(\overline{u}_k^{(\omega_3)i}).$$

In particular, $R_k^{(x)i}(\overline{u}_k^{(\omega_1)i})$ are the gross revenues from outputs creating pollution and $\varphi_k^{(x)i}(\overline{u}_k^{(\omega_1)i})$ are the costs of producing the output. $R_k^{(y)i}(\overline{u}_k^{(\omega_2)i})$ are the gross revenues from outputs leading to damages in the ecosystems and $\varphi_k^{(y)i}(\overline{u}_k^{(\omega_2)i})$ are the costs of producing the output. The production functions of the clean outputs which use the region's realized green capitals $\kappa_k^i z_k$ are $(q_k^{(z)i}(u_k^{IIIi}, \kappa_k^i z_k)$ with non-durable inputs u_k^{IIIi}. $R_k^{(z)i}(q_k^{(z)i}(u_k^{IIIi}, \kappa_k^i z_k))$ are the gross revenues from clean outputs using green technologies. The cost of green technology investments are $\varphi_k^{(z)i}(\overline{u}_k^{(\omega_3)i})$.

The costs of pollution abatements are $c_k^{Ii}(u_k^{Ii})$, the costs of improving the ecosystems are $c_k^{IIi}(u_k^{IIi})$ and the costs of non-durable inputs u_k^{IIIi} are $c_k^{IIIi}(u_k^{IIIi})$. The damages from pollutions are $h_k^i(x_k)$ and the environmental amenities from the ecosystems are $\chi_k^i(y_k)$.

The payoff of region $i \in N$ is

$$\sum_{k=1}^{T}\Big(R_k^{(x)i}(\overline{u}_k^{(\omega_1)i}) - \varphi_k^{(x)i}(\overline{u}_k^{(\omega_1)i}) - c_k^{Ii}(u_k^{Ii}) + R_k^{(y)i}(\overline{u}_k^{(\omega_2)i})$$
$$- \varphi_k^{(y)i}(\overline{u}_k^{(\omega_2)i}) - c_k^{IIi}(u_k^{IIi}) + R_k^{(z)i}(q_k^{(z)i}(u_k^{IIIi}, \kappa_k^i z_k))$$
$$-\varphi_k^{(z)i}(\overline{u}_k^{(\omega_3)i}) - c_k^{IIIi}(u_k^{IIIi}) - h_k^i(x_k) + \chi_k^i(y_k)\Big)\delta_1^k$$
$$+ \Big(Q_{T+1}^{x(i)}(x_{T+1}) + Q_{T+1}^{y(i)}(y_{T+1}) + Q_{T+1}^{z(i)}(\kappa_{T+1}^i z_{T+1}) + \varpi_{T+1}^i\Big)\delta_1^{T+1} \tag{2.4}$$

where $\Big(Q_{T+1}^{x(i)}(x_{T+1}) + Q_{T+1}^{y(i)}(y_{T+1}) + Q_{T+1}^{z(i)}(\kappa_{T+1}^i z_{T+1}) + Q_{T+1}^i\Big)$ is the terminal payoff of region i in stage $T+1$.

To exploit the potential gains from cooperation, the regions agree to act cooperatively and distribute the payoffs among themselves according to an agreed-upon gain sharing optimality principle. To ensure group optimality, the regions cooperate to maximize their joint payoff

$$\sum_{i=1}^{n}\Bigg[\sum_{k=1}^{T}\Big(R_k^{(x)i}(\bar{u}_k^{(\omega_1)i})-\varphi_k^{(x)i}(\bar{u}_k^{(\omega_1)i})-c_k^{Ii}(u_k^{Ii})$$
$$+R_k^{(y)i}(\bar{u}_k^{(\omega_2)i})-\varphi_k^{(y)i}(\bar{u}_k^{(\omega_2)i})-c_k^{IIi}(u_k^{IIi})+R_k^{(z)i}\Big(q_k^{(z)i}(u_k^{IIIi},k_k^i z_k)\Big) \tag{2.5}$$
$$-\varphi_k^{(z)i}(\bar{u}_k^{(\omega_3)i})-c_k^{IIIi}(u_k^{IIIi})-h_k^i(x_k)+\chi_k^i(y_k)\,\delta_1^k\Big)$$
$$+\Big(Q_{T+1}^{x(i)}(x_{T+1})+Q_{T+1}^{y(i)}(y_{T+1})+Q_{T+1}^{z(i)}(\kappa_{T+1}^i z_{T+1})+\varpi_{T+1}^i\Big)\delta_1^{T+1}\Bigg],$$

subject to dynamics (2.1)–(2.3).

2.2 *Group Optimal Solution*

Invoking Theorem 1.1 in Chap. 5, we characterize the group optimal joint payoff of the cooperative game (2.1)–(2.4) as follows.

Corollary 2.1. *Let $W(k,x,y,z;\underline{\overline{u}}_{k-}^{(\omega_1)},\underline{\overline{u}}_{k-}^{(\omega_2)},\underline{\overline{u}}_{k-}^{(\omega_3)})$ be the maximal value of the joint payoff.*

$$\sum_{i=1}^{n}\Bigg[\sum_{t=k}^{T}\Big(R_k^{(x)i}(\overline{u}_t^{(\omega_1)i})-\varphi_t^{(x)i}(\overline{u}_t^{(\omega_1)i})$$
$$-c_t^{Ii}(u_t^{Ii})+R_t^{(y)i}(\overline{u}_t^{(\omega_2)i})-\varphi_t^{(y)i}(\overline{u}_t^{(\omega_2)i})-c_t^{IIi}(u_t^{IIi})$$
$$+R_t^{(z)i}(q_t^{(z)i}(u_t^{IIIi},\kappa_t^i z_t))-\varphi_t^{(z)i}(\overline{u}_t^{(\omega_3)i})-c_t^{IIIi}(u_t^{IIIi})$$
$$-h_t^i(x_t)+\chi_t^i(y_t)\Big)\delta_1^t$$
$$+\Big(Q_{T+1}^{x(i)}(x_{T+1})+Q_{T+1}^{y(i)}(y_{T+1})+Q_{T+1}^{z(i)}(\kappa_{T+1}^i z_{T+1})+\varpi_{T+1}^i\Big)\delta_1^{T+1}\Bigg].$$

for the joint payoff maximization problem (2.1), (2.2), (2.3) and (2.5) starting at stage k with state $(x_k,y_k,z_k)=(x,y,z)$ and previously executed controls $(\underline{\overline{u}}_{k-}^{(\omega_1)},\underline{\overline{u}}_{k-}^{(\omega_2)},\underline{\overline{u}}_{k-}^{(\omega_3)})$, then the function $W(k,x,y,z;\underline{\overline{u}}_{k-}^{(\omega_1)},\underline{\overline{u}}_{k-}^{(\omega_2)},\underline{\overline{u}}_{k-}^{(\omega_3)})$ satisfies the following system of recursive equations:

$$W(T+1,x,y,z;\underline{\overline{u}}_{(T+1)-}^{(\omega_1)},\underline{\overline{u}}_{(T+1)-}^{(\omega_2)},\underline{\overline{u}}_{(T+1)-}^{(\omega_3)})$$
$$=\sum_{i=1}^{n}\Big(Q_{T+1}^{x(i)}(x_{T+1})+Q_{T+1}^{y(i)}(y_{T+1})+Q_{T+1}^{z(i)}(\kappa_{T+1}^i z_{T+1})+Q_{T+1}^i\Big)\delta_1^{T+1}, \tag{2.6}$$

$$W(k,x,y,z;\underline{\overline{u}}_{k-}^{(\omega_1)},\underline{\overline{u}}_{k-}^{(\omega_2)},\underline{\overline{u}}_{k-}^{(\omega_3)})$$

$$= \max_{\substack{u_k^{Ii}, u_k^{IIi}, u_k^{IIIi}, \overline{u}_k^{(\omega_1)i}, \\ \overline{u}_k^{(\omega_2)i}, \overline{u}_k^{(\omega_3)i}, i \in N}} \Bigg\{ \sum_{k=1}^{T} \Big(R_k^{(x)i}(\overline{u}_k^{(\omega_1)i}) - \varphi_k^{(x)i}(\overline{u}_k^{(\omega_1)i}) - c_k^{Ii}(u_k^{Ii})$$

$$+ R_k^{(y)i}(\overline{u}_k^{(\omega_2)i}) - \varphi_k^{(y)i}(\overline{u}_k^{(\omega_2)i}) - c_k^{IIi}(u_k^{IIi}) \Big)$$

$$+ R_k^{(z)i}(q_k^{(z)i}(u_k^{IIIi}, \kappa_k^i z))$$

$$- \varphi_k^{(z)i}(\overline{u}_k^{(\omega_3)i}) - c_k^{IIIi}(u_k^{IIIi}) - h_k^i(x) + \chi_k^i(y))\delta_1^k$$

$$+ W(k+1, x_{k+1}, y_{k+1}, z_{k+1}; \underline{\overline{u}}_{(k+1)-}^{(\omega_1)}, \underline{\overline{u}}_{(k+1)-}^{(\omega_2)}, \underline{\overline{u}}_{(k+1)-}^{(\omega_3)}) \Bigg\},$$

$$\text{for } k \in \{1, 2, \cdots, T\}, \tag{2.7}$$

where $x_{k+1} = x + \sum_{j=1}^{n} m_k^{|k|j}(\overline{u}_k^{(\omega_1)j}) + \sum_{j=1}^{n} \sum_{\tau=k-\omega_1+1}^{k-1} m_k^{|\tau|j}(\overline{u}_\tau^{(\omega_1)j}) - \sum_{j=1}^{n} b_k^j(u_k^{Ij}, x) - \lambda_k^x(x)$,

$$y_{k+1} = y + g_k(y) - \sum_{j=1}^{n} \sigma_k^{|k|j}(\overline{u}_k^{(\omega_2)j}) - \sum_{j=1}^{n} \sum_{\tau=k-\omega_2+1}^{k-1} \sigma_k^{|\tau|j}(\overline{u}_\tau^{(\omega_2)j})$$

$$+ \sum_{j=1}^{n} \varepsilon_k^j(u_k^{IIj}) - \vartheta_k(x) - \lambda_k^y(y),$$

$$z_{k+1} = z + \sum_{j=1}^{n} a_k^{|k|j}(\overline{u}_k^{(\omega_3)j}) + \sum_{j=1}^{n} \sum_{\tau=k-\omega_3+1}^{k-1} a_k^{|\tau|j}(\overline{u}_\tau^{(\omega_3)j}) - \lambda_k^z(z). \tag{2.8}$$

Performing the indicated maximization operator in Corollary 2.1, we obtain the following optimal cooperative conditions.

(i) Region i's pollution creating outputs will be produced up to the levels

$$\frac{\partial R_k^{(x)i}(\overline{u}_k^{(\omega_1)i})}{\partial \overline{u}_k^{(\omega_1)i}} \delta_1^k = \frac{\partial \varphi_k^{(x)i}(\overline{u}_k^{(\omega_1)i})}{\partial \overline{u}_k^{(\omega_1)i}} \delta_1^k + \frac{\partial W_{k+1}}{\partial x_{k+1}} \frac{\partial x_{k+1}}{\partial \overline{u}_k^{(\omega_1)i}}, + \frac{\partial W_{k+1}}{\partial \overline{u}_k^{(\omega_1)i}} \quad \text{for } i \in N \tag{2.9}$$

where W_{k+1} is the short form of $W(k+1, x_{k+1}, y_{k+1}, z_{k+1}; \underline{\overline{u}}_{(k+1)-}^{(\omega_1)}, \underline{\overline{u}}_{(k+1)-}^{(\omega_2)}, \underline{\overline{u}}_{(k+1)-}^{(\omega_3)})$.

In particular, $\frac{\partial R_k^{(x)i}(\overline{u}_k^{(\omega_1)i})}{\partial \overline{u}_k^{(\omega_1)i}} \delta_1^k$ are region i's marginal gross revenues from outputs creating pollution, $\frac{\partial \varphi_k^{(x)i}(\overline{u}_k^{(\omega_1)i})}{\partial \overline{u}_k^{(\omega_1)i}} \delta_1^k$ are the marginal costs of producing the polluting outputs, $+\frac{\partial W_{k+1}}{\partial x_{k+1}} \frac{\partial x_{k+1}}{\partial \overline{u}_k^{(\omega_1)i}}$ are the marginal effects of $\overline{u}_k^{(\omega_1)i}$ on the joint cooperative payoff through the increments in pollution in stage $k+1$ and $\frac{\partial W_{k+1}}{\partial \overline{u}_k^{(\omega_1)i}}$ are the marginal effects of $\overline{u}_k^{(\omega_1)i}$ on the joint cooperative payoff through the increments in pollution in stage $k+1$ to stage $k + \omega_1 - 1$. Thus, in a cooperative optimum, outputs creating pollution will be produced up to the levels where the marginal gross revenues equal the sum of the marginal production costs and the marginal environmental effects from pollution to all regions.

(ii) Region i's ecosystems damaging outputs will be produced up to the levels

$$\frac{\partial R_k^{(y)i}(\overline{u}_k^{(\omega_2)i})}{\partial \overline{u}_k^{(\omega_2)i}}\delta_1^k = \frac{\partial \varphi_k^{(y)i}(\overline{u}_k^{(\omega_2)i})}{\partial \overline{u}_k^{(\omega_2)i}}\delta_1^k + \frac{\partial W_{k+1}}{\partial y_{k+1}}\frac{\partial y_{k+1}}{\partial \overline{u}_k^{(\omega_2)i}} + \frac{\partial W_{k+1}}{\partial \overline{u}_k^{(\omega_2)i}}, \quad \text{for } i \in N \tag{2.10}$$

In particular, $\frac{\partial R_k^{(y)i}(\overline{u}_k^{(\omega_2)i})}{\partial \overline{u}_k^{(\omega_2)i}}\delta_1^k$ are the marginal gross revenues from outputs damaging the ecosystems, $\frac{\partial \varphi_k^{(y)i}(\overline{u}_k^{(\omega_2)i})}{\partial \overline{u}_k^{(\omega_2)i}}\delta_1^k$ are the marginal costs of producing the ecosystems damaging outputs, $+\frac{\partial W_{k+1}}{\partial y_{k+1}}\frac{\partial y_{k+1}}{\partial \overline{u}_k^{(\omega_2)i}}$ are the marginal effects of $\overline{u}_k^{(\omega_2)i}$ on the optimal cooperative payoff through the damage of the ecosystems in stage $k+1$ and $\frac{\partial W_{k+1}}{\partial \overline{u}_k^{(\omega_2)i}}$ are the marginal effects of $\overline{u}_k^{(\omega_2)i}$ on the optimal cooperative payoff through the increments in the damage of the ecosystems in stage $k+1$ to stage $k+\omega_1-1$. Thus, in a cooperative optimum, outputs damaging the ecosystem's pollution will be produced up to the levels where the marginal gross revenues equal the sum of the marginal production costs and the marginal environmental effects to all regions from damages to the ecosystems.

(iii) Region i's investments for building up green technologies will be employed up to the levels

$$\frac{\partial \varphi_k^{(z)i}(\overline{u}_k^{(\omega_3)i})}{\partial \overline{u}_k^{(\omega_3)i}}\delta_1^k = \frac{\partial W_{k+1}}{\partial z_{k+1}}\frac{\partial z_{k+1}}{\partial \overline{u}_k^{(\omega_3)i}} + \frac{\partial W_{k+1}}{\partial \overline{u}_k^{(\omega_3)i}} \quad \text{for } i \in N \tag{2.11}$$

In particular, $\frac{\partial \varphi_k^{(z)i}(\overline{u}_k^{(\omega_3)i})}{\partial \overline{u}_k^{(\omega_3)i}}\delta_1^k$ are the marginal costs of investments for building up green technologies, $\frac{\partial W_{k+1}}{\partial z_{k+1}}\frac{\partial z_{k+1}}{\partial \overline{u}_k^{(\omega_3)i}}$ are the marginal contributions to the optimal cooperative payoff through the increments in green technologies in stage $k+1$ and $\frac{\partial W_{k+1}}{\partial \overline{u}_k^{(\omega_3)i}}$ are the marginal contributions to the optimal cooperative payoff through the increments in green technologies in stage $k+1$ to stage $k+\omega_1-1$.

(iv) Pollution abatement efforts of region i are employed up to the levels where the marginal costs equal the marginal benefits to all regions, that is

$$\frac{\partial c_k^{Ii}(u_k^{Ii})}{\partial u_k^{Ii}}\delta_1^k = \frac{\partial W_{k+1}}{\partial x_{k+1}}\frac{\partial x_{k+1}}{\partial u_k^{Ii}}, \quad \text{for } i \in N \tag{2.12}$$

(v) Ecosystems improvement efforts of region i are employed up to the levels where the marginal costs equal the marginal benefits to all regions, that is

$$\frac{\partial c_k^{IIi}(u_k^{IIi})}{\partial u_k^{IIi}}\delta_1^k = \frac{\partial W_{k+1}}{\partial y_{k+1}}\frac{\partial y_{k+1}}{\partial u_k^{IIi}}, \quad \text{for } i \in N \tag{2.13}$$

(vi) Non-durable inputs of region i used to produce the clean outputs are employed up to the levels where the marginal revenues of the clean outputs equal the marginal costs, that is

$$\frac{\partial R_k^{(z)i}(q_k^{(z)i}(u_k^{IIIi},\kappa_k^i z))}{\partial q_k^{(z)i}(u_k^{IIIi},\kappa_k^i z)}\frac{\partial q_k^{(z)i}(u_k^{IIIi},\kappa_k^i z)}{\partial u_k^{IIIi}} = \frac{\partial c_k^{IIIi}(u_k^{IIIi})}{\partial u_k^{IIIi}} \quad \text{for } i \in N \tag{2.14}$$

If the optimal cooperative conditions in (2.9)-(2.14) satisfy the implicit function theorem, the optimal strategies $(u_k^I, u_k^{II}, u_k^{III}, \overline{u}_k^{(\omega_1)}, \overline{u}_k^{(\omega_2)}, \overline{u}_k^{(\omega_3)})$ can be obtained as explicit functions of $(x, y, z; \overline{u}_{k-}^{(\omega_1)}, \overline{u}_{k-}^{(\omega_2)}, \overline{u}_{k-}^{(\omega_3)})$, Substituting these functions into the stage k equations of (2.7), we can obtain $W(k, x, y, z; \overline{u}_{k-}^{(\omega_1)}, \overline{u}_{k-}^{(\omega_2)}, \overline{u}_{k-}^{(\omega_3)})$, for $k \in \{1, 2, \cdots, T\}$. In particular, $W(k, x, y, z; \overline{u}_{k-}^{(\omega_1)}, \overline{u}_{k-}^{(\omega_2)}, \overline{u}_{k-}^{(\omega_3)})$ reflects the optimal cooperative payoff which depends on the states (x, y, z) and the previously executed durable strategies $(\overline{u}_{k-}^{(\omega_1)}, \overline{u}_{k-}^{(\omega_2)}, \overline{u}_{k-}^{(\omega_3)})$.

We use $(u_k^{I*}, u_k^{II*}, u_k^{III*}, \overline{u}_k^{(\omega_1)*}, \overline{u}_k^{(\omega_2)*}, \overline{u}_k^{(\omega_3)*})$ for $k \in \{1, 2, \cdots, T\}$ to denote the optimal control strategies derived from Corollary 2.1. Substituting these optimal controls into the state dynamics (2.1)-(2.3), we can obtain the dynamics of the optimal cooperative trajectory as:

$$x_{k+1} = x_k + \sum_{j=1}^{n} m_k^{|k|j}(\overline{u}_k^{(\omega_1)j*}) + \sum_{j=1}^{n}\sum_{\tau=k-\omega_1+1}^{k-1} m_k^{|\tau|j}(\overline{u}_\tau^{(\omega_1)j*})$$
$$- \sum_{j=1}^{n} b_k^j(u_k^{Ij*}, x_k) - \lambda_k^x(x_k),$$

$$y_{k+1} = y_k + g_k(y_k) - \sum_{j=1}^{n} \sigma_k^{|k|j}(\overline{u}_k^{(\omega_2)j*}) - \sum_{j=1}^{n}\sum_{\tau=k-\omega_2+1}^{k-1} \sigma_k^{|\tau|j}(\overline{u}_\tau^{(\omega_2)j*})$$
$$+ \sum_{j=1}^{n} \varepsilon_k^j(u_k^{IIj*}) - \vartheta_k(x_k) - \lambda_k^y(y_k),$$

$$z_{k+1} = z_k + \sum_{j=1}^{n} a_k^{|k|j}(\overline{u}_k^{(\omega_3)j*}) + \sum_{j=1}^{n}\sum_{\tau=k-\omega_3+1}^{k-1} a_k^{|\tau|j}(\overline{u}_\tau^{(\omega_3)j*}) - \lambda_k^z(z_k), \tag{2.15}$$

$$k \in \{1, 2, \cdots, T\} \text{ and } (x_1, y_1, z_1) = (x_1^0, y_1^0, z_1^0)$$

We use $\{x_k^*, y_k^*, z_k^*\}_{k=1}^{T+1}$ to denote the solution to the equation system (2.15) which yields the optimal cooperative state trajectory.

2.3 Gain Sharing and Individual Rationality

Sharing the cooperative gain over the cooperation duration in a dynamic stability manner plays an important role in a sustainable cooperation scheme.

Let

$$\begin{aligned}
&\text{Let, } \xi(k, x^*, y^*, z^*; \overline{\underline{u}}_{k-}^{(\omega_1)*}, \overline{\underline{u}}_{k-}^{(\omega_2)*}, \overline{\underline{u}}_{k-}^{(\omega_3)*}) \\
&= [\xi^1(k, x^*, y^*, z^*; \overline{\underline{u}}_{k-}^{(\omega_1)*}, \overline{\underline{u}}_{k-}^{(\omega_2)*}, \overline{\underline{u}}_{k-}^{(\omega_3)*}), \\
&\quad \xi^2(k, x^*, y^*, z^*; \overline{\underline{u}}_{k-}^{(\omega_1)*}, \overline{\underline{u}}_{k-}^{(\omega_2)*}, \overline{\underline{u}}_{k-}^{(\omega_3)*}), \\
&\quad \cdots, \xi(k, x^*, y^*, z^*; \overline{\underline{u}}_{k-}^{(\omega_1)*}, \overline{\underline{u}}_{k-}^{(\omega_2)*}, \overline{\underline{u}}_{k-}^{(\omega_3)*})]
\end{aligned}$$

denote the agreed-upon sharing of cooperative payoffs among the regions in stage $k \in \{1, 2, \cdots, T\}$ along the cooperative path $\{x_k^*, y_k^*, z_k^*\}_{k=1}^{T+1}$ with previously executed controls being $(\overline{\underline{u}}_{k-}^{(\omega_1)*}, \overline{\underline{u}}_{k-}^{(\omega_2)*}, \overline{\underline{u}}_{k-}^{(\omega_3)*})$. To satisfy individual rationality throughout the cooperation duration, the cooperative payoff allotted to region i has to satisfy the following condition

$$\begin{aligned}
&\xi^i(k, x^*, y^*, z^*; \overline{\underline{u}}_{k-}^{(\omega_1)*}, \overline{\underline{u}}_{k-}^{(\omega_2)*}, \overline{\underline{u}}_{k-}^{(\omega_3)*}) \\
&\geq V^i(k, x^*, y^*, z^*; \overline{\underline{u}}_{k-}^{(\omega_1)*}, \overline{\underline{u}}_{k-}^{(\omega_2)*}, \overline{\underline{u}}_{k-}^{(\omega_3)*}) \\
&\text{for } k \in \{1, 2, \cdots, T\} \text{ and } i \in N,
\end{aligned} \tag{2.16}$$

along the cooperative trajectory $\{x_k^*, y_k^*, z_k^*\}_{k=1}^{T+1}$ given the previously executed controls $(\overline{\underline{u}}_{k-}^{(\omega_1)*}, \overline{\underline{u}}_{k-}^{(\omega_2)*}, \overline{\underline{u}}_{k-}^{(\omega_3)*})$.

Condition in (2.16) guarantees the fulfilment of individual rationality throughout the cooperation duration so that the payoff allocated to any region under cooperation will be no less than its non-cooperative payoff. Failure to guarantee individual rationality at any stage would lead to the condition where the concerned regions would deviate from the agreed-upon solution plan and play non-cooperatively.

To satisfy group optimality along the cooperative trajectory $\{x_k^*, y_k^*, z_k^*\}_{k=1}^{T+1}$, it is required that

$$W(k, x^*, y^*, z^*; \overline{\underline{u}}_{k-}^{(\omega_1)*}, \overline{\underline{u}}_{k-}^{(\omega_2)*}, \overline{\underline{u}}_{k-}^{(\omega_3)*})$$

$$= \sum_{j=1}^{n} \xi^i(k, x^*, y^*, z^*; \underline{\overline{u}}_{k-}^{(\omega_1)*}, \underline{\overline{u}}_{k-}^{(\omega_2)*}, \underline{\overline{u}}_{k-}^{(\omega_3)*}),$$

$$\text{for } k \in \{1, 2, \cdots, T\} \tag{2.17}$$

The condition in (2.17) guarantees the maximal joint payoff is distributed to the regions throughout the cooperation duration.

For dynamical consistency to be satisfied, it is required that. $\xi(k, x^*, y^*, z^*; \underline{\overline{u}}_{k-}^{(\omega_1)*}, \underline{\overline{u}}_{k-}^{(\omega_2)*}, \underline{\overline{u}}_{k-}^{(\omega_3)*})$ be maintained in all stages $k \in \{1, 2, \cdots, T\}$ in the cooperation scheme.

3 Imputation Distribution Procedure

Crucial to the analysis is the derivation of an Imputation Distribution Procedure (IDP) leading to the realization of the agreed imputations $\xi(k, x^*, y^*, z^*; \underline{\overline{u}}_{k-}^{(\omega_1)*}, \underline{\overline{u}}_{k-}^{(\omega_2)*}, \underline{\overline{u}}_{k-}^{(\omega_3)*})$ in all stages $k \in \{1, 2, \cdots, T\}$ throughout the cooperation duration. If this is satisfied, no region will have incentive to deviate from the cooperation scheme along its cooperative state trajectory.

To do this, we follow Yeung and Petrosyan (2010, 2016a, 2019) and use $\beta_k^i(x_k^*, y_k^*, z_k^*; \underline{\overline{u}}_{k-}^{(\omega_1)*}, \underline{\overline{u}}_{k-}^{(\omega_2)*}, \underline{\overline{u}}_{k-}^{(\omega_3)*})$ to denote the payment that region i receives in stage k under the cooperative agreement along the cooperative trajectory $\{x_k^*, y_k^*, z_k^*\}_{k=1}^{T+1}$, with durable strategies executed previously but still in effect being $(\underline{\overline{u}}_{k-}^{(\omega_1)*}, \underline{\overline{u}}_{k-}^{(\omega_2)*}, \underline{\overline{u}}_{k-}^{(\omega_3)*})$. The payment scheme involving $\beta_k^i(x_k^*, y_k^*, z_k^*; \underline{\overline{u}}_{k-}^{(\omega_1)*}, \underline{\overline{u}}_{k-}^{(\omega_2)*}, \underline{\overline{u}}_{k-}^{(\omega_3)*})$ constitutes an IDP in the sense that the payoff to region i over the stages from k to $T+1$ satisfies the condition:

$$\xi^i(k, x^*, y^*, z^*; \underline{\overline{u}}_{k-}^{(\omega_1)*}, \underline{\overline{u}}_{k-}^{(\omega_2)*}, \underline{\overline{u}}_{k-}^{(\omega_3)*})$$
$$= \beta_k^i(x_k^*, y_k^*, z_k^*; \underline{\overline{u}}_{k-}^{(\omega_1)*}, \underline{\overline{u}}_{k-}^{(\omega_2)*}, \underline{\overline{u}}_{k-}^{(\omega_3)*})\delta_1^k$$

$$+\left\{\sum_{\zeta=k+1}^{T} \beta_\zeta^i(x_\zeta^*, y_\zeta^*, z_\zeta^*; \underline{\overline{u}}_{\zeta-}^{(\omega_1)*}, \underline{\overline{u}}_{\zeta-}^{(\omega_2)*}, \underline{\overline{u}}_{\zeta-}^{(\omega_3)*})\,\delta_1^\zeta\right.$$
$$\left.+\left(Q_{T+1}^{x(i)}(x_{T+1}) + Q_{T+1}^{y(i)}(y_{T+1}) + Q_{T+1}^{z(i)}(\kappa_{T+1}^i z_{T+1}) + Q_{T+1}^i\right)\delta_1^{T+1}\right\}, \tag{3.1}$$

for $i \in N$ and $k \in \{1, 2, \cdots, T\}$.

Invoking Theorem 2.1 in Chap. 5, the Imputation Distribution Procedure $\beta_k^i(x_k^*, y_k^*, z_k^*; \underline{\overline{u}}_{k-}^{(\omega_1)*}, \underline{\overline{u}}_{k-}^{(\omega_2)*}, \underline{\overline{u}}_{k-}^{(\omega_3)*})$, for $k \in \{1, 2, \cdots, T\}$ and $i \in N$, can be obtained as

$$\beta_k^i(x_k^*, y_k^*, z_k^*; \underline{\overline{u}}_{k-}^{(\omega_1)*}, \underline{\overline{u}}_{k-}^{(\omega_2)*}, \underline{\overline{u}}_{k-}^{(\omega_3)*})$$

$$= (\delta_1^k)^{-1}\left[-\xi^i(k+1, x_{k+1}^*, y_{k+1}^*, z_{k+1}^*; \underline{\overline{u}}_{(k+1)-}^{(\omega_1)*}, \underline{\overline{u}}_{(k+1)-}^{(\omega_2)*}, \underline{\overline{u}}_{(k+1)-}^{(\omega_3)*})\right], \tag{3.2}$$

where $(x_{k+1}^*, y_{k+1}^*, z_{k+1}^*)$ are given in (2.8).

The payment scheme in (3.2) gives rise to the realization of the imputation guided by the agreed-upon optimality principle and constitutes a dynamically consistent payment scheme. More specifically, the payment of $\beta_k^i(x_k^*, y_k^*, z_k^*; \underline{\overline{u}}_{k-}^{(\omega_1)*}, \underline{\overline{u}}_{k-}^{(\omega_2)*}, \underline{\overline{u}}_{k-}^{(\omega_3)*})$ allotted to region $i \in N$ in stage $k \in \{1, 2, \cdots, T\}$ will establish a cooperative plan that matches with the agreed-upon imputation to every region along the cooperative path.

Given that the sizes of regions are asymmetric and their payoffs under non-cooperation can vary greatly, a practicable gain sharing optimality principle is to share the total cooperative payoff proportional to the regions' non-cooperative payoffs. According to this optimality principle, the imputation to region i is

$$\begin{aligned} &\xi^i(k, x^*, y^*, z^*; \underline{\overline{u}}_{k-}^{(\omega_1)*}, \underline{\overline{u}}_{k-}^{(\omega_2)*}, \underline{\overline{u}}_{k-}^{(\omega_3)*}) \\ &= \frac{V^i(k, x_k^*, y_k^*, z_k^*; \underline{\overline{u}}_{k-}^{(\omega_1)*}, \underline{\overline{u}}_{k-}^{(\omega_2)*}, \underline{\overline{u}}_{k-}^{(\omega_3)*})}{\sum_{j=1}^n V^j(k, x_k^*, y_k^*, z_k^*; \underline{\overline{u}}_{k-}^{(\omega_1)*}, \underline{\overline{u}}_{k-}^{(\omega_2)*}, \underline{\overline{u}}_{k-}^{(\omega_3)*})} \\ &\quad W(k, x_k^*, y_k^*, z_k^*; \underline{\overline{u}}_{k-}^{(\omega_1)*}, \underline{\overline{u}}_{k-}^{(\omega_2)*}, \underline{\overline{u}}_{k-}^{(\omega_3)*}), \end{aligned} \tag{3.3}$$

for $i \in N$ at stage $k \in \{1, 2, \cdots, T\}$.

Applying Theorem 2.1 in Chap. 5, a payment

$$\begin{aligned} &\beta_k^i(x_k^*, y_k^*, z_k^*; \underline{\overline{u}}_{k-}^{(\omega_1)*}, \underline{\overline{u}}_{k-}^{(\omega_2)*}, \underline{\overline{u}}_{k-}^{(\omega_3)*}) \\ &= (\delta_1^k)^{-1}\Bigg[\frac{V^i(k, x_k^*, y_k^*, z_k^*; \underline{\overline{u}}_{k-}^{(\omega_1)*}, \underline{\overline{u}}_{k-}^{(\omega_2)*}, \underline{\overline{u}}_{k-}^{(\omega_3)*})}{\sum_{j=1}^n V^j(k, x_k^*, y_k^*, z_k^*; \underline{\overline{u}}_{k-}^{(\omega_1)*}, \underline{\overline{u}}_{k-}^{(\omega_2)*}, \underline{\overline{u}}_{k-}^{(\omega_3)*})} \\ &\quad W(k, x_k^*, y_k^*, z_k^*; \underline{\overline{u}}_{k-}^{(\omega_1)*}, \underline{\overline{u}}_{k-}^{(\omega_2)*}, \underline{\overline{u}}_{k-}^{(\omega_3)*}) \\ &\quad - \frac{V^i(k+1, x_{k+1}^*, y_{k+1}^*, z_{k+1}^*; \underline{\overline{u}}_{(k+1)-}^{(\omega_1)*}, \underline{\overline{u}}_{(k+1)-}^{(\omega_2)*}, \underline{\overline{u}}_{(k+1)-}^{(\omega_3)*})}{\sum_{j=1}^n V^j(k+1, x_{k+1}^*, y_{k+1}^*, z_{k+1}^*; \underline{\overline{u}}_{(k+1)-}^{(\omega_1)*}, \underline{\overline{u}}_{(k+1)-}^{(\omega_2)*}, \underline{\overline{u}}_{(k+1)-}^{(\omega_3)*})} \\ &\quad \times W(k+1, x_{k+1}^*, y_{k+1}^*, z_{k+1}^*; \underline{\overline{u}}_{(k+1)-}^{(\omega_1)*}, \underline{\overline{u}}_{(k+1)-}^{(\omega_2)*}, \underline{\overline{u}}_{(k+1)-}^{(\omega_3)*})\Bigg] \end{aligned} \tag{3.4}$$

will be given to region $i \in N$ at stage $k \in \{1, 2, \cdots, T\}$.

Finally, under cooperation, all regions would use the cooperative strategies and the payoff that region i will directly receive at stage k along the cooperative trajectory $\{x_k^*, y_k^*, z_k^*\}_{k=1}^{T+1}$ with previously executed durable strategies $(\underline{\overline{u}}_{k-}^{(\omega_1)*}, \underline{\overline{u}}_{k-}^{(\omega_2)*}, \underline{\overline{u}}_{k-}^{(\omega_3)*})$ becomes

$$\sigma_k^i(x_k^*, y_k^*, z_k^*; \underline{\overline{u}}_{k-}^{(\omega_1)*}, \underline{\overline{u}}_{k-}^{(\omega_2)*}, \underline{\overline{u}}_{k-}^{(\omega_3)*}) = R_k^{(x)i}(\overline{u}_k^{(\omega_1)i*}) - \varphi_k^{(x)i}(\overline{u}_k^{(\omega_1)i*}) - c_k^{Ii}(u_k^{(Ii*)})$$

$$+R_k^{(y)i}(\overline{u}_k^{(\omega_2)i*}) - \varphi_k^{(y)i}(\overline{u}_k^{(\omega_2)i*}) - c_k^{IIi}(u_k^{IIi*}) + R_k^{(z)i}(q_k^{(z)i}(u_k^{IIIi*}, \kappa_k^i z_k^*))$$

$-\varphi_k^{(z)i}(\overline{u}_k^{(\omega_3)i*}) - c_k^{IIIi}(u_k^{IIIi*}) - h_k^i(x_k^*) + \chi_k^i(y_k^*)$.

However, according to the agreed-upon imputation, region i will receive $\beta_k^i(x_k^*, y_k^*, z_k^*; \underline{\overline{u}}_{k-}^{(\omega_1)*}, \underline{\overline{u}}_{k-}^{(\omega_2)*}, \underline{\overline{u}}_{k-}^{(\omega_3)*})$ at stage k. Therefore, a side-payment

$$\begin{aligned}
&\pi_k^i(x_k^*, y_k^*, z_k^*; \underline{\overline{u}}_{k-}^{(\omega_1)*}, \underline{\overline{u}}_{k-}^{(\omega_2)*}, \underline{\overline{u}}_{k-}^{(\omega_3)*}) \\
&= \beta_k^i(x_k^*, y_k^*, z_k^*; \underline{\overline{u}}_{k-}^{(\omega_1)*}, \underline{\overline{u}}_{k-}^{(\omega_2)*}, \underline{\overline{u}}_{k-}^{(\omega_3)*}) \\
&- \sigma_k^i(x_k^*, y_k^*, z_k^*; \underline{\overline{u}}_{k-}^{(\omega_1)*}, \underline{\overline{u}}_{k-}^{(\omega_2)*}, \underline{\overline{u}}_{k-}^{(\omega_3)*}), \\
&\text{for } k \in \{1, 2, \cdots, T\},
\end{aligned} \tag{3.5}$$

has to be given to region $i \in N$ to yield the cooperative imputation $\xi^i(k, x^*, y^*, z^*; \underline{\overline{u}}_{k-}^{(\omega_1)*}, \underline{\overline{u}}_{k-}^{(\omega_2)*}, \underline{\overline{u}}_{k-}^{(\omega_3)*})$, for $k \in \{1, 2, \cdots, T\}$.

4 An Optimal Environmental Management Game

In this section, we present a representative cooperative dynamic environmental game involving under durable investments with explicit functional forms.

4.1 Game Formulation

Consider the $T-$ stage dynamic environmental game in Sect. 3 of Chap. 4 in which there are n regions that produce three types of outputs. The first type of output $\overline{u}_k^{(\omega_1)i} \in \overline{U}^{(\omega_3)i} \subset R^+$ emits pollutants over ω_1 stages. The second type of output $\overline{u}_k^{(\omega_2)i} \in \overline{U}^{(\omega_2)i} \subset R^+$ causes damages to the ecosystems over ω_2 stages. The level of pollution stocks is $x_k \in X \subset R$ and the dynamics of pollution accumulation is governed by the difference equation:

$$\begin{aligned}
x_{k+1} =& x_k + \sum_{j=1}^{n} m_k^{|k|j} \overline{u}_k^{(\omega_1)j} + \sum_{j=1}^{n} \sum_{\tau=k-\omega_1+1}^{k-1} m_k^{|\tau|j} \overline{u}_\tau^{(\omega_1)j} \\
&- \sum_{j=1}^{n} b_k^j u_k^{Ij} (x_k)^{1/2} - \lambda_k^x x_k, \quad x_1 = x_1^0,
\end{aligned} \tag{4.1}$$

where $m_k^{|k|j}$, $m_k^{|\tau|j}$, b_k^j and $\lambda_k^x(x_k)$ are positive parameters; and $\overline{u}_\tau^{(\omega_1)j} = 0$, for $\tau \in \{0, -1, -2, \cdots, -\omega_1+1\}$ and $j \in N$.

The state of the ecosystem is reflected by $y_k \in Y \subset R$ which conditions generate benefits to the regions. Negative benefits are possible if the state of the ecosystem becomes bad enough. The dynamics governing the evolution of the condition of the ecosystem is

$$y_{k+1} = y_k + g_k - \sum_{j=1}^{n} \sigma_k^{|k|j} \overline{u}_k^{(\omega_2)j} - \sum_{j=1}^{n} \sum_{\tau=k-\omega_2+1}^{k-1} \sigma_k^{|\tau|j} \overline{u}_\tau^{(\omega_2)j}$$
$$+ \sum_{j=1}^{n} \varepsilon_k^j u_k^{IIj} - \vartheta_k x_k - \lambda_k^y y_k,$$
$$y_1 = y_1^0, \tag{4.2}$$

where $\sigma_k^{|\tau|j}$, ε_k^j, g_k, λ_k^y and ϑ_k are positive parameters.

The third type of output is a clean output which is produced by non-durable input $u_k^{IIIi} \in U^{IIIi} \subset R^+$ and green technology $z_k \in Z \subset R^{n_3}$. The dynamics governing the accumulation of green technology is

$$z_{k+1} = z_k + \sum_{j=1}^{n} a_k^{|k|j} \overline{u}_k^{(\omega_3)j} + \sum_{j=1}^{n} \sum_{\tau=k-\omega_3+1}^{k-1} a_k^{|\tau|j} \overline{u}_\tau^{(\omega_3)j}$$
$$- \lambda_k^z z_k, \quad z_1 = z_1^0 \tag{4.3}$$

where $\overline{u}_k^{(\omega_3)j} \in \overline{U}^{(\omega_3)j} \subset R^{m_3^j}$ are the investments in the public green technology by region j in stage k and it would take ω_3 stages to fully transmit the investments into the public technology, $a_k^{|k|j}$, $a_k^{|\tau|j}$ and λ_k^z are positive parameters.

The realized green technologies from the public technologies for region i are $\kappa_k^i z_k$. The region uses a non-durable input $u_k^{IIIi} \in U^{IIIi} \subset R^{m^{III(1)i}}$ together with its realized green technologies to produce clean outputs.

To ensure group optimality, the regions cooperate to maximize their joint payoff

$$\sum_{i=1}^{n} \Bigg[\sum_{k=1}^{T} \Big(R_k^{(x)i} \overline{u}_k^{(\omega_1)i} - \varphi_k^{(x)i} (\overline{u}_k^{(\omega_1)i})^2 - c_k^{Ii} (u_k^{Ii})^2$$
$$+ R_k^{(y)i} u_k^{(\omega_2)i} - \varphi_k^{(y)i} (\overline{u}_k^{(\omega_2)i})^2 - c_k^{IIi} (u_k^{IIi})^2$$
$$+ R_k^{(z)i} (u_k^{IIIi} \kappa_k^i z_k)^{1/2} - \varphi_k^{(z)i} (\overline{u}_k^{(\omega_3)i})^2 - c_k^{IIIi} u_k^{IIIi} - h_k^i x_k + \chi_k^i y_k \Big) \delta^{k-1}$$
$$+ \Big(Q_{T+1}^{x(i)} x_{T+1} + Q_{T+1}^{y(i)} y_{T+1} + Q_{T+1}^{z(i)} \kappa_{T+1}^i z_{T+1} + \varpi_{T+1}^i \Big) \delta^T \Bigg] \tag{4.4}$$

where $R_k^{(x)i}$, $R_k^{(y)i}$, $R_k^{(z)i}$, $\varphi_k^{(x)i}$, $\varphi_k^{(y)i}$, $\varphi_k^{(z)i}$, c_k^{Ii}, c_k^{IIi}, c_k^{IIIi}, h_k^i, χ_k^i, $Q_{T+1}^{y(i)}$, $Q_{T+1}^{z(i)}$ are positive parameters and $Q_{T+1}^{x(i)}$ is negative.

4.2 Group Optimal Solution

Invoking Corollary 2.1, an optimal solution to the joint maximization problem (4.1)–(4.4) can be characterized as below.

Corollary 4.1. *Let $W(k, x, y, z; \underline{\overline{u}}_{k-}^{(\omega_1)}, \underline{\overline{u}}_{k-}^{(\omega_2)}, \underline{\overline{u}}_{k-}^{(\omega_3)})$ be the maximal value of the joint payoff.*

$$\sum_{i=1}^{n}\Bigg[\sum_{t=k}^{T}\Big(R_t^{(x)i}\overline{u}_t^{(\omega_1)i} - \varphi_t^{(x)i}(\overline{u}_t^{(\omega_1)i})^2 - c_t^{Ii}(u_t^{Ii})^2 + R_t^{(y)i}u_t^{(\omega_2)i} - \varphi_t^{(y)i}(\overline{u}_t^{(\omega_2)i})^2$$
$$-c_t^{IIi}(u_t^{IIi})^2 + R_t^{(z)i}(u_t^{IIIi}\kappa_t^i z_t)^{1/2} - \varphi_t^{(z)i}(\overline{u}_t^{(\omega_3)i})^2 - c_t^{IIIi}u_t^{IIIi} - h_t^i x_t + \chi_t^i y_t\Big)\delta^{t-1}$$
$$+\Big(Q_{T+1}^{x(i)}x_{T+1} + Q_{T+1}^{y(i)}y_{T+1} + Q_{T+1}^{z(i)}\kappa_{T+1}^i z_{T+1} + \varpi_{T+1}^i\Big)\delta^T\Bigg]$$

for the joint payoff maximization problem (4.1)–(4.4) starting at stage k with state $(x_k, y_k, z_k) = (x, y, z)$ and previously executed controls $(\underline{\overline{u}}_{k-}^{(\omega_1)}, \underline{\overline{u}}_{k-}^{(\omega_2)}, \underline{\overline{u}}_{k-}^{(\omega_3)})$, then the function $W(k, x, y, z; \underline{\overline{u}}_{k-}^{(\omega_1)}, \underline{\overline{u}}_{k-}^{(\omega_2)}, \underline{\overline{u}}_{k-}^{(\omega_3)})$ satisfies the following system of recursive equations:

$$W(T+1, x, y, z; \underline{\overline{u}}_{(T+1)-}^{(\omega_1)}, \underline{\overline{u}}_{(T+1)-}^{(\omega_2)}, \underline{\overline{u}}_{(T+1)-}^{(\omega_3)})$$
$$= \sum_{j=1}^{n}\Big(Q_{T+1}^{x(j)}x_{T+1} + Q_{T+1}^{y(j)}y_{T+1} + Q_{T+1}^{z(j)}\kappa_{T+1}^j z_{T+1} + Q_{T+1}^j\Big)\delta^T \tag{4.5}$$

$$W(k, x, y, z; \underline{\overline{u}}_{k-}^{(\omega_1)}, \underline{\overline{u}}_{k-}^{(\omega_2)}, \underline{\overline{u}}_{k-}^{(\omega_3)})$$
$$= \max_{\substack{u_k^{Ii}, u_k^{IIi}, u_k^{IIIi}, \overline{u}_k^{(\omega_1)i} \\ \overline{u}_k^{(\omega_2)i}, \overline{u}_k^{(\omega_2)i}, i\in N}} \Bigg\{\sum_{j=1}^{n}\Big(R_k^{(x)j}\overline{u}_k^{(\omega_1)j} - \varphi_k^{(x)j}(\overline{u}_k^{(\omega_1)j})^2$$
$$- c_k^{Ij}(u_k^{Ij})^2 + R_k^{(y)j}u_k^{(\omega_2)j} - \varphi_k^{(y)j}(\overline{u}_k^{(\omega_2)j})^2$$
$$- c_k^{IIj}(u_k^{IIj})^2 + R_k^{(z)j}(u_k^{IIIj}\kappa_k^j z_k)^{1/2}$$
$$- \varphi_k^{(z)j}(\overline{u}_k^{(\omega_3)j})^2 - c_k^{IIIj}u_k^{IIIj} - h_k^j x + \chi_k^j y\Big]\delta^{k-1}$$
$$+ W(k+1, x_{k+1}, y_{k+1}, z_{k+1}; \underline{\overline{u}}_{(k+1)-}^{(\omega_1)}, \underline{\overline{u}}_{(k+1)-}^{(\omega_2)}, \underline{\overline{u}}_{(k+1)-}^{(\omega_3)}\Bigg\}, \tag{4.6}$$

for $k \in \{1, 2, \cdots, T\}$.

Performing the indicated maximization operator in Corollary 4.1, we obtain the following optimal conditions.

(i) The pollution creating output, that is $\overline{u}_k^{(\omega_1)i}$, will be produced up to the level where

$$\overline{u}_k^{(\omega_1)i} = \frac{1}{2\varphi_k^{(x)i}}\Bigg[R_k^{(x)i} + \Bigg(\frac{\partial Wl_{k+1}}{\partial x_{k+1}}m_k^{|k|i} + \frac{\partial W_{k+1}}{\partial u_k^{(\omega_1)i}}\Bigg)\delta^{1-k}\Bigg], \quad \text{for } i \in N \tag{4.7}$$

(ii) The output that damages the ecosystem will be produced up to the level where

$$\overline{u}_k^{(\omega_2)i} = \frac{1}{2\varphi_k^{(y)i}}\left[R_k^{(y)i} + \left(\frac{\partial W_{k+1}}{\partial \overline{u}_k^{(\omega_1)i}} - \frac{\partial W_{k+1}}{\partial y_{k+1}} \sigma_k^{|k|i} \right)\delta^{1-k} \right], \text{ for } i \in N \tag{4.8}$$

(iii) Investments for building up green technologies, that is $\overline{u}_k^{(\omega_3)i}$, will be employed up to the level where

$$\overline{u}_k^{(\omega_3)i} = \frac{1}{2\phi_k^{(z)i}}\left(\frac{\partial W_{k+1}}{\partial z_{k+1}} a_k^{|k|i} + \frac{\partial W_{k+1}}{\partial \overline{u}_k^{(\omega_3)i}} \right)\delta^{1-k}, \quad \text{for } i \in N \tag{4.9}$$

(iv) Pollution abatement effort is employed up to the level where

$$u_k^{Ii} = -\frac{1}{2c_k^{Ii}} \frac{\partial W_{k+1}}{\partial x_{k+1}} b_k^i(x)^{1/2}\delta^{1-k}, \quad \text{for } i \in N \tag{4.10}$$

(v) Ecosystem improvement effort is employed up to the level where

$$u_k^{IIi} = \frac{1}{2c_k^{IIi}} \frac{\partial W_{k+1}}{\partial y_{k+1}} \varepsilon_k^i \delta^{1-k}, \quad \text{for } i \in N \tag{4.11}$$

(vi) Non-durable input used to produce the clean output are employed up to the level where

$$u_k^{IIIi} = \left(\frac{R_k^{(z)i}}{2c_k^{IIIi}} \right)^2 \kappa_k^i z, \quad \text{for } i \in N \tag{4.12}$$

An optimal solution of the cooperative game (4.1)–(4.4) can be solved as:

Proposition 4.1. *System* (4.1)–(4.4) *admits a solution with the optimal cooperative payoff being.*

System (3.5)–(3.6) *admits a solution with the game equilibrium payoff of player i being*

$$W(k, x, y, z; \underline{\overline{u}}_{k-}^{(\omega_1)}, \underline{\overline{u}}_{k-}^{(\omega_2)}, \underline{\overline{u}}_{k-}^{(\omega_3)}) = (A_k x + B_k y + C_k z + D_k)\delta^{k-1}, \tag{4.13}$$

where

$$A_k = -\sum_{i=1}^{n}\left(\frac{(\delta A_{k+1} b_k^i)^2}{4c_k^{Ii}} + h_k^i\right)$$
$$+ \delta A_{k+1}\left(\sum_{j=1}^{n}\frac{\delta A_{k+1}(b_k^j)^2}{2\phi_k^j} + (1-\lambda_k^x)\right) - B_{k+1}\vartheta_k,$$
$$B_k = \sum_{i=1}^{n} \chi_k^i + \delta B_{k+1}(1-\lambda_k^y),$$
$$C_k = \sum_{i=1}^{n} \frac{(R_k^{(z)i})^2 \kappa_k^i}{4(c_k^{IIIi})} + \delta C_{k+1}(1-\lambda_k^z),$$

and D_k is an expression containing the previously executed controls $(\underline{u}_{k-}^{(\omega_1)}, \underline{u}_{k-}^{(\omega_2)}, \underline{u}_{k-}^{(\omega_3)})$,for $k \in \{,1,2,\cdots,T\}$.

The optimal cooperative strategies stage $k \in \{1,2,\cdots,T\}$ can be obtained as:

$$\overline{u}_k^{(\omega_1)i} = \frac{R_k^{(x)i} + \sum_{\tau=k}^{k+\omega_1-1} \delta^{\tau-k+1} A_{\tau+1} m_\tau^{|k|i}}{2\varphi_k^{(x)i}}, \text{ and } A_\tau = 0 \text{ for } \tau > T+1$$
$$\overline{u}_k^{(\omega_2)i} = \frac{R_k^{(y)i} - \sum_{\tau=k}^{k+\omega_2-1} \delta^{\tau-k+1} B_{\tau+1} \sigma_\tau^{|k|i}}{2\varphi_k^{(y)i}}, \text{ and } B_\tau = 0 \text{ for } \tau > T+1$$
$$\overline{u}_k^{(\omega_3)i} = \frac{\sum_{\tau=k}^{k+\omega_3-1} \delta^{\tau-k+1} C_{\tau+1} a_\tau^{|k|i}}{2\varphi_k^{(z)i}}, \text{ and } C_\tau = 0 \text{ for } \tau > T+1$$
$$u_k^{Ii} = -\frac{\delta A_{k+1} b_k^i}{2c_k^{Ii}} x^{1/2}, \quad u_k^{IIi} = \frac{\delta B_{k+1}\varepsilon_k^i}{2c_k^{IIi}},$$
$$u_k^{IIIi} = \frac{(R_k^{(z)i})^2 \kappa_k^i z}{4(c_k^{IIIi})^2} \quad \text{for } i \in N. \tag{4.14}$$

Proof. See Appendix in Sect. 6. ■

Substituting the strategies from (4.14) into the state dynamics (4.1)–(4.3), we can obtain the optimal cooperative dynamics of the state variables. In particular, the optimal cooperative dynamics of pollution accumulation is

$$x_{k+1} = x_k + \sum_{j=1}^{n} m_k^{|k|j} \frac{R_k^{(x)j} + \sum_{\tau=k}^{k+\omega_1-1} \delta^{\tau-k+1} A_{\tau+1} m_\tau^{|k|j}}{2\varphi_k^{(x)i}}$$
$$+ \sum_{j=1}^{n}\sum_{t=k-\omega_1+1}^{k-1} m_k^{|t|j} \frac{R_t^{(x)j} + \sum_{\tau=t}^{t+\omega_1-1} \delta^{\tau-t+1} A_{\tau+1} m_\tau^{|t|j}}{2\varphi_t^{(x)i}}$$
$$+ \sum_{j=1}^{n} b_k^j \frac{\delta A_{k+1} b_k^j}{2c_k^{Ij}} x_k - \lambda_k^x x_k, \quad x_1 = x_1^0, \tag{4.15}$$

To solve (4.15), we first start from the initial stage, that is stage 1, where strategies before stage 1 are zero, that is $\overline{u}_{\tau}^{(\omega_1)j} = 0$, for $\tau \in \{0, -1, -2, \cdots, -\omega_1 + 1\}$. We can obtain x_2 using (4.15). With x_2 and (4.15) we can obtain x_3. Continuing with the process for stage 4 to stage $T + 1$, we can obtain the optimal cooperative path of the pollution stock $\{x_k\}_{k=1}^{T+1}$.

The dynamics governing the evolution of the conditions of the ecosystems in the under cooperation is

$$\begin{aligned} y_{k+1} =& y_k + g_k - \sum_{j=1}^{n} \sigma_k^{|k|j} \frac{R_k^{(y)j} - \sum_{\tau=k}^{k+\omega_2-1} \delta^{\tau-k+1} B_{\tau+1} \sigma_\tau^{|k|j}}{2\varphi_k^{(y)j}} \\ & - \sum_{j=1}^{n} \sum_{t=k-\omega_2+1}^{k-1} \sigma_k^{|t|j} \frac{R_t^{(y)j} - \sum_{\tau=t}^{t+\omega_2-1} \delta^{\tau-t+1} B_{\tau+1} \sigma_\tau^{|t|j}}{2\varphi_t^{(y)j}} \\ & + \sum_{j=1}^{n} \varepsilon_k^j \frac{\delta B_{k+1}^j \varepsilon_k^j}{2c_k^{IIj}} - \vartheta_k x_k - \lambda_k^y y_k, \\ y_1 =& y_1^0 \text{ and } x_1 = x_1^0 \end{aligned} \tag{4.16}$$

We first start from the initial stage, that is stage 1, where strategies before stage 1 are zero, that is $\overline{u}_{\tau}^{(\omega_2)j} = 0$, for $\tau \in \{0, -1, -2, \cdots, -\omega_2 + 1\}$. We can obtain y_2 using (4.16) and x_1. With x_2, y_2 and (4.16) we can obtain y_3. Continuing with the process for stage 4 to stage $T + 1$, we can obtain the optimal cooperative path of the condition of the ecosystems $\{y_k\}_{k=1}^{T+1}$.

The optimal cooperative dynamics governing the accumulation of green technology is

$$\begin{aligned} z_{k+1} =& z_k + \sum_{j=1}^{n} a_k^{|k|j} \frac{\sum_{\tau=k}^{k+\omega_3-1} \delta^{\tau-k+1} C_{\tau+1}^j a_\tau^{|k|j}}{2\varphi_k^{(z)j}} \\ & + \sum_{j=1}^{n} \sum_{t=k-\omega_3+1}^{k-1} a_k^{|\tau|j} \frac{\sum_{\tau=t}^{t+\omega_3-1} \delta^{\tau-t+1} C_{\tau+1}^j a_\tau^{|t|j}}{2\varphi_t^{(z)j}} - \lambda_k^z z_k, \\ z_1 =& z_1^0 \end{aligned} \tag{4.17}$$

Again, we first start from the initial stage, that is stage 1, where strategies before stage 1 are zero, that is $\overline{u}_{\tau}^{(\omega_3)j} = 0$, for $\tau \in \{0, -1, -2, \cdots, -\omega_3 + 1\}$. We can obtain z_2 using (4.17). With z_2 and (4.17) we can obtain z_3. Continuing with the process for stage 4 to stage $T + 1$, we can obtain the optimal cooperative path of the public green technology $\{z_3\}_{k=1}^{T+1}$.

We use (x_k^*, y_k^*, z_k^*) to denote the optimal paths of (x_k, y_k, z_k) and$(\underline{u}_k^{I*}, \underline{u}_k^{II*}, \underline{u}_k^{III*}, \overline{\underline{u}}_k^{(\omega_1)*}, \overline{\underline{u}}_k^{(\omega_2)*}, \overline{\underline{u}}_k^{(\omega_3)*})$ to denote the set of optimal cooperative strategies.

4.3 Sharing of Cooperative Payoff

Since the regions agree to distribute the cooperative payoff proportional to the regions' non-cooperative payoffs. According to this optimality principle, the imputation to region i is

$$\xi^i(k, x^*, y^*, z^*; \underline{\overline{u}}_{k-}^{(\omega_1)*}, \underline{\overline{u}}_{k-}^{(\omega_2)*}, \underline{\overline{u}}_{k-}^{(\omega_3)*})$$
$$= \frac{V^i(k, x_k^*, y_k^*, z_k^*; \underline{\overline{u}}_{k-}^{(\omega_1)*}, \underline{\overline{u}}_{k-}^{(\omega_2)*}, \underline{\overline{u}}_{k-}^{(\omega_3)*})}{\sum_{j=1}^n V^j(k, x_k^*, y_k^*, z_k^*; \underline{\overline{u}}_{k-}^{(\omega_1)*}, \underline{\overline{u}}_{k-}^{(\omega_2)*}, \underline{\overline{u}}_{k-}^{(\omega_3)*})}$$
$$\times W(k, x_k^*, y_k^*, z_k^*; \underline{\overline{u}}_{k-}^{(\omega_1)*}, \underline{\overline{u}}_{k-}^{(\omega_2)*}, \underline{\overline{u}}_{k-}^{(\omega_3)*}), \quad (4.18)$$

for $i \in N$ at stage $k \in \{1, 2, \cdots, T\}$,

Using Proposition 4.1 and (4.15)–(4.17), we obtain

$$\xi^i(k, x^*, y^*, z^*; \underline{\overline{u}}_{k-}^{(\omega_1)*}, \underline{\overline{u}}_{k-}^{(\omega_2)*}, \underline{\overline{u}}_{k-}^{(\omega_3)*})$$
$$= \frac{(A_k^i x_k^* + B_k^i y_k^* + C_k^i z_k^* + D_k^i)}{\sum_{j=1}^n (A_k^j x_k^* + B_k^j y_k^* + C_k^j z_k^* + D_k^j)}$$
$$\times (A_k x_k^* + B_k y_k^* + C_k z_k^* + D_k)\delta^{k-1}, \quad (4.19)$$

for $i \in N$ at stage $k \in \{1, 2, \cdots, T\}$,

Invoking (3.4), a payment

$$\beta_k^i(x_k^*, y_k^*, z_k^*; \underline{\overline{u}}_{k-}^{(\omega_1)*}, \underline{\overline{u}}_{k-}^{(\omega_2)*}, \underline{\overline{u}}_{k-}^{(\omega_3)*})$$
$$= \delta^{1-k}\Bigg[\frac{(A_k^i x_k^* + B_k^i y_k^* + C_k^i z_k^* + D_k^i)}{\sum_{j=1}^n (A_k^j x_k^* + B_k^j y_k^* + C_k^j z_k^* + D_k^j)}$$
$$(A_k x_k^* + B_k y_k^* + C_k z_k^* + D_k)\delta^{k-1}$$
$$- \frac{(A_{k+1}^i x_{k+1}^* + B_{k+1}^i y_{k+1}^* + C_{k+1}^i z_{k+1}^* + D_{k+1}^i)}{\sum_{j=1}^n (A_{k+1}^j x_{k+1}^* + B_{k+1}^j y_{k+1}^* + C_{k+1}^j z_{k+1}^* + D_{k+1}^j)}$$
$$\times (A_{k+1} x_k^* + B_{k+1} y_k^* + C_{k+1} z_k^* + D_{k+1})\delta^k\Bigg] \quad (4.20)$$

will be given to region $i \in N$ at stage $k \in \{1, 2, \cdots, T\}$.

Note that the previously executed controls $(\underline{\overline{u}}_{k-}^{(\omega_1)*}, \underline{\overline{u}}_{k-}^{(\omega_2)*}, \underline{\overline{u}}_{k-}^{(\omega_3)*})$ are embedded in D_k and D_k^i for $i \in N$.

Finally, under cooperation, all regions would use the cooperative strategies and the payoff that region i will directly receive at stage k along the cooperative trajectory $\{x_k^*, y_k^*, z_k^*\}_{k=1}^{T+1}$ with previously executed durable strategies $(\underline{\overline{u}}_{k-}^{(\omega_1)*}, \underline{\overline{u}}_{k-}^{(\omega_2)*}, \underline{\overline{u}}_{k-}^{(\omega_3)*})$ becomes

$$\sigma_k^i(x_k^*, y_k^*, z_k^*; \underline{\overline{u}}_{k-}^{(\omega_1)*}, \underline{\overline{u}}_{k-}^{(\omega_2)*}, \underline{\overline{u}}_{k-}^{(\omega_3)*})$$

$$= R_k^{(x)i}\overline{u}_k^{(\omega_1)i*} - \varphi_k^{(x)i}(\overline{u}_k^{(\omega_1)i*})^2 - c_k^{Ii}(u_k^{Ii*})^2 + R_k^{(y)i}u_k^{(\omega_2)i*}$$
$$- \varphi_k^{(y)i}(\overline{u}_k^{(\omega_2)i*})^2 - c_k^{IIi}(u_k^{IIi*})^2 + R_k^{(z)i}(u_k^{IIIi*}\kappa_k^i z_k^*)^{1/2} - \varphi_k^{(z)i}(\overline{u}_k^{(\omega_3)i*})^2$$
$$- c_k^{IIIi}u_k^{IIIi*} - h_k^i x_k^* + \chi_k^i y_k^*.$$

However, according to the agreed-upon imputation, region i will receive $\beta_k^i(x_k^*, y_k^*, z_k^*; \underline{\overline{u}}_{k-}^{(\omega_1)*}, \underline{\overline{u}}_{k-}^{(\omega_2)*}, \underline{\overline{u}}_{k-}^{(\omega_3)*})$ at stage k. Therefore, a side-payment

$$\pi_k^i(x_k^*, y_k^*, z_k^*; \underline{\overline{u}}_{k-}^{(\omega_1)*}, \underline{\overline{u}}_{k-}^{(\omega_2)*}, \underline{\overline{u}}_{k-}^{(\omega_3)*})$$
$$= \beta_k^i(x_k^*, y_k^*, z_k^*; \underline{\overline{u}}_{k-}^{(\omega_1)*}, \underline{\overline{u}}_{k-}^{(\omega_2)*}, \underline{\overline{u}}_{k-}^{(\omega_3)*})$$
$$- \sigma_k^i(x_k^*, y_k^*, z_k^*; \underline{\overline{u}}_{k-}^{(\omega_1)*}, \underline{\overline{u}}_{k-}^{(\omega_2)*}, \underline{\overline{u}}_{k-}^{(\omega_3)*}), \quad \text{for } k \in \{1, 2, \cdots, T\} \tag{4.21}$$

has to be given to region $i \in N$ to yield the cooperative imputation $\xi^i(k, x^*, y^*, z^*; \underline{\overline{u}}_{k-}^{(\omega_1)*}, \underline{\overline{u}}_{k-}^{(\omega_2)*}, \underline{\overline{u}}_{k-}^{(\omega_3)*})$ for $k \in \{1, 2, \cdots, T\}$ in (4.19).

5 Efficiency Gain and Sustainable Imputation

As pointed out in Chap. 5 (see also Koutsoupias and Papadimitriou (2009)), the efficiency of a system degrades due to self-seeking optimization especially when externalities exist. Given the potentially large number of regions and severity of externalities in environmental analysis, self-seeking optimization by individual regions has caused serious damages to the ecosystems and established a continual trend of eco-degradation. Cooperative optimization represents the best way of obtaining solutions with maximum efficiency to the problem. It identifies the strategies that maximize the joint payoff collectively via internalizing the externalities.

Invoking Proposition 3.1 in Chap. 4 and Proposition 4.1 above we note that the value of A_k, which the marginal damage of pollution to all the n regions, is always more negative than A_k^i, which measures the magnitude of the marginal damage of pollution to region i, for $i \in N$ and $k \in \{, 1, 2, \cdots, T + 1\}$. The negative externalities of pollution are internalized in the optimal cooperative solution by taking into consideration of damages to all regions rather than just the damage to a single region. This leads to

$$\overline{u}_k^{(\omega_1)i*} = \frac{R_k^{(x)i} + \sum_{\tau=k}^{k+\omega_1-1}\delta^{\tau-k+1}A_{\tau+1}m_\tau^{|k|i}}{2\varphi_k^{(x)i}} < \overline{u}_k^{(\omega_1)i**}$$
$$= \frac{R_k^{(x)i} + \sum_{\tau=k}^{k+\omega_1-1}\delta^{\tau-k+1}A_{\tau+1}^i m_\tau^{|k|i}}{2\varphi_k^{(x)i}} \tag{5.1}$$

which reflects that the production of pollution creating outputs under cooperation would be less than that under non-cooperation. In addition, the pollution abatement efforts under cooperation are larger than the pollution abatement efforts under non-cooperation, that is

$$u_k^{Ii*} = -\frac{\delta A_{k+1} b_k^i}{2c_k^{Ii}} x^{1/2} > u_k^{Ii**} = -\frac{\delta A_{k+1}^i b_k^i}{2c_k^{Ii}} x^{1/2}, \tag{5.2}$$

Hence, the pollution stock under cooperation is lower than that under non-cooperation in every stage $k \in \{2, 3, \cdots, T+1\}$.

Again, invoking Proposition 3.1 in Chap. 4 and Proposition 4.1 above we note that the value of B_k, which measures the marginal benefit of the condition of the ecosystems to all the n regions, is always larger than B_k^i, which measures marginal benefit of the condition of the ecosystems to region i, for $i \in N$ and $k \in \{, 1, 2, \cdots, T+1\}$. The positive externalities of the condition of the ecosystems are internalized in the optimal cooperative solution by taking into consideration of the benefits to all regions rather than just the benefit to a single region. This leads to

$$\begin{aligned}\overline{u}_k^{(\omega_2)i*} &= \frac{R_k^{(y)i} - \sum_{\tau=k}^{k+\omega_2-1} \delta^{\tau-k+1} B_{\tau+1} \sigma_\tau^{|k|i}}{2\varphi_k^{(y)i}} < \overline{u}_k^{(\omega_2)i**} \\ &= \frac{R_k^{(y)i} - \sum_{\tau=k}^{k+\omega_2-1} \delta^{\tau-k+1} B_{\tau+1}^i \sigma_\tau^{|k|i}}{2\varphi_k^{(y)i}},\end{aligned} \tag{5.3}$$

which reflects that the production of ecosystem damaging outputs under cooperation would be less than that under non-cooperation. In addition, the ecosystems improvement efforts under cooperation are larger than the ecosystems improvement efforts under non-cooperation, that is

$$u_k^{IIi*} = \frac{\delta B_{k+1} \varepsilon_k^i}{2c_k^{IIi}} > u_k^{IIi**} = \frac{\delta B_{k+1}^i \varepsilon_k^i}{2c_k^{IIi}} \tag{5.4}$$

Hence, the condition of the ecosystems under cooperation is better than that under non-cooperation in every stage $k \in \{2, 3, \cdots, T+1\}$.

Invoking again Proposition 3.1 in Chap. 4 and Proposition 4.1 above we note that the value of C_k, which measures the marginal benefit of the green technology to all the n regions, is always larger than C_k^i, which measures marginal benefit of the green technology to region i, for $i \in N$ and $k \in \{, 1, 2, \cdots, T+1\}$. The positive externalities of the green technology are internalized in the optimal cooperative solution by taking into consideration of the benefits to all regions rather than just the benefit to a single region. This leads to

$$\overline{u}_k^{(\omega_3)i*} = \frac{\sum_{\tau=k}^{k+\omega_3-1} \delta^{\tau-k+1} C_{\tau+1} a_\tau^{|k|i}}{2\varphi_k^{(z)i}} > \overline{u}_k^{(\omega_3)i**}$$

$$= \frac{\sum_{\tau=k}^{k+\omega_3-1} \delta^{\tau-k+1} C_{\tau+1}^{i} a_{\tau}^{|k|i}}{2\varphi_k^{(z)i}}, \tag{5.5}$$

which reflects that the investments in green technology under cooperation are larger than that under the investments in green technology under non-cooperation. Hence, the condition of the green technology level under cooperation is higher than that under non-cooperation in every stage $k \in \{2, 3, \cdots, T+1\}$. In addition, the inputs employed to produce clean outputs under cooperation would be larger than the inputs employed to produce clean outputs under non-cooperation, that is

$$u_k^{IIIi*} = \frac{(R_k^{(z)i})^2 \kappa_k^i z_k^*}{4(c_k^{IIIi})^2} > u_k^{IIIi**} = \frac{(R_k^{(z)i})^2 \kappa_k^i z_k^{**}}{4(c_k^{IIIi})^2} \tag{5.6}$$

This leads to the situation where the clean outputs produced by each region under cooperation is higher than the clean outputs produced by each region under non-cooperation, that is

$$\begin{aligned} R_k^{(z)j} (u_k^{IIIi*})^{1/2} (\kappa_k^i z_k^*)^{1/2} &= \frac{(R_k^{(z)i})^2 \kappa_k^i z_k^*}{2c_k^{IIIi}} \\ > R_k^{(z)j} (u_k^{IIIi**})^{1/2} (\kappa_k^i z_k^{**})^{1/2} &= \frac{(R_k^{(z)i})^2 \kappa_k^i z_k^{**}}{2c_k^{IIIi}} \end{aligned} \tag{5.7}$$

Under the optimal cooperative environmental management, which maximizes efficiency, clean outputs will be higher, pollution creating outputs and ecosystems damaging outputs will be lower. In addition, investments in green technology and the level of green technology will be higher.

Moreover, to calibrate how poor an environmental outcome one can get in a non-cooperative situation where regions act in their own interests (fulfilling individuality) versus the Pareto optimum that fulfils optimality collectively, we consider the Price of Anarchy (PoA). It measures the ratio of the most efficient (Pareto optimal) sum of regions' payoffs to the sum of the regions' payoffs in a non-cooperative equilibrium, which can be expressed as:

$$\begin{aligned} &PoA(k, x_k^*, y_k^*, z_k^*; \overline{\underline{u}}_{k-}^{(\omega_1)*}, \overline{\underline{u}}_{k-}^{(\omega_2)*}, \overline{\underline{u}}_{k-}^{(\omega_3)*}) \\ &= \frac{W(k, x_k^*, y_k^*, z_k^*; \overline{\underline{u}}_{k-}^{(\omega_1)*}, \overline{\underline{u}}_{k-}^{(\omega_2)*}, \overline{\underline{u}}_{k-}^{(\omega_3)*})}{\sum_{j=1}^{n} V^j(k, x_k^*, y_k^*, z_k^*; \overline{\underline{u}}_{k-}^{(\omega_1)*}, \overline{\underline{u}}_{k-}^{(\omega_2)*}, \overline{\underline{u}}_{k-}^{(\omega_3)*})} \\ &= \frac{(A_k x_k^* + B_k y_k^* + C_k z_k^* + D_k)}{\sum_{j=1}^{n} (A_k^j x_k^* + B_k^j y_k^* + C_k^j z_k^* + D_k^j)} \end{aligned} \tag{5.8}$$

The optimal cooperative joint payoff $W(k, x_k^*, y_k^*, z_k^*; \overline{\underline{u}}_{k-}^{(\omega_1)*}, \overline{\underline{u}}_{k-}^{(\omega_2)*}, \overline{\underline{u}}_{k-}^{(\omega_3)*})$ which takes into consideration the externalities (in pollution, ecosystems conditions and green technology involving of a large number of regions) could be substantially higher than the sum of the regions' non-cooperative payoffs. Hence, the magnitude of PoA can be sizeable.

In addition, the percentage of efficiency gain brought about by Pareto cooperative optimization if the regions maintain cooperation at any stage $k \in \{1, 2, \cdots, T\}$ along the cooperative trajectory (x_k^*, y_k^*, z_k^*) can be obtained as:

$$\frac{W(k, x_k^*, y_k^*, z_k^*; \overline{\underline{u}}_{k-}^{(\omega_1)*}, \overline{\underline{u}}_{k-}^{(\omega_2)*}, \overline{\underline{u}}_{k-}^{(\omega_3)*}) - \sum_{j=1}^{n} V^j(k, x_k^*, y_k^*, z_k^*; \overline{\underline{u}}_{k-}^{(\omega_1)*}, \overline{\underline{u}}_{k-}^{(\omega_2)*}, \overline{\underline{u}}_{k-}^{(\omega_3)*})}{\sum_{j=1}^{n} V^j(k, x_k^*, y_k^*, z_k^*; \overline{\underline{u}}_{k-}^{(\omega_1)*}, \overline{\underline{u}}_{k-}^{(\omega_2)*}, \overline{\underline{u}}_{k-}^{(\omega_3)*})}$$

$$= \frac{(A_k x_k^* + B_k y_k^* + C_k z_k^* + D_k) - \sum_{j=1}^{n} (A_k^j x_k^* + B_k^j y_k^* + C_k^j z_k^* + D_k^j)}{\sum_{j=1}^{n} (A_k^j x_k^* + B_k^j y_k^* + C_k^j z_k^* + D_k^j)} = PoA - 1 \tag{5.9}$$

In the process of capturing efficiency gain in a dynamic system with durable strategies, the delay in cooperation can play a significant role. The cost of delay from deferring the cooperative scheme from stage 1 to stage $\tau + 1$ is

$$\Bigg\{\Big[W(1, x_1^*, y_1^*, z_1^*; \overline{\underline{u}}_{k-}^{(\omega_1)*}, \overline{\underline{u}}_{k-}^{(\omega_2)*}, \overline{\underline{u}}_{k-}^{(\omega_3)*})$$
$$-W(\tau, x_\tau^*, y_\tau^*, z_\tau^*; \overline{\underline{u}}_{k-}^{(\omega_1)*}, \overline{\underline{u}}_{k-}^{(\omega_2)*}, \overline{\underline{u}}_{k-}^{(\omega_3)*})\Big]$$
$$-\Bigg[\sum_{j=1}^{n} V^j(1, x_1^{**}, y_1^{**}, z_1^{**}; \overline{\underline{u}}_{k-}^{(\omega_1)**}, \overline{\underline{u}}_{k-}^{(\omega_2)**}, \overline{\underline{u}}_{k-}^{(\omega_3)**})$$
$$-\sum_{j=1}^{n} V^j(\tau, x_\tau^{**}, y_\tau^{**}, z_\tau^{**}; \overline{\underline{u}}_{k-}^{(\omega_1)**}, \overline{\underline{u}}_{k-}^{(\omega_2)**}, \overline{\underline{u}}_{k-}^{(\omega_3)**})\Bigg]\Bigg\}$$
$$+\Big\{W(\tau+1, x_{\tau+1}^*, y_{\tau+1}^*, z_{\tau+1}^*; \overline{\underline{u}}_{k-}^{(\omega_1)*}, \overline{\underline{u}}_{k-}^{(\omega_2)*}, \overline{\underline{u}}_{k-}^{(\omega_3)*})$$
$$-W(\tau+1, x_{\tau+1}^{**}, y_{\tau+1}^{**}, z_{\tau+1}^{**}; \overline{\underline{u}}_{k-}^{(\omega_1)**}, \overline{\underline{u}}_{k-}^{(\omega_2)**}, \overline{\underline{u}}_{k-}^{(\omega_3)**})\Big\}. \tag{5.10}$$

The term inside the first set of curly brackets yields the loss in payoff from stage 1 to stage τ for acting non-cooperatively rather than cooperatively. The term inside the second set of curly brackets reflects the loss in the cooperative joint payoff from stage $\tau + 1$ to the end of the game if cooperation starts in stage $\tau + 1$ rather than stage 1.

The delay in cooperation can cause very significant loss because of (i) the state $(x_{\tau+1}^{**}, y_{\tau+1}^{**}, z_{\tau+1}^{**})$ along the game equilibrium path can reach a potentially devastating condition and (ii) the previously executed strategies $(\overline{\underline{u}}_{k-}^{(\omega_1)**}, \overline{\underline{u}}_{k-}^{(\omega_2)**}, \overline{\underline{u}}_{k-}^{(\omega_3)**})$ under non-cooperation from stage 1 to stage τ can generate a heavy burden to the future payoffs and state dynamics.

Finally, since the asymmetric regions agree to share the optimal cooperative payoff proportional to the regions' non-cooperative payoffs, the imputation to region $i \in N$ at stage $k \in \{1, 2, \cdots, T\}$ is

$$\xi^i(k, x^*, y^*, z^*; \overline{\underline{u}}_{k-}^{(\omega_1)*}, \overline{\underline{u}}_{k-}^{(\omega_2)*}, \overline{\underline{u}}_{k-}^{(\omega_3)*})$$
$$= \frac{V^i(k, x_k^*, y_k^*, z_k^*; \overline{\underline{u}}_{k-}^{(\omega_1)*}, \overline{\underline{u}}_{k-}^{(\omega_2)*}, \overline{\underline{u}}_{k-}^{(\omega_3)*})}{\sum_{j=1}^n V^j(k, x_k^*, y_k^*, z_k^*; \overline{\underline{u}}_{k-}^{(\omega_1)*}, \overline{\underline{u}}_{k-}^{(\omega_2)*}, \overline{\underline{u}}_{k-}^{(\omega_3)*})}$$
$$\times W(k, x_k^*, y_k^*, z_k^*; \overline{\underline{u}}_{k-}^{(\omega_1)*}, \overline{\underline{u}}_{k-}^{(\omega_2)*}, \overline{\underline{u}}_{k-}^{(\omega_3)*}) \quad (5.11)$$

The imputation in (5.11) can be expressed as

$$\xi^i(k, x_k^*, y_k^*, z_k^*; \overline{\underline{u}}_{k-}^{(\omega_1)*}, \overline{\underline{u}}_{k-}^{(\omega_2)*}, \overline{\underline{u}}_{k-}^{(\omega_3)*})$$
$$= \frac{W(k, x_k^*, y_k^*, z_k^*; \overline{\underline{u}}_{k-}^{(\omega_1)*}, \overline{\underline{u}}_{k-}^{(\omega_2)*}, \overline{\underline{u}}_{k-}^{(\omega_3)*})}{\sum_{j=1}^n V^j(k, x_k^*, y_k^*, z_k^*; \overline{\underline{u}}_{k-}^{(\omega_1)*}, \overline{\underline{u}}_{k-}^{(\omega_2)*}, \overline{\underline{u}}_{k-}^{(\omega_3)*})}$$
$$\times V^j(k, x_k^*, y_k^*, z_k^*; \overline{\underline{u}}_{k-}^{(\omega_1)*}, \overline{\underline{u}}_{k-}^{(\omega_2)*}, \overline{\underline{u}}_{k-}^{(\omega_3)*})$$
$$= PoA(k, x_k^*, y_k^*, z_k^*; \overline{\underline{u}}_{k-}^{(\omega_1)*}, \overline{\underline{u}}_{k-}^{(\omega_2)*}, \overline{\underline{u}}_{k-}^{(\omega_3)*})$$
$$\times V^j(k, x_k^*, y_k^*, z_k^*; \overline{\underline{u}}_{k-}^{(\omega_1)*}, \overline{\underline{u}}_{k-}^{(\omega_2)*}, \overline{\underline{u}}_{k-}^{(\omega_3)*})$$

The cooperative payoff to an individual region equals its non-cooperative payoff times the Price of Anarchy. Hence, individual regions would obtain the proportion of efficiency gain measured by PoA along the cooperative state trajectory. Given that each region would receive a cooperative payoff that exploits full efficiency and is guided by PoA along the cooperative path (x_k^*, y_k^*, z_k^*), the scheme is dynamically stable with the agreed-upon optimality principle maintained at any stage. Moreover, the previously executed strategies $(\overline{\underline{u}}_{k-}^{(\omega_1)*}, \overline{\underline{u}}_{k-}^{(\omega_2)*}, \overline{\underline{u}}_{k-}^{(\omega_3)*})$ are also taken into consideration in obtaining the imputation to the regions.

6 Appendix: Proof of Proposition 4.1.

Using (4.5), we have

$$A_{T+1} = \sum_{i=1}^n Q_{T+1}^{x(i)}, \; B_{T+1} = \sum_{i=1}^n Q_{T+1}^{y(i)},$$
$$C_{T+1} = \sum_{i=1}^n Q_{T+1}^{z(i)}, \; D_{T+1} = \sum_{i=1}^n \varpi_{T+1}^i \quad (6.1)$$

Using the technique of backward induction, we consider first the last operational stage T. Using Proposition 4.1 and (4.7)–(4.12), we can express the optimal cooperative strategies in stage T as:

$$\overline{u}_T^{(\omega_1)i} = \frac{R_T^{(x)i} + \delta A_{T+1} m_T^{|T|i}}{2\varphi_T^{(x)i}}, \; \overline{u}_T^{(\omega_2)i} = \frac{R_T^{(y)i} - \delta B_{T+1} \sigma_T^{|T|i}}{2\varphi_T^{(y)i}},$$

$$\overline{u}_T^{(\omega_3)i} = \frac{\delta C_{T+1} a_T^{|T|i}}{2\varphi_T^{(z)i}}, \quad u_T^{Ii} = -\frac{\delta A_{T+1} b_T^i}{2c_T^{Ii}} x^{1/2}$$

$$u_T^{IIi} = \frac{\delta B_{T+1} \varepsilon_T^i}{2c_T^{IIi}}, \quad u_T^{IIIi} = \frac{(R_T^{(z)i})^2 \kappa_T^i z}{4(c_T^{IIIi})^2}, \quad \text{for } i \in N \tag{6.2}$$

Substituting the optimal strategies in (6.2) into the stage T equation of (4.6) we obtain

$$\begin{aligned}
&W(T, x, y, z; \underline{\overline{u}}_{T-}^{(\varpi_1)}, \underline{\overline{u}}_{T-}^{(\omega_2)}, \underline{\overline{u}}_{T-}^{(\omega_3)}) = (A_T x + B_T y + C_T z + D_T)\delta^{T-1} \\
&= \sum_{i=1}^{n} \Bigg(\frac{(R_T^{(x)i} - \delta A_{T+1} m_T^{|T|i})(R_T^{(x)i} + \delta A_{T+1} m_T^{|T|i})}{4\varphi_T^{(x)i}} - \frac{(\delta A_{T+1} b_T^i)^2}{4c_T^{Ii}} x \\
&+ \frac{(R_T^{(y)i} + \delta B_{T+1} \sigma_T^{|T|i})(R_T^{(y)i} - \delta B_{T+1} \sigma_T^{|T|i})}{4\varphi_T^{(y)i}} - \frac{(\delta B_{T+1} \varepsilon_T^i)^2}{4c_T^{IIi}} \\
&+ \frac{(R_T^{(z)i})^2 \kappa_T^i z}{4(c_T^{IIIi})} - \frac{(\delta C_{T+1} a_T^{|T|i})^2}{4\varphi_T^{(z)i}} - h_T^i x + \chi_T^i y \Bigg) \delta^{T-1} \\
&+ \Bigg[A_{T+1} \Bigg(x + \sum_{j=1}^{n} m_T^{|T|j} \frac{R_T^{(x)j} + \delta A_{T+1} m_T^{|T|j}}{2\varphi_T^{(x)j}} \\
&+ \sum_{j=1}^{n} \sum_{\tau=T-\omega_1+1}^{T-1} m_T^{|\tau|j} \overline{u}_\tau^{(\omega_1)j} + \sum_{j=1}^{n} \frac{\delta A_{T+1} (b_T^j)^2}{2c_T^{Ij}} x - \lambda_T^x x \Bigg) \\
&+ B_{T+1} \Bigg(y + g_T - \sum_{j=1}^{n} \sigma_T^{|T|j} \frac{R_T^{(y)j} - \delta B_{T+1} \sigma_T^{|T|j}}{2\varphi_T^{(y)j}} \\
&- \sum_{j=1}^{n} \sum_{\tau=T-\omega_2+1}^{T-1} \sigma_T^{|\tau|j} \overline{u}_\tau^{(\omega_2)j} + \sum_{j=1}^{n} \varepsilon_T^j \frac{\delta B_{T+1} \varepsilon_T^j}{2c_T^{IIj}} - \vartheta_T x - \lambda_T^y y \Bigg) \\
&+ C_{T+1} \Bigg(z + \sum_{j=1}^{n} a_T^{|T|j} \frac{\delta C_{T+1} a_T^{|T|j}}{2\varphi_T^{(z)j}} + \sum_{j=1}^{n} \sum_{\tau=T-\omega_3+1}^{T-1} a_T^{|\tau|j} \overline{u}_\tau^{(\omega_3)j} - \lambda_T^z z \Bigg) \\
&+ \varpi_{T+1}^i \Bigg] \delta^T, \quad \text{for } i \in N
\end{aligned} \tag{6.3}$$

System (6.3) yield an equation in which right-hand-side and left-hand-side are linear functions of x, y and z. For (6.3) to hold, it is required that:

$$A_T x = -\sum_{i=1}^{n} \left(\frac{(\delta A_{T+1}^i b_T^i)^2}{4c_T^{Ii}} x + h_T^i x \right) + \delta A_{T+1} \left(\sum_{j=1}^{n} \frac{\delta A_{T+1} (b_T^j)^2}{2c_T^{Ij}} + (1 - \lambda_T^x) \right) x$$
$$- B_{T+1} \vartheta_T x,$$

$$B_T y = \sum_{i=1}^{n} \chi_T^i y + \delta B_{T+1} (1 - \lambda_T^y) y,$$

$$C_T z = \sum_{i=1}^{n} \frac{(R_T^{(z)i})^2 \kappa_T^i z}{4(c_T^{IIIi})} + \delta C_{T+1}(1 - \lambda_T^z)z \quad \text{for } i \in N \tag{6.4}$$

From (6.4) we obtain:

$$A_T^i = -\sum_{i=1}^{n} \left(\frac{(\delta A_{T+1}^i b_T^i)^2}{4c_T^{Ii}} + h_T^i \right) + \delta A_{T+1} \left(\sum_{j=1}^{n} \frac{\delta A_{T+1}(b_T^j)^2}{2c_T^{Ij}} + (1 - \lambda_T^x) \right) - B_{T+1}\vartheta_T$$

$$B_T^i = \sum_{i=1}^{n} \chi_T^i + \delta B_{T+1}(1 - \lambda_T^y)$$

$$C_T^i = \sum_{i=1}^{n} \frac{(R_T^{(z)i})^2 \kappa_T^i}{4(c_T^{IIIi})} + \delta C_{T+1}(1 - \lambda_T^z), \quad \text{for } i \in N. \tag{6.5}$$

In addition,

$$\begin{aligned} D_T = \sum_{i=1}^{n} \Bigg(& \frac{(R_T^{(x)i} - \delta A_{T+1} m_T^{|T|i})(R_T^{(x)i} + \delta A_{T+1} m_T^{|T|i})}{4\varphi_T^{(x)i}} \\ & + \frac{(R_T^{(y)i} + \delta B_{T+1} \sigma_T^{|T|i})(R_T^{(y)i} - \delta B_{T+1} \sigma_T^{|T|i})}{4\varphi_T^{(y)i}} \\ & - \frac{(\delta B_{T+1} \varepsilon_T^i)^2}{4c_T^{IIi}} - \frac{(\delta C_{T+1} a_T^{|T|i})^2}{4\varphi_T^{(z)i}} \Bigg) \\ & + \Bigg[A_{T+1} \left(\sum_{j=1}^{n} m_T^{|T|j} \frac{R_T^{(x)j} + \delta A_{T+1} m_T^{|T|j}}{2\varphi_T^{(x)j}} + \sum_{j=1}^{n} \sum_{\tau=T-\omega_1+1}^{T-1} m_T^{|\tau|j} \overline{u}_\tau^{(\omega_1)j} \right) \\ & + B_{T+1} \Bigg(g_T - \sum_{j=1}^{n} \sigma_T^{|T|j} \frac{R_T^{(y)j} - \delta B_{T+1} \sigma_T^{|T|j}}{2\varphi_T^{(y)j}} \\ & - \sum_{j=1}^{n} \sum_{\tau=T-\omega_2+1}^{T-1} \sigma_T^{|\tau|j} \overline{u}_\tau^{(\omega_2)j} + \sum_{j=1}^{n} \varepsilon_T^j \frac{\delta B_{T+1} \varepsilon_T^j}{2c_T^{IIj}} \Bigg) \\ & + C_{T+1} \left(\sum_{j=1}^{n} a_T^{|T|j} \frac{\delta C_{T+1} a_T^{|T|j}}{2\varphi_T^{(z)j}} + \sum_{j=1}^{n} \sum_{\tau=T-\omega_3+1}^{T-1} a_T^{|\tau|j} \overline{u}_\tau^{(\omega_3)j} \right) + \varpi_{T+1}^i \Bigg] \delta, \end{aligned}$$

for $i \in N$ (6.6)

which is a function of previously executed controls $(\underline{\overline{u}}_{T-}^{(\omega_1)}, \underline{\overline{u}}_{T-}^{(\omega_2)}, \underline{\overline{u}}_{T-}^{(\omega_3)})$.

Then we move to stage $T-1$.

Using $W(T, x, y, z; \underline{\overline{u}}_{T-}^{(\omega_1)}, \underline{\overline{u}}_{T-}^{(\omega_2)}, \underline{\overline{u}}_{T-}^{(\omega_3)}) = (A_T x + B_T y + C_T z + D_T)\delta^{T-1}$ derived in (6.5)–(6.6) and the stage $T-1$ equation in (4.6), the optimal cooperative strategies in stage $T-1$ can be expressed as:

$$\overline{u}_{T-1}^{(\omega_1)i} = \frac{R_{T-1}^{(x)i} + \delta A_T m_{T-1}^{|T-1|i} + \delta^2 A_{T+1} m_T^{|T-1|i}}{2\varphi_{T-1}^{(x)i}},$$

$$\overline{u}_{T-1}^{(\omega_2)i} = \frac{R_{T-1}^{(y)i} - \delta B_T \sigma_{T-1}^{|T-1|i} - \delta^2 B_{T+1} \sigma_T^{|T-1|i}}{2\varphi_{T-1}^{(y)i}},$$

$$\overline{u}_{T-1}^{(\omega_3)i} = \frac{\delta C_T a_{T-1}^{|T-1|i} + \delta^2 C_{T+1} a_T^{|T-1|i}}{2\varphi_{T-1}^{(z)i}}, \quad u_{T-1}^{Ii} = -\frac{\delta A_T b_{T-1}^i}{2c_{T-1}^{Ii}} x^{1/2}$$

$$u_{T-1}^{IIi} = \frac{\delta B_T \varepsilon_{T-1}^i}{2c_{T-1}^{IIi}}, \quad u_{T-1}^{IIIi} = \frac{(R_{T-1}^{(z)i})^2 \kappa_{T-1}^i z}{4(c_{T-1}^{IIIi})^2} \quad \text{for } i \in N \tag{6.7}$$

Substituting the optimal strategies in (6.7) into the stage $T-1$ equation of (4.6) we obtain

$$
\begin{aligned}
& W(T-1, x, y, z; \underline{\overline{u}}_{(T-1)-}^{(\omega_1)}, \underline{\overline{u}}_{(T-1)-}^{(\omega_2)}, \underline{\overline{u}}_{(T-1)-}^{(\omega_3)}) \\
& = (A_{T-1}x + B_{T-1}y + C_{T-1}z + D_{T-1})\delta^{T-2} \\
& = \sum_{i=1}^{n} \left(\frac{(R_{T-1}^{(x)i} - \delta A_T m_{T-1}^{|T-1|i} - \delta^2 A_{T+1} m_T^{|T-1|i|})(R_{T-1}^{(x)i} + \delta A_T m_{T-1}^{|T-1|i} + \delta^2 A_{T+1} m_T^{|T-1|i})}{4\varphi_{T-1}^{(x)i}} \right. \\
& - \frac{(\delta A_T b_{T-1}^i)^2}{4c_{T-1}^{Ii}} x \\
& + \frac{(R_T^{(y)i} + \delta B_T \sigma_{T-1}^{|T-1|i} + \delta^2 B_{T+1} \sigma_T^{|T-1|i})(R_{T-1}^{(y)i} - \delta B_T \sigma_{T-1}^{|T-1|i} - \delta^2 B_{T+1} \sigma_T^{|T-1|i})}{4\varphi_{T-1}^{(y)i}} \\
& - \frac{(\delta B_T \varepsilon_{T-1}^i)^2}{4c_{T-1}^{IIi}} + \frac{(R_{T-1}^{(z)i})^2 \kappa_{T-1}^i z}{4(c_{T-1}^{IIIi})} \\
& \left. - \frac{(\delta C_T a_{T-1}^{|T-1|i} + \delta^2 C_{T+1} a_T^{|T-1|i})^2}{4\varphi_{T-1}^{(z)i}} - h_{T-1}^i x + \chi_{T-1}^i y \right) \delta^{T-2} \\
& + \left[A_T \left(x + \sum_{j=1}^{n} m_{T-1}^{|T-1|j} \frac{R_{T-1}^{(x)j} + \delta A_T m_{T-1}^{|T-1|j} + \delta^2 A_{T+1} m_T^{|T-1|j}}{2\varphi_{T-1}^{(x)j}} \right. \right. \\
& \left. + \sum_{j=1}^{n} \sum_{\tau=T-1-\omega_1+1}^{T-2} m_{T-1}^{|\tau|j} \overline{u}_\tau^{(\omega_1)j} + \sum_{j=1}^{n} \frac{\delta A_T (b_{T-1}^j)^2}{2c_{T-1}^{Ij}} x - \lambda_{T-1}^x x \right)
\end{aligned}
$$

$$+ B_T\left(y + g_{T-1} - \sum_{j=1}^{n} \sigma_T^{|T-1|j} \frac{R_{T-1}^{(y)j} - \delta B_T \sigma_{T-1}^{|T-1|j} - \delta^2 B_{T+1}\sigma_T^{|T-1|j}}{2\varphi_{T-1}^{(y)j}}\right.$$

$$\left. - \sum_{j=1}^{n} \sum_{\tau=T-1-\omega_2+1}^{T-2} \sigma_{T-1}^{|\tau|j} \overline{u}_\tau^{(\omega_2)j} + \sum_{j=1}^{n} \varepsilon_{T-1}^{j} \frac{\delta B_T \varepsilon_{T-1}^{j}}{2c_{T-1}^{IIj}} - \vartheta_{T-1}x - \lambda_{T-1}^{y} y \right.$$

$$+ C_T\left(z + \sum_{j=1}^{n} a_{T-1}^{|T-1|j} \frac{\delta C_T a_{T-1}^{|T-1|j} + \delta^2 C_{T+1} a_T^{|T-1|j}}{2\varphi_{T-1}^{(z)j}}\right.$$

$$\left.\left. + \sum_{j=1}^{n} \sum_{\tau=T-1-\omega_3+1}^{T-2} a_{T-1}^{|\tau|j} \overline{u}_\tau^{(\omega_3)j} - \lambda_{T-1}^{z} z\right) + D_T^{***}\right]\delta^{T-1}, \quad \text{for } i \in N, \tag{6.8}$$

where D_{T+1}^{***} is D_T in (6.6) with

$$\overline{u}_{T-1}^{(\omega_1)j} = \frac{R_{T-1}^{(x)j} + \delta A_T m_{T-1}^{|T-1|j} + \delta^2 A_{T+1} m_T^{|T-1|j}}{2\varphi_{T-1}^{(x)j}},$$

$$\overline{u}_{T-1}^{(\omega_2)j} = \frac{R_{T-1}^{(y)j} - \delta B_T \sigma_{T-1}^{|T-1|j} - \delta^2 B_{T+1}\sigma_T^{|T-1|j}}{2\varphi_{T-1}^{(y)j}},$$

$$\overline{u}_{T-1}^{(\omega_3)j} = \frac{\delta C_T a_{T-1}^{|T-1|j} + \delta^2 C_{T+1} a_T^{|T-1|j}}{2\varphi_{T-1}^{(z)j}}.$$

System (6.8) yields an equation in which right-hand-side and left-hand-side are linear functions of x, y and z. For (6.8) to hold, it is required that:

$$A_{T-1}^{i} = -\sum\nolimits_{i=1}^{n}\left(\frac{(\delta A_T b_{T-1}^{i})^2}{4c_{T-1}^{Ii}} + h_{T-1}^{i}\right)$$

$$+ \delta A_T\left(\sum\nolimits_{j=1}^{n} \frac{\delta A_T (b_{T-1}^{j})^2}{2c_{T-1}^{Ij}} + (1 - \lambda_{T-1}^{x})\right) - B_T\vartheta_{T-1},$$

$$B_{T-1} = \sum\nolimits_{i=1}^{n} \chi_{T-1}^{i} + \delta B_T(1 - \lambda_{T-1}^{y}),$$

$$C_{T-1} = \sum\nolimits_{i=1}^{n} \frac{(R_{T-1}^{(z)i})^2 \kappa_{T-1}^{i}}{4(c_{T-1}^{IIIi})} + \delta C_T(1 - \lambda_{T-1}^{z}), \quad \text{for } i \in N. \tag{6.9}$$

In addition,

$$D_{T-1}$$

$$= \sum_{i=1}^{n}\left(\frac{(R_{T-1}^{(x)i} - \delta A_T m_{T-1}^{|T-1|i} - \delta^2 A_{T+1} m_T^{|T-1|i|})(R_{T-1}^{(x)i} + \delta A_T m_{T-1}^{|T-1|i} + \delta^2 A_{T+1} m_T^{|T-1|i})}{4\varphi_{T-1}^{(x)i}}\right.$$

$$+\frac{(R_T^{(y)i}+\delta B_T\sigma_{T-1}^{|T-1|i}+\delta^2 B_{T+1}\sigma_T^{|T-1|i})(R_{T-1}^{(y)i}-\delta B_T\sigma_{T-1}^{|T-1|i}-\delta^2 B_{T+1}\sigma_T^{|T-1|i})}{4\varphi_{T-1}^{(y)i}}$$

$$-\frac{(\delta B_T\varepsilon_{T-1}^i)^2}{4c_{T-1}^{IIi}}$$

$$-\frac{(\delta C_T a_{T-1}^{|T-1|i}+\delta^2 C_{T+1}a_T^{|T-1|i})^2}{4\varphi_{T-1}^{(z)i}}\Bigg).$$

$$+\Bigg[A_T\Bigg(\sum_{j=1}^n m_{T-1}^{|T-1|j}\frac{R_{T-1}^{(x)j}+\delta A_T m_{T-1}^{|T-1|j}+\delta^2 A_{T+1}m_T^{|T-1|j}}{2\varphi_{T-1}^{(x)j}}$$

$$+\sum_{j=1}^n\sum_{\tau=T-1-\omega_1+1}^{T-2} m_{T-1}^{|\tau|j}\overline{u}_\tau^{(\omega_1)j}\Bigg)$$

$$+B_T\Bigg(g_{T-1}-\sum_{j=1}^n\sigma_{T-1}^{|T-1|j}\frac{R_{T-1}^{(y)j}-\delta B_T\sigma_{T-1}^{|T-1|j}-\delta^2 B_{T+1}\sigma_T^{|T-1|j}}{2\varphi_{T-1}^{(y)j}}$$

$$-\sum_{j=1}^n\sum_{\tau=T-1-\omega_2+1}^{T-2}\sigma_{T-1}^{|\tau|j}\overline{u}_\tau^{(\omega_2)j}+\sum_{j=1}^n\varepsilon_{T-1}^j\frac{\delta B_T\varepsilon_{T-1}^j}{2c_{T-1}^{IIj}}\Bigg)$$

$$+C_T\Bigg(\sum_{j=1}^n a_{T-1}^{|T-1|j}\frac{\delta C_T a_{T-1}^{|T-1|j}+\delta^2 C_{T+1}a_T^{|T-1|j}}{2\varphi_{T-1}^{(z)j}}$$

$$+\sum_{j=1}^n\sum_{\tau=T-1-\omega_3+1}^{T-2} a_{T-1}^{|\tau|j}\overline{u}_\tau^{(\omega_3)j}\Bigg)+D_T^{***}\Bigg]\delta,\quad \text{for } i\in N \tag{6.10}$$

which is a function of previously executed controls $(\underline{\overline{u}}_{(T-1)-}^{(\omega_1)}, \underline{\overline{u}}_{(T-1)-}^{(\omega_2)}, \underline{\overline{u}}_{(T-1)-}^{(\omega_3)})$.

Then, we move to stage $T-2$.

Using $W(T-1,x,y,z;\underline{\overline{u}}_{(T-1)-}^{(\omega_1)}, \underline{\overline{u}}_{(T-1)-}^{(\omega_2)}, \underline{\overline{u}}_{(T-1)-}^{(\omega_3)}) = (A_{T-1}x + B_{T-1}y + C_{T-1}z + D_{T-1})\delta^{T-2}$ derived in (6.9)–(6.10) and the stage $T-2$ equation in (4.6), the optimal cooperative strategies in stage $T-2$ can be expressed as:

$$\overline{u}_{T-2}^{(\omega_1)i}=\frac{R_{T-2}^{(x)i}+\delta A_{T-1}m_{T-2}^{|T-2|i}+\delta^2 A_T m_{T-1}^{|T-2|i}+\delta^3 A_{T+1}m_T^{|T-2|i}}{2\varphi_{T-2}^{(x)i}},$$

$$\overline{u}_{T-2}^{(\omega_2)i}=\frac{R_{T-2}^{(y)i}-\delta B_{T-1}\sigma_{T-2}^{|T-2|i}-\delta^2 B_T\sigma_{T-1}^{|T-2|i}-\delta^3 B_{T+1}\sigma_T^{|T-2|i}}{2\varphi_{T-2}^{(x)i}},$$

$$\overline{u}_{T-2}^{(\omega_3)i} = \frac{\delta C_{T-1} a_{T-2}^{|T-2|i} + \delta^2 C_T a_{T-1}^{|T-2|i} + \delta^3 C_{T+1} a_T^{|T-2|i}}{2\varphi_{T-2}^{(z)i}},$$

$$u_{T-2}^{Ii} = -\frac{\delta A_{T-1} b_{T-2}^i}{2c_{T-2}^{Ii}} x^{1/2},$$

$$u_{T-2}^{IIi} = \frac{\delta B_{T-1} \varepsilon_{T-2}^i}{2c_{T-2}^{IIi}} u_{T-2}^{IIIi} = \frac{(R_{T-2}^{(z)i})^2 \kappa_{T-2}^i z}{4(c_{T-2}^{IIIi})^2} \quad \text{for } i \in N. \tag{6.11}$$

Substituting the optimal strategies in (6.11) into the stage $T-2$ equation of (4.6) we obtain a system of equations in which right-hand-side and left-hand-side are linear functions of x, y and z. For (6.11) to hold, it requires that:

$$A_{T-2} = -\sum_{i=1}^{n} \left(\frac{(\delta A_{T-1} b_{T-2}^i)^2}{4c_{T-2}^{Ii}} + h_{T-2}^i \right)$$
$$+ \delta A_{T-1} \left(\sum_{j=1}^{n} \frac{\delta A_{T-1} (b_{T-2}^j)^2}{2c_{T-2}^{Ij}} + (1 - \lambda_{T-2}^x) \right) - B_{T-1} \vartheta_{T-2},$$

$$B_{T-2} = \sum_{i=1}^{n} \chi_{T-2}^i + \delta B_{T-1}^i (1 - \lambda_{T-2}^y),$$

$$C_{T-2} = \sum_{i=1}^{n} \frac{(R_{T-2}^{(w)i})^2 \kappa_{T-2}^i}{4(c_{T-2}^{(w)i})} + \delta C_{T-1} (1 - \lambda_{T-2}^z), \quad \text{for } i \in N. \tag{6.12}$$

In addition, D_{T-2} is an expression that contains the previously executed controls $(\overline{u}_{(T-2)-}^{(\omega_1)}, \overline{u}_{(T-2)-}^{(\omega_2)}, \overline{u}_{(T-2)-}^{(\omega_3)})$.

Following the above analysis for stage, the optimal strategies in stage $k \in \{T, T-1, \cdots, 1\}$ can be expressed as:

$$\overline{u}_k^{(\omega_1)i} = \frac{R_k^{(x)i} + \sum_{\tau=k}^{k+\omega_1-1} \delta^{\tau-k+1} A_{\tau+1} m_\tau^{|k|i}}{2\varphi_k^{(x)i}}, \quad \text{and } A_\tau = 0 \text{ for } \tau > T+1,$$

$$\overline{u}_k^{(\omega_2)i} = \frac{R_k^{(y)i} - \sum_{\tau=k}^{k+\omega_2-1} \delta^{\tau-k+1} B_{\tau+1} \sigma_\tau^{|k|i}}{2\varphi_k^{(y)i}} \quad \text{and } B_\tau = 0 \text{ for } \tau > T+1$$

$$\overline{u}_k^{(\omega_3)i} = \frac{\sum_{\tau=k}^{k+\omega_3-1} \delta^{\tau-k+1} C_{\tau+1} a_\tau^{|k|i}}{2\varphi_k^{(z)i}}, \quad \text{and } C_\tau = 0 \text{ for } \tau > T+1$$

$$u_k^{Ii} = -\frac{\delta A_{k+1} b_k^i}{2c_k^{Ii}} x^{1/2}, \; u_k^{IIi} = \frac{\delta B_{k+1} \varepsilon_k^i}{2c_k^{IIi}}, \; u_k^{IIIi} = \frac{(R_k^{(z)i})^2 \kappa_k^i z}{4(c_k^{IIIi})^2} \quad \text{for } i \in N \tag{6.13}$$

Substituting the optimal strategies in (6.13) into the stage k equation of (4.6) we obtain

$$W(k,x,y,z;\overline{\underline{u}}_{k-}^{(\omega_1)},\overline{\underline{u}}_{k-}^{(\omega_2)},\overline{\underline{u}}_{k-}^{(\omega_3)}) = (A_k x + B_k y + C_k z + D_k)\delta^{k-1}$$

where

$$A_k = -\sum_{i=1}^{n}\left(\frac{(\delta A_{k+1} b_k^i)^2}{4c_k^{Ii}} + h_k^i\right)$$
$$+\delta A_{k+1}\left(\sum_{j=1}^{n}\frac{\delta A_{k+1}(b_k^j)^2}{2\phi_k^j} + (1-\lambda_k^x)\right) - B_{k+1}\vartheta_k,$$

$$B_k = \sum_{i=1}^{n}\chi_k^i + \delta B_{k+1}(1-\lambda_k^y),$$

$$C_k = \sum_{i=1}^{n}\frac{(R_k^{(z)i})^2\kappa_k^i}{4(c_k^{IIIi})} + \delta C_{k+1}(1-\lambda_k^z),$$

and D_k is an expression which contains the previously executed controls

$$(\overline{\underline{u}}_{k-}^{(\omega_1)},\overline{\underline{u}}_{k-}^{(\omega_2)},\overline{\underline{u}}_{k-}^{(\omega_3)}),\quad \text{for } i\in N \tag{6.14}$$

In particular, the term involving $\overline{\underline{u}}_{k-}^{(\omega_1)}$ in D_k is

$$\delta A_{k+1}\sum_{j=1}^{n}\sum_{\tau=k-\omega_1+1}^{k-1} m_k^{|\tau|j}\overline{u}_\tau^{(\omega_1)j} + \delta^2 A_{k+2}\sum_{j=1}^{n}\sum_{\tau=k-\omega_1+2}^{k-1} m_{k+1}^{|\tau|j}\overline{u}_\tau^{(\omega_1)j}$$
$$+\delta^3 A_{k+3}\sum_{j=1}^{n}\sum_{\tau=k-\omega_1+3}^{k-1} m_{k+2}^{|\tau|j}\overline{u}_\tau^{(\omega_1)j}$$
$$+\delta^4 A_{k+4}\sum_{j=1}^{n}\sum_{\tau=k-\omega_1+4}^{k-1} m_{k+3}^{|\tau|j}\overline{u}_\tau^{(\omega_1)j} + \cdots + \delta^{\omega_1-1} A_{k+\omega_1-1}\sum_{j=1}^{n}\sum_{\tau=k-1}^{k-1} m_{k+\omega_2-2}^{|\tau|j}\overline{u}_\tau^{(\omega_1)j}. \tag{6.15}$$

The term involving $\overline{\underline{u}}_{k-}^{(\omega_2)}$ in D_k is

$$-\Bigg(\delta B_{k+1}\sum_{j=1}^{n}\sum_{\tau=k-\omega_2+1}^{k-1}\sigma_k^{|\tau|j}\overline{u}_\tau^{(\omega_2)j} + \delta^2 B_{k+2}\sum_{j=1}^{n}\sum_{\tau=k-\omega_2+2}^{k-1}\sigma_{k+1}^{|\tau|j}\overline{u}_\tau^{(\omega_2)j}$$
$$+\delta^3 B_{k+3}\sum_{j=1}^{n}\sum_{\tau=k-\omega_2+3}^{k-1}\sigma_{k+2}^{|\tau|j}\overline{u}_\tau^{(\omega_2)j}$$

$$+\delta^4 B_{k+4} \sum_{j=1}^{n} \sum_{\tau=k-\omega_2+4}^{k-1} \sigma_{k+3}^{|\tau|j} \overline{u}_{\tau}^{(\omega_2)j} + \cdots + \delta^{\omega_2-1}$$

$$B_{k+\omega_2-1} \sum_{j=1}^{n} \sum_{\tau=k-1}^{k-1} \sigma_{k+\omega_2-2}^{|\tau|j} \overline{u}_{\tau}^{(\omega_3)j} \Bigg). \tag{6.16}$$

The term involving $\underline{\overline{u}}_{k-}^{(\omega_3)}$ in D_k is

$$\delta C_{k+1} \sum_{j=1}^{n} \sum_{\tau=k-\omega_3+1}^{k-1} a_k^{|\tau|j} \overline{u}_{\tau}^{(\omega_3)j} + \delta^2 C_{k+2} \sum_{j=1}^{n} \sum_{\tau=k-\omega_3+2}^{k-1} a_{k+1}^{|\tau|j} \overline{u}_{\tau}^{(\omega_3)j}$$

$$+ \delta^3 C_{k+3} \sum_{j=1}^{n} \sum_{\tau=k-\omega_3+3}^{k-1} a_{k+2}^{|\tau|j} \overline{u}_{\tau}^{(\omega_3)j}$$

$$+\delta^4 C_{k+4} \sum_{j=1}^{n} \sum_{\tau=k-\omega_3+4}^{k-1} a_{k+3}^{|\tau|j} \overline{u}_{\tau}^{(\omega_3)j} + \cdots + \delta^{\omega_3-1} C_{k+\omega_3-1} \sum_{j=1}^{n} \sum_{\tau=k-1}^{k-1} a_{k+\omega_3-2}^{|\tau|j} \overline{u}_{\tau}^{(\omega_3)j}. \tag{6.17}$$

7 Chapter Notes

This chapter presents a general class of cooperative dynamic environmental games involving anthropogenic eco-degradation and green technology development under durable strategies lags. Cooperative differential games in environmental control had been studied by Dockner and Long (1993), Jørgensen and Zaccour (2001), Fredj et al. (2004), Breton et al. (2005, 2006), Petrosyan and Zaccour (2003), Yeung (2007, 2008) and Yeung and Petrosyan (2008). Cooperative differential games that have identified dynamically consistent solutions can be found in Jørgensen and Zaccour (2001), Petrosyan and Zaccour (2003), Yeung (2007) and Yeung and Petrosyan (2004, 2005, 2006a, b, 2008), Yeung et al. (2017) and Yeung et al. (2021). Yeung (2014) presented a dynamic cooperative game of environmental management in which discrete choices of production techniques are available. Yeung and Petrosyan (2016b) provided the first cooperative dynamic environmental game with technology switching and clean technology development. This chapter enriches the Yeung and Petrosyan (2016b) game with the incorporation of natural resources and their effects on the evolution of the pollution stock. Studies on using green technology to alleviate environmental damages had been presented in Billatos and Basaly (1997), Ane (2000), Kazi and Makhija (2013), Eis et al. (2016), Yeung and Petrosyan (2016b) and Yeung et al. (2017).

Again, the terminal payoffs of the regions can be extended to include previously executed strategies that have effects after stage T. In particular, the terminal payoff of region i in stage $T+1$ can be expressed as

$$\Big(Q_{T+1}^{x(i)}(x_{T+1}) + Q_{T+1}^{y(i)}(y_{T+1}) + Q_{T+1}^{z(i)}(\kappa_{T+1}^{i} z_{T+1}) + v_{T+1}^{x(i)}(u_{(T+1)-}^{(\omega_1)i})$$

$$+v_{T+1}^{y(i)}(\overline{u}_{(T+1)-}^{(\omega_2)i}) + v_{T+1}^{z(i)}(\overline{u}_{(T+1)-}^{(\omega_3)}) + Q_{T+1}^{i}\Big), \quad \text{for } i \in N.$$

An exercise is presented in Problem 1 of Sect. 7.

In addition, we can allow the ecosystem degrading outputs to have two types of durable damages to the environment. The first type of damage involves the damage to the ecosystem and the second type of damages involves environmental effects which affect the regions payoffs directly. Both damages would be in effect for ω_2 stages.

8 Problems

1. (Warmup exercise) Consider a 2-region dynamic environmental game in which each region produces two types of outputs. The first type of output $\overline{u}_k^{(2)i}$ emits pollutants over 2 stages. The second type of output is a clean output u_k^{Ii}.

The pollution stock is $x_k \in X \subset R$ and the dynamics of pollution accumulation is governed by the dynamical equation:

$$x_{k+1} = x_k + \sum_{j=1}^{2} \overline{u}_k^{(2)j} + \sum_{j=1}^{2} 0.2\overline{u}_{k-1}^{(2)j} - \sum_{j=1}^{2} 0.3u_k^{IIj}(x_k)^{1/2} - 0.1\, x_k, \quad x_1 = 10,$$

where u_k^{IIi} is the pollution abatement effort of region i.

The cost of pollution abatement is $2(u_k^{IIi})^2$. The damages from pollutions is $2x_k$ to region 1 and $0.5x_k$ to region 2.

The payoff of region 1 is

$$\sum_{k=1}^{5} \Big(5\overline{u}_k^{(2)1} - 2(\overline{u}_k^{(2)1})^2 + 3u_k^{I1} - (u_k^{I1})^2 - 2(u_k^{II1})^2 - 2x_k\Big)0.9^{k-1}$$
$$+ (-10x_6 + 100)0.9^5.$$

(i) Derive the non-cooperative game equilibrium payoffs of the region. (ii) Obtain the cooperative payoff if the regions agree to cooperate. (iii) Compute the Price of Anarchy (PoA) at stage 1.

2. Consider the Game (3.1)–(3.4) in Sect. 3. The Terminal payoff of Region i in Stage $T+1$ is Replaced by

$$\left(Q_{T+1}^{x(i)} x_{T+1} + Q_{T+1}^{y(i)} y_{T+1} + Q_{T+1}^{z(i)} \kappa_{T+1}^{i} z_{T+1} + \sum_{\tau=T+1-\omega_1+1}^{T} v_{T+1}^{|\tau|x(i)} \overline{u}_{\tau}^{(\omega_1)i} \right.$$

$$\left. + \sum_{\tau=T+1-\omega_2+1} v_{T+1}^{|\tau|y(i)} \overline{u}_{\tau}^{(\omega_2)i} + \sum_{\tau=T+1-\omega_3+1}^{T} {}^{|\tau|z(i)}_{T+1} \overline{u}_{\tau}^{(\omega_3)i} + \varpi_{T+1}^{i} \right),$$

where $Q_{T+1}^{x(i)}$ and $v_{T+1}^{|\tau|x(i)}$ are negative and $Q_{T+1}^{y(i)}$, $Q_{T+1}^{z(i)}$, ϖ_{T+1}^{i}, $v_{T+1}^{|\tau|y(i)}$ and $v_{T+1}^{|\tau|z(i)}$ are positive.

The regions agree to maximize their joint payoff and share the cooperative payoff proportional to their non-cooperative payoff.

Derive (i) the optimal cooperative strategies, (ii) the optimal cooperative joint payoff, (iii) the cooperative state trajectories and the cooperative payoffs of region $i \in N$.

3. Consider a 3-player 8-stage dynamic environmental game in which region $i \in \{1, 2, 3\}$ produces three types of outputs. The first type of output $\overline{u}_k^{(\omega_1)i}$ emits pollutants over 2 stages. The second type of output $\overline{u}_k^{(\omega_2)i}$ causes damages to the ecosystems over 4 stages. The third type of output is a clean output that uses green technologies.

The pollution stock is $x_k \in X \subset R$ and the dynamics of pollution accumulation is governed by the dynamical equation:

$$x_{k+1} = x_k + \sum_{j=1}^{3} \overline{u}_k^{(\omega_1)j} + \sum_{j=1}^{3} \sum_{\tau=k-1}^{k-1} 0.5\overline{u}_{\tau}^{(\omega_1)j} - \sum_{j=1}^{3} 0.1 u_k^{Ij} (x_k)^{1/2} - 0.2 x_k, \quad x_1 = 10$$

where u_k^{Ii} is the pollution abatement effort of region i.

The dynamics governing the evolution of the condition of the ecosystems $y_k \in Y \subset R$ is $y_{k+1} = y_k + 80 - \sum_{j=1}^{3} 1.5\overline{u}_k^{(\omega_2)j} - \sum_{j=1}^{3} \sum_{\tau=k-3}^{k-1} \overline{u}_{\tau}^{(\omega_2)j} + \sum_{j=1}^{3} 0.5 u_k^{IIj} - x_k - 0.05 y_k$, $y_1 = 8000$, where u_k^{IIi} is ecosystem improvement effort.

There is a public green technology $z_k \in Z \subset R$ used to produce the third type of outputs. The dynamics governing the built up of green technology is

$$z_{k+1} = z_k + 0.4 \sum_{j=1}^{3} \overline{u}_k^{(\omega_3)j} + \sum_{j=1}^{3} \sum_{\tau=k-2}^{k-1} 0.2\overline{u}_{\tau}^{(\omega_3)j} - 0.1 z_k \quad z_1 = 100,$$

where $u_k^{(\omega_3)i}$ is the physical investment in green technology which takes three stages to have it fully converted into green technology.

The realized green technology from the public technology for region 1 is $\kappa^1 z_k = 0.8 z_k$, for region 2 is $\kappa^2 z_k = 0.5 z_k$ and for region 3 is $\kappa^3 z_k = 0.1 z_k$. The regions use non-durable inputs c_k^{IIIj}, for $j \in \{1, 2, 3\}$ together with their green technologies to produce clean outputs.

The cost of pollution abatement is $(u_k^{Ii})^2$, the cost of improving the ecosystems is $2({}_k^{IIi})^2$, the non-durable input in green technology production is $0.5 u_k^{IIIi}$ and the investment cost in the public green technology is $3(\overline{u}_k^{(\omega_2)i})^2$. The damages from pollutions is $2x_k$ and the environmental amenities from the ecosystems is $500y_k$.

The payoff of region $i \in \{1, 2, 3\}$ is

$$\sum_{k=1}^{8}\Bigg(8\overline{u}_k^{(\omega_1)i} - 2(\overline{u}_k^{(\omega_1)i})^2 + 5\overline{u}_k^{(\omega_2)i} - (\overline{u}_k^{(\omega_2)i})^2 + 4(u_k^{IIIi})^{1/2}(\kappa_k^i z_k)^{1/2}$$

$$-(u_k^{Ii})^2 - 2(_k^{IIi})^2 - 0.5u_k^{IIIi} - 3(\overline{u}_k^{(\omega_3)i})^2 - 2x_k + 500y_k \Bigg)0.9^{k-1}$$

$$+\Bigg(- 20x_{T+1} + 100y_{T+1} + 5\kappa_{T+1}^i z_{T+1} + 750 \Bigg)0.9^8.$$

Compute the Price of Anarchy (PoA) and the cost of delay from deferring the cooperative scheme from stage 1 to stage $\tau + 1$.

References

Ane RP (2000) Stimulating innovation in green technology. Policy Alter Oppor, Am Behav Sci 44(2):188–212

Billatos S, Basaly N (1997) Green technology and design for the environment. CRC Press, p 312. ISBN 9781560324607

Breton M, Zaccour G, Zahaf M (2005) A differential game of joint implementation of environmental projects. Automatica 41(10):1737–1749

Breton M, Zaccour G, Zahaf M (2006) A game-theoretic formulation of joint implementation of environmental projects. Eur J Oper Res 168(1):221–239

Dockner E, Long NV (1993) International pollution control: cooperative versus noncooperative strategies. J Environ Econ Manag 24:13–29

Eis J, Bishop R, Gradwell P (2016) Galvanising low-carbon innovation. In: A new climate economy working paper for seizing the global opportunity: partnerships for better growth and a better climate. London, New Climate Economy, pp 1–28

Fredj K, Martín-Herrán G, Zaccour G (2004) Slowing deforestation pace through subsidies: a differential game. Automatica 40:301–309. https://doi.org/10.1016/j.automatica.2003.10.020

Jørgensen S, Zaccour G (2001) Time consistent side payments in a dynamic game of downstream pollution. J Econ Dyn Control 25:1973–1987

Kazi AG, Makhija A (2013) Green technology & protection of environment. Published on May 28, 2013. Published in: Technology, Business

Koutsoupias E, Papadimitriou C (2009) Worst-case equilibria. Comput Sci Rev 3(2):65–69. Archived from the original on 2016-03-13. Retrieved 2010-09-12

Petrosyan LA, Zaccour G (2003) Time-consistent shapley value allocation of pollution cost reduction. J Econ Dyn Control 27:381–398

Yeung DWK (2007) Dynamically consistent cooperative solution in a differential game of transboundary industrial pollution. J Optim Theory Appl 134:143–160

Yeung DWK (2014) Dynamically consistent collaborative environmental management with production technique choices. Ann Oper Res 220(1):181–204. https://doi.org/10.1007/s10479-011-0844-0

Yeung DWK, Petrosyan LA (2004) Subgame consistent cooperative solution in stochastic differential games. J Optim Theory Appl 120(3):651–666
Yeung DWK, Petrosyan LA (2005) Subgame consistent solution of a cooperative stochastic differential game with nontransferable payoffs. J Optim Theory Appl 124(3):701–724
Yeung DWK, Petrosyan LA (2006) Dynamically stable corporate joint ventures. Automatica 42:365–370
Yeung DWK, Petrosyan LA (2008) A cooperative stochastic differential game of transboundary industrial pollution. Automatica 44(6):1532–1544
Yeung DWK, Petrosyan LA (2010) subgame consistent solutions for cooperative stochastic dynamic games. J Optim Theory Appl 145:579–596
Yeung DWK, Petrosyan LA (2016a) A cooperative dynamic environmental game of subgame consistent clean technology development. Int Game Theory Rev 18(2):164008.01-164008.23
Yeung DWK, Petrosyan LA (2016b) Subgame consistent cooperation: a comprehensive treatise. Springer
Yeung DWK, Petrosyan LA (2019) Cooperative dynamic games with control lags. Dyn Games Appl 9(2):550–567. https://doi.org/10.1007/s13235-018-0266-6
Yeung DWK, Zhang YX, Bai HT, Islam S (2021) Collaborative environmental management for transboundary air pollution problems: a differential Levies Game. J Ind Manag Optim 17(2):517–531. https://doi.org/10.3934/jimo.2019121
Yeung DWK, Petrosyan LA (2006b) Cooperative stochastic differential games. New York, Springer-Verlag, pp 242. ISBN-10: 0-387-27620-3, e-ISBN: 0-387-27622-X, ISBN-13: 987-0387-27260-5
Yeung DWK, Petrosyan LA, Zhang YX, Cheung F (2017) BEST SCORES solution to the catastrophe bound environment, Nova Science, New York, ISBN 978-1-53610-924-5
Yeung DWK (2008) Dynamically consistent solution for a pollution management game in collaborative abatement with uncertain future payoffs, special issue on frontiers in game theory: in honour of John F. Nash. Int Game Theory Rev 10(4):517–538

Chapter 7
Random Horizon Dynamic Games with Durable Strategies

This chapter presents a class of dynamic games which incorporates two frequently observed real-life phenomena—durable strategies and uncertain horizon. In the presence of durable strategies and random horizon, significant modification of the dynamic optimization techniques is required to accommodate these phenomena. A dynamic optimization theorem is developed and a set of equations characterizing a non-cooperative game equilibrium is derived. A subgame consistent solution for the cooperative game counterpart is obtained with a new theorem for the derivation of a payoff distribution procedure under random horizon and durable strategies. The organization of the Chapter is as follows. Random horizon decision problems with durable strategies are reviewed in Sect. 1. In Section 2, a dynamic optimization theorem for solving problems with durable strategies and random horizon is presented. A general class of general class of non-cooperative dynamic games with random horizon and durable-strategies is formulated in Sect. 3. A computational illustration using an interactive dynamic investments game is given in Sect. 4. A general class of cooperative random horizon durable-strategies dynamic games and their subgame consistent solutions are provided. Computational illustrations for the cooperative game solution are provided in Sect. 6. Mathematical appendices are provided in Sect. 7 and chapter notes are given in Sect. 8.

1 Random Horizon Decision Problems with Durable Strategies

Durable strategies and random horizon are frequently observed in real-life practical decision involving strategic interactions. Revenue generating investments, toxic waste disposal, durable goods, business contracts, taxes, regulations, coalition agreements, diffusion of knowledge, advertising and investments to build up physical capital are vivid examples of durable strategies. In many cases, both the decision-maker's payoff and the evolution of the state dynamics can be subjected to effects

D. W. K. Yeung and L. A. Petrosyan, *Durable-Strategies Dynamic Games*, Theory and Decision Library C 50, https://doi.org/10.1007/978-3-030-92742-4_7

arising from durable strategies. For instance, durable goods and revenue generating investments affect the decision-maker's payoffs for more than one stage. Tariffs have delayed impacts on the economy's growth dynamics. Conversion of investments into physical capital stock involves a certain number of stages and the corresponding costs may be paid by instalment over a certain period of time. Advertising is well-known for having promotion lags under which reputation takes time to establish. Examples can be found in Mela et al. (1997), Tellis (2006) and Ruffino (2008). Sarkar and Zhang (2015) demonstrated implementation lags in investment processes. Agliardi and Koussisc (2013) showed the impacts of time-to-build characteristics in investment on capital formation. Tsuruga and Shota (2019) and Sarkar and Zhang (2013) studied implementation lags which generate lagged effects on fiscal policy and investment decision. Christiano et al. (2005) analyzed the rigidity and dynamic effects of monetary shocks. Shibata et al. (2019) studied current account dynamics under information rigidity and imperfect capital mobility. The existence of at least some durable strategies in the decision-maker's strategy set is a rule rather than an exception in practical analysis.

In addition, the terminal time of many decision problems is not known with certainty. Commonly observed examples of horizontal uncertainties include the terms of offices of elected authorities, contract renewal, continuation of agreements subjected to periodic negotiations and the life span of business firms. A classic case is that business firms are normally hypothesized to have an indefinitely long life expectancy but in reality the average life span of today's multinational Fortune 500-size corporation is just around fifty years. Moreover, the demise of firms is often unpredictable.

Due to the common presence of durable strategies and uncertain horizon in real world interactions, practical situations of random horizon durable strategies dynamic games are prevalent. For instance:

(i) *Climate Change Management*

Damages to the ecosystem and climate change have been a major concern in the international policy arena. Anthropogenic impacts on the environment are not only confined to pollution emission but also ecosystem degradation. Industrial activities damaging the ecosystem in the current stage—like, toxic wastes released, deforestation and biodiversity loss—would continue to produce harmful effects in many subsequent stages. These industrial activities can produce continual effects to the evolution dynamics of the state of the ecosystem and the accumulation process of the pollutant stock in many stages afterwards. On the other hand, the build up of green technology to alleviate the environment problem is subjected to lengthy innovation processes. At the same time, industrial activities can produce direct durable and non-durable impacts on the payoffs of the agents. In addition, the current regime (game structure) would cease when new regulatory measures are introduced in the future. The Kyoto Protocol, Bali Road Map, Copenhagen Accord, Cancun Agreements, Doha Climate Gateway and Warsaw Outcomes were examples of regimes with uncertain horizons. Even the existing Paris Agreement has an uncertain horizon.

Modelling climate change policy in a random horizon durable strategies dynamic game framework would yield new and potentially significant results.

(ii) *Political Unions*

Political unions, like the EU, are vivid examples of random horizon durable strategies dynamic games. The EU is an international union which focuses on governing through common economic, social and security policies. Tariff agreements, different kinds of investments and common currency arrangement are durable economic policy strategies. Citizenships, people mobility, education, legal system, refugee quotas and immigration services are durable social policy strategies. Border definition and defense system are durable security policy strategies. The growth in economic state variables, like capital stocks, production plants, technology and long-term investments, are subjected to the effects of durable strategies. The transformation in social state variables, like education and population of immigrants, are subjected to the impacts of durable strategies. The change in security and defense state variables, like arm forces and weapon stock, are subjected to durable controls. Nevertheless, the horizon of a political union can be subjected to random elements. Brexit is an example of an unpredictable exit of a union member which would put a halt on the continuation of the current structure of the regime.

(iii) *International Trade*

Global trade of goods and services are worth trillions of dollars annually. The benefits from international trade coming from gains in specialization, technology spillover and economies of scale occupy a large proportion of income generated globally. Under globalization, physical capital buildup, knowledge capital formation, technology transfer, public capital enlargement are state dynamics which are affected by durable investments. Company goodwill, brand reputation and client base are state variables which evolution are affected by durable controls like customer relation building, advertisement and social responsibilities promotion. Debt financing and income generating investments under global trade are durable strategies which affect the payoff of trading corporations. Trade wars and trade disputes would alter the existing global trade environment. New trade pacts would result in new business relationships and trading networks. International trade can indeed be regarded as a random horizon durable strategies dynamic games.

(iv) *Public Knowledge-based Capital Provision*

The economic growth of many nations came largely from knowledge-based economies. Many knowledge-based capitals and technologies have a certain degree of public goods property in that other parties can benefit (at least partially) from the technologies developed by the originators. Since diffusion of knowledge cannot be completed instantaneously, many investments in knowledge-based capital may take time to be completely converted into applicable technologies. There also exist unpredictable future events that would end the current structure of operation. These events include the discovery of a new technology, an agreement to develop the technology cooperatively and the filing of patents that restrict the use the spillover part of certain knowledge-based technology.

(v) *Disarmament Initiatives*

Disarmament or arms control attempts to impose restrictions upon the development, production, stockpiling, proliferation of conventional weapons and weapons of mass destruction. The strategies involved in arm race are often durable. Weapon development takes a considerably long time to transform into usable military equipment. Military threats can prevail for a lengthy time. Heavy spending on arms development would lead to financial burden in the future while the resulting damage to the environment can be lasting. The state variables include a vector of stockpiling of weapons which accumulation is affected by durable investments. A nation's weapon stocks yield defense security to itself and its allies over a period of time. But such weapon stocks would cause threats to confronting nations. Moreover, random ending of arms control agreement is not uncommon. The Intermediate-range Nuclear Forces (INF) ended recently due to the failure of the coming up of a US-Russian agreement to keep it alive. On the other hand, the possibility of new military confrontations would cause the life span of the existing military equilibrium to be a random element.

2 Random Horizon Dynamic Optimization under Durable Strategies

In this section, we develop a dynamic optimization technique for solving intertemporal problems with random horizon and durable strategies which will serve as the foundation of solving the game problem. Consider a durable strategies dynamic optimization problem with a random horizon of $\hat{T}$ stages, where $\hat{T}$ is a random variable with range $\{1, 2, \cdots, T\}$ and corresponding probabilities $\{\theta_1, \theta_2, \cdots, \theta_T\}$. Conditional upon the reaching of stage τ, the probability of the game would last up to stages $\tau, \tau+1, \cdots, T$ becomes respectively.

$$\frac{\theta_\tau}{\sum_{\zeta=\tau}^{T} \theta_\zeta}, \frac{\theta_{\tau+1}}{\sum_{\zeta=\tau}^{T} \theta_\zeta}, \cdots, \frac{\theta_T}{\sum_{\zeta=\tau}^{T} \theta_\zeta}.$$

There exist multiple durable strategies of different lag durations which affect the payoff and the state dynamics or both. We use $u_k \in U \subset R^m$ to denote the set of non-durable control strategies. We use $\overline{u}_k = (\overline{u}_k^{(2)}, \overline{u}_k^{(3)}, \cdots, \overline{u}_k^{(\omega)})$ to denote the set of durable control strategies, where $\overline{u}_k^{(\zeta)} \in \overline{U}^\zeta \subset R^{m_\zeta}$ for $\zeta \in \{2, 3, \cdots, \omega\}$. In particular, the strategies $\overline{u}_k^{(2)}$ are durable strategies that have effects in stages k and $k+1$. The strategies $\overline{u}_k^{(3)}$ are durable strategies that have effects within stages k, $k+1$ and $k+2$. The strategies $u_k^{(\omega)}$ are durable strategies that have effects within the duration from stages k to stage $k+\omega-1$. The single-stage payoff received in stage k can then be expressed as $g_k(x_k, u_k, \overline{u}_k; \overline{u}_{k-})$, where $x_k \in X \subset R^m$ is the state at stage k and $\overline{u}_{k-}$ is the set of durable controls which are executed before stage k but still in effect in stage k.

If the plan ends after stage $\hat{T}$, the decision-maker will receive a terminal payment $q_{\hat{T}+1}(x_{\hat{T}+1}; \overline{u}_{(\hat{T}+1)-})$ in stage $\hat{T}+1$, which can be zero, positive (a salvage value) or negative (a penalty).

The state dynamics is.

$$x_{k+1} = f_k(x_k, u_k, \overline{u}_k; \overline{u}_{k-}), \quad x_1 = x_1^0. \tag{2.1}$$

The expected payoff to be maximized becomes

$$E\left\{\sum_{k=1}^{\hat{T}} g_k(x_k, u_k, \overline{u}_k; \overline{u}_{k-})\delta_1^k + q_{\hat{T}+1}(x_{\hat{T}+1}; \overline{u}_{(\hat{T}+1)-})\delta_1^{\hat{T}+1}\right\}$$
$$= \sum_{\hat{T}=1}^{T} \theta_{\hat{T}} \left\{\sum_{k=1}^{\hat{T}} g_k(x_k, u_k, \overline{u}_k; \overline{u}_{k-})\delta_1^k + q^i_{\hat{T}+1}(x_{\hat{T}+1}; \overline{u}_{(\hat{T}+1)-})\delta_1^{\hat{T}+1}\right\}, \tag{2.2}$$

where δ_1^k is the discount factor from stage 1 to stage k.

The controls executed before the start of the operation in stage 1, that is u_{1-}, are known and some or all of them can be zeros. The functions $g_k(x_k, u_k, \overline{u}_k; \overline{u}_{k-})$, $f_k(x_k, u_k, \overline{u}_k; \overline{u}_{k-})$ and $q_{\hat{T}+1}(x_{\hat{T}+1}; \overline{u}_{(\hat{T}+1)-})$ are differentiable functions.

Now consider the case when stage τ has arrived with the state being x_τ and the previously executed durable strategies being $\overline{u}_{\tau-}$. The problem can be formulated as the maximization of the payoff:

$$E\left\{\sum_{k=\tau}^{\hat{T}} g_k(x_k, u_k, \overline{u}_k; \overline{u}_{k-})\delta_1^k + q_{\hat{T}+1}(x_{\hat{T}+1}; \overline{u}_{(\hat{T}+1)-})\delta_1^{\hat{T}+1}\right\}$$
$$= \sum_{\hat{T}=\tau}^{T} \frac{\theta_{\hat{T}}}{\sum_{\zeta=\tau}^{T}\theta_\zeta} \left\{\sum_{k=\tau}^{\hat{T}} g_k(x_k, u_k, \overline{u}_k; \overline{u}_{k-})\delta_1^k + q_{\hat{T}+1}(x_{\hat{T}+1}; \overline{u}_{(\hat{T}+1)-})\delta_1^{\hat{T}+1}\right\}, \tag{2.3}$$

subject to the dynamics

$$x_{k+1} = f_k(x_k, u_k, \overline{u}_k; \overline{u}_{k-}), \quad \text{for } k \in \{\tau, \tau+1, \cdots, T\}. \tag{2.4}$$

A theorem characterizing an optimal solution to the random horizon problem (2.1)–(2.2) is provided below.

Theorem 2.1. *Let the value function $V(\tau, x; \overline{u}_{\tau-})$ denote the maximal value of the expected payoffs*

$$E\left\{\sum_{k=\tau}^{\hat{T}} g_k(x_k, u_k, \overline{u}_k; \overline{u}_{k-})\delta_1^k + q_{\hat{T}+1}(x_{\hat{T}+1}; \overline{u}_{(\hat{T}+1)-})\delta_1^{\hat{T}+1}\right\}$$

$$= \sum_{\hat{T}=\tau}^{T} \frac{\theta_{\hat{T}}}{\sum_{\zeta=\tau}^{T}\theta_\zeta}\left\{\sum_{k=\tau}^{\hat{T}} g_k(x_k, u_k, \overline{u}_k; \overline{u}_{k-})\delta_1^k + q_{\hat{T}+1}(x_{\hat{T}+1}; \overline{u}_{(\hat{T}+1)-})\delta_1^{\hat{T}+1}\right\},$$

for problem (2.3)–(2.4) *starting at stage* τ *with state* $x_\tau = x$ *and previously executed controls* $u_{\tau-}$. *Then the function* $V(\tau, x; \overline{u}_{\tau-})$ *satisfies the following system of recursive equations*:

$$V(T+1, x; \overline{u}_{(T+1)-}) = q_{T+1}(x; \overline{u}_{(T+1)-})\delta_1^{T+1}, \tag{2.5}$$

$$\begin{aligned} V(T, x; \overline{u}_{T-}) = \max_{u_T, \overline{u}_T}\big\{&g_T(x, u_T, \overline{u}_T; \overline{u}_{T-})\delta_1^T \\ &+q_{T+1}[f_T(x, u_T, \overline{u}_T; \overline{u}_{T-}); \overline{u}_T, \overline{u}_{(T+1)-} \cap \overline{u}_{T-}]\delta_1^{T+1}\big\}, \end{aligned} \tag{2.6}$$

$$\begin{aligned} V(\tau, x; \overline{u}_{\tau-}) = \max_{u_\tau, \overline{u}_\tau}\Big\{&g_\tau(x, u_\tau, \overline{u}_\tau; \overline{u}_{\tau-})\delta_1^\tau \\ &+ \frac{\theta_\tau}{\sum_{\zeta=\tau}^{T}\theta_\zeta} q_{\tau+1}[f_T(x, u_\tau, \overline{u}_\tau; \overline{u}_{\tau-}); \overline{u}_\tau, \overline{u}_{(\tau+1)-} \cap \overline{u}_{\tau-}]\delta_1^{\tau+1} \\ &+ \frac{\sum_{\zeta=\tau+1}^{T}\theta_\zeta}{\sum_{\zeta=\tau}^{T}\theta_\zeta} V(\tau+1, f_\tau(x, u_\tau, \overline{u}_\tau; \overline{u}_{\tau-}); \overline{u}_\tau, \overline{u}_{(\tau+1)-} \cap \overline{u}_{\tau-}]\Big\}, \\ &\text{for } \tau \in \{1, 2, \cdots, T-1\}. \end{aligned} \tag{2.7}$$

Proof See Appendix I. ■

Theorem 2.1 yields a novel set of Bellman equations for the random horizon optimization problems under durable strategies. According to the Bellman (1957) standard dynamic programming technique, the controls executed in stage k will affect the state x_{k+1} in stage $k+1$ through the dynamic equation. In Theorem 2.1, the current state x_k and previously executed controls $\overline{u}_{k-}$ appear as given in the stage k optimization problem. However, the previously executed controls $\overline{u}_{k-}$ have no transition equations governing their transition from one stage to another, but they will last for some finite stages. They act like a vector of idiosyncratic state variables which affects the state dynamics (2.1) and the payoff function (2.2) when their impacts are still effective. In addition, the theorem also takes care of the combined effects of random horizon in the problem.

3 Random Horizon Durable Strategies Dynamic Games

In this section, we first formulate a general class of random horizon dynamic games with durable strategies. Then, we characterize the non-cooperative game equilibrium.

3.1 Game Formulation

Consider the $n-$ person dynamic game with $\hat{T}$ stages where $\hat{T}$ is a random variable with range $\{1, 2, \cdots, T\}$ and corresponding probabilities $\{\theta_1, \theta_2, \cdots, \theta_T\}$. Conditional upon the reaching of stage τ, the probability of the game would last up to stages $\tau, \tau+1, \cdots, T$ becomes respectively

$$\frac{\theta_\tau}{\sum_{\zeta=\tau}^{T}\theta_\zeta}, \frac{\theta_{\tau+1}}{\sum_{\zeta=\tau}^{T}\theta_\zeta}, \cdots, \frac{\theta_T}{\sum_{\zeta=\tau}^{T}\theta_\zeta}.$$

There exist durable and non-durable strategies affecting the players' payoffs and the state dynamics. We use $u_k^i \in U^i \subset R^{m^i}$ to denote the set of non-durable control strategies of player i. We use $\overline{u}_k^i = (\overline{u}_k^{(2)i}, \overline{u}_k^{(3)i}, \cdots, \overline{u}_k^{(\omega_i)i})$ to denote the set of durable strategies of player i, where $\overline{u}_k^{(\zeta)i} \in \overline{U}^{(\zeta)i} \subset R^{m_{(\zeta)i}}$ for $\zeta \in \{2, 3, \cdots, \omega_i\}$. In particular, $\overline{u}_k^{(2)i}$ are non-durable strategies that have effects in stages k and $k+1$. The strategies $\overline{u}_k^{(3)i}$ are durable strategies that have effects within stage k to stage $k+2$. The strategies $\overline{u}_k^{(\omega)i}$ are durable strategies that have effects within stages k, $k+1, \cdots, k+\omega-1$. The state at stage k is $x_k \in X \subset R^m$ and the state space is common for all players. The single-stage payoff of player i in stage k is

$$g_k^i(x_k, \underline{u}_k, \underline{\overline{u}}_k; \underline{\overline{u}}_{k-}), \text{for } k \in \{1, 2, \cdots, T\} \text{ and } i \in \{1, 2, \cdots, n\} \equiv N,$$

where $\underline{u}_k = (u_k^1, u_k^2, \cdots, u_k^n)$ is the set of durable strategies of all the n players, $\underline{\overline{u}}_k = (\overline{u}_k^1, \overline{u}_k^2, \cdots, \overline{u}_k^n)$ is the set of durable strategies of all the n players and $\underline{\overline{u}}_{k-} = (\overline{u}_{k-}^1, \overline{u}_{k-}^2, \cdots, \overline{u}_{k-}^n)$ is the set of strategies which are executed before stage k by all players but still in effect in stage k.

If the game ends after stage $\hat{T}$, player i will receive a terminal payment $q_{\hat{T}+1}^i(x_{\hat{T}+1}; \overline{u}_{(\hat{T}+1)-})$ in stage $\hat{T}+1$, which can be zero, positive (a salvage value) or negative (a penalty).

The expected payoff of player i is

$$E\left\{\sum_{k=1}^{\hat{T}} g_k^i(x_k, \underline{u}_k, \underline{\overline{u}}_k; \underline{\overline{u}}_{k-})\delta_1^k + q_{\hat{T}+1}^i(x_{\hat{T}+1}; \underline{\overline{u}}_{(\hat{T}+1)-})\delta_1^{\hat{T}+1}\right\}$$

$$= \sum_{\hat{T}=1}^{T} \theta_{\hat{T}} \left\{ \sum_{k=1}^{\hat{T}} g_k^i(x_k, \underline{u}_k, \overline{u}_k; \overline{u}_{k-}) \delta_1^k + q_{\hat{T}+1}^i(x_{\hat{T}+1}; \overline{u}_{(\hat{T}+1)-}) \delta_1^{\hat{T}+1} \right\}, \quad \text{for } i \in N, \tag{3.1}$$

where δ_1^k is the discount factor from stage 1 to stage k.

The state dynamics is characterized by a vector of difference equations:

$$x_{k+1} = f_k(x_k, \underline{u}_k, \overline{u}_k; \overline{u}_{k-}), \quad x_1 = x_1^0, \tag{3.2}$$

for $k \in \{1, 2, \cdots, T\}$.

The controls executed before the start of the operation in stage 1, that is $\overline{u}_{1-}$, are known and some or all of them can be zeros. The $g_k^i(x_k, \underline{u}_k, \overline{u}_k; \overline{u}_{k-})$, $f_k^i(x_k, \underline{u}_k, \overline{u}_k; \overline{u}_{k-})$ and $q_{\hat{T}+1}^i(x_{\hat{T}+1}; \overline{u}_{(\hat{T}+1)-})$ are continuously differentiable functions.

Now consider the case when stage τ has arrived with the state being x_τ and the previously executed durable strategies $\overline{u}_{\tau-}$. Then it becomes a game in which the payoff of player i is

$$E\left\{ \sum_{k=\tau}^{\hat{T}} g_k^i(x_k, \underline{u}_k, \overline{u}_k; \overline{u}_{k-}) \delta_1^k + q_{\hat{T}+1}^i(x_{\hat{T}+1}; \overline{u}_{(\hat{T}+1)-}) \delta_1^{\hat{T}+1} \right\}$$

$$= \sum_{\hat{T}=\tau}^{T} \frac{\theta_{\hat{T}}}{\sum_{\zeta=\tau}^{T} \theta_\zeta} \left\{ \sum_{k=\tau}^{\hat{T}} g_k^i(x_k, \underline{u}_k, \overline{u}_k; \overline{u}_{k-}) \delta_1^k + q_{\hat{T}+1}^i(x_{\hat{T}+1}; \overline{u}_{(\hat{T}+1)-}) \delta_1^{\hat{T}+1} \right\}, \quad \text{for } i \in N, \tag{3.3}$$

and the state dynamics are

$$x_{k+1} = f_k(x_k, \underline{u}_k, \overline{u}_k; \overline{u}_{k-}), \text{ for } k = \{\tau, \tau+1, \cdots, T\}, \; x_\tau = x. \tag{3.4}$$

The information set of every player includes the knowledge in

(i) all the possible moves by himself and other players, that is $(u_k^i, \overline{u}_k^i)$, for $k \in \{1, 2, \cdots, T\}$ and $i \in N$;
(ii) the set of controls which are executed before stage k by all players but still in effect in stage k, that is $\overline{u}_{k-} = (\overline{u}_{k-}^1, \overline{u}_{k-}^2, \cdots, \overline{u}_{k-}^n)$, for $k \in \{1, 2, \cdots, T\}$;
(iii) the state dynamics $x_{k+1} = f_k(x_k, \underline{u}_k, \overline{u}_k; \overline{u}_{k-})$ and the values of present and past states $(x_k, x_{k-1}, \cdots, x_1)$;
(iv) the payoff functions of all players $g_k^i(x_k, \underline{u}_k, \overline{u}_k; \overline{u}_{k-})$, for $i \in N$ and $k \in \{1, 2, \cdots, T\}$;
(v) the knowledge of the random variable $\hat{T}$ with range $\{1, 2, \cdots, T\}$ and corresponding probabilities $\{\theta_1, \theta_2, \cdots, \theta_T\}$; and
(vi) the terminal payment $q_{\hat{T}+1}^i(x_{\hat{T}+1}; \overline{u}_{(\hat{T}+1)-})$ in stage $\hat{T}+1$, for $i \in N$, if the game terminates after stage $\hat{T}$.

3.2 Non-cooperative Equilibrium

In this subsection, we investigate the non-cooperative outcome of the random horizon durable strategies dynamic game (3.1)–(3.2). In particular, a feedback Nash equilibrium of the game can be characterized by the following theorem.

Theorem 3.1. *Let* $\{\underline{u}_\tau^{**}, \overline{u}_\tau^{**}\}$ *denote the set of the players' feedback Nash equilibrium strategies and* $V^i(\tau, x; \underline{\overline{u}}_{\tau-}^{**})$ *denote the feedback Nash equilibrium payoff of player* i *in the non-cooperative game* (3.3)–(3.4), *then the function* $V^i(\tau, x; \underline{\overline{u}}_{\tau-}^{**})$ *satisfies the following system of recursive equations*

$$V^i(T+1, x; \underline{\overline{u}}_{(T+1)-}^{**}) = q_{T+1}^i(x; \underline{\overline{u}}_{(T+1)-}^{**})\delta_1^{T+1}, \tag{3.5}$$

$$\begin{aligned} &V^i(T, x; \underline{\overline{u}}_{T-}^{**}) \\ &= \max_{u_T^i, \overline{u}_T^i} \Big\{ g_T^i(x, u_T^i, \overline{u}_T^i, \underline{u}_T^{(\neq i)**}, \overline{u}_T^{(\neq i)**}; \underline{\overline{u}}_{T-}^{**})\delta_1^T \\ &\quad + q_{T+1}^i[f_T(x, u_T^i, \overline{u}_T^i, \underline{u}_T^{(\neq i)**}, \overline{u}_T^{(\neq i)**}; \underline{u}_{T-}^{**}); \overline{u}_T^i, \overline{u}_T^{(\neq i)**}, \underline{\overline{u}}_{(T+1)-} \cap \underline{\overline{u}}_{T-}^{**}]\delta_1^{T+1} \Big\}, \end{aligned} \tag{3.6}$$

$$\begin{aligned} &V^i(\tau, x; \underline{\overline{u}}_{\tau-}^{**}) \\ &= \max_{u_\tau^i, \overline{u}_\tau^i} \Big\{ g_\tau^i(x, u_\tau^i, \overline{u}_\tau^i, \underline{u}_\tau^{(\neq i)**}, \overline{u}_\tau^{(\neq i)**}; \underline{\overline{u}}_{\tau-}^{**})\delta_1^\tau \\ &\quad + \frac{\theta_\tau}{\sum_{\zeta=\tau}^T \theta_\zeta} q_{\tau+1}^i[f_\tau(x, u_T^i, \overline{u}_T^i, \underline{u}_T^{(\neq i)**}, \overline{u}_T^{(\neq i)**}; \underline{u}_{\tau-}^{**}); \overline{u}_\tau^i, \overline{u}_\tau^{(\neq i)**}, \underline{\overline{u}}_{(\tau+1)-}^{**} \cap \underline{\overline{u}}_{\tau-}^{**}]\delta_1^{\tau+1} \\ &\quad + \frac{\sum_{\zeta=\tau+1}^T \theta_\zeta}{\sum_{\zeta=\tau}^T \theta_\zeta} V^i[\tau+1, f_\tau(x, u_\tau^i, \overline{u}_\tau^i, \underline{u}_\tau^{(\neq i)**}, \overline{u}_\tau^{(\neq i)**}; \underline{u}_{\tau-}^{**}); \overline{u}_\tau^i, \overline{u}_\tau^{(\neq i)**}, \underline{\overline{u}}_{(\tau+1)-}^{**} \cap \underline{\overline{u}}_{\tau-}^{**}] \Big\}, \\ &\text{for } \tau \in \{1, 2, \cdots, T-1\}, \end{aligned} \tag{3.7}$$

where $\underline{u}_\tau^{(\neq i)**} = \underline{u}_\tau^{**} \backslash u_\tau^{i**}$ and $\overline{u}_\tau^{(\neq i)**} = \overline{u}_\tau^{**} \backslash \overline{u}_\tau^{i**}$.

Proof The conditions in (3.5)–(3.7) shows that the optimal random horizon dynamic optimization result under durable strategies in Theorem 2.1 holds for each player given other players' equilibrium strategies $(\underline{u}_\tau^{(\neq i)**}, \overline{u}_\tau^{(\neq i)**})$, for $\tau \in \{1, 2, \cdots, T-1\}$. Hence the conditions of a Nash (1951) equilibrium are satisfied and Theorem 3.1 follows. ∎

Theorem 3.1 is a novel solution technique for the characterization of a feedback Nash equilibrium in a dynamic game. The set of equations in (3.5)–(3.7) represents the random horizon durable strategies analogue of the Isaacs-Bellman equations in a feedback Nash game equilibrium.

Substituting the set of feedback Nash equilibrium strategies $(\underline{u}_k^{**}, \overline{u}_k^{**})$, for $k \in \{1, 2, \cdots, T\}$ from Theorem 3.1 into state dynamics (3.1) yields the game equilibrium dynamics

$$x_{k+1} = f_k(x_k, \underline{u}_k^{**}; \underline{u}_{k-}^{**}), \quad x_1 = x_1^0. \tag{3.8}$$

Substituting the set of feedback Nash equilibrium strategies $\underline{u}_k^{**}$, for $k \in \{1, 2, \cdots, T\}$ into the player i's payoff yields

$$V^i(\tau, x; \underline{u}_{\tau-}^{**})$$
$$= E\left\{\sum_{k=\tau}^{\hat{T}} g_k^i(x_k, \underline{u}_k^{**}, \overline{u}_k^{**}; \overline{u}_{k-}^{**})\delta_1^k + q_{\hat{T}+1}^i(x_{\hat{T}+1}; \overline{u}_{(\hat{T}+1)-}^{**})\delta_1^{\hat{T}+1}\right\}$$
$$= \sum_{\hat{T}=\tau}^{T} \frac{\theta_{\hat{T}}}{\sum_{\zeta=\tau}^{T} \theta_\zeta} \left\{\sum_{k=\tau}^{\hat{T}} g_k^i(x_k, \underline{u}_k^{**}, \overline{u}_k^{**}; \overline{u}_{k-}^{**})\delta_1^k + q_{\hat{T}+1}^i(x_{\hat{T}+1}; \overline{u}_{(\hat{T}+1)-}^{**})\delta_1^{\hat{T}+1}\right\},$$
$$\text{for } i \in N. \tag{3.9}$$

The value function $V^i(\tau, x; \overline{u}_{\tau-}^{**})$ gives the expected game equilibrium payoff to player i from stage τ to the end of the game.

4 A Computational Illustration in Dynamic Investments

In this section, we present a computational illustration in interactive investments with technology spillover to demonstrate the computation of the non-cooperative equilibrium in a dynamic game under random horizon with durable strategies.

4.1 Game Formulation

Consider the case of two economic agents (regions or firms) which are given a certain lease to trade with each other. The lease for trade has to be renewed after each stage (year or decade) for up to a maximum of 10 stages. At stage 1, it is known that the probabilities that the lease will last for $1, 2, 3, \cdots, 10$ stages with respective probabilities $\theta_1, \theta_2, \theta_3, \cdots, \theta_{10}$. Conditional upon the reaching of stage $\tau > 1$, the probability of the game would last up to stages $\tau, \tau + 1$, to 10 are

$$\frac{\theta_\tau}{\sum_{\zeta=\tau}^{10} \theta_\zeta}, \frac{\theta_{\tau+1}}{\sum_{\zeta=\tau}^{10} \theta_\zeta}, \frac{\theta_{\tau+2}}{\sum_{\zeta=\tau}^{10} \theta_\zeta}, \cdots, \frac{\theta_{10}}{\sum_{\zeta=\tau}^{10} \theta_\zeta}.$$

Each agent has a private production capital $x_k^i \in X^i \subset R$, for $i \in \{1, 2\}$. Investment in private capital takes three stages to complete its transformation into the agent's capital stock and we use $u_k^{(3)i}$ to denote the investment in the private capital by agent i in stage k. The increment in the private capital in stage k brought about by agent i's investment $\overline{u}_k^{(3)i}$ in stage $t \in \{k, k-1, k-2\}$ is $\varepsilon_k^{|t|i} \overline{u}_t^{(3)i}$. Technology spillover is a common phenomenon among firms (see Madsen (2007)). Cellini and Lambertini (2003, 2009) were among the first in analyzing technology spillover in a dynamic game framework. In this illustration, we model technology spillover in a way such that the private capital of an agents can positively affect the increment of the other agent's private capital. The spillover of agent j's private capital to the increment of agent i's private capital stock is $\gamma_k^{(i)j} x_k^j$. The terms $\varepsilon_k^{|\tau|i}$ and $\gamma_k^{(i)j}$ are positive parameters.

The accumulation process of the private capital stock of agent i is governed by the dynamical equation

$$\begin{aligned} x_{k+1}^i &= x_k^i + \varepsilon_k^{|k|i} \overline{u}_k^{(3)i} + \sum\nolimits_{t=k-2}^{k-1} \varepsilon_k^{|t|i} \overline{u}_t^{(3)i} - \lambda_k^i x_k^i + \gamma_k^{(i)j} x_k^j, \\ x_1^i &= x_1^{i(0)}, \; i, j \in \{1, 2\} \text{ and } i \neq j, \end{aligned} \tag{4.1}$$

where λ_k^i is the depreciation rate of capital x_k^i.

Each firm also has a revenue generating investment which yields an income stream over four stages. We use $\overline{u}_k^{(4)i}$ to denote agent i's revenue generating investment in stage k. The income generated by agent i's stage k generating investment is $p_t^{|k|i} \overline{u}_k^{(4)i}$ in stage $t \in \{k, k+1, k+2, k+3\}$. The cost of the revenue generating investment $\overline{u}_k^{(4)i}$ is $\phi_k^i (\overline{u}_k^{(4)i})^2$, with ϕ_k^i being a positive parameter. The cost of the capital investment $\overline{u}_k^{(3)i}$ is $\varphi_k^{(x)i} (\overline{u}_k^{(3)i})^2$, with $\varphi_k^{(x)i}$ being a positive parameter.

Agent i uses a non-durable strategy u_k^i along with its private capital x_k^i to produce an output. The unit cost of the non-durable strategy is c_k^i. The total operating profit of agent i from trade in stage k is $R_k^i (u_k^i x_k^i)^{1/2} - c_k^i u_k^i$, where $(u_k^i x_k^i)^{1/2}$ is the production function of the output.

If the agent is still in operation in stage k, the single-stage profit of agent i is

$$R_k^i (u_k^i x_k^i)^{1/2} - c_k^i u_k^i - \varphi_k^{(x)i} (\overline{u}_k^{(3)i})^2 - \phi_k^i (\overline{u}_k^{(4)i})^2 + \sum\nolimits_{t=k-3}^{k} p_k^{|t|i} \overline{u}_t^{(4)i}. \tag{4.2}$$

If the agents remain in operation until stage 10, agent $i \in \{1, 2\}$ will receive a terminal payoff in stage 11 equalling

$$q^i(11, x; \underline{\overline{u}}_{11-}^{(3)}, \underline{\overline{u}}_{11-}^{(4)}) = \left(Q_{11}^{(i)i} x_{11}^i + \sum_{t=11-2}^{10} \overline{v}_{11}^{|t|i} \overline{u}_t^{(3)i} + \sum_{t=11-3}^{10} \overline{p}_{11}^{|t|i} \overline{u}_t^{(4)i} + \varpi_{11}^i \right),$$

where $\underline{\overline{u}}_{11-}^{(3)} = (\overline{u}_{11-}^{(3)1}, \overline{u}_{11-}^{(3)2})$, $\dot{\underline{u}}_{11-}^{(4)} = (\overline{u}_{11-}^{(4)1}, \overline{u}_{11-}^{(4)2})$, $\overline{v}_{11}^{|t|i} \overline{u}_t^{(3)i}$ is the reward in stage 11 paid to capital investment $\overline{u}_t^{(3)i}$ and $\overline{p}_{11}^{|t|i} \overline{u}_t^{(4)i}$ is the return in stage 11 from the revenue generating investment $\overline{u}_t^{(4)i}$.

If the game terminates after stage τ, agent $i \in \{1, 2\}$ will receive a terminal payoff in stage $\tau + 1$ equalling

$$q^i(\tau+1, x; \underline{\overline{u}}^{(3)}_{(\tau+1)-}, \underline{\overline{u}}^{(4)}_{(\tau+1)-})$$
$$= \left(Q^{(i)i}_{\tau+1} x^i_{\tau+1} + \sum_{t=\tau+1-2}^{\tau} \overline{v}^{|t|i}_{\tau+1} \overline{u}^{(3)i}_t + \sum_{t=\tau+1-3}^{\tau} \overline{p}^{|t|i}_{\tau+1} \overline{u}^{(4)i}_t + \varpi^i_{\tau+1} \right).$$

In particular, $Q^{(i)i}_{\tau+1} x^i_{\tau+1}$ is the value (or salvage price) of agent i private capital $x^i_{\tau+1}$ in stage $\tau+1$. The term $\sum_{t=\tau+1-3}^{\tau} \overline{p}^{|t|i}_{\tau+1} \overline{u}^{(4)i}_t$ is the value of the payments for previously procured revenue generating investments. Finally, $\sum_{t=\tau+1-2}^{\tau} \overline{v}^{|t|i}_{\tau+1} \overline{u}^{(3)i}_t$ is the payment to agent i's previously executed investments in private capital. Some or all of the terms $\overline{p}^{|t|i}_{\tau+1}$ and $\overline{v}^{|t|i}_{\tau+1}$ can be zero.

The expected payoff of agent i to be maximized is

$$\sum_{\hat{T}=1}^{10} \theta_{\hat{T}} \Bigg\{ \sum_{k=1}^{\hat{T}} \left(R^i_k (u^i_k x^i_k)^{1/2} - c^i_k u^i_k - \varphi^{(x)i}_k (\overline{u}^{(3)i}_k)^2 - \phi^i_k (\overline{u}^{(4)i}_k)^2 + \sum_{t=k-3}^{k} p^{|t|i}_k \overline{u}^{(4)i}_t \right) \delta^{k-1}$$
$$+ \left(Q^{(i)i}_{\hat{T}+1} x^i + \sum_{t=\hat{T}+1-2}^{\hat{T}} \overline{v}^{|t|i}_{\hat{T}+1} \overline{u}^{(3)i}_t + \sum_{t=\hat{T}+1-3}^{\hat{T}} \overline{p}^{|t|i}_{\hat{T}+1} \overline{u}^{(4)i}_t + \varpi^i_{\hat{T}+1} \right) \delta^{\hat{T}} \Bigg\}, \tag{4.3}$$

where δ is the discount factor and the controls executed before the start of the operation in stage 1, that is $(\underline{\overline{u}}^{(3)}_{1-}, \underline{\overline{u}}^{(4)}_{-})$, are known to can be zeros.

4.2 Game Equilibrium Outcome

Invoking Theorem 3.1, the feedback Nash equilibrium of the game (4.1)–(4.3) can be characterized as follows.

Corollary 4.1. *Let $\{\underline{\overline{u}}^{(3)**}_{\tau}, \underline{\overline{u}}^{(4)**}_{\tau}\}$ denote the set of the agents' feedback Nash equilibrium strategies and $V^i(\tau, x; \underline{\overline{u}}^{(3)**}_{\tau-}, \underline{\overline{u}}^{(4)**}_{\tau-})$ denote the feedback Nash equilibrium payoff of agent i in the non-cooperative game (3.3)–(3.4), then the function $V^i(\tau, x; \underline{\overline{u}}^{(3)**}_{\tau-}, \underline{\overline{u}}^{(4)**}_{\tau-})$ satisfies the following system of recursive equations*

$$V^i(11, x; \underline{\overline{u}}^{(3)**}_{11-}, \underline{\overline{u}}^{(4)**}_{11-})$$
$$= \left(Q^{(i)i}_{11} x^i_{11} + \sum_{t=11-2}^{10} \overline{v}^{|t|i}_{11} \overline{u}^{(3)i**}_t + \sum_{t=11-3}^{10} \overline{p}^{|t|i}_{11} \overline{u}^{(4)i**}_t + \varpi^i_{11} \right) \delta^{10}, \tag{4.4}$$

$$V^i(10, x; \underline{\overline{u}}^{(3)**}_{10-}, \underline{\overline{u}}^{(4)**}_{10-})$$
$$= \max_{u^i_{10}, \overline{u}^{(3)i}_{10}, \overline{u}^{(4)i}_{10}} \Big\{ \Big(R^i_{10} (u^i_{10} x^i)^{1/2} - c^i_{10} u^i_{10} - \varphi^{(x)i}_{10} (\overline{u}^{(3)i}_{10})^2 - \phi^i_{10} (\overline{u}^{(4)i}_{10})^2$$

$$+p_{10}^{|10|i}\overline{u}_{10}^{(4)i} + \sum_{t=10-3}^{9} p_{10}^{|t|i}\overline{u}_t^{(4)i**}\Bigg)\delta^9$$

$$+\Bigg(Q_{11}^{(i)i}[x^i + \varepsilon_{10}^{|10|i}\overline{u}_{10}^{(3)i} + \sum_{t=10-2}^{9} \varepsilon_{10}^{|t|i}\overline{u}_t^{(3)i**} - \lambda_{10}^{i}x^i + \gamma_{10}^{(i)j}x^j]$$

$$+\overline{v}_{11}^{|10|i}\overline{u}_{10}^{(3)i} + \sum_{t=10-2}^{9} \overline{v}_{11}^{|t|i}\overline{u}_t^{(3)i**} + \overline{p}_{11}^{|10|i}\overline{u}_{10}^{(4)i} + \sum_{t=11-3}^{9} \overline{p}_{11}^{|t|i}\overline{u}_t^{(4)i**} + \varpi_{11}^{i}\Bigg)\delta^{10}\Bigg\},\tag{4.5}$$

$$V^i(\tau, x; \underline{\overline{u}}_{\tau-}^{(3)**}, \underline{\overline{u}}_{\tau-}^{(4)**})$$
$$= \max_{u_\tau^i, \overline{u}_\tau^{(3)i}, \overline{u}_\tau^{(4)i}} \Big\{\Big((R_\tau^i(u_\tau^i x^i)^{1/2} - c_\tau^i u_\tau^i - \varphi_\tau^{(x)i}(\overline{u}_\tau^{(3)i})^2 - \phi_\tau^i(\overline{u}_\tau^{(4)i})^2$$
$$+p_\tau^{|\tau|i}\overline{u}_\tau^{(4)i} + \sum_{t=\tau-3}^{\tau} p_\tau^{|t|i}\overline{u}_t^{(4)i**}\Big)\delta^{\tau-1}$$
$$+\frac{\theta_\tau}{\sum_{\zeta=\tau}^{T}\theta_\zeta}\Bigg(Q_{\tau+1}^{(i)i}x_{\tau+1}^{i***} + \overline{v}_{\tau+1}^{|\tau|i}\overline{u}_\tau^{(3)i} + \sum_{t=\tau-2}^{\tau-1}\overline{v}_{\tau+1}^{|t|i}\overline{u}_t^{(3)i**}$$
$$+\overline{p}_{\tau+1}^{|\tau|i}\overline{u}_\tau^{(4)i} + \sum_{t=\tau+1-3}^{\tau-1}\overline{p}_{\tau+1}^{|t|i}\overline{u}_t^{(4)i**} + \varpi_{\tau+1}^{i}\Bigg)\delta^\tau$$
$$+\frac{\sum_{\zeta=\tau+1}^{T}\theta_\zeta}{\sum_{\zeta=\tau}^{T}\theta_\zeta}V^i[\tau+1, x_{\tau+1}^{i***}, x_{\tau+1}^{j***}; \overline{u}_\tau^{(3)i}, \overline{u}_\tau^{(4)i}, \underline{\overline{u}}_{(\tau+1)-}^{(3)i**} \cap \underline{\overline{u}}_{\tau-}^{(3)i**},$$
$$\underline{\overline{u}}_{(\tau+1)-}^{(4)i**} \cap \underline{\overline{u}}_{\tau-}^{(4)i**}, \underline{\overline{u}}_{(\tau+1)-}^{(3)j**}, \underline{\overline{u}}_{(\tau+1)-}^{(4)j**}]\Big\},$$
$$\text{for } \tau \in \{1, 2, \cdots, 9\},\tag{4.6}$$

where $x_{\tau+1}^{i***} = x^i + \varepsilon_\tau^{|\tau|i}\overline{u}_\tau^{(3)i} + \sum_{t=\tau-2}^{\tau-1}\varepsilon_\tau^{|t|i}\overline{u}_t^{(3)i**} - \lambda_\tau^i x^i + \gamma_\tau^{(i)j}x^j$,
$x_{\tau+1}^{j***} = x^j + \varepsilon_\tau^{|\tau|j}\overline{u}_\tau^{(3)j**} + \sum_{t=\tau-2}^{\tau-1}\varepsilon_\tau^{|t|j}\overline{u}_t^{(3)j**} - \lambda_\tau^j x^j + \gamma_\tau^{(j)i}x^i$, for $i, j \in \{1, 2\}$ and $i \neq j$ ■

Performing the indicated maximization operator in (4.5)–(4.6) of Corollary 4.1, we obtain a feedback Nash equilibrium solution of the game (3.1)–(3.3) as follows.

Proposition 4.1. *System* (4.4)–(4.5) *admits a solution with the game equilibrium payoff of firm i being*
$V^i(\tau, x; \underline{\overline{u}}_{\tau-}^{(3)**}, \underline{\overline{u}}_{\tau-}^{(4)**}) = (A_\tau^{(i)i}x^i + A_\tau^{(i)j}x^j + C_\tau^i)\delta^{\tau-1}$, *for* $\tau \in \{1, 2, \cdots, 11\}$,
where

$$A_{11}^{(i)i} = Q_{11}^{(i)i},\ A_{11}^{(i)j} = 0,\ C_{11}^i = \sum_{t=11-2}^{10}\overline{v}_{11}^{|t|i}\overline{u}_t^{(3)i**} + \sum_{t=11-3}^{10}\overline{p}_{11}^{|t|i}\overline{u}_t^{(4)i**} + \varpi_{11}^i,$$
$$A_{10}^{(i)i} = \frac{(R_{10}^i)^2}{4c_{10}^i} + \delta A_{11}^{(i)i}(1-\lambda_{10}^i),\ A_{10}^{(i)j} = \delta A_{11}^{(i)i}\gamma_T^{(i)j},$$

$$A_\tau^{(i)i} = \frac{(R_\tau^i)^2}{4c_\tau^i} + \frac{\theta_\tau}{\sum_{\zeta=\tau}^{10}\theta_\zeta}\delta Q_{\tau+1}^{(i)i}(1-\lambda_\tau^i) + \frac{\sum_{\zeta=\tau+1}^{10}\theta_\zeta}{\sum_{\zeta=\tau}^{10}\theta_\zeta}\left(A_{\tau+1}^{(i)i}(1-\lambda_\tau^i) + A_{\tau+1}^{(i)j}\gamma_\tau^{(j)i}\right)\delta,$$

$$A_\tau^{(i)j} = \frac{\theta_\tau}{\sum_{\zeta=\tau}^{10}\theta_\zeta}\left(A_{\tau+1}^{(i)i}\gamma_\tau^{(i)j} + A_{\tau+1}^{(i)j}(1-\lambda_\tau^j)\right)\delta, \tag{4.7}$$

C_τ^i is an expression with previously executed controls $(\overline{u}_{\tau-}^{(3)**}, \overline{u}_{\tau-}^{(4)**})$, for $\tau \in \{1, 2, \cdots, 9\}$ and $i, j \in \{1, 2\}$ and $i \neq j$.

Proof See Appendix II. ∎

The values of $A_\tau^{(i)i}$ and $A_\tau^{(i)j}$, for $\tau \in \{1, 2, \cdots, 10\}$ and $i, j \in \{1, 2\}$ and $i \neq j$, in Proposition 4.1 can be solved backwardly from $\tau = 10$ to $\tau = 1$. Using Corollary 4.1 and Proposition 4.1, the game equilibrium strategies of the agent $i \in \{1, 2\}$ can be obtained as (see derivation details in Appendix II):

$$u_\tau^{i**} = \frac{(R_\tau^i)^2}{4(c_\tau^i)^2}x^i, \qquad \text{for } \tau \in \{1, 2, \cdots, 10\};$$

$$\overline{u}_{10}^{(3)i**} = \frac{(A_{11}^{(i)i}\varepsilon_{10}^{|10|i} + \overline{v}_{11}^{|10|i})\delta}{2\varphi_{10}^{(x)i}},$$

$$\overline{u}_9^{(3)i**} = \frac{1}{2\varphi_9^{(x)i}}\left[\frac{\theta_9}{\sum_{\zeta=9}^{10}\theta_\zeta}\left(Q_{10}^{(i)i}\varepsilon_9^{|9|i}\delta + \overline{v}_{10}^{|9|i}\delta\right) + \frac{\sum_{\zeta=10}^{10}\theta_\zeta}{\sum_{\zeta=9}^{10}\theta_\zeta}\left(A_{10}^{(i)i}\varepsilon_9^{|9|i}\delta + A_{11}^{(i)i}\varepsilon_{10}^{|9|i}\delta^2 + \overline{v}_{11}^{|9|i}\delta^2\right)\right],$$

$$\overline{u}_8^{(3)i**} = \frac{1}{2\varphi_8^{(x)i}}\left[\frac{\theta_8}{\sum_{\zeta=8}^{10}\theta_\zeta}\left(Q_9^{(i)i}\varepsilon_8^{|8|i}\delta + \overline{v}_9^{|8|i}\delta\right) + \frac{\sum_{\zeta=9}^{10}\theta_\zeta}{\sum_{\zeta=8}^{10}\theta_\zeta}\left(A_9^{(i)i}\varepsilon_8^{|8|i}\delta + A_{11}^{(i)i}\varepsilon_{10}^{|8|i}\delta^3 + \overline{v}_{11}^{|8|i}\delta^3\right)\right],$$

$$\overline{u}_\tau^{(3)i**} = \frac{1}{2\varphi_\tau^{(x)i}}\left[\frac{\theta_\tau}{\sum_{\zeta=\tau}^{10}\theta_\zeta}\left(Q_{\tau+1}^{(i)i}\varepsilon_\tau^{|\tau|i}\delta + \overline{v}_{\tau+1}^{|\tau|i}\delta\right) + \frac{\sum_{\zeta=\tau+1}^{10}\theta_\zeta}{\sum_{\zeta=\tau}^{10}\theta_\zeta}\left(\sum_{t=\tau}^{\tau+2}A_{t+1}^{(i)i}\varepsilon_t^{|\tau|i}\delta^{t+1-\tau}\right)\right],$$

for $\tau \in \{1, 2, \cdots, 7\}$;

$$\overline{u}_{10}^{(4)i**} = \frac{p_{10}^{|10|i} + \delta\overline{p}_{11}^{|10|i}}{2\phi_{10}^i},$$

$$\overline{u}_9^{(4)i**} = \frac{1}{2\phi_9^i}\left[p_9^{|9|i} + \frac{\theta_9}{\sum_{\zeta=9}^{10}\theta_\zeta}\overline{p}_{10}^{|9|i}\delta + \frac{\sum_{\zeta=10}^{10}\theta_\zeta}{\sum_{\zeta=9}^{10}\theta_\zeta}\left(p_{10}^{|9|i}\delta + \overline{p}_{11}^{|9|i}\delta^2\right)\right],$$

$$\overline{u}_8^{(4)i**} = \frac{1}{2\phi_8^i}\left[p_8^{|8|i} + \frac{\theta_8}{\sum_{\zeta=8}^{10}\theta_\zeta}\overline{p}_9^{|8|i}\delta + \frac{\sum_{\zeta=9}^{10}\theta_\zeta}{\sum_{\zeta=8}^{10}\theta_\zeta}\left(p_9^{|8|i}\delta + +p_{10}^{|8|i}\delta^2 + \overline{p}_{11}^{|8|i}\delta^3\right)\right],$$

$$\overline{u}_\tau^{(4)i**} = \frac{1}{2\phi_\tau^i}\left[p_\tau^{|\tau|i} + \frac{\theta_\tau}{\sum_{\zeta=\tau}^{10}\theta_\zeta}\overline{p}_{\tau+1}^{|\tau|i}\delta + \frac{\sum_{\zeta=\tau+1}^{10}\theta_\zeta}{\sum_{\zeta=\tau}^{10}\theta_\zeta}\left(\sum_{t=\tau}^{\tau+2} p_{t+1}^{|\tau|i}\delta^{t+1-\tau}\right)\right],$$

$$\text{for } \tau \in \{1, 2, \cdots, 7\}. \tag{4.8}$$

With the values of $A_\tau^{(i)i}$, for $\tau \in \{1, 2, \cdots, 10\}$ and $i \in \{1, 2\}$, the game equilibrium capital investment strategies $\overline{u}_\tau^{(3)i**}$ and $\overline{u}_\tau^{(3)j**}$ in (4.8) can be obtained explicitly. Substituting the game equilibrium capital investment strategies into the accumulation processes of the private capital stock of agents in (4.1) yields

$$x_{k+1}^1 = x_k^1 + \varepsilon_k^{|k|1}\overline{u}_k^{(3)1**} + \sum_{t=k-2}^{k-1}\varepsilon_k^{|t|1}\overline{u}_t^{(3)1**} - \lambda_k^1 x_k^1 + \gamma_k^{(1)2} x_k^2, \quad x_1^1 = x_1^{1(0)},$$

$$x_{k+1}^2, = x_k^2 + \varepsilon_k^{|k|2}\overline{u}_k^{(3)2**} + \sum_{t=k-2}^{k-1}\varepsilon_k^{|t|2}\overline{u}_t^{(3)2**} - \lambda_k^2 x_k^2 + \gamma_k^{(2)1} x_k^1, \quad x_1^2 = x_1^{2(0)}. \tag{4.9}$$

In initial stage 1, $\overline{u}_{\tau-}^{(3)1**}$ and $\overline{u}_{\tau-}^{(3)2**}$ are zeros. Given $x_1^1 = x_1^{1(0)}$ and $x_1^2 = x_1^{2(0)}$, the values of x_2^1 and x_2^2 can be obtained. With x_2^1 and x_2^2, the values of x_3^1 and x_3^2 can be obtained. Performing the analysis repeatedly, the values of x_t^1 and x_t^2 for $t \in \{4, 5, \cdots, 11\}$ can be obtained. We use $\{x_t^{1**}, x_t^{2**}\}_{t=1}^{T+1}$ to denote the game equilibrium trajectory of private capitals.

Novel results in dynamic game analysis are identified in the following remarks.

Remark 4.1. The terminal values of the agents' previously executed durable strategies may be low or equal to zeros in a premature termination of the game. For instance, the termination return of the income generating investment $(\overline{p}_{\tau+1}^{|t|i})$ may be substantially below the regular return $(p_{\tau+1}^{|t|i})$ and the termination compensation for previously executed durable capital investments (that is $\overline{v}_{\tau+1}^{|t|i}$) may be negligible or even zero. Loss of the values of previously executed durable strategies in a premature termination is a new outcome in dynamic game analysis. ■

Remark 4.2. In formulating durable investment strategies, the agents have to take into consideration the durable effects and the random termination effects.

In particular, in determining the durable income-earning investment $\overline{u}_\tau^{(4)i**}$ in (4.8), the agent i has to consider

(i) the probability that the game would terminate in the next stage, that is $\theta_\tau / \sum_{\zeta=\tau}^{10}\theta_\zeta$; and the probability that the game would continue onward, that is $\sum_{\zeta=\tau+1}^{10}\theta_\zeta / \sum_{\zeta=\tau}^{10}\theta_\zeta$; and

(ii) the future return of the investment if the game would terminate in the next stage, that is $\overline{p}_{\tau+1}^{(\tau)i}\delta$; and the future returns of the investment if the game would continue onward, that is $\left(\sum_{t=\tau}^{\tau+2} p_{t+1}^{|\tau|i}\delta^{t+1-\tau}\right)$.

In determining the durable capital investment strategies $\overline{u}_{\tau}^{(3)i**}$ in (4.8), the agent has to consider

(i) the probability that the game would terminate in the next stage, that is $\theta_\tau / \sum_{\zeta=\tau}^{10}\theta_\zeta$; and the probability that the game would continue onward, that is $\sum_{\zeta=\tau+1}^{10}\theta_\zeta / \sum_{\zeta=\tau}^{10}\theta_\zeta$; and
(ii) the future return of capital investment if the game would terminate in the next stage, that is $\left(Q_{\tau+1}^{(i)i}\varepsilon_{\tau}^{|\tau|i}\delta + \overline{v}_{\tau+1}^{|\tau|i}\delta\right)$; and the future benefits of capital investment if the game would continue onward, that is $\left(\sum_{t=\tau}^{\tau+2} A_{t+1}^{(i)i}\varepsilon_{t}^{|\tau|i}\delta^{t+1-\tau}\right)$.

Taking into consideration of the durable effects and the random termination effects in formulating strategies is a new outcome in dynamic game analysis. ■

Remark 4.3. The presence of random horizon has put a significant discounting effect on the benefits of durable strategies. Consider agent i's investment in private capital

$$\overline{u}_{\tau}^{(3)i**} = \frac{1}{2\varphi_{\tau}^{(x)i}}\left[\frac{\theta_\tau}{\sum_{\zeta=\tau}^{10}\theta_\zeta}\left(Q_{\tau+1}^{(i)i}\varepsilon_{\tau}^{|\tau|i}\delta + \overline{v}_{\tau+1}^{|\tau|i}\delta\right) + \frac{\sum_{\zeta=\tau+1}^{10}\theta_\zeta}{\sum_{\zeta=\tau}^{10}\theta_\zeta}\left(\sum_{t=\tau}^{\tau+2} A_{t+1}^{(i)i}\varepsilon_{t}^{|\tau|i}\delta^{t+1-\tau}\right)\right].$$

If the terminal payoff $\left(Q_{\tau+1}^{(i)i}\varepsilon_{\tau}^{|\tau|i}\delta + \overline{v}_{\tau+1}^{|\tau|i}\delta\right)$ in stage $\tau+1$ is negligible, the benefit of agent i's investment in private capital $\left(\sum_{t=\tau}^{\tau+2} A_{t+1}^{(i)i}\varepsilon_{t}^{|\tau|i}\delta^{t+1-\tau}\right)$ is discounted by the factor $\sum_{\zeta=\tau+1}^{10}\theta_\zeta / \sum_{\zeta=\tau}^{10}\theta_\zeta$.

Moreover, there may exist cases where there is no redemption values for previously executed durable capital investment strategies when operations stop prematurely. This brings about the situation where durable investments are subject to the risk of early termination. ■

The outcomes identified in remarks 4.1, 4.2 and 4.3 are outcomes which are novel in dynamic game analysis. However, these outcomes are not uncommon in real-life operations. Random horizon durable strategies dynamic games indeed enrich game theory in analyzing decision making.

Remark 4.4. In the computational illustration, we consider spillover involving increments (dynamics) of technology build up. However, the payoffs of the agents are affected by technology level x_k^i as:

$$\sum_{\hat{T}=1}^{10}\theta_{\hat{T}}\left\{\sum_{k=1}^{\hat{T}}\left(R_k^i(u_k^i x_k^i)^{1/2} - c_k^i u_k^i - \varphi_k^{(x)i}(\overline{u}_k^{(3)i})^2 - \phi_k^i(u_k^{(4)i})^2 + \sum_{t=k-3}^{k} p_k^{|t|i}u_t^{(4)i}\right)\delta^{k-1}\right.$$
$$\left. + \left(Q_{\hat{T}+1}^{(i)i}x^i + \sum_{t=\hat{T}+1-2}^{\hat{T}} \overline{v}_{\hat{T}+1}^{|t|i}\overline{u}_t^{(3)i} + \sum_{t=\hat{T}+1-3}^{\hat{T}} \overline{p}_{\hat{T}+1}^{|t|i}\overline{u}_t^{(4)i} + \varpi_{\hat{T}+1}^i\right)\delta^{\hat{T}}\right\}.$$

The spillover effects are indirectly affecting the payoffs of the agents through the additional increment in technology level brought about by the other agent's capital stock. ■

Remark 4.5. The accumulation process of private capital stocks with technology spillover can take on various forms of specification. For instance, the increment of agent i's private capital can be affected by the capital investment of agent j. The capital accumulation dynamics then become

$$x_{k+1}^{i} = x_{k}^{i} + \varepsilon_{k}^{|k|i} \overline{u}_{k}^{(3)i} + \sum_{t=k-2}^{k-1} \varepsilon_{k}^{|t)|i} \overline{u}_{t}^{(3)i} + \upsilon_{k}^{|k|i} \overline{u}_{k}^{(3)j} + \sum_{t=k-2}^{k-1} \upsilon_{k}^{|t|i} \overline{u}_{t}^{(3)j} - \lambda_{k}^{i} x_{k}^{i},$$

for $i, j \in \{1, 2\}$ and $i \neq j$ and $\upsilon_{k}^{|k|i} \overline{u}_{k}^{(3)j} + \sum_{t=k-2}^{k-1} \upsilon_{k}^{|t|i} \overline{u}_{t}^{(3)j}$ is the effect of agent j's capital investment on the increment of agent i's private capital in stage k.

We do not specify the investments in capital to be dependent upon the level of capital stock to avoid the problem of starting up at zero capital stock. Hence, we are considering the case where the amounts of physical capital investments are the sole determinant of increments of capital. ■

5 Dynamic Cooperation under Random Horizon and Strategies Lags

Now we consider the case where the players agree to cooperate and distribute the cooperative payoff among themselves according to an optimality principle. Two crucial properties that a cooperative scheme has to satisfy are group optimality and individual rationality. In addition to these crucial properties, the scheme has to satisfy the property of subgame consistency to guarantee dynamical stability. In this section first examine the properties of group optimality and individual rationality and then consider subgame consistent cooperative solution under random horizon and durable strategies.

5.1 *Group Optimality and Individual Rationality*

Group optimality requires the maximization of the players' expected joint payoff so that all gains are captured. To obtain group optimality the players have to solve the discrete-time dynamic optimization problem of maximizing their expected joint payoff

$$E\left\{\sum_{j=1}^{n}\left(\sum_{k=1}^{\hat{T}} g_{k}^{j}(x_{k}, \underline{u}_{k}, \overline{u}_{k}; \underline{\overline{u}}_{k-})\delta_{1}^{k} + q_{\hat{T}+1}^{j}(x_{\hat{T}+1}; \underline{\overline{u}}_{(\hat{T}+1)-})\delta_{1}^{\hat{T}+1}\right)\right\}$$

$$= \sum_{\hat{T}=1}^{T} \theta_{\hat{T}} \left\{ \sum_{j=1}^{n} \left(\sum_{k=1}^{\hat{T}} g_k^j(x_k, \underline{u}_k, \overline{u}_k; \overline{u}_{k-}) \delta_1^k + q_{\hat{T}+1}^j(x_{\hat{T}+1}; \overline{u}_{(\hat{T}+1)-}) \delta_1^{\hat{T}+1} \right) \right\}, \tag{5.1}$$

subject to dynamics

$$x_{k+1} = f_k(x_k, \underline{u}_k, \overline{u}_k; \overline{u}_{k-}), \quad x_1 = x_1^0. \tag{5.2}$$

Invoking the random horizon durable strategies dynamic optimization method in Theorem 2.1 we can characterize an optimal solution to the problem (5.1)–(5.2) as

Theorem 5.1. *Let the value function* $W(\tau, x; \overline{u}_{\tau-})$ *denote the maximal value of the expected joint payoffs*

$$E\left\{ \sum_{j=1}^{n} \left(\sum_{k=\tau}^{\hat{T}} g_k^j(x_k, \underline{u}_k, \overline{u}_k; \overline{u}_{k-}) \delta_1^k + q_{\hat{T}+1}^j(x_{\hat{T}+1}; \overline{u}_{(\hat{T}+1)-}) \delta_1^{\hat{T}+1} \right) \right\}$$

$$= \sum_{\hat{T}=\tau}^{T} \frac{\theta_{\hat{T}}}{\sum_{\zeta=\tau}^{T} \theta_\zeta} \left\{ \sum_{j=1}^{n} \left(\sum_{k=\tau}^{\hat{T}} g_k^j(x_k, \underline{u}_k, \overline{u}_k; \overline{u}_{k-}) \delta_1^k + q_{\hat{T}+1}^j(x_{\hat{T}+1}; \overline{u}_{(\hat{T}+1)-}) \delta_1^{\hat{T}+1} \right) \right\},$$

for problem (5.1)–(5.2) *starting at stage* τ *with state* $x_\tau = x$ *and previously executed controls* $\underline{u}_{\tau-}$. *Then the function* $W(\tau, x_\tau; \overline{u}_{\tau-})$ *satisfies the following system of recursive equations*:

$$W(T+1, x; \overline{u}_{(T+1)-}) = \sum_{j=1}^{j} q_{T+1}^j(x; \overline{u}_{(T+1)-}) \delta_1^{T+1}, \tag{5.3}$$

$$W(T, x; \overline{u}_{T-}) = \max_{\underline{u}_T, \overline{u}_T} \left\{ \sum_{j=1}^{n} (g_T^j(x, \underline{u}_T, \overline{u}_T; \overline{u}_{T-}) \delta_1^T \right.$$
$$\left. + q_{T+1}^j[f_T(x, \underline{u}_T, \overline{u}_T; \overline{u}_{T-}); \overline{u}_T, \overline{u}_{(T+1)-} \cap \overline{u}_{T-}] \delta_1^{T+1}) \right\}, \tag{5.4}$$

$$W(\tau, x; \overline{u}_{\tau-}) = \max_{\underline{u}_\tau, \overline{u}_\tau} \left\{ \sum_{j=1}^{n} \left(g_\tau^j(x, \underline{u}_\tau, \overline{u}_\tau; \overline{u}_{\tau-}) \delta_1^\tau \right. \right.$$
$$\left. + \frac{\theta_\tau}{\sum_{\zeta=\tau}^{T} \theta_\zeta} q_{\tau+1}^j[f_\tau(x, \underline{u}_\tau, \overline{u}_\tau; \overline{u}_{\tau-}); \overline{u}_\tau, \overline{u}_{(\tau+1)-} \cap \overline{u}_{\tau-}] \delta_1^{\tau+1} \right)$$
$$\left. + \frac{\sum_{\zeta=\tau+1}^{T} \theta_\zeta}{\sum_{\zeta=\tau}^{T} \theta_\zeta} W[\tau+1, f_\tau(x, \underline{u}_\tau, \overline{u}_\tau; \overline{u}_{\tau-}); \overline{u}_\tau, \overline{u}_{(\tau+1)-} \cap \overline{u}_{\tau-}] \right\},$$
$$\text{for } \tau \in \{1, 2, \cdots, T-1\}. \tag{5.5}$$

Proof The conditions in (5.3)–(5.5) satisfy the optimal random horizon dynamic optimization result under durable strategies in Theorem 2.1. Hence Theorem 5.1 follows. ■

We use $(\underline{u}_\tau^*, \overline{u}_\tau^*)$, for $\tau \in \{1, 2, \cdots, T\}$ to denote the set of group optimal strategies derived in Theorem 5.1. Substituting the optimal strategies $(\underline{u}_\tau^*, \overline{u}_\tau^*)$ into the state dynamics (5.1), one can obtain the dynamics of the cooperative trajectory as:

$$x_{k+1} = f_k(x_k, \underline{u}_k^*, \overline{u}_k^*; \overline{u}_{k-}^*), \quad x_1 = x_1^0. \tag{5.6}$$

We use $\{x_k^*\}_{k=1}^{T+1}$ to denote the solution generated by dynamics (5.6). Using the set of optimal strategies $(\underline{u}_\tau^*, \overline{u}_\tau^*)$ one can obtain the expected joint cooperative payoff as

$$W(\tau, x_\tau^*; \overline{u}_{\tau-}^*)$$
$$= E\left\{\sum_{j=1}^{n}\left(\sum_{k=\tau}^{\hat{T}} g_k^j(x_k^*, \underline{u}_k^*, \overline{u}_k^*; \overline{u}_{k-}^*)\delta_1^k + q_{\hat{T}+1}^j(x_{\hat{T}+1}^*; \overline{u}_{(\hat{T}+1)-}^*)\delta_1^{\hat{T}+1}\right)\right\}$$
$$= \sum_{\hat{T}=\tau}^{T} \frac{\theta_{\hat{T}}}{\sum_{\zeta=\tau}^{T}\theta_\zeta}\left\{\sum_{j=1}^{n}\left(\sum_{k=\tau}^{\hat{T}} g_k^j(x_k^*, \underline{u}_k^*, \overline{u}_k^*; \overline{u}_{k-}^*)\delta_1^k + q_{\hat{T}+1}^j(x_{\hat{T}+1}^*; \overline{u}_{(\hat{T}+1)-}^*)\delta_1^{\hat{T}+1}\right)\right\}, \tag{5.7}$$

for $\tau \in \{1, 2, \cdots, T\}$.

We then move to the property of individual rationality. The players have to agree to an optimality principle in distributing the total cooperative payoff among themselves. For individual rationality to be upheld, the imputations (see von Neumann and Morgenstern (1944)) under the agreed-upon optimality principle received by the players have to be no less than their expected non-cooperative payoff along the cooperative state trajectory. Let

$$\xi(\tau, x_\tau; \overline{u}_{\tau-}) = [\xi^1(\tau, x_\tau; \overline{u}_{\tau-}), \xi^2(\tau, x_\tau; \overline{u}_{\tau-}), \cdots, \xi^n(\tau, x_\tau; \overline{u}_{\tau-})]$$

denote the agreed-upon imputation vector guiding the distribution of the total cooperative payoff in stage $\tau \in \{1, 2, \cdots, T\}$. For individual rationality to be maintained throughout all the stages $\tau \in \{1, 2, \cdots, T\}$ along the optimal cooperative trajectory $\{x_k^*\}_{k=1}^{T+1}$, it is required that:

$\xi^i(\tau, x_\tau^*; \overline{u}_{\tau-}^*) \geq V^i(\tau, x_\tau^*; \overline{u}_{\tau-}^*)$, for $i \in N$ and $\tau \in \{1, 2, \cdots, T\}$,along the cooperative trajectory $\{x_k^*\}_{k=1}^{T}$ for given previously executed strategies $\overline{u}_{\tau-}^*$.

Finally, to satisfy group optimality, the imputation vector has to satisfy $W(\tau, x_\tau^*; \underline{u}_{\tau-}^*) = \sum_{j=1}^n \xi^j(\tau, x_\tau^*; \underline{u}_{\tau-}^*)$, for $\tau \in \{1, 2, \cdots, T\}$.

5.2 *Subgame Consistent Solutions and Payment Mechanism*

To guarantee dynamical stability in a dynamic cooperation scheme, the solution has to satisfy the property of subgame consistency. A cooperative solution is subgame consistent if an extension of the solution policy to a subgame starting at a later time with a state along the optimal cooperative trajectory would remain optimal. In particular, subgame consistency ensures that as the game proceeds players are guided by the same optimality principle at each stage of the game and hence do not possess incentives to deviate from the previously adopted optimal behaviour. Therefore, for subgame consistency to be satisfied, the agreed-upon imputation

$$\xi(\tau, x_\tau^*; \underline{u}_{\tau-}^*) = [\xi^1(\tau, x_\tau^*; \underline{u}_{\tau-}^*), \xi^2(\tau, x_\tau^*; \underline{u}_{\tau-}^*), \cdots, \xi^n(\tau, x_\tau^*; \underline{u}_{\tau-}^*)],$$
$$\tau \in \{1, 2, \cdots, T\}, \tag{5.8}$$

has to be maintained along the cooperative trajectory $\{x_k^*\}_{k=1}^T$ for given previously executed strategies $\underline{u}_{\tau-}^*$.

Crucial to the analysis is the formulation of a payment mechanism so that the imputation in (5.8) can be realized as the game proceeds. Following the analysis of Yeung and Petrosyan (2010, 2011), we formulate a discrete-time random horizon Payoff Distribution Procedure (PDP) so that the agreed-upon imputations (4.1) can be realized. Let $\beta_k^i(x_k^*; \underline{u}_{k-}^*)$ denote the payment that player i will received at stage k under the cooperative agreement.

The payment scheme involving $\beta_k^i(x_k^*; \underline{u}_{k-}^*)$ constitutes a PDP in the sense that along the cooperative trajectory $\{x_k^*\}_{k=1}^T$ with previously executed durable strategies $\underline{u}_{k-}^*$ the imputation to player i over the stages from k to T can be expressed as:

$$\xi^i(\tau, x_\tau^*; \underline{u}_{\tau-}^*) = E\left\{\sum_{k=\tau}^{\hat{T}} \beta_k^i(x_k^*; \underline{u}_{k-}^*)\delta_1^k + q_{\hat{T}+1}^i(x_{\hat{T}+1}^*; \underline{u}_{(\hat{T}+1)-}^*)\delta_1^{\hat{T}+1}\right\}$$
$$= \sum_{\hat{T}=\tau}^{T} \frac{\theta_{\hat{T}}}{\sum_{\zeta=\tau}^{T} \theta_\zeta} \left\{\sum_{k=\tau}^{\hat{T}} \beta_k^i(x_k^*; \underline{u}_{k-}^*)\delta_1^k + q_{\hat{T}+1}^i(x_{\hat{T}+1}^*; \underline{u}_{(\hat{T}+1)}^*)\delta_1^{\hat{T}+1}\right\}, \tag{5.9}$$

for $i \in N$ and $\tau \in \{1, 2, \cdots, T\}$.

A theorem characterizing a computational formula for $\beta_\tau^i(x_\tau^*; \underline{u}_{\tau-}^*)$, for $\tau \in \{1, 2, \cdots, T\}$ and $i \in N$, which yields (5.9) is provided below.

Theorem 5.2. *A payment equalling*

$$\beta_\tau^i(x_\tau^*; \underline{\overline{u}}_{\tau-}^*) = (\delta_1^\tau)^{-1} \left(\xi^i(\tau, x_\tau^*; \underline{\overline{u}}_{\tau-}^*) - \frac{\theta_\tau}{\sum_{\zeta=\tau}^T \theta_\zeta} q_{\tau+1}^i(x_{\tau+1}^*; \underline{\overline{u}}_{(\tau+1)-}^*) \delta_1^{\tau+1} \right.$$
$$\left. - \frac{\sum_{\zeta=\tau+1}^T \theta_\zeta}{\sum_{\zeta=\tau}^T \theta_\zeta} \xi^i(\tau+1, x_{\tau+1}^*; \underline{\overline{u}}_{(\tau+1)-}^*) \right), \tag{5.10}$$

given to player $i \in N$ at stage $\tau \in \{1, 2, \cdots, T\}$ would lead to the realization of the imputation vector $\xi(\tau, x_\tau^*; \underline{\overline{u}}_{\tau-}^*)$ in (5.8).

Proof See Appendix III. ■

Note that the payoff distribution procedure $\beta_\tau^i(x_\tau^*; \underline{\overline{u}}_{\tau-}^*)$ in Theorem 5.2 would give rise to the agreed-upon imputation vector in (5.8). Therefore, the property of subgame consistency is satisfied.

When all players are using the cooperative strategies, the payoff that player i will directly receive at stage $\tau \in \{1, 2, \cdots, T\}$ is $g_\tau(x, \underline{u}_\tau^*, \underline{\overline{u}}_\tau^*; \underline{\overline{u}}_{\tau-}^*)$. However, according to the agreed-upon imputation, player i is to receive $\beta_\tau^i(x_\tau^*; \underline{\overline{u}}_{\tau-}^*)$ at stage τ. Therefore, a side-payment equalling

$$\beta_\tau^i(x_\tau^*; \underline{\overline{u}}_{\tau-}^*) - g_\tau(x, \underline{u}_\tau^*, \underline{\overline{u}}_\tau^*; \underline{\overline{u}}_{\tau-}^*)$$

will be given to player $i \in N$ in stage $\tau \in \{1, 2, \cdots, T\}$.

6 A Random Horizon Cooperative Durable Strategies Game

In this section, we present a random horizon cooperative durable strategies dynamic game of interactive investments to demonstrate the computation of the cooperative solution.

6.1 Game Formulation and Group Payoff Maximization

Consider the game in Sect. 4. Now, the agents agree to cooperate, maximize their expected joint profit and adopt an optimality principle which would allot the agents' cooperative payoffs proportional to the sizes of their non-cooperative payoff. To maximize the expected joint payoff, the players have to solve the discrete-time dynamic programming problem of maximizing

$$\sum_{i=1}^{2} \left\{ \sum_{\hat{T}=1}^{10} \theta_{\hat{T}} \left[\sum_{k=1}^{\hat{T}} \left(R_k^i (u_k^i x_k^i)^{1/2} - c_k^i u_k^i - \varphi_k^{(x)i} (\overline{u}_k^{(3)i})^2 - \phi_k^i (\overline{u}_k^{(4)i})^2 + \sum_{t=k-3}^{k} p_k^{|t|i} u_t^{(4)i} \right) \delta^{k-1} \right. \right.$$

$$+\left(Q_{\hat{T}+1}^{(i)i}x^i + \sum_{t=\hat{T}+1-2}^{\hat{T}} \overline{v}_{\hat{T}+1}^{|t|i}u_t^{(3)i} + \sum_{t=\hat{T}+1-3}^{\hat{T}} \overline{p}_{\hat{T}+1}^{|t|i}u_t^{(4)i} + \varpi_{\hat{T}+1}^i\right)\delta^{\hat{T}}\Bigg]\Bigg\}, \quad (6.1)$$

subject to

$$x_{k+1}^i = x_k^i + \varepsilon_k^{|k|i}\overline{u}_k^{|3|i} + \sum\nolimits_{t=k-2}^{k-1} \varepsilon_k^{|t|i}\overline{u}_t^{(3)i} - \lambda_k^i x_k^i + \gamma_k^{(i)j}x_k^j,$$
$$x_1^i = x_1^{i(0)},\ i, j \in \{1, 2\} \text{ and } i \neq j. \quad (6.2)$$

Invoking Theorem 5.1, the expected joint payoff of the cooperative game (6.1)–(6.2) can be characterized as follows.

Corollary 6.1. *Let the value function $W(\tau, x; \underline{\overline{u}}_{\tau-}^{(3)}, \underline{\overline{u}}_{\tau-}^{(4)})$ denote the maximal value of the expected joint payoffs of the game (6.1)–(6.2) starting at stage $\tau \in \{1, 2, \cdots, 10\}$ with state $x_\tau = x$ and previously executed controls $(\underline{u}_{\tau-}^{(3)}, \underline{u}_{\tau-}^{(4)})$. Then the function $W(\tau, x; \underline{\overline{u}}_{\tau-}^{(3)}, \underline{\overline{u}}_{\tau-}^{(4)})$ satisfies the following system of recursive equations:*

$$W(11, x; \underline{\overline{u}}_{11-}^{(3)}, \underline{\overline{u}}_{11-}^{(4)})$$
$$= \sum_{i=1}^{2}\left(Q_{11}^{(i)i}x_{11}^i + \sum_{t=11-2}^{10} \overline{v}_{11}^{|t|i}\overline{u}_t^{(3)i} + \sum_{t=11-3}^{10} \overline{p}_{11}^{|t|i}\overline{u}_t^{(4)i} + \varpi_{11}^i\right)\delta^{10}, \quad (6.3)$$

$$W(10, x; \underline{\overline{u}}_{10-}^{(3)}, \underline{\overline{u}}_{10-}^{(4)})$$
$$= \max_{\substack{u_{10}^i, \overline{u}_{10}^{(3)i}, \overline{u}_{10}^{(4)i} \\ i\in\{12\}}} \Bigg\{\sum_{i=1}^{2}\Bigg[\Bigg(R_{10}^i(u_{10}^i x^i)^{1/2} - c_{10}^i u_{10}^i - \varphi_{10}^{(x)i}(\overline{u}_{10}^{(3)i})^2 - \phi_{10}^i(\overline{u}_{10}^{(4)i})^2$$
$$+ p_{10}^{|10|)i}\overline{u}_{10}^{(4)i} + \sum_{t=10-3}^{9} p_{10}^{|t|i}\overline{u}_t^{(4)i}\Bigg)\delta^9$$
$$+\Bigg(Q_{11}^{(i)i}[x^i + \varepsilon_{10}^{|10|i}\overline{u}_{10}^{(3)i} + \sum_{t=10-2}^{9} \varepsilon_{10}^{|t|i}\overline{u}_t^{(3)i} - \lambda_{10}^i x^i + \gamma_{10}^{(i)j}x^j]$$
$$+\overline{v}_{11}^{|10|i}\overline{u}_{10}^{(3)i} + \sum_{t=10-2}^{9} \overline{v}_{11}^{|t|i}\overline{u}_t^{(3)i} + \overline{p}_{11}^{|10|i}\overline{u}_{10}^{(4)i} + \sum_{t=11-3}^{9} \overline{p}_{11}^{|t|i}\overline{u}_t^{(4)i} + \varpi_{11}^i\Bigg)\delta^{10}\Bigg]\Bigg\}, \quad (6.4)$$

$$W(\tau, x; \underline{\overline{u}}_{\tau-}^{(3)}, \underline{\overline{u}}_{\tau-}^{(4)})$$
$$= \max_{\substack{u_\tau^i, \overline{u}_\tau^{(3)i}, \overline{u}_\tau^{(4)i} \\ i\in\{1,2\}}} \Bigg\{\sum_{i=1}^{2}\Big(R_\tau^i(u_\tau^i x^i)^{1/2} - c_\tau^i u_\tau^i - \varphi_\tau^{(x)i}(\overline{u}_\tau^{(3)i})^2 - \phi_\tau^i(\overline{u}_\tau^{(4)i})^2$$

$$+p_\tau^{|\tau|i}u_\tau^{(4)i} + \sum_{t=\tau-3}^{\tau} p_\tau^{|t|i}\overline{u}_t^{(4)i}\Bigg)\delta^{\tau-1}$$

$$+\sum_{i=1}^{2}\frac{\theta_\tau}{\sum_{\zeta=\tau}^{T}\theta_\zeta}\Bigg(Q_{\tau+1}^{(i)i}x_{\tau+1}^i + \overline{v}_{\tau+1}^{|\tau|i}\overline{u}_\tau^{(3)i} + \sum_{t=\tau-2}^{\tau-1}\overline{v}_{\tau+1}^{|t|i}\overline{u}_t^{(3)i} + \overline{p}_{\tau+1}^{|\tau|i}\overline{u}_\tau^{(4)i} +$$

$$\sum_{t=\tau+1-3}^{\tau-1}\overline{p}_{\tau+1}^{|t|i}\overline{u}_t^{(4)i} + \varpi_{\tau+1}^i\Bigg)\delta^\tau$$

$$+\frac{\sum_{\zeta=\tau+1}^{T}\theta_\zeta}{\sum_{\zeta=\tau}^{T}\theta_\zeta}W[\tau+1, x_{\tau+1}; \overline{u}_\tau^{(3)}, \underline{\overline{u}}_{(\tau+1)-}^{(3)} \cap \underline{\overline{u}}_{\tau-}^{(3)}, \overline{u}_\tau^{(4)}, \underline{\overline{u}}_{(\tau+1)-}^{(4)} \cap \underline{\overline{u}}_{\tau-}^{(4)}]\Bigg\},$$

$$\text{for } \tau \in \{1, 2, \cdots, 9\}, \tag{6.5}$$

where $x_{\tau+1}$ is a vector containing two elements
$x^1 + \varepsilon_\tau^{|\tau|1}\overline{u}_\tau^{(3)1} + \sum_{t=\tau-2}^{\tau-1}\varepsilon_\tau^{|t|1}\overline{u}_t^{(3)1} - \lambda_\tau^1 x^1 + \gamma_\tau^{(1)2}x^2$ and
$x^2 + \varepsilon_\tau^{|\tau|2}\overline{u}_\tau^{(3)2} + \sum_{t=\tau-2}^{\tau-1}\varepsilon_\tau^{|t|2}\overline{u}_t^{(3)2} - \lambda_\tau^2 x^2 + \gamma_\tau^{(2)1}x^1$. ■

Performing the indicated maximization operator in (6.4)–(6.5) of Corollary 6.1, we obtain an optimal cooperative solution of the game (6.1)–(6.2) as follows.

Proposition 6.1. *System* (6.3)–(6.5) *admits an optimal cooperative solution with the expected joint payoff being*

$$W(\tau, x; \underline{u}_{\tau-}^{(3)}, \underline{u}_{\tau-}^{(4)}) = (A_\tau^1 x^1 + A_\tau^2 x^2 + C_\tau)\delta^{\tau-1}, \quad \text{for } \tau \in \{1, 2, \cdots, 11\}, \tag{6.6}$$

where

$$A_{11}^1 = Q_{T+1}^{(1)1},\ A_{11}^2 = Q_{T+1}^{(2)2},\ C_{11} = \sum_{i=1}^{2}\left(\sum_{t=11-2}^{10}\overline{v}_{11}^{|t|i}\overline{u}_t^{(3)i} + \sum_{t=11-3}^{10}\overline{p}_{11}^{|t|i}\overline{u}_t^{(4)i} + \varpi_{11}^i\right);$$

$$A_{10}^1 = \sum_{i=1}^{2}\frac{(R_{10}^i)^2}{4c_{10}^i} + \delta A_{11}^1(1-\lambda_{10}^1) + \delta A_{11}^2\gamma_{10}^{(2)1},$$

$$A_{10}^2 = \sum_{i=1}^{2}\frac{(R_{10}^i)^2}{4c_{10}^i} + \delta A_{11}^2(1-\lambda_{10}^2) + \delta A_{11}^1\gamma_{10}^{(1)2};$$

$$A_\tau^1 = \frac{(R_\tau^1)^2}{4c_\tau^1} + \frac{\theta_\tau}{\sum_{\zeta=\tau}^{10}\theta_\zeta}\left(Q_{\tau+1}^{(1)1}(1-\lambda_\tau^1) + Q_{\tau+1}^{(2)2}\gamma_\tau^{(2)1}\right)\delta$$
$$+\frac{\sum_{\zeta=\tau+1}^{10}\theta_\zeta}{\sum_{\zeta=\tau}^{10}\theta_\zeta}(A_{\tau+1}^1(1-\lambda_\tau^1) + A_{\tau+1}^2\gamma_\tau^{(2)1})\delta,$$

$$A_\tau^2 = \frac{(R_\tau^2)^2}{4c_\tau^2} + \frac{\theta_\tau}{\sum_{\zeta=\tau}^{10}\theta_\zeta}\left(Q_{\tau+1}^{(2)2}(1-\lambda_\tau^2) + Q_\tau^{(1)1}\gamma_\tau^{(1)2}\right)\delta$$

$$+\frac{\sum_{\zeta=\tau+1}^{10}\theta_\zeta}{\sum_{\zeta=\tau}^{10}\theta_\zeta}(A_{\tau+1}^2(1-\lambda_\tau^2)+A_{\tau+1}^1\gamma_\tau^{(1)2})\delta, \text{ for } \tau\in\{1,2,\cdots,9\}. \quad (6.7)$$

In addition, C_τ is an expression containing previously executed controls $(\underline{\overline{u}}_{\tau-}^{(3)}, \underline{\overline{u}}_{\tau-}^{(4)})$, $\tau\in\{1,2,\cdots,11\}$.

Proof See Appendix IV. ■

The values of A_τ^1 and A_τ^2, for $\tau\in\{1,2,\cdots,10\}$, in Proposition 6.1 can be solved backwardly from $\tau=10$ to $\tau=1$. In addition, using Corollary 6.1 and Proposition 6.1 and the optimal conditions derived in Corollary 6.1, the optimal cooperative strategies can be obtained as:

$$u_\tau^{i*}=\frac{(R_\tau^i)^2}{4(c_\tau^i)}x^i, \quad \text{for } i\in\{1,2\} \text{ and } \tau\in\{1,2,\cdots,10\};$$

$$\overline{u}_{10}^{(3)i*}=\frac{A_{11}^i\varepsilon_{10}^{|10|i}+\overline{v}_{11}^{|10|i}}{2\varphi_{10}^{(x)i}}\delta,$$

$$\overline{u}_9^{(3)i*}$$
$$=\frac{1}{2\varphi_9^{(x)i}}\left[\frac{\theta_9}{\sum_{\zeta=9}^{10}\theta_\zeta}\left(Q_{10}^{(i)i}\varepsilon_9^{|9|i}\delta+\overline{v}_{10}^{|9|i}\delta\right)+\frac{\sum_{\zeta=10}^{10}\theta_\zeta}{\sum_{\zeta=9}^{10}\theta_\zeta}\left(A_{10}^i\varepsilon_9^{|9|i}\delta+A_{11}^i\varepsilon_{10}^{|9|i}\delta^2+\overline{v}_{11}^{|9|i}\delta^2\right)\right],$$

$$\overline{u}_8^{(3)i*}=\frac{1}{2\varphi_8^{(x)i}}\left[\frac{\theta_8}{\sum_{\zeta=8}^{10}\theta_\zeta}\left(Q_9^{(i)i}\varepsilon_8^{|8|i}\delta+\overline{v}_9^{|8|i}\delta\right)\right.$$
$$\left.+\frac{\sum_{\zeta=9}^{10}\theta_\zeta}{\sum_{\zeta=8}^{10}\theta_\zeta}\left(A_9^i\varepsilon_8^{|8|i}\delta+A_{10}^i\varepsilon_9^{|8|i}\delta^2+A_{11}^i\varepsilon_{10}^{|8|i}\delta^3+\overline{v}_{11}^{|8|i}\delta^3\right)\right],$$

$$\overline{u}_\tau^{(3)i*}=\frac{1}{2\varphi_\tau^{(x)i}}\left[\frac{\theta_\tau}{\sum_{\zeta=\tau}^{10}\theta_\zeta}\left(Q_{\tau+1}^{(i)i}\varepsilon_\tau^{|\tau|i}\delta+\overline{v}_{\tau+1}^{|\tau|i}\delta\right)+\frac{\sum_{\zeta=\tau+1}^{10}\theta_\zeta}{\sum_{\zeta=\tau}^{10}\theta_\zeta}\left(\sum_{t=\tau}^{\tau+2}A_{t+1}^i\varepsilon_t^{|\tau|i}\delta^{t+1-\tau}\right)\right],$$

for $i\in\{1,2\}$ and $\tau\in\{1,2,\cdots,7\}$;

$$\overline{u}_{10}^{(4)i*}=\frac{p_{10}^{|10|i}+\delta\overline{p}_{11}^{|10|i}}{2\varphi_{10}^i},$$

$$\overline{u}_9^{(4)i*}=\frac{1}{2\phi_9^i}\left[p_9^{|9|i}+\frac{\theta_9}{\sum_{\zeta=9}^{10}\theta_\zeta}\overline{p}_{10}^{|9|i}\delta+\frac{\sum_{\zeta=10}^{10}\theta_\zeta}{\sum_{\zeta=9}^{10}\theta_\zeta}\left(p_{10}^{|9|i}\delta+\overline{p}_{11}^{|9|i}\delta^2\right)\right],$$

$$\overline{u}_8^{(4)i*}=\frac{1}{2\phi_8^i}\left[p_8^{|8|i}+\frac{\theta_8}{\sum_{\zeta=8}^{10}\theta_\zeta}\overline{p}_9^{|8|i}\delta+\frac{\sum_{\zeta=9}^{10}\theta_\zeta}{\sum_{\zeta=8}^{10}\theta_\zeta}\left(p_9^{|8|i}\delta+p_{10}^{|8|i}\delta^2+\overline{p}_{11}^{|8|i}\delta^3\right)\right],$$

$$\overline{u}_\tau^{(4)i*}=\frac{1}{2\phi_\tau^i}\left[p_\tau^{|\tau|i}+\frac{\theta_\tau}{\sum_{\zeta=\tau}^{10}\theta_\zeta}\overline{p}_{\tau+1}^{|\tau|i}\delta+\frac{\sum_{\zeta=\tau+1}^{10}\theta_\zeta}{\sum_{\zeta=\tau}^{10}\theta_\zeta}\left(\sum_{t=\tau}^{\tau+2}p_{t+1}^{|\tau|i}\delta^{t+1-\tau}\right)\right],$$

for $i \in \{1,2\}$ and $\tau \in \{1,2,\cdots,7\}$. (6.8)

Comparing Proposition 6.1 with Proposition 4.1, one can obtain that $A_\tau^1 > A_\tau^{(1)1}$ and $A_\tau^2 > A_\tau^{(2)2}$, for $\tau \in \{1,2,\cdots,11\}$. Therefore, private capital investments under cooperation in (6.8) are greater than private capital investments under non-cooperation in (4.8). Substituting the optimal cooperative capital investment strategies into the accumulation processes of the private capital stock of agents in (6.2) yields

$$x_{k+1}^1 = x_k^1 + \varepsilon_k^{|k|1}\overline{u}_k^{(3)1*} + \sum_{t=k-2}^{k-1} \varepsilon_k^{|t|1}\overline{u}_t^{(3)1*} - \lambda_k^1 x_k^1 + \gamma_k^{(1)2} x_k^2, \quad x_1^1 = x_1^{1(0)},$$

$$x_{k+1}^2 = x_k^2 + \varepsilon_k^{|k|2}\overline{u}_k^{(3)2*} + \sum_{t=k-2}^{k-1} \varepsilon_k^{|t|2}\overline{u}_t^{(3)2*} - \lambda_k^2 x_k^2 + \gamma_k^{(2)1} x_k^1, \quad x_1^2 = x_1^{2(0)}. \quad (6.9)$$

In initial stage 1, $\overline{u}_{1-}^{(3)1}$ and $\overline{u}_{1-}^{(3)2}$ are zeros. Given $x_1^1 = x_1^{1(0)}$ and $x_1^2 = x_1^{2(0)}$, the values of x_2^1 and x_2^2 can be obtained. With x_2^1 and x_2^2, the values of x_3^1 and x_3^2 can be obtained. Performing the analysis repeatedly, the values of x_t^1 and x_t^2 for $t \in \{4,5,\cdots,11\}$ can be obtained. We use $\{x_t^{1*}, x_t^{2*}\}_{t=1}^{T+1}$ to denote the optimal cooperative trajectory of private capitals. With private capital investments under cooperation being greater than private capital investments under non-cooperation, we have $x_t^{1*} > x_t^{1**}$ and $x_t^{2*} > x_t^{2**}$ and outputs produced using private capitals will be higher under cooperation.

6.2 Subgame Consistent Payoff Distribution

Let $\xi(\tau, x; \underline{\overline{u}}_{\tau-}^{(3)}, \underline{\overline{u}}_{\tau-}^{(4)}) = [\xi^1(\tau, x; \underline{\overline{u}}_{\tau-}^{(3)}, \underline{\overline{u}}_{\tau-}^{(4)}), \xi^2(\tau, x; \underline{\overline{u}}_{\tau-}^{(3)}, \underline{\overline{u}}_{\tau-}^{(4)})]$ denote the agreed-upon imputation vector guiding the distribution of the total cooperative payoff in stage $\tau \in \{1,2,\cdots,10\}$. Since the agents agree to distribute the cooperative payoff proportional to their non-cooperative payoffs, the imputation to agent i is

$$\begin{aligned}
&\xi^i(\tau, x^*; \underline{\overline{u}}_{\tau-}^{(3)*}, \underline{\overline{u}}_{\tau-}^{(4)*}) \\
&= \frac{V^i(\tau, x^*; \underline{\overline{u}}_{\tau-}^{(3)*}, \underline{\overline{u}}_{\tau-}^{(4)*})}{\sum_{\ell=1}^2 V^\ell(\tau, x^*; \underline{\overline{u}}_{\tau-}^{(3)*}, \underline{\overline{u}}_{\tau-}^{(4)*})} W(\tau, x^*; \underline{\overline{u}}_{\tau-}^{(3)*}, \underline{\overline{u}}_{\tau-}^{(4)*}) \\
&= \frac{A_\tau^{(i)i} x_\tau^{i*} + A_\tau^{(i)j} x_\tau^{j*} + C_\tau^i}{\sum_{\ell=1}^2 (A_\tau^{(\ell)\ell} x_\tau^{\ell*} + A_\tau^{(\ell)j} x_\tau^{j*} + C_\tau^\ell)} (A_\tau^1 x_\tau^{1*} + A_\tau^2 x_\tau^{2*} + C_\tau)\delta^{\tau-1}, \quad (6.10)
\end{aligned}$$

for $i \in \{1,2\}$ at stage $\tau \in \{1,2,\cdots,10\}$.

Invoking Theorem 5.2, a payment equalling

$$\beta_\tau^i(x_\tau^*; \underline{\overline{u}}_{\tau-}^{(3)*}, \underline{\overline{u}}_{\tau-}^{(4)*})$$
$$= \delta^{1-\tau}\left(\xi^i(\tau, x_\tau^*; \underline{\overline{u}}_{\tau-}^{(3)*}, \underline{\overline{u}}_{\tau-}^{(4)*}) - \frac{\theta_\tau}{\sum_{\zeta=\tau}^T \theta_\zeta} q_{\tau+1}^i(x_{\tau+1}^*; \underline{\overline{u}}_{(\tau+1)-}^{(3)*}, \underline{\overline{u}}_{(\tau+1)-}^{(4)*})\delta^\tau \right.$$
$$\left. - \frac{\sum_{\zeta=\tau+1}^T \theta_\zeta}{\sum_{\zeta=\tau}^T \theta_\zeta} \xi^i(\tau+1, x_{\tau+1}^*; \underline{\overline{u}}_{(\tau+1)-}^{(3)*}, \underline{\overline{u}}_{(\tau+1)-}^{(4)*})\right), \tag{6.11}$$

given to player $i \in \{1, 2\}$ at stage $\tau \in \{1, 2, \cdots, 10\}$ would lead to the realization of the imputation vector $\xi(\tau, x^*; \underline{\overline{u}}_{\tau-}^{(3)*}, \underline{\overline{u}}_{\tau-}^{(4)*})$ in (6.10).

Using Proposition 6.1, the payment $\beta_\tau^i(x_\tau^*; \underline{\overline{u}}_{\tau-}^{(3)*}, \underline{\overline{u}}_{\tau-}^{(4)*})$ in (6.11) can be expressed as

$$\beta_\tau^i(x_\tau^*; \underline{\overline{u}}_{\tau-}^{(3)*}, \underline{\overline{u}}_{\tau-}^{(4)*})$$
$$= \frac{A_\tau^{(i)i} x_\tau^{i*} + A_\tau^{(i)j} x_\tau^{j*} + C_\tau^i}{\sum_{\ell=1}^2 (A_\tau^{(\ell)\ell} x_\tau^{\ell*} + A_\tau^{(\ell)j} x_\tau^{j*} + C_\tau^\ell)} (A_\tau^1 x_\tau^{1*} + A_\tau^2 x_\tau^{2*} + C_\tau)$$
$$- \frac{\theta_\tau}{\sum_{\zeta=\tau}^T \theta_\zeta}\left(Q_{\tau+1}^{(i)i} x_{\tau+1}^{i*} + \overline{v}_{\tau+1}^{(\tau)i} \overline{u}_\tau^{(3)i*}\right.$$
$$\left. + \sum_{t=\tau-2}^{\tau-1} \overline{v}_{\tau+1}^{(t)i} \overline{u}_t^{(3)i*} + \overline{p}_{\tau+1}^{(\tau)i} \overline{u}_\tau^{(4)i*} + \sum_{t=\tau+1-3}^{\tau-1} \overline{p}_{\tau+1}^{(t)i} \overline{u}_t^{(4)i*} + \varpi_{\tau+1}^i \right)\delta$$
$$- \frac{\sum_{\zeta=\tau+1}^T \theta_\zeta}{\sum_{\zeta=\tau}^T \theta_\zeta}\left(\frac{A_{\tau+1}^{(i)i} x_{\tau+1}^{i*} + A_{\tau+1}^{(i)j} x_{\tau+1}^{j*} + C_{\tau+1}^i}{\sum_{\ell=1}^2 (A_{\tau+1}^{(\ell)\ell} x_{\tau+1}^{\ell*} + A_{\tau+1}^{(\ell)j} x_{\tau+1}^{j*} + C_{\tau+1}^\ell)}\right.$$
$$\left. \cdot (A_{\tau+1}^1 x_{\tau+1}^{1*} + A_{\tau+1}^2 x_{\tau+1}^{2*} + C_{\tau+1})\delta\right). \tag{6.12}$$

Note that previously executed durable strategies $(\underline{\overline{u}}_{\tau-}^{(3)*}, \underline{\overline{u}}_{\tau-}^{(4)*})$ are embedded in $\beta_\tau^i(x_\tau^*; \underline{\overline{u}}_{\tau-}^{(3)*}, \underline{\overline{u}}_{\tau-}^{(4)*})$. Under cooperation, all agents would use the cooperative strategies and the single-stage payment that agent i will directly receive at stage τ along the cooperative trajectory $\{x_\tau^*\}_{\tau=1}^T$ with previously executed durable strategies $(\underline{\overline{u}}_{\tau-}^{(3)*}, \underline{\overline{u}}_{\tau-}^{(4)*})$ becomes

$$\pi_\tau^i(x_\tau^*; \underline{\overline{u}}_{\tau-}^{(3)*}, \underline{\overline{u}}_{\tau-}^{(4)*}) = R_\tau^i(u_\tau^{i*} x_\tau^{i*})^{1/2} - c_\tau^i u_\tau^{i*} - \phi_\tau^{(x)i}(\overline{u}_\tau^{(3)i*})^2 - \varphi_\tau^i(\overline{u}_\tau^{(4)i*})^2$$
$$+ p_{k|}^{|k|i} \overline{u}_k^{(4)i} + \sum_{t=\tau-3}^{\tau-1} p_\tau^{|t|i} \overline{u}_t^{(4)i*}.$$

However, according to the agreed-upon imputation, agent i will receive $\beta_\tau^i(x_\tau^*; \underline{\overline{u}}_{\tau-}^{(3)*}, \underline{\overline{u}}_{\tau-}^{(4)*})$ at stage τ. Therefore, a side-payment equalling

$$\beta_\tau^i(x_\tau^*; \underline{\overline{u}}_{\tau-}^{(3)*}, \underline{\overline{u}}_{\tau-}^{(4)*}) - \pi_\tau^i(x_\tau^*; \underline{\overline{u}}_{\tau-}^{(3)*}, \underline{\overline{u}}_{\tau-}^{(4)*}) \tag{6.13}$$

has to be given to agent $i \in \{1, 2\}$ in stage $\tau \in \{1, 2, \cdots, 10\}$ to yield the cooperative imputation $\xi(\tau, x^*; \underline{\overline{u}}_{\tau-}^{(3)*}, \underline{\overline{u}}_{\tau-}^{(4)*})$ in (6.10).

A couple of novel results in cooperative dynamic game analysis is identified in the following remarks.

Remark 6.1. The positive spillover effect from agent i's capital investment to agent j's private capital stock could be substantial and cooperation would enhance the effects of the agents' durable capital investments to the expected joint payoff. In formulating the durable capital investment strategies $\overline{u}_\tau^{(3)i}$ in (6.8), the agents have to consider

(i) the potential cooperative gains

$$\left(\sum_{t=\tau}^{\tau+2} A_{t+1}^i \varepsilon_t^{|\tau|i} \delta^{t+1-\tau}\right);$$

(ii) the probability that the game would terminate in the next stage, that is $\theta_\tau / \sum_{\zeta=\tau}^{10} \theta_\zeta$; and the probability that the game would continue onward, that is $\sum_{\zeta=\tau+1}^{10} \theta_\zeta / \sum_{\zeta=\tau}^{10} \theta_\zeta$; and

(iii) the future return of capital investments if the game would terminate in the next stage, that is

$$\sum_{i=1}^{2} \left(Q_{\tau+1}^{(i)i} \varepsilon_\tau^{|\tau|i} \delta + \overline{v}_{\tau+1}^{|\tau|i} \delta\right).$$

■

Remark 6.2. Under cooperation, the maximization of the expected joint payoff would take into consideration the terminal payoffs of the agents jointly if the game stops prematurely after stage τ, that is

$$\left(Q_{\tau+1}^{(i)i} x_{\tau+1}^i + \overline{v}_{\tau+1}^{|\tau|i} u_\tau^{(3)i} + \sum_{t=\tau-2}^{\tau-1} \overline{v}_{\tau+1}^{|t|i} \overline{u}_t^{(3)i} + \overline{p}_{\tau+1}^{|\tau|i} \overline{u}_\tau^{(4)i} + \sum_{t=\tau+1-3}^{\tau-1} \overline{p}_{\tau+1}^{|t|i} \overline{u}_t^{(4)i} + \varpi_{\tau+1}^i\right),$$

for $i \in \{1, 2\}$ and $\tau \in \{1, 2, \cdots, 9\}$. ■

The outcomes identified in Remarks 6.1 and 6.2 reflect outcomes which are novel in cooperative dynamic game analysis.

7 Appendices

7.1 Appendix I: Proof of Theorem 2.1

According to (2.5), the value function at stage $T+1$ is

$$V(T+1, x; \overline{u}_{(T+1)-}) = q_{T+1}(x_{T+1}; \overline{u}_{(T+1)-})\delta_1^{T+1}$$
$$= q_{T+1}(x_{T+1}; \overline{u}_T, \overline{u}_{(T+1)-} \cap \overline{u}_{T-})\delta_1^{T+1}. \quad (7.1)$$

We first consider the case when the last stage T has been reached. The optimization problem then becomes

$$V(T, x; \overline{u}_{T-}) = \max_{u_T, \overline{u}_T}\{g_T(x, u_T, \overline{u}_T; \overline{u}_{T-})\delta_1^T$$
$$+ q_{T+1}(f_T(x, u_T, \overline{u}_T; \overline{u}_{T-}); \overline{u}_T, \overline{u}_{(T+1)-} \cap \overline{u}_{T-})\delta_1^{T+1}\}. \quad (7.2)$$

The maximization operator in stage T involves $(u_T, \overline{u}_T)$ only and $\overline{u}_{(T+1)-} \cap \overline{u}_{T-}$ is a subset of $\overline{u}_{T-}$. The current state x and the previously executed controls u_{T-} appear in the stage T maximization problem as given parameters. If the first order conditions of the maximization problem in (7.2) satisfy the implicit function theorem, one can obtain the optimal controls $(u_T, \overline{u}_T)$ as functions of x and u_{T-}. Substituting these optimal controls into the function on the right-hand-side of (7.2) yields the function $V(T, x; \overline{u}_{T-})$, which satisfies the optimal conditions of a maximum for given x and $\overline{u}_{T-}$.

Now consider the problem in stage $\tau \in \{T-1, T-2, \cdots, 1\}$, according to (2.3) one has to maximize the expected payoff

$$\sum_{\hat{T}=\tau}^{T} \frac{\theta_{\hat{T}}}{\sum_{\zeta=\tau}^{T} \theta_\zeta} \left\{ \sum_{k=\tau}^{\hat{T}} g_k(x_k, u_k, \overline{u}_k; \overline{u}_{k-})\delta_1^k + q_{\hat{T}+1}(x_{\hat{T}+1}; \overline{u}_{(\hat{T}+1)-})\delta_1^{\hat{T}+1} \right\}. \quad (7.3)$$

The expected payoff in (7.3) can be expressed as

$$g_\tau(x_\tau, u_\tau, \overline{u}_\tau; \overline{u}_{\tau-})\delta_1^\tau + \frac{\theta_\tau}{\sum_{\zeta=\tau}^{T} \theta_\zeta} q_{\tau+1}(x_{\tau+1}; \overline{u}_{(\tau+1)-})\delta_1^{\tau+1}$$
$$+ \frac{\sum_{\hat{T}=\tau+1}^{T} \theta_{\hat{T}}}{\sum_{\zeta=\tau}^{T} \theta_\zeta} \left\{ \sum_{k=\tau+1}^{\hat{T}} g_k(x_k, u_k, \overline{u}_k; \overline{u}_{k-})\delta_1^k + q_{\hat{T}+1}(x_{\hat{T}+1}; \overline{u}_{(\hat{T}+1)-})\delta_1^{\hat{T}+1} \right\}$$
$$= g_\tau(x_\tau, u_\tau, \overline{u}_\tau; \overline{u}_{\tau-})\delta_1^\tau + \frac{\theta_\tau}{\sum_{\zeta=\tau}^{T} \theta_\zeta} q_{\tau+1}(x_{\tau+1}; \overline{u}_{(\tau+1)-})\delta_1^{\tau+1}$$
$$+ \frac{\sum_{\zeta=\tau+1}^{T} \theta_\zeta}{\sum_{\zeta=\tau}^{T} \theta_\zeta} \left[\frac{\sum_{\hat{T}=\tau+1}^{T} \theta_{\hat{T}}}{\sum_{\zeta=\tau+1}^{T} \theta_\zeta} \left\{ \sum_{k=\tau+1}^{\hat{T}} g_k(x_k, u_k, \overline{u}_k; \overline{u}_{k-})\delta_1^k + q_{\hat{T}+1}(x_{\hat{T}+1}; \overline{u}_{(\hat{T}+1)-})\delta_1^{\hat{T}+1} \right\} \right].$$
$$(7.4)$$

Note that from (2.3), the term inside the squared brackets of (7.4) can be expressed as

$$\left[\frac{\sum_{\hat{T}=\tau+1}^{T}\theta_{\hat{T}}}{\sum_{\zeta=\tau+1}^{T}\theta_{\zeta}}\left\{\sum_{k=\tau+1}^{\hat{T}} g_k(x_k, u_k, \overline{u}_k; \overline{u}_{k-})\delta_1^k + q_{\hat{T}+1}(x_{\hat{T}+1}; \overline{u}_{(\hat{T}+1)-})\delta_1^{\hat{T}+1}\right\}\right]$$

$$= \sum_{\hat{T}=\tau+1}^{T} \frac{\theta_{\hat{T}}}{\sum_{\zeta=\tau+1}^{T}\theta_{\zeta}}\left\{\sum_{k=\tau+1}^{\hat{T}} g_k(x_k, u_k, \overline{u}_k; \overline{u}_{k-})\delta_1^k + q_{\hat{T}+1}(x_{\hat{T}+1}; \overline{u}_{(\hat{T}+1)-})\delta_1^{\hat{T}+1}\right\}$$

$$= E\left\{\sum_{k=\tau+1}^{\hat{T}} g_k(x_k, u_k, \overline{u}_k; \overline{u}_{k-})\delta_1^k + q_{\hat{T}+1}(x_{\hat{T}+1}; \overline{u}_{(\hat{T}+1)-})\delta_1^{\hat{T}+1}\right\}. \tag{7.5}$$

Invoking Theorem 2.1, $V(\tau + 1, x; \overline{u}_{(\tau+1)-})$ is the maximal value of

$$E\left\{\sum_{k=\tau+1}^{\hat{T}} g_k(x_k, u_k, \overline{u}_k; \overline{u}_{k-})\delta_1^k + q_{\hat{T}+1}(x_{\hat{T}+1}; \overline{u}_{(\hat{T}+1)-})\delta_1^{\hat{T}+1}\right\}.$$

Therefore, the expected payoff to be maximized in stage τ, as specified in (7.4), can be expressed as

$$g_\tau(x, u_\tau, \overline{u}_\tau; \overline{u}_{\tau-})\delta_1^\tau + \frac{\theta_\tau}{\sum_{\zeta=\tau}^{T}\theta_\zeta} q_{\tau+1}[f_\tau(x, u_\tau, \overline{u}_\tau; \overline{u}_{\tau-}), \overline{u}_\tau, \overline{u}_{(\tau+1)-} \cap \overline{u}_{\tau-}]\delta_1^{\tau+1}$$
$$+ \frac{\sum_{\zeta=\tau+1}^{T}\theta_\zeta}{\sum_{\zeta=\tau}^{T}\theta_\zeta} V[\tau + 1, f_\tau(x, u_\tau, \overline{u}_\tau; \overline{u}_{\tau-}); \overline{u}_\tau, \overline{u}_{(\tau+1)-} \cap \overline{u}_{\tau-}]. \tag{7.6}$$

The τth equation in (2.7) of Theorem 2.1 indeed leads to

$$\begin{aligned} &V(\tau, x; \overline{u}_{\tau-}) \\ &= \max_{u_\tau, \overline{u}_\tau}\Big\{g_\tau(x, u_\tau, \overline{u}_\tau; \overline{u}_{\tau-})\delta_1^\tau \\ &\quad + \frac{\theta_\tau}{\sum_{\zeta=\tau}^{T}\theta_\zeta} q_{\tau+1}[f_\tau(x, u_\tau, \overline{u}_\tau; \overline{u}_{\tau-}); \overline{u}_\tau, \overline{u}_{(\tau+1)-} \cap \overline{u}_{\tau-}]\delta_1^{\tau+1} \\ &\quad + \frac{\sum_{\zeta=\tau+1}^{T}\theta_\zeta}{\sum_{\zeta=\tau}^{T}\theta_\zeta} V[\tau + 1, f_\tau(x, u_\tau, \overline{u}_\tau; \overline{u}_{\tau-}); \overline{u}_\tau, \overline{u}_{(\tau+1)-} \cap \overline{u}_{\tau-}]\Big\}, \\ &\text{for } \tau \in \{1, 2, \cdots, T-2\}. \end{aligned} \tag{7.7}$$

The maximization operator involves $(u_\tau, \overline{u}_\tau)$ and $\overline{u}_{(\tau+1)-} \cap \overline{u}_{\tau-}$ is a subset of $\overline{u}_{\tau-}$. Again, the current state x and the previously executed controls $\overline{u}_{\tau-}$ appear in the stage τ optimization problem. If the first order conditions of the maximization problem in (7.7) satisfy the implicit function theorem, one can obtain the optimal controls $(u_\tau, \overline{u}_\tau)$ as functions of x and $\overline{u}_{\tau-}$. Substituting these optimal controls into the function on the right-hand-side of (7.7) yields the function $V(\tau, x; \overline{u}_{\tau-})$.

Hence Theorem 2.1 follows. ■

7.2 Appendix II: Proof of Proposition 4.1

Using (4.4), we have

$$A_{11}^{(i)i} = Q_{11}^{(i)i},\ A_{11}^{(i)j} = 0,\ C_{11}^{i} = \sum_{t=11-2}^{10} \overline{v}_{11}^{|t|i} \overline{u}_t^{(3)i**} + \sum_{t=11-3}^{10} \overline{p}_{11}^{|t|i} \overline{u}_t^{(4)i**} + \varpi_{11}^{i}, \tag{7.8}$$

for $i, j \in \{1, 2\}$ and $i \neq j$.

Invoking the technique of backward induction, we consider first the last operational stage 10. Using Proposition 4.1 and (4.4), we can express the game equilibrium strategies in stage 10 as:

$$u_{10}^{i**} = \frac{(R_{10}^{i})^2}{4(c_{10}^{i})^2} x^i,\ \overline{u}_{10}^{(3)i**} = \frac{(A_{11}^{(i)i} \varepsilon_{10}^{|10|i} + \overline{v}_{11}^{|10|i})\delta}{2\varphi_{10}^{(x)i}},\ \overline{u}_{10}^{(4)i**} = \frac{p_{10}^{|10|i} + \delta \overline{p}_{11}^{|10|i}}{2\phi_{10}^{i}}, \quad \text{for } i \in \{1, 2\}. \tag{7.9}$$

Substituting the optimal strategies in (7.9) into the stage 10 equation of (4.5) we obtain

$$\begin{aligned}
&V^i(10, x; \overline{u}_{10-}^{(3)**}, \overline{u}_{10-}^{(4)**}) \\
&= (A_{10}^{(i)i} x^i + A_{10}^{(i)j} x^j + C_{10}^{i})\delta^9 \\
&= \left(\frac{(R_{10}^{i})^2}{4c_{10}^{i}} x^i - \frac{(A_{11}^{(i)i} \varepsilon_{10}^{|10|i} + \overline{v}_{11}^{|10|i})^2 \delta^2}{4\varphi_{10}^{(x)i}} - \frac{(p_{10}^{|10|i} + \delta \overline{p}_{11}^{|10|i})^2}{4\phi_{10}^{i}} \right. \\
&\quad \left. + \frac{p_{10}^{|10|i}(p_{10}^{|10|i} + \delta \overline{p}_{11}^{|10|i})}{2\phi_{10}^{i}} + \sum_{t=10-3}^{9} p_{10}^{|t|i} \overline{u}_t^{(4)i**} \right) \delta^9 \\
&\quad + \left(A_{11}^{(i)i} [x^i + \varepsilon_{10}^{|10|i} \frac{(A_{11}^{(i)i} \varepsilon_{10}^{|10|i} + \overline{v}_{11}^{|10|i})\delta}{2\varphi_{10}^{(x)i}} + \sum_{t=10-2}^{9} \varepsilon_{10}^{|t|i} \overline{u}_t^{(3)i**} \right. \\
&\quad - \lambda_{10}^{i} x^i + \gamma_{10}^{(i)j} x^j] + \overline{v}_{11}^{(10)i} \frac{(A_{11}^{(i)i} \varepsilon_{10}^{|10|i} + \overline{v}_{11}^{|10|i})\delta}{2\varphi_{10}^{(x)i}} \\
&\quad + \sum_{t=10-2}^{9} \overline{v}_{11}^{|t|i} \overline{u}_t^{(3)i**} + \frac{\overline{p}_{11}^{|10|i}(p_{10}^{|10|i} + \delta \overline{p}_{11}^{|10|i})}{2\phi_{10}^{i}}
\end{aligned}$$

$$+\sum_{t=11-3}^{9}\overline{p}_{11}^{|t|i}\overline{u}_t^{(4)i**}+\varpi_{11}^{i}\Bigg)\delta^{10},\quad i,j\in\{1,2\}\text{ and } i\neq j. \tag{7.10}$$

System (7.10) yields a system of equations which right-hand-side and left-hand-side are linear functions of (x^1, x^2). For (7.10) to hold, it is required that:

$$A_{10}^{(i)i}x^i=\frac{(R_{10}^i)^2x^i}{4c_{10}^i}+\delta A_{11}^{(i)i}(1-\lambda_{10}^i)x^i,$$
$$A_{10}^{(i)j}x^j=\delta A_{11}^{(i)i}\gamma_T^{(i)j}x^j. \tag{7.11}$$

From (7.11) we obtain:

$$A_{10}^{(i)i}=\frac{(R_{10}^i)^2}{4c_{10}^i}+\delta A_{11}^{(i)i}(1-\lambda_{10}^i),$$
$$A_{10}^{(i)j}=\delta A_{11}^{(i)i}\gamma_T^{(i)j}. \tag{7.12}$$

In addition,

$$C_{10}^i$$
$$=-\frac{(A_{11}^{(i)i}\varepsilon_{10}^{|10|i}+\overline{v}_{11}^{|10|i})^2\delta^2}{4\varphi_{10}^{(x)i}}-\frac{(p_{10}^{|10|i}+\delta\overline{p}_{11}^{|10|i})^2}{4\phi_{10}^i}$$
$$+\frac{p_{10}^{|10|i}(p_{10}^{|10|i}+\delta\overline{p}_{11}^{|10|i})}{2\phi_{10}^i}+\sum_{t=10-3}^{9}p_{10}^{|t|i}\overline{u}_t^{(4)i**}\Bigg)$$
$$+\Bigg(A_{11}^{(i)i}[\varepsilon_{10}^{|10|i}\frac{(A_{11}^{(i)i}\varepsilon_{10}^{|10|i}+\overline{v}_{11}^{|10|i})\delta}{2\varphi_{10}^{(x)i}}+\sum_{t=10-2}^{9}\varepsilon_{10}^{|t|i}\overline{u}_t^{(3)i**}]$$
$$+\overline{v}_{11}^{(10)i}\frac{(A_{11}^{(i)i}\varepsilon_{10}^{|10|i}+\overline{v}_{11}^{|10|i})\delta}{2\varphi_{10}^{(x)i}}$$
$$+\sum_{t=10-2}^{9}\overline{v}_{11}^{|t|i}\overline{u}_t^{(3)i**}+\frac{\overline{p}_{11}^{|10|i}(p_{10}^{|10|i}+\delta\overline{p}_{11}^{|10|i})}{2\phi_{10}^i}+\sum_{t=11-3}^{9}\overline{p}_{11}^{|t|i}\overline{u}_t^{(4)i**}+\varpi_{11}^i\Bigg)\delta,$$
$$\text{for } i,j\in\{1,2\}\text{ and } i\neq j. \tag{7.13}$$

which is an expression containing previously executed controls $(\underline{\overline{u}}_{10-}^{(3)**}, \underline{\overline{u}}_{10-}^{(4)**})$.

Then we move to stage 9.

Using $V^i(10, x; \underline{\overline{u}}_{10-}^{(3)**}, \underline{\overline{u}}_{10-}^{(4)**}) = (A_{10}^{(i)i}x^i + A_{10}^{(i)j}x^j + C_{10}^i)\delta^9$ derived in (7.10)–(7.13) and the stage 9 equation in (4.6), the game equilibrium strategies in stage 9 can be expressed as:

$u_9^{i**} = \frac{(R_9^i)^2}{4(c_9^i)^2}x^i$,

$$\overline{u}_9^{(3)i**} = \frac{1}{2\varphi_9^{(x)i}}\Bigg[\frac{\theta_9}{\sum_{\zeta=9}^{10}\theta_\zeta}\left(Q_{10}^{(i)i}\varepsilon_9^{|9|i}\delta + \overline{v}_{10}^{|9|i}\delta\right)$$
$$+\frac{\sum_{\zeta=10}^{10}\theta_\zeta}{\sum_{\zeta=9}^{10}\theta_\zeta}\left(A_{10}^{(i)i}\varepsilon_9^{|9|i}\delta + A_{11}^{(i)i}\varepsilon_{10}^{|9|i}\delta^2 + \overline{v}_{11}^{|9|i}\delta^2\right)\Bigg],$$

$$u_9^{(4)i**} = \frac{1}{2\varphi_9^i}\left[p_9^{|9|i} + \frac{\theta_9}{\sum_{\zeta=9}^{10}\theta_\zeta}\overline{p}_{10}^{|9|i}\delta + \frac{\sum_{\zeta=10}^{10}\theta_\zeta}{\sum_{\zeta=9}^{10}\theta_\zeta}\left(p_{10}^{|9|i}\delta + \overline{p}_{11}^{|9|i}\delta^2\right)\right], \tag{7.14}$$

for $i \in \{1, 2\}$.

Substituting the optimal strategies in (7.14) into the stage 9 equation of (4.6) we obtain

$$V^i(9, x; \underline{\overline{u}}_{9-}^{(3)**}, \underline{\overline{u}}_{9-}^{(4)**})$$
$$= (A_9^{(i)i}x^i + A_9^{(i)j}x^j + C_9^i)\delta^8$$
$$= \left(\frac{(R_9^i)^2}{4c_9^i}x^i - \varphi_9^{(x)i}(\overline{u}_9^{(3)i**})^2 - \phi_9^i(\overline{u}_9^{(4)i**})^2 + p_9^{|9|i}\overline{u}_9^{(4)i**} + \sum_{t=9-3}^{8} p_9^{|t|i}\overline{u}_t^{(4)i**}\right)\delta^8$$
$$+\frac{\theta_9}{\sum_{\zeta=9}^{10}\theta_\zeta}\Bigg(Q_{10}^{(i)i}[x^i + \varepsilon_9^{|9|i}\overline{u}_9^{(3)i**} + \sum_{t=9-2}^{8}\varepsilon_9^{|t|i}\overline{u}_t^{(3)i**} - \lambda_9^i x^i + \gamma_9^{(i)j}x^j]$$
$$+\overline{v}_{10}^{|9|i}\overline{u}_9^{(3)i**} + \sum_{t=9-2}^{8}\overline{v}_{10}^{|t|i}\overline{u}_t^{(3)i**} + \overline{p}_{10}^{|9|i}\overline{u}_9^{(4)i**} + \sum_{t=9+1-3}^{8}\overline{p}_{10}^{|t|i}\overline{u}_t^{(4)i**} + \varpi_{10}^i\Bigg)\delta^9$$
$$+\frac{\sum_{\zeta=10}^{10}\theta_\zeta}{\sum_{\zeta=9}^{10}\theta_\zeta}\Bigg[A_{10}^{(i)i}[x^i + \varepsilon_9^{|9|i}\overline{u}_9^{(3)i**} + \sum_{t=9-2}^{8}\varepsilon_9^{|t|i}\overline{u}_t^{(3)i**} - \lambda_9^i x^i + \gamma_9^{(i)j}x^j]$$
$$+A_{10}^{(i)j}[x^j + \varepsilon_9^{|9|j}\overline{u}_9^{(3)j**} + \sum_{t=9-2}^{8}\varepsilon_9^{|t|j}\overline{u}_t^{(3)j**} - \lambda_9^j x^j + \gamma_9^{(j)i}x^i]$$
$$+\Bigg(\frac{p_{10}^{|10|i}(p_{10}^{|10|i} + \delta\overline{p}_{11}^{|10|i})}{2\phi_{10}^i}$$
$$-\frac{(A_{11}^{(i)i}\varepsilon_{10}^{|10|i} + \overline{v}_{11}^{|10|i})^2\delta^2}{4\varphi_{10}^{(x)i}} - \frac{(p_{10}^{|10|i} + \delta\overline{p}_{11}^{|10|i})^2}{4\phi_{10}^i} + p_{10}^{|9|i}\overline{u}_9^{(4)i**} + \sum_{t=10-3}^{8} p_{10}^{|t|i}\overline{u}_t^{(4)i**}$$
$$+\Bigg(A_{11}^{(i)i}[\varepsilon_{10}^{|10|i}\frac{(A_{11}^{(i)i}\varepsilon_{10}^{|10|i} + \overline{v}_{11}^{|10|i})\delta}{2\varphi_{10}^{(x)i}} + \varepsilon_{10}^{|9|i}\overline{u}_9^{(3)i**} + \sum_{t=10-2}^{8}\varepsilon_{10}^{|t|i}\overline{u}_t^{(3)i**}]$$
$$+\overline{v}_{11}^{|10|i}\frac{(A_{11}^{(i)i}\varepsilon_{10}^{|10|i} + \overline{v}_{11}^{|10|i})\delta}{2\varphi_{10}^{(x)i}} + \overline{v}_{11}^{|9|i}\overline{u}_9^{(3)i**} + \sum_{t=10-2}^{8}\overline{v}_{11}^{|t|i}\overline{u}_t^{(3)i**} + \frac{\overline{p}_{11}^{|10|i}(p_{10}^{|10|i} + \delta\overline{p}_{11}^{|10|i})}{2\phi_{10}^i}$$
$$+\overline{p}_{11}^{|9|i}u_9^{(4)i**} + \sum_{t=11-3}^{8}\overline{p}_{11}^{|t|i}u_t^{(4)i**} + \varpi_{11}^i\Bigg)\delta\Bigg]\delta^9, \quad i, j \in \{1, 2\} \text{ and } i \neq j, \tag{7.15}$$

where $u_9^{(3)i**}$, $\overline{u}_9^{(3)j**}$, $\overline{u}_9^{(4)i**}$ and $\overline{u}_9^{(4)j**}$ are given in (7.14).

System (7.15) yields a system of equations which right-hand-side and left-hand-side are linear functions of (x^1, x^2). For (7.15) to hold, it is required that:

$$A_9^{(i)i} x^i = \frac{(R_9^i)^2 x^i}{4c_9^i} + \frac{\theta_9}{\sum_{\zeta=9}^{10} \theta_\zeta} \delta Q_{10}^{(i)i} (1 - \lambda_9^i) x^i + \frac{\sum_{\zeta=10}^{10} \theta_\zeta}{\sum_{\zeta=9}^{10} \theta_\zeta} \left(A_{10}^{(i)i} (1 - \lambda_9^i) x^i + A_{10}^{(i)j} \gamma_9^{(j)i} x^i \right) \delta,$$

$$A_9^{(i)j} x^j = \frac{\theta_9}{\sum_{\zeta=9}^{10} \theta_\zeta} \left(A_{10}^{(i)i} \gamma_9^{(i)j} x^j + A_{10}^{(i)j} (1 - \lambda_9^j) x^j \right) \delta,$$

$$i, j \in \{1, 2\} \text{ and } i \neq j. \tag{7.16}$$

From (7.16) we obtain:

$$A_9^{(i)i} = \frac{(R_9^i)^2}{4c_9^i} + \frac{\theta_9}{\sum_{\zeta=9}^{10} \theta_\zeta} \delta Q_{10}^{(i)i} (1 - \lambda_9^i) + \frac{\sum_{\zeta=10}^{10} \theta_\zeta}{\sum_{\zeta=9}^{10} \theta_\zeta} \left(A_{10}^{(i)i} (1 - \lambda_9^i) + A_{10}^{(i)j} \gamma_9^{(j)i} \right) \delta,$$

$$A_9^{(i)j} = \frac{\theta_9}{\sum_{\zeta=9}^{10} \theta_\zeta} \left(A_{10}^{(i)i} \gamma_9^{(i)j} + A_{10}^{(i)j} (1 - \lambda_9^j) \right) \delta, \tag{7.17}$$

In addition, C_9^i is an expression including all the terms on the right-hand-side of (7.15) which do not involve x^1 or x^2. It is a function of previously executed controls $(\overline{u}_{9-}^{(3)**}, \overline{u}_{9-}^{(4)**})$.

Then, we move to stage 8.

Using $V^i(9, x; \overline{u}_{9-}^{(3)**}, \overline{u}_{9-}^{(4)**}) = (A_9^{(i)i} x^i + A_9^{(i)j} x^j + C_9^i)\delta^9$ derived in (7.15)–(7.17) and the stage 8 equation in (4.6), the game equilibrium strategies in stage 8 can be expressed as:

$$u_8^{i**} = \frac{(R_8^i)^2}{4(c_8^i)^2} x^i,$$

$$\overline{u}_8^{(3)i**} = \frac{1}{2\varphi_8^{(x)i}} \left[\frac{\theta_8}{\sum_{\zeta=8}^{10} \theta_\zeta} \left(Q_9^{(i)i} \varepsilon_8^{|8|i} \delta + \overline{v}_9^{|8|i} \delta \right) + \frac{\sum_{\zeta=9}^{10} \theta_\zeta}{\sum_{\zeta=8}^{10} \theta_\zeta} \left(A_9^{(i)i} \varepsilon_8^{|8|i} \delta + A_{10}^{(i)i} \varepsilon_9^{|8|i} \delta^2 + A_{11}^{(i)i} \varepsilon_{10}^{|8|i} \delta^3 + \overline{v}_{11}^{|8|i} \delta^3 \right) \right],$$

$$\overline{u}_8^{(4)i**} = \frac{1}{2\phi_8^i} \left[p_8^{|8|i} + \frac{\theta_8}{\sum_{\zeta=8}^{10} \theta_\zeta} \overline{p}_9^{|8|i} \delta + \frac{\sum_{\zeta=9}^{10} \theta_\zeta}{\sum_{\zeta=8}^{10} \theta_\zeta} \left(p_9^{|8|i} \delta + p_{10}^{|8|i} \delta^2 + \overline{p}_{11}^{|8|i} \delta^3 \right) \right], \tag{7.18}$$

for $i \in \{1, 2\}$.

Substituting the optimal strategies in (7.18) into the stage 8 equation of (4.6) we obtain

$$V^i(8, x; \underline{u}_{8-}^{(3)**}, \underline{u}_{8-}^{(4)**})$$

$$= (A_8^{(i)i} x^i + A_8^{(i)j} x^j + C_8^i)\delta^7$$

$$= \left(\frac{(R_8^i)^2}{4c_8^i} x^i - \varphi_8^{(x)i} (\overline{u}_8^{(3)i**})^2 - \phi_8^i (\overline{u}_8^{(4)i**})^2 + p_8^{|8|i} \overline{u}_8^{(4)i**} + \sum_{t=8-3}^{7} p_8^{|t|i} \overline{u}_t^{(4)i**} \right) \delta^7$$

$$+ \frac{\theta_8}{\sum_{\zeta=8}^{10} \theta_\zeta} \left(Q_9^{(i)i} [x^i + \varepsilon_8^{|8|i} \overline{u}_8^{(3)i**} + \sum_{t=8-2}^{7} \varepsilon_8^{|t|i} \overline{u}_t^{(3)i**} - \lambda_8^i x^i + \gamma_8^{(i)j} x^j] \right.$$

$$\left. + \overline{v}_9^{|8|i} \overline{u}_8^{(3)i**} + \sum_{t=8-2}^{7} \overline{v}_8^{|t|i} \overline{u}_t^{(3)i**} + \overline{p}_9^{|8|i} \overline{u}_8^{(4)i**} + \sum_{t=8+1-3}^{7} \overline{p}_9^{|t|i} \overline{u}_t^{(4)i**} + \varpi_9^i \right) \delta^8$$

$$+ \frac{\sum_{\zeta=9}^{10} \theta_\zeta}{\sum_{\zeta=8}^{10} \theta_\zeta} \left[A_9^{(i)i} [x^i + \varepsilon_8^{|8|i} \overline{u}_8^{(3)i**} + \sum_{t=8-2}^{7} \varepsilon_8^{|t|i} \overline{u}_t^{(3)i**} - \lambda_8^i x^i + \gamma_8^{(i)j} x^j] \right.$$

$$\left. + A_9^{(i)j} [x^j + \varepsilon_8^{|8|j} \overline{u}_8^{(3)j**} + \sum_{t=8-2}^{7} \varepsilon_8^{|t|j} \overline{u}_t^{(3)j**} - \lambda_8^j x^j + \gamma_8^{(j)i} x^i] + C_9^i \right] \delta^8, \quad (7.19)$$

$i, j \in \{1, 2\}$ and $i \neq j$,
where $\overline{u}_8^{(3)i**}$, $\overline{u}_8^{(3)j**}$, $\overline{u}_8^{(4)i**}$ and $\overline{u}_8^{(4)j**}$ are given in (7.18).

System (7.18) yields a system of equations which right-hand-side and left-hand-side are linear functions of (x^1, x^2). For (7.19) to hold, it is required that:

$$A_8^{(i)i} x^i = \frac{(R_8^i)^2 x^i}{4c_8^i} + \frac{\theta_8}{\sum_{\zeta=8}^{10} \theta_\zeta} \delta Q_9^{(i)i} (1 - \lambda_8^i) x^i$$
$$+ \frac{\sum_{\zeta=9}^{10} \theta_\zeta}{\sum_{\zeta=8}^{10} \theta_\zeta} \left(A_9^{(i)i} (1 - \lambda_8^i) x^i + A_9^{(i)j} \gamma_8^{(j)i} x^i \right) \delta,$$

$$A_8^{(i)j} x^j = \frac{\theta_8}{\sum_{\zeta=8}^{10} \theta_\zeta} \left(A_9^{(i)i} \gamma_8^{(i)j} x^j + A_9^{(i)j} (1 - \lambda_8^j) x^j \right) \delta,$$
$$i, j \in \{1, 2\} \text{ and } i \neq j. \quad (7.20)$$

From (7.20) we obtain:

$$A_8^{(i)i} = \frac{(R_8^i)^2}{4c_8^i} + \frac{\theta_8}{\sum_{\zeta=8}^{10} \theta_\zeta} \delta Q_9^{(i)i} (1 - \lambda_8^i) + \frac{\sum_{\zeta=9}^{10} \theta_\zeta}{\sum_{\zeta=8}^{10} \theta_\zeta} \left(A_9^{(i)i} (1 - \lambda_8^i) + A_9^{(i)j} \gamma_8^{(j)i} \right) \delta,$$

$$A_8^{(i)j} = \frac{\theta_8}{\sum_{\zeta=8}^{10} \theta_\zeta} \left(A_9^{(i)i} \gamma_8^{(i)j} + A_9^{(i)j} (1 - \lambda_8^j) \right) \delta, \quad (7.21)$$

$i, j \in \{1, 2\}$ and $i \neq j$.

In addition, C_8^i is an expression including all the terms on the right-hand-side of (7.19) which do not involve x^1 or x^2. It is an expression containing previously executed controls $(\overline{\underline{u}}_{8-}^{(3)**}, \overline{\underline{u}}_{8-}^{(4)**})$.

Repeating the process for stages $\tau \in \{7, 6, \cdots, 1\}$, we obtain

$$u_\tau^{i**} = \frac{(R_\tau^i)^2}{4(c_\tau^i)^2} x^i,$$

$$\overline{u}_\tau^{(3)i**} = \frac{1}{2\varphi_\tau^{(x)i}} \left[\frac{\theta_\tau}{\sum_{\zeta=\tau}^{10} \theta_\zeta} \left(Q_{\tau+1}^{(i)i} \varepsilon_\tau^{|\tau|i} \delta + \overline{v}_{\tau+1}^{|\tau|i} \delta \right) + \frac{\sum_{\zeta=\tau+1}^{10} \theta_\zeta}{\sum_{\zeta=\tau}^{10} \theta_\zeta} \left(\sum_{t=\tau}^{\tau+2} A_{t+1}^{(i)i} \varepsilon_t^{|\tau|i} \delta^{t+1-\tau} \right) \right],$$

$$\overline{u}_\tau^{(4)i**} = \frac{1}{2\phi_\tau^i} \left[p_\tau^{|\tau|i} + \frac{\theta_\tau}{\sum_{\zeta=\tau}^{10} \theta_\zeta} \overline{p}_{\tau+1}^{|\tau|i} \delta + \frac{\sum_{\zeta=\tau+1}^{10} \theta_\zeta}{\sum_{\zeta=\tau}^{10} \theta_\zeta} \left(\sum_{t=\tau}^{\tau+2} p_{t+1}^{|\tau|i} \delta^{t+1-\tau} \right) \right], \quad (7.22)$$

for $i \in \{1, 2\}$.

The game equilibrium payoff of agent i for stage $\tau \in \{7, 6, \cdots, 1\}$ can be obtained as

$$V^i(\tau, x; \overline{\underline{u}}_{\tau-}^{(3)**}, \overline{\underline{u}}_{\tau-}^{(4)**}) = (A_\tau^{(i)i} x^i + A_\tau^{(i)j} x^j + C_\tau^i) \delta^{\tau-1},$$

where

$$A_\tau^{(i)i} = \frac{(R_\tau^i)^2}{4c_\tau^i} + \frac{\theta_\tau}{\sum_{\zeta=\tau}^{10} \theta_\zeta} \delta Q_{\tau+1}^{(i)i} (1 - \lambda_\tau^i) + \frac{\sum_{\zeta=\tau+1}^{10} \theta_\zeta}{\sum_{\zeta=\tau}^{10} \theta_\zeta} \left(A_{\tau+1}^{(i)i} (1 - \lambda_\tau^i) + A_{\tau+1}^{(i)j} \gamma_\tau^{(j)i} \right) \delta,$$

$$A_\tau^{(i)j} = \frac{\theta_\tau}{\sum_{\zeta=\tau}^{10} \theta_\zeta} \left(A_{\tau+1}^{(i)i} \gamma_\tau^{(i)j} + A_{\tau+1}^{(i)j} (1 - \lambda_\tau^j) \right) \delta, \quad (7.23)$$

$i, j \in \{1, 2\}$ and $i \neq j$.

In addition, C_τ^i is an expression containing previously executed controls $(\overline{\underline{u}}_{\tau-}^{(3)**}, \overline{\underline{u}}_{\tau-}^{(4)**})$. ■

7.3 Appendix III: Proof of Theorem 5.2

Using (5.9) we obtain

$$\xi^i(\tau, x_\tau^*; \overline{\underline{u}}_{\tau-}^*) = \beta_\tau^i(x_\tau^*; \overline{\underline{u}}_{\tau-}^*) \delta_1^\tau + \frac{\theta_\tau}{\sum_{\zeta=\tau}^T \theta_\zeta} q_{\tau+1}^i(x_{\tau+1}^*; \overline{\underline{u}}_{(\tau+1)-}^*) \delta_1^{\tau+1}$$
$$+ \frac{\sum_{\zeta=\tau+1}^T \theta_{\hat{T}}}{\sum_{\zeta=\tau}^T \theta_\zeta} \left\{ \sum_{k=\tau+1}^{\hat{T}} \beta_k^i(x_k^*; \overline{\underline{u}}_{k-}^*) \delta_1^k + q_{\hat{T}+1}^i(x_{\hat{T}+1}^*; \overline{\underline{u}}_{(\hat{T}+1)-}^*) \delta_1^{\hat{T}+1} \right\}. \quad (7.24)$$

The term $\xi^i(\tau, x_\tau^*; \underline{u}_{\tau-}^*)$ in (7.24) can be expressed as

$$\xi^i(\tau, x_\tau^*; \underline{u}_{\tau-}^*)$$
$$= \beta_\tau^i(x_\tau^*; \underline{u}_{\tau-}^*)\delta_1^\tau + \frac{\theta_\tau}{\sum_{\zeta=\tau}^T \theta_\zeta} q_{\tau+1}^i(x_{\tau+1}^*; \underline{u}_{(\tau+1)-}^*)\delta_1^{\tau+1}$$
$$+ \frac{\sum_{\zeta=\tau+1}^T \theta_\zeta}{\sum_{\zeta=\tau}^T \theta_\zeta} \frac{\sum_{\zeta=\tau+1}^T \theta_{\hat{T}}}{\sum_{\zeta=\tau+1}^T \theta_\zeta} \left\{ \sum_{k=\tau+1}^{\hat{T}} \beta_k^i(x_k^*; \underline{u}_{k-}^*)\delta_1^k + q_{\hat{T}+1}^i(x_{\hat{T}+1}^*; \underline{u}_{(\hat{T}+1)-}^*)\delta_1^{\hat{T}+1} \right\}. \tag{7.25}$$

Invoking (5.9), we can express $\xi^i(\tau+1, x_{\tau+1}^*; \underline{u}_{(\tau+1)-}^*)$ as follows:

$$\xi^i(\tau+1, x_{\tau+1}^*; \underline{u}_{(\tau+1)-}^*)$$
$$= E\left\{ \sum_{k=\tau+1}^{\hat{T}} \beta_k^i(x_k^*; \underline{u}_{k-}^*)\delta_i^k + q_{\hat{T}+1}^i(x_{\hat{T}+1}^*; \underline{u}_{(\hat{T}+1)-}^*)\delta_1^{\hat{T}+1} \right\}$$
$$= \frac{\sum_{\hat{T}=\tau+1}^T \theta_{\hat{T}}}{\sum_{\zeta=\tau+1}^T \theta_\zeta} \left\{ \sum_{k=\tau+1}^{\hat{T}} \beta_k^i(x_k^*; \underline{u}_{k-}^*)\delta_1^k + q_{\hat{T}+1}^i(x_{\hat{T}+1}^*; \underline{u}_{(\hat{T}+1)}^*)\delta_1^{\hat{T}+1} \right\}. \tag{7.26}$$

Therefore, $\xi^i(\tau, x_\tau^*; \underline{u}_{\tau-}^*)$ in (7.25) can be expressed as:

$$\xi^i(\tau, x_\tau^*; \underline{u}_{\tau-}^*) = \beta_\tau^i(x_\tau^*; \underline{u}_{\tau-}^*)\delta_1^\tau + \frac{\theta_\tau}{\sum_{\zeta=\tau}^T \theta_\zeta} q_{\tau+1}^i(x_{\tau+1}^*; \underline{u}_{(\tau+1)-}^*)\delta_1^{\tau+1}$$
$$+ \frac{\sum_{\zeta=\tau+1}^T \theta_\zeta}{\sum_{\zeta=\tau}^T \theta_\zeta} \xi^i(\tau+1, x_{\tau+1}^*; \underline{u}_{(\tau+1)-}^*). \tag{7.27}$$

The term $\beta_\tau^i(x_\tau^*; \underline{u}_{\tau-}^*)$ can be obtained as

$$\beta_\tau^i(x_\tau^*; \underline{u}_{\tau-}^*) = (\delta_1^\tau)^{-1} \Bigg(\xi^i(\tau, x_\tau^*; \underline{u}_{\tau-}^*) - \frac{\theta_\tau}{\sum_{\zeta=\tau}^T \theta_\zeta} q_{\tau+1}^i(x_{\tau+1}^*; \underline{u}_{(\tau+1)-}^*)\delta_1^{\tau+1}$$
$$- \frac{\sum_{\zeta=\tau+1}^T \theta_\zeta}{\sum_{\zeta=\tau}^T \theta_\zeta} \xi^i(\tau+1, x_{\tau+1}^*; \underline{u}_{(\tau+1)-}^*) \Bigg). \tag{7.28}$$

Hence Theorem 5.2 follows. ■

7.4 Appendix IV: Proof of Proposition 6.1

Using (6.3), we have

$$A_{11}^1 = Q_{T+1}^{(1)1},\ A_{11}^2 = Q_{T+1}^{(2)2},\ C_{11} = \sum\nolimits_{i=1}^{2} \left(\sum\nolimits_{t=11-2}^{10} \overline{v}_{11}^{(t)i} \overline{u}_t^{(3)i} + \sum\nolimits_{t=11-3}^{10} \overline{p}_{11}^{(t)i} \overline{u}_t^{(4)i} + \varpi_{11}^i \right). \tag{7.29}$$

Invoking the technique of backward induction, we consider first the last operational stage 10. Using Proposition 6.1 and (6.4), we can express the optimal cooperative strategies in stage 10 as:

$$u_{10}^i = \frac{(R_{10}^i)^2}{4(c_{10}^i)^2} x^i,\ \overline{u}_{10}^{(3)i} = \frac{A_{11}^i \varepsilon_{10}^{|10|i} + \overline{v}_{11}^{|10|i}}{2\varphi_{10}^{(x)i}} \delta,\ \overline{u}_{10}^{(4)i} = \frac{p_{10}^{|10|i} + \delta \overline{p}_{11}^{|10|i}}{2\phi_{10}^i}, \quad \text{for } i \in \{1, 2\}. \tag{7.30}$$

Substituting the optimal strategies in (7.9) into the stage 10 equation of (6.4) we obtain

$$\begin{aligned}
& W(10, x; \underline{\overline{u}}_{10-}^{(3)}, \underline{\overline{u}}_{10-}^{(4)}) \\
&= (A_{10}^1 x^1 + A_{10}^2 x^2 + C_{10}) \delta^9 \\
&= \sum_{i=1}^{2} \left(\frac{(R_{10}^i)^2}{4c_{10}^i} x^i - \frac{(A_{11}^i \varepsilon_{10}^{|10|i} + \overline{v}_{11}^{|10|i})^2 \delta^2}{4\varphi_{10}^{(x)i}} - \frac{(p_{10}^{|10|i} + \delta \overline{p}_{11}^{|10|i})^2}{4\phi_{10}^i} \right. \\
&\quad \left. + \frac{p_{10}^{|10|i} (p_{10}^{|10|i} + \delta \overline{p}_{11}^{|10|)i})}{2\phi_{10}^i} + \sum_{t=10-3}^{9} p_{10}^{|t|i} \overline{u}_t^{(4)i} \right) \delta^9 \\
&\quad + \left[A_{11}^1 [x^1 + \varepsilon_{10}^{|10|1} \frac{(A_{11}^1 \varepsilon_{10}^{|10|1} + \overline{v}_{11}^{|10|1}) \delta}{2\varphi_{10}^{(x)1}} + \sum_{t=10-2}^{9} \varepsilon_{10}^{|t|1} \overline{u}_t^{(3)1} - \lambda_{10}^1 x^1 + \gamma_{10}^{(1)2} x^2] \right. \\
&\quad + A_{11}^2 [x^2 + \varepsilon_{10}^{|10|2} \frac{(A_{11}^2 \varepsilon_{10}^{|10|2} + \overline{v}_{11}^{|10|2}) \delta}{2\varphi_{10}^{(x)2}} + \sum_{t=10-2}^{9} \varepsilon_{10}^{|t|2} \overline{u}_t^{(3)2} - \lambda_{10}^2 x^2 + \gamma_{10}^{(2)1} x^1] \\
&\quad + \sum_{i=1}^{2} \left(\overline{v}_{11}^{|10|i} \frac{(A_{11}^i \varepsilon_{10}^{|10|i} + \overline{v}_{11}^{|10|i}) \delta}{2\varphi_{10}^{(x)i}} + \sum_{t=10-2}^{9} \overline{v}_{11}^{|t|i} \overline{u}_t^{(3)i} + \frac{\overline{p}_{11}^{|10|i} (p_{10}^{|10|i} + \delta \overline{p}_{11}^{|10|i})}{2\phi_{10}^i} \right. \\
&\quad \left. \left. + \sum_{t=11-3}^{9} \overline{p}_{11}^{|t|i} \overline{u}_t^{(4)i} + \varpi_{11}^i \right) \right] \delta^{10}.
\end{aligned} \tag{7.31}$$

System (7.31) is an equation which right-hand-side and left-hand-side are linear functions of (x^1, x^2). For (7.31) to hold, it is required that:

$$A_{10}^1 x^1 = \frac{(R_{10}^1)^2 x^1}{4c_{10}^1} + \delta A_{11}^1 (1 - \lambda_{10}^1) x^1 + \delta A_{11}^2 \gamma_{10}^{(2)1} x^1,$$

$$A_{10}^2 x^2 = \frac{(R_{10}^2)^2 x^2}{4c_{10}^2} + \delta A_{11}^2 (1 - \lambda_{10}^2) x^2 + \delta A_{11}^1 \gamma_{10}^{(1)2} x^2. \tag{7.32}$$

From (7.32) we obtain:

$$A_{10}^1 = \sum_{i=1}^{2} \frac{(R_{10}^i)^2}{4c_{10}^i} + \delta A_{11}^1 (1 - \lambda_{10}^1) + \delta A_{11}^2 \gamma_{10}^{(2)1},$$

$$A_{10}^2 = \sum_{i=1}^{2} \frac{(R_{10}^i)^2}{4c_{10}^i} + \delta A_{11}^2 (1 - \lambda_{10}^2) + \delta A_{11}^1 \gamma_{10}^{(1)2}. \tag{7.33}$$

In addition,

$$\begin{aligned} C_{10} = \sum_{i=1}^{2} \Bigg(& \frac{p_{10}^{|10|i} (p_{10}^{|10|i} + \delta \overline{p}_{11}^{|10|)i})}{2\phi_{10}^i} + \sum_{t=10-3}^{9} p_{10}^{|t|i} \overline{u}_t^{(4)i} \\ & - \frac{(A_{11}^i \varepsilon_{10}^{|10|i} + \overline{v}_{11}^{|10|i})^2 \delta^2}{4\varphi_{10}^{(x)i}} - \frac{(p_{10}^{|10|i} + \delta \overline{p}_{11}^{|10|i})^2}{4\phi_{10}^i} \Bigg) \\ & + \Bigg[A_{11}^1 [\varepsilon_{10}^{|10|1} \frac{(A_{11}^1 \varepsilon_{10}^{|10|1} + \overline{v}_{11}^{|10|1})\delta}{2\varphi_{10}^{(x)1}} + \sum_{t=10-2}^{9} \varepsilon_{10}^{|t|1} \overline{u}_t^{(3)1}] \\ & + A_{11}^2 [\varepsilon_{10}^{|10|2} \frac{(A_{11}^2 \varepsilon_{10}^{|10|2} + \overline{v}_{11}^{|10|2})\delta}{2\varphi_{10}^{(x)2}} + \sum_{t=10-2}^{9} \varepsilon_{10}^{|t|2} \overline{u}_t^{(3)2}] \\ & + \sum_{i=1}^{2} \Bigg(\overline{v}_{11}^{|10|i} \frac{(A_{11}^i \varepsilon_{10}^{|10|i} + \overline{v}_{11}^{|10|i})\delta}{2\varphi_{10}^{(x)i}} + \sum_{t=10-2}^{9} \overline{v}_{11}^{|t|i} \overline{u}_t^{(3)i} + \frac{\overline{p}_{11}^{|10|i} (p_{10}^{|10|i} + \delta \overline{p}_{11}^{|10|i})}{2\phi_{10}^i} \\ & + \sum_{t=11-3}^{9} \overline{p}_{11}^{|t|i} \overline{u}_t^{(4)i} + \varpi_{11}^i \Bigg) \Bigg] \delta, \end{aligned} \tag{7.34}$$

which is an expression containing previously executed controls $(\underline{\overline{u}}_{10-}^{(3)}, \underline{\overline{u}}_{10-}^{(4)})$. Then we move to stage 9.

Using $W(10, x; \underline{\overline{u}}_{10-}^{(3)}, \underline{\overline{u}}_{10-}^{(4)}) = (A_{10}^1 x^1 + A_{10}^2 x^2 + C_{10})\delta^9$ derived in (7.31)–(7.34) and the stage 9 equation in (6.5), the optimal cooperative strategies in stage 9 can be expressed as:

$$u_9^i = \frac{(R_9^i)^2}{4(c_9^i)^2} x^i,$$

$$\overline{u}_9^{(3)i} = \frac{1}{2\varphi_9^{(x)i}}\left[\frac{\theta_9}{\sum_{\zeta=9}^{10}\theta_\zeta}\left(Q_{10}^{(i)i}\varepsilon_9^{|9|i}\delta + \overline{v}_{10}^{|9|i}\delta\right)\right.$$
$$\left.+\frac{\sum_{\zeta=10}^{10}\theta_\zeta}{\sum_{\zeta=9}^{10}\theta_\zeta}\left(A_{10}^{i}\varepsilon_9^{|9|i}\delta + A_{11}^{i}\varepsilon_{10}^{|9|i}\delta^2 + \overline{v}_{11}^{|9|i}\delta^2\right)\right],$$

$$\overline{u}_9^{(4)i} = \frac{1}{2\varphi_9^{i}}\left[p_9^{|9|i} + \frac{\theta_9}{\sum_{\zeta=9}^{10}\theta_\zeta}\overline{p}_{10}^{|9|i}\delta + \frac{\sum_{\zeta=10}^{10}\theta_\zeta}{\sum_{\zeta=9}^{10}\theta_\zeta}\left(p_{10}^{(9)i}\delta + \overline{p}_{11}^{(9)i}\delta^2\right)\right], \quad (7.35)$$

for $i \in \{1, 2\}$.

Substituting the optimal strategies in (7.35) into the stage 9 equation of (6.5) we obtain

$$W(9, x; \overline{u}_{9-}^{(3)}, \overline{u}_{9-}^{(4)})$$
$$= (A_9^1 x^1 + A_9^2 x^2 + C_9)\delta^8$$
$$= \sum_{i=1}^{2}\left(\frac{(R_9^i)^2}{4c_9^i}x^i - \varphi_9^{(x)i}(\overline{u}_9^{(3)i})^2 - \phi_9^i(\overline{u}_9^{(4)i})^2 + p_9^{|9|i}\overline{u}_9^{(4)i} + \sum_{t=9-3}^{8} p_9^{|t|i}\overline{u}_t^{(4)i}\right)\delta^8$$
$$+\sum_{i=1}^{2}\frac{\theta_9}{\sum_{\zeta=9}^{10}\theta_\zeta}\left(Q_{10}^{(i)i}[x^i + \varepsilon_9^{|9|i}\overline{u}_9^{(3)i} + \sum_{t=9-2}^{8}\varepsilon_9^{|t|i}\overline{u}_t^{(3)i} - \lambda_9^i x^i + \gamma_9^{(i)j}x^j]\right.$$
$$\left.+\overline{v}_{10}^{|9|i}\overline{u}_9^{(3)i} + \sum_{t=9-2}^{8}\overline{v}_{10}^{|t|i}\overline{u}_t^{(3)i} + \overline{p}_{10}^{|9|i}\overline{u}_9^{(4)i} + \sum_{t=9+1-3}^{8}\overline{p}_{10}^{|t|i}\overline{u}_t^{(4)i} + \varpi_{10}^i\right)\delta^9$$
$$+\frac{\sum_{\zeta=10}^{10}\theta_\zeta}{\sum_{\zeta=9}^{10}\theta_\zeta}\left(A_{10}^1[x^1 + \varepsilon_9^{|9|1}\overline{u}_9^{(3)1} + \sum_{t=9-2}^{8}\varepsilon_9^{|t|1}\overline{u}_t^{(3)1} - \lambda_9^1 x^1 + \gamma_9^{(1)2}x^2]\right.$$
$$\left.+A_{10}^2[x^2 + \varepsilon_9^{|9|2}\overline{u}_9^{(3)2} + \sum_{t=9-2}^{8}\varepsilon_9^{|t|2}\overline{u}_t^{(3)2} - \lambda_9^2 x^2 + \gamma_9^{(2)1}x^1] + C_{10}\right)\delta^9, \quad (7.36)$$

where $\overline{u}_9^{(3)1}$, $\overline{u}_9^{(3)2}$, $\overline{u}_9^{(4)1}$ and $\overline{u}_9^{(4)2}$ are given in (7.35).

Equation (7.36) is an equation which right-hand-side and left-hand-side are linear functions of (x^1, x^2). For (7.36) to hold, it is required that:

$$A_9^1 x^1 = \frac{(R_9^1)^2 x^1}{4c_9^1} + \frac{\theta_9}{\sum_{\zeta=9}^{10}\theta_\zeta}\left(Q_{10}^{(1)1}(1-\lambda_9^1)x^1 + Q_{10}^{(2)2}\gamma_9^{(2)1}x^1\right)\delta$$
$$+\frac{\sum_{\zeta=10}^{10}\theta_\zeta}{\sum_{\zeta=9}^{10}\theta_\zeta}\left(A_{10}^1(1-\lambda_9^1)x^1 + A_{10}^2\gamma_9^{(2)1}x^1\right)\delta,$$

$$A_9^2 x^2 = \frac{(R_9^2)^2 x^2}{4c_9^2} + \frac{\theta_9}{\sum_{\zeta=9}^{10}\theta_\zeta}\left(Q_{10}^{(2)2}(1-\lambda_9^2)x^2 + Q_{10}^{(1)1}\gamma_9^{(1)2}x^2\right)\delta$$

$$+\frac{\sum_{\zeta=10}^{10}\theta_\zeta}{\sum_{\zeta=9}^{10}\theta_\zeta}\left(A_{10}^2(1-\lambda_9^2)x^2+A_{10}^1\gamma_9^{(1)2}x^2\right)\delta. \tag{7.37}$$

From (7.37) we obtain:

$$A_9^1 = \frac{(R_9^1)^2}{4c_9^1} + \frac{\theta_9}{\sum_{\zeta=9}^{10}\theta_\zeta}\left(Q_{10}^{(1)1}(1-\lambda_9^1) + Q_{10}^{(2)2}\gamma_9^{(2)1}\right)\delta$$
$$+\frac{\sum_{\zeta=10}^{10}\theta_\zeta}{\sum_{\zeta=9}^{10}\theta_\zeta}\left(A_{10}^1(1-\lambda_9^1)+A_{10}^2\gamma_9^{(2)1}\right)\delta,$$

$$A_9^2 = \frac{(R_9^2)^2}{4c_9^2} + \frac{\theta_9}{\sum_{\zeta=9}^{10}\theta_\zeta}\left(Q_{10}^{(2)2}(1-\lambda_9^2) + Q_{10}^{(1)1}\gamma_9^{(1)2}\right)\delta$$
$$+\frac{\sum_{\zeta=10}^{10}\theta_\zeta}{\sum_{\zeta=9}^{10}\theta_\zeta}\left(A_{10}^2(1-\lambda_9^2)+A_{10}^1\gamma_9^{(1)2}\right)\delta. \tag{7.38}$$

In addition, C_9 is an expression including all the terms on the right-hand-side of (7.15) which do not involve x^1 or x^2. It is an expression containing previously executed controls $(\underline{u}_{9-}^{(3)}, \underline{u}_{9-}^{(4)})$.

Then, we move to stage 8.

Using $W(9, x; \underline{u}_{9-}^{(3)}, \underline{u}_{9-}^{(4)}) = (A_9^1 x^1 + A_9^2 x^2 + C_9)\delta^8$ derived in (7.36)–(7.38) and the stage 8 equation in (6.5), the optimal cooperative strategies in stage 8 can be expressed as:

$$u_8^i = \frac{(R_8^i)^2}{4(c_8^i)^2}x^i,$$

$$\overline{u}_8^{(3)i} = \frac{1}{2\varphi_8^{(x)i}}\left[\frac{\theta_8}{\sum_{\zeta=8}^{10}\theta_\zeta}\left(Q_9^{(i)i}\varepsilon_8^{|8|i}\delta+\overline{v}_9^{|8|i}\delta\right)\right.$$
$$\left.+\frac{\sum_{\zeta=9}^{10}\theta_\zeta}{\sum_{\zeta=8}^{10}\theta_\zeta}\left(A_9^i\varepsilon_8^{|8|i}\delta + A_{10}^i\varepsilon_9^{|8|i}\delta^2 + A_{10}^i\varepsilon_9^{|8|i}\delta^2 + A_{11}^i\varepsilon_{10}^{|8|i}\delta^3 + \overline{v}_{11}^{|8|i}\delta^3\right)\right],$$

$$\overline{u}_8^{(4)i} = \frac{1}{2\phi_8^i}\left[p_8^{|8|i} + \frac{\theta_8}{\sum_{\zeta=8}^{10}\theta_\zeta}\overline{p}_9^{|8|i}\delta + \frac{\sum_{\zeta=9}^{10}\theta_\zeta}{\sum_{\zeta=8}^{10}\theta_\zeta}\left(p_9^{|8|i}\delta + p_{10}^{|8|i}\delta^2 + \overline{p}_{11}^{(8)i}\delta^3\right)\right], \tag{7.39}$$

for $i \in \{1, 2\}$.

Substituting the optimal strategies in (7.39) into the stage 8 equation of (6.5) we obtain

$$W(8, x; \underline{u}_{8-}^{(3)}, \underline{u}_{8-}^{(4)})$$

$$= (A_8^1 x^1 + A_8^2 x^2 + C_8)\delta^7$$

$$= \sum_{i=1}^{2}\left(\frac{(R_8^i)^2}{4c_8^i}x^i - \varphi_8^{(x)i}(\overline{u}_8^{(3)i})^2 - \phi_8^i(\overline{u}_8^{(4)i})^2 + p_8^{|8|i}\overline{u}_8^{(4)i} + \sum_{t=8-3}^{7} p_8^{|t|i}\overline{u}_t^{(4)i}\right)\delta^7$$

$$+\sum_{i=1}^{2}\frac{\theta_8}{\sum_{\zeta=8}^{10}\theta_\zeta}\left(Q_9^{(i)i}[x^i + \varepsilon_8^{|8|i}\overline{u}_8^{(3)i} + \sum_{t=8-2}^{7}\varepsilon_8^{|t|i}\overline{u}_t^{(3)i} - \lambda_8^i x^i + \gamma_8^{(i)j}x^j]\right.$$

$$\left.+\overline{v}_9^{|8|i}\overline{u}_8^{(3)i} + \sum_{t=8-2}^{7}\overline{v}_8^{|t|i}\overline{u}_t^{(3)i} + \overline{p}_9^{|8|i}\overline{u}_8^{(4)i} + \sum_{t=8+1-3}^{7}\overline{p}_9^{|t|i}\overline{u}_t^{(4)i} + \varpi_9^i\right)\delta^8$$

$$+\frac{\sum_{\zeta=9}^{10}\theta_\zeta}{\sum_{\zeta=8}^{10}\theta_\zeta}\left[A_9^1[x^1 + \varepsilon_8^{|8|1}\overline{u}_8^{(3)1} + \sum_{t=8-2}^{7}\varepsilon_8^{|t|1}\overline{u}_t^{(3)1} - \lambda_8^1 x^1 + \gamma_8^{(1)2}x^2]\right.$$

$$\left.+A_9^2[x^2 + \varepsilon_8^{|8|2}\overline{u}_8^{(3)2} + \sum_{t=8-2}^{7}\varepsilon_8^{|t|2}\overline{u}_t^{(3)2} - \lambda_8^2 x^2 + \gamma_8^{(2)1}x^1] + C_9\right]\delta^8, \tag{7.40}$$

where $\overline{u}_8^{(3)1}$, $\overline{u}_8^{(3)2}$, $\overline{u}_8^{(4)1}$ and $\overline{u}_8^{(4)2}$ are given in (7.39).

System (7.40) yields a system of equations which right-hand-side and left-hand-side are linear functions of (x^1, x^2). For (7.40) to hold, it is required that:

$$A_8^1 = \frac{(R_8^1)^2}{4c_8^1} + \frac{\theta_8}{\sum_{\zeta=8}^{10}\theta_\zeta}\left(Q_9^{(1)1}(1-\lambda_8^1) + Q_9^{(2)2}\gamma_8^{(2)1}\right)\delta$$
$$+\frac{\sum_{\zeta=9}^{10}\theta_\zeta}{\sum_{\zeta=8}^{10}\theta_\zeta}\left(A_9^1(1-\lambda_8^1) + A_9^2\gamma_8^{(2)1}\right)\delta,$$

$$A_8^2 = \frac{(R_8^2)^2}{4c_8^2} + \frac{\theta_8}{\sum_{\zeta=8}^{10}\theta_\zeta}\left(Q_9^{(2)2}(1-\lambda_8^2) + Q_9^{(1)1}\gamma_8^{(1)2}\right)\delta$$
$$+\frac{\sum_{\zeta=9}^{10}\theta_\zeta}{\sum_{\zeta=8}^{10}\theta_\zeta}\left(A_9^2(1-\lambda_8^2) + A_9^1\gamma_8^{(1)2}\right)\delta. \tag{7.41}$$

In addition, C_8 is an expression including all the terms on the right-hand-side of (7.40) which do not involve x^1 or x^2. It is an expression containing previously executed controls $(\overline{u}_{8-}^{(3)}, \overline{u}_{8-}^{(4)})$.

Repeating the process for stages $\tau \in \{7, 6, \cdots, 1\}$, we obtain the optimal cooperative strategies in stage τ as:

$$u_\tau^i = \frac{(R_\tau^i)^2}{4(c_\tau^i)}x^i,$$

$$\overline{u}_\tau^{(3)i} = \frac{1}{2\varphi_\tau^{(x)i}}\left[\frac{\theta_\tau}{\sum_{\zeta=\tau}^{10}\theta_\zeta}\left(Q_{\tau+1}^{(i)i}\varepsilon_\tau^{|\tau|i}\delta + \overline{v}_{\tau+1}^{|\tau|i}\delta\right) + \frac{\sum_{\zeta=\tau+1}^{10}\theta_\zeta}{\sum_{\zeta=\tau}^{10}\theta_\zeta}\left(\sum_{t=\tau}^{\tau+2} A_{t+1}^i\varepsilon_t^{|\tau|i}\delta^{t+1-\tau}\right)\right],$$

$$\overline{u}_{\tau}^{(4)i} = \frac{1}{2\phi_{\tau}^{i}}\left[p_{\tau}^{|\tau|i} + \frac{\theta_{\tau}}{\sum_{\zeta=\tau}^{10}\theta_{\zeta}}\overline{p}_{\tau+1}^{|\tau|i}\delta + \frac{\sum_{\zeta=\tau+1}^{10}\theta_{\zeta}}{\sum_{\zeta=\tau}^{10}\theta_{\zeta}}\left(\sum_{t=\tau}^{\tau+2} p_{t+1}^{|\tau|i}\delta^{t+1-\tau}\right)\right], \quad (7.42)$$

for $i \in \{1, 2\}$.

The optimal expected cooperative joint payoff at stage $\tau \in \{7, 6, \cdots, 1\}$ can be obtained as

$$W(\tau, x; \underline{\overline{u}}_{\tau-}^{(3)}, \underline{\overline{u}}_{\tau-}^{(4)}) = (A_{\tau}^{1}x^{1} + A_{\tau}^{2}x^{2} + C_{\tau})\delta^{\tau-1},$$

where

$$\begin{aligned}
A_{\tau}^{1} =& \frac{(R_{\tau}^{1})^{2}}{4c_{\tau}^{1}} + \frac{\theta_{\tau}}{\sum_{\zeta=\tau}^{10}\theta_{\zeta}}\left(Q_{\tau+1}^{(1)1}(1-\lambda_{\tau}^{1}) + Q_{\tau+1}^{(2)2}\gamma_{\tau}^{(2)1}\right)\delta \\
&+ \frac{\sum_{\zeta=\tau+1}^{10}\theta_{\zeta}}{\sum_{\zeta=\tau}^{10}\theta_{\zeta}}(A_{\tau+1}^{1}(1-\lambda_{\tau}^{1}) + A_{\tau+1}^{2}\gamma_{\tau}^{(2)1})\delta, \\
A_{\tau}^{2} =& \frac{(R_{\tau}^{2})^{2}}{4c_{\tau}^{2}} + \frac{\theta_{\tau}}{\sum_{\zeta=\tau}^{10}\theta_{\zeta}}\left(Q_{\tau+1}^{(2)2}(1-\lambda_{\tau}^{2}) + Q_{\tau}^{(1)1}\gamma_{\tau}^{(1)2}\right)\delta \\
&+ \frac{\sum_{\zeta=\tau+1}^{10}\theta_{\zeta}}{\sum_{\zeta=\tau}^{10}\theta_{\zeta}}(A_{\tau+1}^{2}(1-\lambda_{\tau}^{2}) + A_{\tau+1}^{1}\gamma_{\tau}^{(1)2})\delta. \quad (7.43)
\end{aligned}$$

In addition, C_{τ} is an expression containing previously executed controls $(\underline{\overline{u}}_{\tau-}^{(3)}, \underline{\overline{u}}_{\tau-}^{(4)})$. ■

8 Chapter Notes

Random horizon and durable strategies posts an interesting challenge to the study of dynamic game theory. The presence of durable strategies and uncertainty in planning horizon requires significant modification of the dynamic optimization techniques to solve problems under these phenomena. Petrosyan and Murzov (1966) developed the Isaacs-Bellman equations under random horizon for zero-sum differential games. Petrosyan and Shevkoplyas (2003) presented an analysis on dynamically consistent solutions for cooperative differential games with random duration. Shevkoplyas (2011) considered Shapley value in cooperative differential games with random horizon. Yeung and Petrosyan (2011) developed discrete-time dynamic games with random horizon. Subgame consistent solution for cooperative stochastic dynamic games with random horizon was provided in Yeung and Petrosyan (2012). Yeung and Petrosyan (2014) studied subgame consistent cooperative solutions for randomly furcating stochastic dynamic games with uncertain horizon. Existing works on lagged control optimization involved lags in the state dynamics can be found in Bellman

(1957), Arthur (1977), Burdet and Sethi (1976), Hartl and Sethi (1984), Brandt-Pollmann et al. (2008), Huschto et al. (2011), Sethi and Mcguire (1977), Winkler et al. (2003) and Yeung and Petrosyan (2019). Bokov (2011) considered continuous-time optimal control with control lags in the payoff and time-lags in the state dynamics. Petrosyan and Yeung (2020) derived discrete-time dynamic optimization with durable controls affecting both the payoffs and dynamics. Dynamic optimization techniques with random horizon in continuous time can be found in Shevkoplyas (2011). Dynamic optimization techniques with random horizon in discrete time were developed by Yeung and Petrosyan (2011, 2012 and 2014). This chapter develops a class of dynamic games with durable strategies and random game horizon. Non-trivial modification of the dynamic optimization techniques is required to accommodate the phenomena of durable strategies and uncertain horizon. A general class of non-cooperative random horizon durable-strategies dynamic games is presented and a novel set of equations characterizing an equilibrium is derived. Finally, a general class of cooperative random horizon durable-strategies dynamic games is presented. The techniques presented in this chapter were developed by Yeung and Petrosyan (2021).

9 Problems

1. (Warmup exercise) Consider the case of two economic agents (regions or firms) which are given a certain lease to trade with each other. The lease for trade has to be renewed after each stage (year or decade) for up to a maximum of five stages. At stage 1, it is known that the probabilities that the lease will last for 1, 2, 3, 4, 5 stages with respective probabilities 0.1, 0.3, 0.3, 0.2, 0.1. Conditional upon the reaching of stage $\tau > 1$, the probability of the game would last up to stages $\tau, \tau + 1$, to 6 are

$$\frac{\theta_\tau}{\sum_{\zeta=\tau}^{5}\theta_\zeta}, \frac{\theta_{\tau+1}}{\sum_{\zeta=\tau}^{5}\theta_\zeta}, \frac{\theta_{\tau+2}}{\sum_{\zeta=\tau}^{5}\theta_\zeta}, \ldots, \frac{\theta_5}{\sum_{\zeta=\tau}^{5}\theta_\zeta}.$$

For example, if stage 2 has arrived, the probability of the game would last up to stages 2 is $\frac{\theta_2}{\sum_{\zeta=2}^{5}\theta_\zeta} = \frac{0.3}{0.9} = \frac{1}{3}$, the probability of the game would last up to stages 3 is $\frac{0.3}{\sum_{\zeta=2}^{5}\theta_\zeta} = \frac{1}{3}$, the probability of the game would last up to stages 4 is $\frac{0.2}{\sum_{\zeta=2}^{5}\theta_\zeta} = \frac{2}{9}$ and the probability of the game would last up to stages 5 is $\frac{0.1}{\sum_{\zeta=\tau}^{5}\theta_\zeta} = \frac{1}{9}$.

Each agent has a private production capital $x_k^i \in X^i \subset R$, for $i \in \{1, 2\}$. Investment in private capital takes two stages to complete its transformation into the agent's capital stock and we use $\overline{u}_k^{(2)i}$ to denote the investment in the private capital by agent $i \in \{1, 2\}$ i in stage k.

The accumulation process of the private capital stock of agent 1 is governed by the dynamical equation

$$x_{k+1}^1 = x_k^1 + 2\overline{u}_k^{(2)1} + \overline{u}_{k-1}^{(2)1} - 0.1x_k^1 + 0.05x_k^2, \quad x_1^1 = 70,$$

where the depreciation rate of capital x_k^1 is 0.1 and $0.05x_k^2$ is the technology spillover from agent 2's capital.

The accumulation process of the private capital stock of agent 2 is governed by the dynamical equation

$$x_{k+1}^2, = x_k^2 + 3\overline{u}_k^{(2)2} + \overline{u}_{k-1}^{(2)1} - 0.1x_k^2 + 0.08x_k^1, \quad x_1^2 = 50,$$

where the depreciation rate of capital x_k^1 is 0.1 and $0.05x_k^2$ is the technology spillover from agent 1's capital.

The cost of $\overline{u}_k^{(2)1}$ is $2(\overline{u}_k^{(2)1})^2$ and the cost of $\overline{u}_k^{(2)2}$ is $(\overline{u}_k^{(2)2})^2$.

The expected payoff of agent 1 to be maximized is

$$\sum_{\hat{T}=1}^{5} \theta_{\hat{T}} \left\{ \sum_{k=1}^{\hat{T}} \left(10x_k^1 - 2(\overline{u}_k^{(2)1})^2\right)(0.9)^{k-1} + \left(8x_{\hat{T}+1}^1 + 20\right)(0.9)^{\hat{T}} \right\},$$

The expected payoff of agent 2 to be maximized is

$$\sum_{\hat{T}=1}^{5} \theta_{\hat{T}} \left\{ \sum_{k=1}^{\hat{T}} \left(8x_k^2 - (\overline{u}_k^{(2)2})^2\right)(0.9)^{k-1} + \left(7x_{\hat{T}+1}^2 + 30\right)(0.9)^{\hat{T}} \right\},$$

Derive a non-cooperative equilibrium.

2. Consider the game in Sect. 4. In particular, when the game ends the agents would go bankrupt, so that the terminal payoff in stage $\tau + 1$ becomes

$$q^i(\tau+1, x; \underline{\overline{u}}_{(\tau+1)-}^{(3)}, \underline{\overline{u}}_{(\tau+1)-}^{(4)})$$
$$= \left(Q_{\tau+1}^{(i)i} x_{\tau+1}^i + \sum_{t=\tau+1-2}^{\tau} \overline{v}_{\tau+1}^{|t|i} \overline{u}_t^{(3)i} + \sum_{t=\tau+1-3}^{\tau} \overline{p}_{\tau+1}^{|t|i} \overline{u}_t^{(4)i} + \varpi_{\tau+1}^i \right) = 0,$$

for $\tau = \{1, 2, \cdots, 10\}$.

Therefore, the expected payoff of agent i to be maximized is

$$\sum_{\hat{T}=1}^{10} \theta_{\hat{T}} \left\{ \sum_{k=1}^{\hat{T}} \left(R_k^i (u_k^i x_k^i)^{1/2} - c_k^i u_k^i - \phi_k^{(x)i} (\overline{u}_k^{(3)i})^2 - \varphi_k^i (\overline{u}_k^{(4)i})^2 + \sum_{t=k-3}^{k} p_k^{|t|i} \overline{u}_t^{(4)i} \right) \delta^{k-1} \right\}.$$

The accumulation process of the private capital stock of agent i is governed by the dynamical equation

$$x_{k+1}^{i} = x_{k}^{i} + \varepsilon_{k}^{|k|i} \overline{u}_{k}^{(3)i} + \sum_{t=k-2}^{k-1} \varepsilon_{k}^{|t|i} \overline{u}_{t}^{(3)i} - \lambda_{k}^{i} x_{k}^{i} + \gamma_{k}^{(i)j} x_{k}^{j},$$

$$x_{1}^{i} = x_{1}^{i(0)}, \; i, j \in \{1, 2\}, \text{ and } i \neq j.$$

Derive a non-cooperative equilibrium.

3. Consider the case where the agents agree to cooperate, maximize their expected joint profit and adopt an optimality principle which would allot the agents' cooperative payoffs proportional to the sizes of their non-cooperative payoff. Characterize a subgame consistent cooperative solution.

References

Agliardi E, Koussisc N (2013) Optimal capital structure and the impact of time-to-build. Financ Res Lett 10:124–130

Arthur WB (1977) Control of linear processes with distributed lags using dynamic programming from first principles. J Optim Theory Appl 23:429–443

Bellman R (1957) Terminal control, time lags, and dynamic programming. Proc Natl Acad Sci USA 43:927–930

Bokov GV (2011) Pontryagin's maximum principle of optimal control problems with time-delay. J Math Sci 172:623–634

Brandt-Pollmann U, Winkler R, Sager S, Moslener U, Schlöder J (2008) Numerical solution of optimal control problems with constant control delays. Comput Econ 31:181–206

Burdet CA, Sethi SP (1976) On the maximum principle for a class of discrete dynamical systems with lags. J Optim Theory Appl 19:445–454

Cellini R, Lambertini L (2003) Advertising with spillover effects in a differential oligopoly game with differentiated goods. CEJOR 11:409–423

Cellini R, Lambertini L (2009) Dynamic R&D with spillovers: competition vs cooperation. J Econ Dyn Control 33:568–582

Christiano LJ, Eichenbaum M, Evans CL (2005) Nominal rigidities and the dynamic effects of a shock to monetary policy. J Polit Econ 113:1–45

Hartl RF, Sethi SP (1984) Optimal control of a class of systems with continuous lags: dynamic programming approach and economic interpretations. J Optim Theory Appl 43(1):73–88

Huschto T, Feichtinger G, Hartl RF, Kort PM, Sager S (2011) Numerical solution of a conspicuous consumption model with constant control delay. Automatica 47:1868–1877

Madsen JB (2007) Technology spillover through trade and TFP convergence: 135 years of evidence for the OECD countries. J Int Econ 72:464–480

Mela CF, Gupta S, Lehmann DR (1997) The long-term impact of promotion and advertising on consumer brand choice. J Mark Res 34:248–261

Nash JF Jr (1951) Noncooperative games. Ann Math 54:286–295

Petrosyan LA, Shevkoplyas EV (2003) Cooperative solutions for games with random duration. Game theory and applications, vol IX, pp 125–139

Petrosyan LA, Murzov NV (1966) Game theoretic problems in mechanics. Litvsk Math Sb N6:423–433

Petrosyan LA, Yeung DWK (2020) Cooperative dynamic games with durable controls: theory and application. Dyn Games Appl 10:872–896. https://doi.org/10.1007/s13235-019-00336-w

Ruffino CC (2008) Lagged effects of TV advertising on sales of an intermittently advertised product. Bus Econ Rev 18:1–12

Sarkar S, Zhang H (2013) Implementation lag and the investment decision. Econ Lett 119:136–140

Sarkar S, Zhang C (2015) Investment policy with time-to-build. J Bank Finance 55:142–156

Sethi SP, Mcguire TW (1977) Optimal skill mix: an application of the maximum principle for systems with retarded controls. J Optim Theory Appl 23:245–275

Shevkoplyas EV (2011) The shapley value in cooperative differential games with random duration. Ann Dyn Games 11:359–373

Shibata A, Shintani M, Tsurugac T (2019) Current account dynamics under information rigidity and imperfect capital mobility. J Int Money Financ 92:153–167

Tellis GJ (2006) Optimal data interval for estimating advertising response. Mark Sci 25:217–229

Tsuruga T, Shota W (2019) Money-financed fiscal stimulus: the effects of implementation lag. J Eco Dyn Cont 104:132–151

von Neumann J, Morgenstern O (1944) Theory of games and economic behavior. Princeton University Press

Winkler R, Brandt-Pollmann U, Moslener U, Schlöder J (2003) Time-lags in capital accumulation. In: Ahr D, Fahrion R, Oswald M, Reinelt G (eds) Operations research proceedings, pp. 451–458

Yeung DWK, Petrosyan LA (2012) Subgame consistent economic optimization: an advanced cooperative dynamic game analysis. Boston, Birkhäuser, pp 395. ISBN 978-0-8176-8261-3

Yeung DWK, Petrosyan LA (2010) Subgame consistent solutions for cooperative stochastic dynamic games. J Optim Theory Appl 145:579–596

Yeung DWK, Petrosyan LA (2011) Subgame consistent cooperative solution of dynamic games with random horizon. J Optim Theory Appl 150:78–97

Yeung DWK, Petrosyan LA (2014) Subgame consistent cooperative solutions for randomly furcating stochastic dynamic games with uncertain horizon. Int Game Theo Rev 16(2):1440012.01-1440012.29

Yeung DWK, Petrosyan LA (2019) Cooperative dynamic games with control lags. Dyn Games Appl 9(2):550–567. https://doi.org/10.1007/s13235-018-0266-6

Yeung DWK, Petrosyan LA (2021) Generalized dynamic games with durable strategies under uncertain planning horizon. J Comp Appl Math 395:113595

Chapter 8
Asynchronous Horizons Durable-Strategies Dynamic Games

In many game situations, the players' time horizons differ. This may arise from different life spans, different entry and exit times in different markets and the different durations of leases and contracts. Asynchronous horizons game situations occur frequently in economic and social activities. Examples include political activities involving elected governments, resource extraction by companies holding leases of exploitation expiring at different times and the situation of dominant firms facing 'hit and run' fringe competitors. In this chapter, we consider durable-strategies dynamic games in which the players' game horizons are asynchronous. Durable strategies, when combined with asynchronous horizons, complicated the analysis of strategic behaviour over time for players entering the game at different stages. In Sect. 1, we first begin with the formulation of a durable-strategies dynamic game in which the players' game horizons are asynchronous. The non-cooperative equilibrium is also characterized in Sect. 2. An illustrative example is provided in Sect. 3. In Section 4, we consider asynchronous horizons cooperative game in the presence durable strategies. In Sect. 5, a theoretical compensation scheme is proposed to handle the imputations under cooperation to players with different game horizons. Section 6 considers the problem of the tragedy of the cross-generational environmental commons under durable strategies. Mathematical appendices are relegated to Sect. 7 with Chapter notes provided in Sect. 8.

1 Game Formulation

In this section, we consider a general class of n—person, T—stage asynchronous durable- strategies dynamic game in which the players enter the game at different stages. There exist multiple durable strategies of different lag durations which affect

D. W. K. Yeung and L. A. Petrosyan, *Durable-Strategies Dynamic Games*,
Theory and Decision Library C 50, https://doi.org/10.1007/978-3-030-92742-4_8

the payoff and the state dynamics. Player $i \in \{1, 2, \ldots, n\} \equiv N$ has a set of non-durable and durable strategies $u_k^i = \left(u_k^{(1)i}, \bar{u}_k^{(2)i}, \bar{u}_k^{(3)i}, \ldots, \bar{u}_k^{(\omega)i)}\right) \in U^i$. In particular, $u_k^{(1)i}$ are non-durable strategies of player i that have effects in only one stage, that is stage k. The strategies $\bar{u}_k^{(2)i}$ are durable strategies that have effects in stages k and $k+1$. The strategies $\bar{u}_k^{(\omega)i}$ are durable strategies that have effects within the duration from stages k to stage $k+\omega-1$. We use $\bar{u}_k^i = (\bar{u}_k^{(2)i}, \bar{u}_k^{(3)i}, \ldots, \bar{u}_k^{(\omega)i})$ to denote the subset of durable strategies of player i at stage k. The term $\underline{u}_k = (u_k^1, u_k^2, \ldots, u_k^n)$ is used to denote the set of strategies of all players and $\underline{\bar{u}}_k = (\bar{u}_k^1, \bar{u}_k^2, \ldots, \bar{u}_k^n)$ to denote the subset of durable strategies of all players at stage k. The single-stage payoff of player i received in stage k is $g_k^i(x_k, \underline{u}_k, \underline{\bar{u}}_k, ; \underline{\bar{u}}_{k-})$, where $x_k \in X \subset R^m$ is the state at stage k and $\underline{\bar{u}}_{k-}$ is the set of durable strategies of all players which are executed before stage k but still in effect in stage k.

The game horizon of player $i \in \{1, 2, \ldots, n\}$ is $\bar{T}^i = \left[T_{En}^i, T_{Ex}^i\right]$, where T_{En}^i represents the stage when he enters the game and T_{Ex}^i is the stage when he exits the game. Player i's game horizon is partitioned into z_i segments. A distinct set of players exists in each of the game segments. For notational convenience in subsequent exposition, we denote the game horizon of player i as $\bar{T}^i = \left[T_{En}^i, T_{Ex}^i\right] = \left[T_1^i, T_{z,1}^i - 1\right]$. Therefore, the z_i game segments in player i's game horizon can be expressed as $[T_1^i, T_2^i - 1], [T_2^i, T_3^i - 1], \cdots, [T_{z_i-1}^i, T_{z_i}^i - 1], [T_{z_i}^i, T_{z_i+1}^i - 1]$. We use $S^{[T_t^i, T_{t+1}^i - 1]}$ to denote the set of players in the game interval $[T_t^i, T_{t+1}^i - 1]$ and $(\underline{u}_k^{[T_t^i, T_{t+1}^i - 1]}, \underline{\bar{u}}_k^{[T_t^i, T_{t+1}^i - 1]})$ to denote the set of strategies of the players in the game segment $[T_t^i, T_{t+1}^i - 1]$ executed in stage k, for $k \in [T_t^i, T_{t+1}^i - 1]$, for $t \in \{1, 2, \cdots, z_i\}$. The interval $[T_t^i, T_{t+1}^i - 1]$ is used interchangeably with the interval $\{T_t^i, T_t^i + 1, T_t^i + 2, \cdots, T_{t+1}^i - 1\}$. Furthermore, we denote the set of players (excluding player i) which would exit after stage $T_{t+1}^i - 1$ by $S^{[T_t^i, T_{t+1}^i - 1]A} \subset S^{[T_t^i, T_{t+1}^i - 1]}$. We denote the set of players (excluding player i) which would stay in the game after stage $T_{t+1}^i - 1$ by $S^{[T_t^i, T_{t+1}^i - 1]B} \subset S^{[T_t^i, T_{t+1}^i - 1]}$.

The present values (viewed at initial stage 1) of the payoffs of player i is.

$$\sum_{t=1}^{z_i} \sum_{k=T_t^i}^{T_{t+1}^i - 1} g_k^i\left(x_k, \underline{u}_k^{[T_t^i, T_{t+1}^i - 1]}, \underline{\bar{u}}_k^{[T_t^i, T_{t+1}^i - 1]}; \underline{\bar{u}}_{k-}\right)\delta_1^k + q_{T_{z_i+1}^i}^i\left(x_{T_{z_i+1}^i}; \underline{u}_{(T_{z_i+1}^i)-}\right)\delta_1^{T_{z_i+1}^i},$$

$$\text{for } i \in N. \tag{1.1}$$

The state dynamics is characterized by the difference equations:

$$x_{k+1} = f_k\left(x_k, \underline{u}_k^{[T_t^i, T_{t+1}^i - 1]}, \underline{\bar{u}}_k^{[T_t^i, T_{t+1}^i - 1]}; \underline{\bar{u}}_{k-}\right),$$

$$\text{for } k \in \left[T_t^i, T_{t+1}^i - 1\right] \text{ and } t \in (1, 2, \cdots, z_i). \tag{1.2}$$

2 Non-cooperative Equilibrium

A theorem characterizing a feedback Nash equilibrium of the asynchronous horizon dynamic game with durable strategies (1.1)–(1.2) is provided below.

Theorem 2.1 *Let $(u_k^{i**}, \bar{u}_k^{i**})$, for $k \in \bar{T}^i$ and $i \in N$, be the set of feedback Nash equilibrium strategies and $V^i(k, x; \underline{\bar{u}}_{k-}^{**})$ be the feedback Nash equilibrium payoff of player i at stage $k \in \bar{T}^i$ in the asynchronous horizon dynamic game (1.1)–(1.2), then the function $V^i(k, x; \underline{\bar{u}}_{k-}^{**})$, for $k \in \bar{T}^i$ and $i \in N$, satisfies the following recursive equations:*

$$
\begin{aligned}
&V^i\left(T_{z_i+1}^i, x; \underline{\bar{u}}_{(T_{z_i+1}^i)-}^{**}\right) = q_{T_{zi}^i+1}^i\left(x, \underline{\bar{u}}_{(T_{z_i+1}^i)-1}^{**}\right)\delta_1^{T_{z_i+1}^i},\\
&V^i\left(T_{t+1}^i, x, \underline{\bar{u}}_{(T_{t+1}^i)-}^{**}\right) = V^i\left(T_{t+1}^i, x, \underline{\bar{u}}_{(T_{t+1}^i)-}^{**}\right),\\
&V^j\left(T_{t+1}^i, x, \underline{\bar{u}}_{(T_{t+1}^i)-}^{**}\right) = q_{T_{t+1}^i}^j\left(x; \underline{\bar{u}}_{(T_{t+1}^i)-}^{**}\right)\delta_1^{T_{t+1}}, \quad \text{for } j \in S^{[T_t^i, T_{T+1}^i-1]A},\\
&V^\ell\left(T_{t+1}^\ell, x, \underline{\bar{u}}_{(T_{t+1}^i)-}^{**}\right) = V^\ell\left(T_{t+1}^\ell, x; \underline{\bar{u}}_{(T_{t+1}^i)-}^{**}\right), \quad \text{for } \ell \in S^{[T_t^i, T_{t+1}^i-1]B};
\end{aligned}
\tag{2.1}
$$

$$
\begin{aligned}
V^i(k, x, \bar{u}_{k-}^{**}) = \max_{u_k^i, \bar{n}_k^i}\Big\{&g_k^i\left(x, u_k^i, \bar{u}_k^i, \underline{u}_k^{[T_t^i, {}_{t+1}^i-1](\neq i)^{**}}, \bar{u}_k^{[T_t^i, T_{t+1}^i-1](\neq i)^{**}}; \bar{u}_{k-}^{**}\right)\delta_1^k\\
&+V^i\left[k+1, x_{k+1}^{(\neq i)**}; \bar{u}_k^i, \underline{u}_k^{[T_i^i, T_{i+1}^i-1](\neq i)^{**}}, \underline{u}_{(k+1)-}^{**} \cap \underline{\bar{u}}_{k-}^{**}\right]\Big\}\\
\text{where } x_{k+1}^{(\neq i)^{**}} &= f_k\left(x, u_k^i, \bar{u}_k^i, \underline{u}_k^{[T_t^i, T_{t+1}^i-1](\neq i)^{**}}, \underline{\bar{u}}_k^{[T_t^i, T_{t+1}^i-1](\neq i)^{**}}; \underline{u}_{k-}^{**}\right);
\end{aligned}
\tag{2.2}
$$

$$
\begin{aligned}
V^j(k, x, \bar{u}_{k-}^{**}) &= \max_{u_k^j, \bar{u}_k^j}\Big\{g_k^j\left(x, u_k^j, \bar{u}_k^j, \underline{u}_k^{[T_t^i, T_{t+1}^i-1](\neq j)^{**}}, \underline{\bar{u}}_k^{[T_t^i, T_{t+1}^i-1](\neq j)^{**}}; \underline{\bar{u}}_{k-}^{**}\right)\delta_1^k\\
&+V^j\left[k+1, x_{k+1}^{(\neq j)**}; \bar{u}_k^j, \underline{u}_k^{[T_i^T, T_{i+1}-1](\neq j)^{**}}, \underline{\bar{u}}_{(k+1)-}^{**} \cap \underline{\bar{u}}_{k-}^{**}\right]\Big\}, \text{ for } j \in S^{[T_t^i, T_{i+1}^i-1]A},
\end{aligned}
$$

Where

$$
x_{k+1}^{(\neq j)**} = f_k\left(x, u_k^j, \bar{u}_k^j, \underline{u}_k^{[T_t^i, T_{t+1}^i-1](\neq j)^{**}}, \underline{u}_k^{[T_t^i, T_{t+1}^i-1](\neq j)^{**}}; \underline{u}_{k-1}^{**}\right);
\tag{2.3}
$$

$$
\begin{aligned}
V^\ell(k, x, \bar{u}_{k-}^{**}) &= \max_{u_k^\ell, \bar{u}_k^\ell}\Big\{g_k^\ell\left(x, u_k^\ell, \bar{u}_k^\ell, \underline{u}_k^{[T_t^i, T_{t+1}^i-1](\neq \ell)^{**}}, \underline{\bar{u}}_k^{[T_t^i, T_{t+1}^i-1](\neq \ell)^{***}}; \underline{\bar{u}}_{k-}^{**}\right)\delta_1^k\\
&+V^\ell\left[k+1, x_{k+1}^{(\neq \ell)^{**}}; \bar{u}_k^i, \bar{u}_k^{[T_t^i, T_{i+1}^i-1](\neq \ell)^{**}}, \underline{u}_{(k+1)-}^{**} \cap \underline{u}_{k-}^{**}\right]\Big\}, \text{ for } \ell \in S^{[T_t^i, T_{t+1}^i-1]B},
\end{aligned}
$$

where

$$
x_{k+1}^{(\neq \ell)^{**}} = f_k\left(x, u_k^\ell, \bar{u}_k^\ell, \underline{u}_k^{[T_t^i, T_{t+1}^i-1](\neq \ell)^{**}}, \underline{\bar{u}}_k^{[T_t^i, T_{t+1}^i-1](\neq \ell)^{**}}; \bar{u}_{k-}^{**}\right);
\tag{2.4}
$$

for $k \in [T_t^i, T_t^i - 1]$ and $t \in \{1, 2, \cdots, z_i\}$.

Proof See Appendix I. ■

Substituting the game equilibrium strategies from Theorem 2.1 into the state dynamics in (1.2), the game equilibrium state dynamics can be obtained as:

$$x_{k+1} = f_k\left(x_k, \underline{u}_k^{[T_t^i, T_{t+1}^i - 1]^{**}}, \bar{\underline{u}}_k^{[T_t^i, T_{t+1}^i - 1]^{**}}; \bar{\underline{u}}_{k-}^{**}\right),$$

$$\text{for } k \in \left[T_t^i, T_{t+1}^i - 1\right] \text{ and } t \in (1, 2, \cdots, z_i). \quad (2.5)$$

The solution to (2.5) yields the game equilibrium state trajectory, which we denote as $\left\{x_k^{**}\right\}_{k=1}^{T+1}$.

Theorem 2.1 yields is a novel set of Hamilton–Jacobi-Bellman equations specifically for asynchronous horizon durable-strategies dynamic games. This class of equations differs from the standard Hamilton–Jacobi-Bellman equations in the following ways:

(i) Durable strategies executed by players would continue to affect the state and payoffs of other players even after they have exited the game.

(ii) Players entering the game after some players have already been in the game would have to share the consequences (positive or negative) of the actions taken by players who already left the game.

3 A Computational Illustration in Eco-degradation

In this section, we provide an illustration of cross-generation durable-strategies environmental games with explicit functional forms.

3.1 Game Formulation

We consider a T—stage n—player/region cross-generational durable strategies environmental game. The game horizon of player $i \in \{1, 2, \cdots, n\}$ is $\bar{T}^i = \left[T_{En}^i, T_{Ex}^i\right]$, where T_{En}^i is the stage when he enters the game and T_{Ex}^i is the stage when he exits the game. Player i's game horizon is partitioned into z_i segments. A distinct set of players exists in each of the game segments. For notational convenience in subsequent exposition, we denote $\bar{T}^i = \left[T_{En}^i, T_{Ex}^i\right] = \left[T_1^i, T_{z_i+1}^i - 1\right]$. Therefore, the z_i game segments in player i's game horizon can be expressed as $[T_1^i, T_2^i - 1]$, $[T_2^i, T_3^i - 1], \cdots, [T_{z_i-1}^i, T_{z_i}^i - 1], [T_{z_i}^i, T_{z_i+1}^i - 1]$. We use $(\underline{u}_k^{[T_t^i, T_{t+1}^i - 1]}, \bar{\underline{u}}_k^{[T_t^i, T_{t+1}^i - 1]})$ to denote the set of strategies of the players in the game segment $[T_t^i, T_{t+1}^i - 1]$ in stage $k \in [T_t^i, T_{t+1}^i - 1]$, for $t \in \{1, 2, \cdots, z_i\}$.

There exist an eco-degrading output produced by player i, denoted by $\bar{u}_k^i \in \bar{U}^i \subset R$, which directly damages the payoffs of the players and add to the state of environmental degradation over ω stages. The increase in the state of environmental degradation created in stage k by $\bar{u}_k^i$ in stage k is $m_k^{|k|i}\bar{u}_k^i$. The increase in the state of environmental degradation created by $\bar{u}_k^i$ in stage $\tau \in \{k+1, k+2, \cdots, k+\omega-1\}$ is $m_\tau^{|k|i}\bar{u}_k^i$.

The evolution of the state of environmental degradation is represented by the dynamical equation:

$$x_{k+1}^1 = x_k^1 + \sum_{\theta \in S^{[T_t^i, T_{t+1}^i-1]}} m_k^{|k|\theta}\bar{u}_k^\theta + \sum_{\theta \in S^{[k-]}} \sum_{\tau=k-\omega+1}^{k-1} m_k^{|\tau|\theta}\bar{u}_\tau^\theta$$
$$- \sum_{\vartheta \in S^{[T_t^i, T_{t+1}^i-1]}} b_k^\theta u_k^{I\theta}(x_k^1)^{1/2} - \lambda_k^1 x_k^1$$

$$\text{for } k \in \left[T_t^i, T_{t+1}^i - 1\right] \text{ and } t \in (1, 2, \cdots, z_i), \tag{3.1}$$

where $S^{[T_t^i, T_{t+1}^i-1]}$ is the set of players existing in the game segment $k \in [T_t^i, T_{t+1}^i-1]$, for $t \in (1, 2, \cdots, z_i)$;

$S^{[k-]}$ is the set of players who had durable strategies executed before stage k but are still effective in stage k;

$m_k^{|\tau|\theta}\bar{u}_\tau^\theta$ is the incremental environment degradation created in stage k by output $\bar{u}_\tau^\theta$ of player $\theta \in S^{[k-]}$ produced in stage $\tau \in \{k-1, k-2, \cdots, k-\omega+1\}$;

$u_k^{I\theta}$ is the level of non-durable environmental improvement effort from player θ in stage k and $b_k^\theta u_k^{I\theta}(x_k^1)^{1/2}$ is the improvement in the state of environment by $u_k^{I\theta}$; and λ_k is the natural regeneration rate of the state of the environment.

In addition, the eco-degrading output $\bar{u}_k^i$ also creates a direct and durable negative impact on the payoffs of all the players. Specifically, the negative impact that player i would experience in stage $k \in [T_t^i, T_{t+1}^i - 1]$ is

$$- \sum_{\theta \in S^{\left[T_t^i - T_{t+1}^i-1\right]}} v_k^{i|k|\theta}\bar{u}_k^\theta - \sum_{\theta \in S^{[k-]}} \sum_{\tau=k-\omega+1}^{k-1} v_k^{i|\tau|\theta}\bar{u}_\tau^\theta, \tag{3.2}$$

where $v_k^{i|k|\theta}\bar{u}_k^\theta$ is the direct damaging impact of $\bar{u}_k^\theta$, for $\theta \in S^{[T_t^i, T_{t+1}^i-1]}$, on player i's payoff and $v_k^{i|\tau|\theta}\bar{u}_\tau^\theta$ is the direct damaging impact on player i's payoff by $\bar{u}_\tau^\theta$, for $\theta \in S^{[k-]}$ and $\tau \in \{k-1, k-2, \cdots, k-\omega+1\}$

There exists a knowledge-based green technology that is accessible to all players and is used to produce a green product. The evolution of green technology is governed by the dynamics

$$x_{k+1}^2 = x_k^2 + \sum_{\theta \in S^{[T_t^i, T_{t+1}^i-1]}} a_k^\theta u_k^{II\theta} - \lambda_k^2 x_k^2$$

$$\text{for } k \in \left[T_t^i, T_{t+1}^i - 1\right] \text{ and } t \in (1, 2, \cdots, z_i), \tag{3.3}$$

where

$u_k^{II\theta} \in U^{II\theta} \subset R$ is the investment in public green technology by player θ in stage k,

$a_k^{|k|\theta}(u_k^{II\theta})$ is the addition to the public green technology in stage k by player θ's technology investment in stage k and $\lambda_k^2 x_k^2$ measures the obsolescence in the green technology.

The extent of realization of the green technology by player i is $\tau_k^i x_k^2$. The production function of the clean outputs from the region's realized green capitals is $u_k^{IIIi}(\tau_k^i x_k^2)^{1/2}$, where u_k^{IIIi} is the related non-durable input. The cost of u_k^{IIIi} is $c_k^{IIIi}(u_k^{IIIi})^2$ and the price of the green output is P_k^i. The net economic gain from the green output at stage k is $P_k^i u_k^{IIIi}\left(\tau_k^i x_k^2\right)^{1/2} - c_k^{IIIi}\left(u_k^{IIIi}\right)^2$. The gain-maximizing choice of the non-durable input u_k^{IIIi} equals $\frac{P_k^i(\tau_k^i x_k^2)^{1/2}}{2c_k^{IIIi}}$. The net economic gain from the green output at stage k of player i can then be obtained as $\frac{(P_k^i)^2\tau_k^i x_k^2}{4c_k^{IIIi}}$. Denoting $\frac{(P_k^i)^2\tau_k^i}{4c_k^{IIIi}}$ by κ_k^i, the net economic gain at stage k can be expressed as $\kappa_k^i x_k^2$ for simplicity in exposition.

The single-stage gain of player $i \in N$ is

$$R_k^i \bar{u}_k^i - \phi_k^i\left(\bar{u}_k^i\right)^2 - c_k^{Ii}\left(u_k^{Ii}\right)^2 + \kappa_k^i x_k^2 - c_k^{IIi}\left(u_k^{IIF}\right)^2 - h_k^i x_k^1$$

$$- \sum_{\theta \in S\left[T_t^i, x_{t+1}^i - 1\right]} v_k^{i|k|\theta} \bar{u}_k^{\theta} - \sum_{\theta \in S^{[k-]}} \sum_{\tau=k-\omega+1}^{k-1} v_k^{i|\tau|\theta} \bar{u}_\tau^{\theta}$$

In particular, $R_k^i \bar{u}_k^i$ is the gross revenue from the eco-degrading output, $\phi_k^i(\bar{u}_k^i)^2$ is the costs of producing the output and the cost of environmental degradation abatement is $c_k^{Ii}(u_k^{Ii})^2$ and $c_k^{IIi} u_k^{IIi}$ is the cost of investment in green technology. The damage brought by the state of environmental degradation is $h_k^i x_k^1$.

The present values (viewed at initial stage 1) of the payoffs of player $i \in N$ can be expressed as:

$$\sum_{t=1}^{z_i} \sum_{k=T_t^i}^{T_{t+1}^i - 1} \left(R_k^i \bar{u}_k^i - \phi_k^i\left(\bar{u}_k^i\right)^2 - c_k^{Ii}\left(u_k^{Ii}\right)^2 + \kappa_k^i x_k^2 - c_k^{IIi}\left(u_k^{IIi}\right)^2 - h_k^i x_k^1 - \sum_{\theta \in S\left[T_t^i, T_{t+1}^i - 1\right]} v_k^{i|k|\theta} \bar{u}_k^{\theta} \right.$$
$$\left. - \sum_{\theta \in S^{[k-]}} \sum_{\tau=k-\omega+1}^{k-1} v_k^{i|\tau|\theta} \bar{u}_\tau^{\theta} \right) \delta^{k-1} + \left(Q_{T_{z_i+1}^i}^{(1)i} x_{T_{z_i+1}^i}^1 + Q_{T_{z_i+1}^i}^{(2)i} x_{T_{z_i+1}^i}^2 + \varpi_{T_{z_i+1}^i}^i \right) \delta^{T_{z_i+1}^i - 1} \tag{3.4}$$

where $\left(Q_{T_{z_i+1}^i}^{(1)i} x_{T_{z_i+1}^i}^1 + Q_{T_{z_i+1}^i}^{(2)i} x_{T_{z_i+1}^i}^2 + \varpi_{T_{z_i+1}^i}^i \right)$ is the terminal value of player i after it exits the game in stage $T_{z_i+1}^i - 1$.

3.2 Non-cooperative Equilibrium

For notational convenience, we denote $x = (x^1, x^2)$. Invoking Theorem 2.1 we can characterize a non-cooperative game equilibrium of the game (3.1)–(3.4) in the Corollary below.

Corollary 3.1 *Let* $(u_k^{Ii**}, u_k^{IIi**}, \bar{u}_k^{i**})$, *for* $k \in \bar{T}^i$ *and* $i \in N$, *be the set of feedback Nash equilibrium strategies and* $V^i(k, x; \underline{\bar{u}}_{k-}^{**})$ *be the feedback Nash equilibrium payoff of player* i *at stage* $k \in \bar{T}^i$ *in the cross-generational dynamic game* (3.1)–(3.4) *then the function* $V^i(k, x; \underline{\bar{u}}_{k-}^{**})$, *for* $k \in \bar{T}^i$ *and* $i \in N$, *satisfies the following recursive equations:*

$$V^i\left(T^i_{z_i+1}, x; \underline{\bar{u}}^{**}_{(T^i_{z_i+1})-}\right) = \left(Q^{(1)i}_{T^i_{z_i+1}} x^1 + Q^{(2)i}_{T^i_{z_i+1}} x^2 + \varpi^i_{T^i_{z_i+1}}\right)\delta^{T^i_{z_i+1}-1},$$

$$V^i\left(T^i_{t+1}, x, \underline{\bar{u}}^{**}_{(T^i_{t+1})-}\right) = V^i\left(T^i_{t+1}, x; \underline{\bar{u}}^{**}_{(T^i_{t+1})-}\right),$$

$$V^j\left(T^i_{t+1}, x; \underline{\bar{u}}^{**}_{(T^i_{t+1})-}\right) = \left(Q^{(1)j}_{T^i_{t+1}} x^1 + Q^{(2)j}_{T^i_{t+1}} x^2 + \varpi^j_{T^i_{t+1}}\right)\delta^{T^i_{t+1}-1}, \quad \text{for } j \in S^{\left[T^i_t, T^i_{t+1}-1\right]A},$$

$$V^l\left(T^i_{t+1}, x; \underline{\bar{u}}^{**}_{(T^i_{t+1})-}\right) = V^\ell\left(T^i_{t+1}, x; \underline{\bar{u}}^{**}_{(T^i_{t+1})-}\right) \quad \text{for} \quad l \in S^{[T^i_t, T^T_{t+1}-1]B} \quad ; \tag{3.5}$$

$$V^i(k, x; \underline{\bar{u}}^{**}_{k-}) = \max_{u_k^{Ii}, u_k^{IIi}, \bar{u}_k^i} \Bigg\{ \Bigg(R^i_k \bar{u}^i_k - \varphi^i_k\left(\bar{u}^i_k\right)^2 - c^{Ii}_k\left(u^{Ii}_k\right)^2 + \kappa^i_k x^2 - c^{IIi}_k\left(u^{\Pi i}_k\right)^2 - h^i_k x^1$$

$$- v_k^{i|k|i}\bar{u}^i_k - \sum_{\theta \in S^{\left[T^i_t, T^i_{t+1}-1\right]}} v_k^{i|k|\theta}\bar{u}_k^{\theta**} - \sum_{\theta \in S^{[k-]}} \sum_{\tau=k=\omega+1}^{k-1} v_k^{i|\tau|\theta}\bar{u}_\tau^{\theta**}\Bigg)\delta^{k-1}$$

$$+ V^i\left(k+1, x_{k+1}^{1\left[T^i_t, T^i_{t+1}-1\right](\neq i)***}, x_{k+1}^{2\left[T^i_t, T^i_{t+1}-1\right](\neq i)**}; \bar{u}^i_k, \underline{\bar{u}}_k^{\left[T^i_t, T^i_{t+1}-1\right](\neq i)**}, \underline{\bar{u}}^{**}_{(k+1)-} \cap \underline{\bar{u}}^{**}_{k-}\right)\Bigg\}; \tag{3.6}$$

$$V^j(k, x; \underline{\bar{u}}^{**}_{k-}) = \max_{u_k^{Ij}, u_k^{IIj}, \bar{u}_k^j} \Bigg\{ \Bigg(R^j_k \bar{u}^j_k - \varphi^j_k\left(\bar{u}^j_k\right)^2 - c^{Ij}_k\left(u^{Ij}_k\right)^2 + \kappa^j_k x^2 - c^{IIj}_k\left(u^{IIj}_k\right)^2 - h^j_k x^1$$

$$- v_k^{j|k|j}\bar{u}^j_k - \sum_{\substack{\theta \in S^{\left[T^i_t, T^i_{t+1}-1\right]} \\ \theta \neq j}} v_k^{j|k|\theta}\bar{u}_k^{\theta**} - \sum_{\theta \in S^{[k-]}} \sum_{\tau=k-\omega+1}^{k-1} v_k^{j|\tau|\theta}\bar{u}_\tau^{\theta**}\Bigg)\delta^{k-1}$$

$$+ V^j\left(k+1, x_{k+1}^{1\left[T^i_i, T^i_{t+1}-1\right](\neq j)**}, x_{k+1}^{2\left[T^i_t, T^i_{t+1}-1\right](\neq j)**}; \bar{u}^j_k, \underline{u}_k^{\left[T^i_t, T^i_{t+1}-1\right](\neq j)***}, \underline{\bar{u}}^{***}_{(k+1)-} \cap \underline{\bar{u}}^{**}_{k-}\right)\Bigg\},$$

$$\text{for } j \in S^{[T^i_t, T^i_{t+1}-1]A}; \tag{3.7}$$

$$V^{\ell}(k, x; \bar{u}_{k-}^{**}) = \max_{u_k^{I\ell}, u_k^{II\ell}, \bar{u}_k^{\ell}} \left\{ \left(R_k^{\ell} \bar{u}_k^{\ell} - \varphi_k^{\ell}\left(\bar{u}_k^{\ell}\right)^2 - c_k^{Ilj}\left(u_k^{Ilj}\right)^2 + \kappa_k^{\ell} x^2 - c_k^{IIJj}\left(u_k^{IIJj}\right)^2 - h_k^{\ell} x^1 \right.\right.$$

$$\left. - v_k^{\ell|k|\ell} \bar{u}_k^{\ell} - \sum_{\substack{\theta \in S^{[T_t^i, T_{t+1}^i - 1]} \\ \theta \neq l}} v_k^{\ell|k|\theta} \bar{u}_k^{\theta **} - \sum_{\theta \in S^{[k-]}} \sum_{\tau = k-\omega+1}^{k-1} v_k^{\ell|\tau|\theta} \bar{u}_\tau^{\theta **} \right) \delta^{k-1}$$

$$\left. + V^{\ell}\left(k+1, x_{k+1}^{1\left[T_T^i, T_{t+1}^i - 1\right](\neq \ell)**}, x_{k+1}^{2\left[T_t^i, T_{t+1}^i - 1\right](\neq l)**}; \bar{u}_k^{\ell}, \underline{\bar{u}}_k^{\left[T_t^i, T_{t+1}^i - 1\right](\neq \ell)**}, \underline{\bar{u}}_{(k+1)-}^{**} \cap \underline{\bar{u}}_{k-}^{**}\right) \right\},$$

$$\text{for } \ell \in S^{[T_t^i, T_{t+1}^i - 1]B}; \tag{3.8}$$

where

$$x_{k+1}^{1[T_t^i, T_{t+1}^i - 1](\neq \varsigma)**} = x^1 + m_k^{|k|\varsigma} \bar{u}_k^{\varsigma} + \sum_{\substack{\theta \in S^{[T_t^i, T_{t+1}^i - 1]} \\ \theta \neq \varsigma}} m_k^{|k|\theta} \bar{u}_k^{\theta **} + \sum_{\theta \in S^{[k-]x^1}} \sum_{\tau = k-\omega+1}^{k-1} m_k^{|\tau|\theta} \bar{u}_\tau^{\theta **}$$

$$- b_k^{\varsigma} u_k^{I\varsigma} (x^1)^{1/2} - \sum_{\substack{\theta \in S^{[T_t^i, T_{t+1}^i - 1]} \\ \theta \neq \varsigma}} b_k^{\theta} u_k^{I\theta **} (x^1)^{1/2} - \lambda_k^1 x^1$$

$$x_{k+1}^{2[T_t^i, T_{t+1}^i - 1](\neq \varsigma)**} = x^2 + a_k^{\varsigma} u_k^{II\varsigma} + \sum_{\substack{\theta \in S^{[T_t^i, T_{t+1}^i - 1]} \\ \theta \neq \varsigma}} a_k^{\theta} u_k^{II\theta **} - \lambda_k^2 x^2,$$

for $\varsigma \in \{i, j, \ell\}$, $k \in [T_t^i, T_{t+1}^i - 1]$ and $t \in \{1, 2, \cdots, z_i\}$. ■

Performing the indicated maximization operator in Corollary 3.1, the value function $V^i(k, x; \underline{\bar{u}}_{k-}^{**})$, which reflects the payoff of the player i at stage k as follows, can be obtained as follows.

Proposition 3.1 *System* (3.5)–(3.8) *admits a game equilibrium solution with the payoffs of player* $i \in N$ *being*

$$V^i\left(T_{z_i+1}^i, x; \bar{u}_{(z_i+1)-}^{**}\right) = \left(Q_{T_{z_i+1}^i}^{(1)i} x^1 + Q_{T_{z_i+1}^i}^{(2)i} x^2 + \varpi_{T_{z_i+1}^i}^i\right) \delta^{T_{z_i+1}^i - 1},$$

$$V^i\left(k, x; \underline{\bar{u}}_{k-}^{**}\right) = \left(A_k^{(1)i} x^1 + A_k^{(2)i} x^2 + C_k^i\right) \delta^{k-1}, \quad k \in \left[T_{z_i}^i, T_{z_i+1}^i - 1\right], \tag{3.9}$$

where $A_{T_{z_i+1}^i}^{(1)i} = Q_{T_{z_i+1}^i}^{(1)i}$, $A_{T_{z_i+1}^i}^{(2)i} = Q_{T_{z_i+1}^i}^{(2)i}$ *and* $C_{T_{z_i+1}^i}^i = \varpi_{T_{z_i+1}^i}^i$;

$$A_k^{(1)i} = -\frac{\left(A_{k+1}^{(1)i} b_k^i \delta\right)^2}{4c_k^{Ii}} - h_k^i + \delta A_{k+1}^{(1)i}\left(1 + \sum_{\theta \in S^{\left[T_{z_i}^i, T_{z_i+1}^i - 1\right]}} b_k^\theta \frac{A_{k+1}^{(1)\theta} b_k^\theta \delta}{2c_k^{I\theta}} - \lambda_k^1\right)$$

and

$$A_k^{(2)i} = \kappa_k^i + \delta A_{k+1}^{(2)i}(1 - \lambda_k^2),$$

and the corresponding C_k^i *is an expression that includes previously executed durable strategies* $\underline{\bar{u}}_{k-}^{**}$, for $k \in [T_{z_i}^i, T_{z_i+1}^i - 1]$;

$$V^i(k, x; \bar{u}_{k-}^{*k}) = \left(A_k^{(1)i} x^1 + A_k^{(2)i} x^2 + C_k^i\right)\delta^{k-1},$$

$$\text{for } k \in \left[T_t^i, T_{t+1}^i - 1\right] \text{ and } t \in \{1, 2, \cdots, z_i - 1\}, \tag{3.10}$$

where

$$A_k^{(1)i} = -\frac{\left(A_{k+1}^{(1)i} b_k^i \delta\right)^2}{4c_k^{Ii}} - h_k^i + \delta A_{k+1}^{(1)i}\left(1 + \sum_{\theta \in S^{[T_t^i, T_t^i - 1]}} b_k^\theta \frac{A_{k+1}^{(1)\theta} b_k^\theta \delta}{2c_k^{I\theta}} - \lambda_k^1\right)$$

$$A_k^{(2)i} = \kappa_k^i + \delta A_{k+1}^{(2)i}(1 - \lambda_k^2),$$

and the corresponding C_k^i *is an expression that includes previously executed durable strategies* $\underline{\bar{u}}_{k-}^{**}$, *for* $k \in [T_t^i, T_{t+1}^i - 1]$.

Proof See Appendix II. ■

The game equilibrium strategies of player $i \in N$ can be obtained as (see derivation details in Appendix II):

$$u_k^{Ii**} = \frac{-\delta A_{k+1}^{(1)i} b_k^i}{2c_k^{Ii}} (x^1)^{1/2}, \; u_k^{IIi**} = \frac{\delta A_{k+1}^{(2)i} a_k^i}{2c_k^{IIi}},$$

$$\bar{u}_k^{i**} = \frac{R_k^i - \sum_{\eta=k}^{k+\omega-1} \delta^{\eta-k} v_\eta^{i|k|i} + \sum_{\eta=k}^{k+\omega-1} \delta^{\eta-k+1} A_{\eta+1}^{(1)i} m_\eta^{|k|i}}{2\phi_k^i},$$

$$\text{with } v_\eta^{i|\eta|i} = A_{\eta+1}^{(1)i} = 0, \text{ for } \eta > T_{z_i+1}^i - 1 \tag{3.11}$$

for $k \in \{T_1^i, T_1^i + 1, T_1^i + 2, \cdots, T_{z_i+1}^i - 1\}$.
The game equilibrium environmental degradation dynamics is.

$$x_{k+1}^{1} = x_k^1 + \sum_{\theta \in S^{[T_t^i, T_{t+1}^i - 1]}} m_k^{|k|\theta} \bar{u}_k^{\theta **} + \sum_{\theta \in S^{[k-]}} \sum_{\tau = k-\omega+1}^{k-1} m_k^{|\tau|\theta} \bar{u}_\tau^{\theta **} - \sum_{\vartheta \in S^{[T_t^i, T_{t+1}^i - 1]}} b_k^\theta u_k^{I\theta **} (x_k^1)^{1/2} - \lambda_k^1 x_k^1$$

$$\text{for } k \in \left[T_t^i, T_{t+1}^i - 1\right] \text{ and } t \in (1, 2, \cdots, z_i). \tag{3.12}$$

The evolution of the green technology is governed by the dynamics

$$x_{k+1}^2 = x_k^2 + \sum_{\theta \in S^{\left[T_t^i, T_{t+1}^i - 1\right]}} a_k^\theta u_k^{II\theta **} - \lambda_k^2 x_k^2,$$

$$\text{for } k \in \left[T_t^i, T_{t+1}^i - 1\right] \text{ and } t \in (1, 2, \cdots, z_i). \tag{3.13}$$

We use $\{x_k^{1**}, x_k^{2**}\}_{k=1}^{T+1}$ to denote the solution to (3.12) and (3.13), which is game equilibrium trajectory of environmental degradation and green technology.

Remark 3.1 One of the novel results is that environmental degradation may continue to worsen even if all the players stop their eco-degrading production, that is

$$x_{k+1}^1 = x_k^1 + \sum_{\theta \in S^{[k-]}} \sum_{\tau = k-\omega+1}^{k-1} m_k^{|\tau|\theta} \bar{u}_\tau^{\theta **} - \sum_{\vartheta \in S^{[T_t^i, T_{t+1}^i - 1]}} b_k^\theta u_k^{I\theta **} (x_k^1)^{1/2} - \lambda_k^1 x_k^1$$

is still positive even $\sum_{\theta \in S^{[T_t^i, T_{t+1}^i - 1]}} m_k^{|k|\theta} \bar{u}_k^{\theta **} = 0.$

Remark 3.2 The negative impacts of eco-degrading outputs by players entering the game before would continue to affect the payoffs of the players entering subsequently. In particular, according to (3.4), the negative impact on player i's payoff at stage k amounts to $-\sum_{\theta \in S^{[k-]}} \sum_{\tau=k-\omega+1}^{k-1} v_k^{i|\tau|\theta} \bar{u}_\tau^{\theta **}$

Another novel result is that even if all the players cease their eco-degrading production at stage k, they would still suffer from previously produced eco-degrading outputs.

The previously executed durable strategies by players who have left the game will continue to affect the state dynamics and payoffs of the players entering the game after their exits. In a non-cooperative equilibrium, environmental degradation can continue to worsen over time. As a result, a *dynamic tragedy of the cross-generational environmental commons* seems to be unavoidable unless effective global cooperative efforts are made available.

4 Asynchronous Horizon Cooperation under Durable Strategies

It is well known that non-cooperative behaviours among participants would, in general, lead to an outcome that is not Pareto optimal. In an asynchronous horizons game situation with durable strategies highly undesirable outcomes could appear for players entering the game in later stages. A cooperative scheme acceptable to all players would bring about the possibility of a socially optimal and group efficient solution to all players. To exploit the potential gains from cooperation, the players agree to act cooperatively and distribute the payoffs among themselves according to an agreed-upon gain sharing optimality principle.

4.1 Game Formulation and Group Optimality

To exploit the potential gains from cooperation, the players agree to act cooperatively and distribute the payoffs among themselves according to an agreed-upon gain sharing optimality principle. We partition the entire game horizon $[1, T]$ into z segments. A distinct set of players exists in each of the game segments. For notational convenience in subsequent exposition, we adopt the notations T_z as the terminal stage T and T_1 as the initial stage 1. Therefore, the z game segments in the entire game horizon can be expressed as $[T_1, T_2 - 1]$, $[T_2, T_3 - 1]$, $\big[T_3, T_{4-1}\big], \cdots\cdots, \big[T_{z-1}, T_z - 1\big], \big[T_z, T_{z+1} - 1\big] \equiv \big[T_z, T\big]$. We use $(\underline{u}_k^{[T_t,T_{t+1}-1]}, \bar{\underline{u}}_k^{[T_t,T_{t+1}-1]})$ to denote the set of strategies of the players in the game segment $[T_t, T_{t+1} - 1]$ in stage $k \in [T_t, T_{t+1} - 1]$, for $t \in (1, 2, \cdots, z)$.

The state dynamics is characterized by the difference equations:

$$x_{k+1} = f_k\Big(x_k, \underline{u}_k^{[T_t,T_{t+1}-1]}, \bar{\underline{u}}_k^{[T_t,T_{t+1}-1]}; \bar{\underline{u}}_{k-}\Big),$$
$$\text{for } k \in \big[T_t, T_{t+1} - 1\big] \text{ and } t \in (1, 2, \cdots, z). \tag{4.1}$$

We use $S^{[T_t,T_{t+1}-1]}$ to denote the set of players existing in the game segment $k \in [T_t, T_{t+1} - 1]$, for $t \in (1, 2, \cdots, z)$. We use $S^{[T_t,T_{t+1}-1]C}$ to denote the set of players who will exit the game after stage $T_{t+1}^i - 1$ and receive a terminal payoff in stage T_{t+1}, for $t \in (1, 2, \cdots, z)$. To achieve group optimality, the players will solve the dynamic optimization problem which maximizes their joint payoff in all game intervals $[T_t, T_{t+1} - 1]$, for $t \in (1, 2, \cdots, z)$

$$\sum_{t=1}^{z} \sum_{i\in S^{[T_t,T_{t+1}-1]}} \Bigg[\sum_{k=T_t}^{T_{t+1}-1} g_k^i(x_k, \underline{u}_k^{[T_t,T_{t+1}^i-1]}, \bar{\underline{u}}_k^{[T_t,T_{t+1}^i-1]}; \bar{\underline{u}}_{k-})\delta_1^k$$
$$+ \sum_{j\in S^{[T_t,T_{t+1}-1]C}} q_{T_{t+1}}^j(x_{T_{t+1}}; \bar{\underline{u}}_{(T_{t+1}^i)-})\,\delta_1^{T_{t+1}} \Bigg] \tag{4.2}$$

subject to dynamics (4.1).

The optimization problem (4.1)–(4.2) is analogous to a dynamic optimization problem in which the sets of control strategies are different in different time intervals of the problem horizon. An optimal solution to the joint maximization problem (4.1)–(4.2) can be characterized by the theorem below.

Theorem 4.1 *Let $W(k, x; \bar{\underline{u}}_{k-})$ denote the maximal value of the joint payoff at stage k with state $x_k = x$ and previously executed durables strategies that are still effective $\bar{\underline{u}}_{k-}$, then the function $W(k, x; \underline{u}_{k-})$ satisfies the following system of recursive equations:*

$$W(T+1, x, \bar{\underline{u}}_{(T+1)-}) = \sum_{i \in S^{[T_z,T]}} q^i_{T+1}\big(x; \bar{\underline{u}}^i_{(T+1)-}\big)\delta_1^{T+1}, \tag{4.3}$$

$$W(k, x; \bar{\underline{u}}_{k-}) = \max_{\substack{u^i_k, \bar{u}^i_k \\ i \in S^{[T_z,T]}}} \Bigg\{ \sum_{i \in S^{[T_z,T]}} g^i_k(x, \underline{u}_k^{[T_z,T]}, \bar{\underline{u}}_k^{[T_z,T]}; \bar{\underline{u}}_{k-})\delta_1^k$$

$$+W\Big[k+1, f_k\Big(x, \underline{u}_k^{[T_z,T]}, \bar{\underline{u}}_k^{[T_z,T]}; \bar{\underline{u}}_{k-}\Big); \bar{\underline{u}}_k^{[T_z,T]}, \bar{\underline{u}}_{(k+1)} - \cap \bar{\underline{u}}_{k-}\Big]\Bigg\}$$

$$\text{for } k \in \big[T_z, T\big]; \tag{4.4}$$

$$W\Big(T_{t+1} - 1, x; \bar{\underline{u}}_{(T_{t+1}-1)-}\Big) = \max_{\substack{u^i_{T+1-1-1}, \bar{u}^i_{t+1-1} - 1 \\ i \in S^{[T_t, T_{t+1}-1]}}}$$

$$\Bigg\{ \sum_{i \in S^{[T_t,T_{t+1}-1]}} g^i_{T_{t+1}-1}\Big(x, \underline{u}_{T_{t+1}-1}^{[T_t,T_{t+1}-1]}, \bar{\underline{u}}_{T_{t+1}-1}^{[T_t,T_{t+1}-1]}; \bar{\underline{u}}_{(T_{t+1}-1)-}\Big)\delta_1^{T_{t+1}-1}$$

$$+ \sum_{j \in S^{[T_t,T_{t+1}-1]C}} q^j_{T_{t+1}}[f_{T_{t+1}-1}(x, \underline{u}_{T_{t+1}-1}^{[T_t,T_{t+1}-1]}, \bar{\underline{u}}_{T_{t+1}-1}^{[T_t,T_{t+1}-1]};$$

$$\bar{\underline{u}}_{(T_{t+1}-1)-}); \bar{\underline{u}}_{T_{t+1}-1}^{[T_t,T_{t+1}-1]}, \bar{\underline{u}}_{(T_{t+1})-} \cap \bar{\underline{u}}_{(T_{t+1}-1)-}]\delta_1^{T_{t+1}}$$

$$+W\Big[T_{t+1}, f_{T_{t+1}-1}\Big(x, \underline{u}_{T_{t+1}-1}^{[T_t,T_{t+1}-1]}, \bar{\underline{u}}_{T_{t+1}-1}^{[T_t,T_{t+1}-1]}, \bar{\underline{u}}_{(T_{t+1}-1)-}\Big);$$

$$\bar{\underline{u}}_{T_{t+1}-1}^{[T_t,T_{t+1}-1]}, \bar{\underline{u}}_{(T_{t+1})-} \cap \bar{\underline{u}}_{(T_{t+1}-1)-}\Big]\Bigg\} \tag{4.5}$$

$$W(k, x; \bar{\underline{u}}_{k-}) = \max_{\substack{u^i_k, \bar{u}^i_k \\ i \in S^{[T_t,T_{t+1}-1]}}} \Bigg\{ \sum_{i \in S^{[T_t,T_{t+1}-1]}} g^i_k(x, \underline{u}_k^{[T_t,T_{t+1}-1]}, \bar{\underline{u}}_k^{[T_t,T_{t+1}-1]}; \bar{\underline{u}}_{k-})\delta_1^k$$

$$+W\Big[k+1, f_k\Big(x, \underline{u}_k^{[T_t,T_{t+1}-1]}, \bar{\underline{u}}_k^{[T_t,T_{t+1}-1]}; \bar{\underline{u}}_{k-}\Big); \bar{u}_k^{[T_t,T_{t+1}-1]}, \bar{u}_{(k+1)-} \cap \bar{u}_{k-}\Big]\Bigg\},$$

$$\text{for } k \in [T_t, T_{t+1} - 2] \text{ and } t \in \{1, 2, \cdots, z - 1\}. \tag{4.6}$$

Proof
See Appendix III. ■

We use $(\underline{u}_k^*, \bar{\underline{u}}_k^*)$ to denote the optimal control strategies derived from Theorem 4.1. Substituting the optimal strategies into the state dynamics in (1.2), the optimal state dynamics can be obtained as:

$$x_{k+1} = f_k\Big(x_k, \underline{u}_k^{[T_t, T_{t+1}-1]^*}, \bar{\underline{u}}_k^{[T_t, T_{t+1}-1]^*}; \bar{\underline{u}}_{k-}^*\Big),$$
$$\text{for } k \in [T_t, T_{t+1} - 1] \text{ and } t \in (1, 2, \cdots, z). \tag{4.7}$$

The solution to (4.7) yields the optimal trajectory which we denote as $\{x_k^*\}_{k=1}^{T+1}$.

4.2 *Individual Rationality and Group Optimality*

We let

$$[\xi^1(T_1^1, x_1^0; \bar{\underline{u}}_{T_1^1-}), \xi^2(T_1^2, x_1^0; \bar{\underline{u}}_{T_1^2-}), \cdots, \xi^n(T_1^n, x_1^0; \bar{\underline{u}}_{T_1^n-})] \tag{4.8}$$

denote (in present value) the agreed-upon distribution of cooperative payoffs among the players. In particular, player i would receive $\xi^i(T_1^i, x_1^0; \underline{u}_{T_1^i-}^*)$ over the cooperation duration, for $i \in N$

To satisfy group optimality in the cooperative scheme, it is required to satisfy.

$$W(1, x_1^0; \bar{\underline{u}}_{1-}) = \sum_{j=1}^{n} \xi^j\Big(T_1^j, x_1^0; \bar{\underline{u}}_{T_1^j-}\Big). \tag{4.9}$$

This condition guarantees the maximal joint payoff is distributed to the players. Individual rationality is required to hold so that the payoff allocated to any player under cooperation will be no less than his non-cooperative payoff. Failure to guarantee individual rationality leads to the condition where the concerned participants would deviate from the agreed-upon solution plan and play non-cooperatively. To satisfy individual rationality, the payoffs received by the players under cooperation must be no less than their non-cooperative payoffs along the cooperative state trajectory.

For individual rationality to be maintained, one of the conditions is that the payoff of player i under cooperation must be no less than his non-cooperative payoff in the outset of the game, that is.

$$\xi^i\Big(T_1^i, x_1^0; \bar{\underline{u}}_{T_1^i-}\Big) \geq V^i\Big(T_1^i, x_1^0; \bar{\underline{u}}_{T_1^i-}\Big), \quad \text{for } i \in N. \tag{4.10}$$

If condition (4.10) is not satisfied, no agreement can be reached.

5 A Theoretical Compensatory Solution

In this section, we present a theoretical cooperative solution to alleviate the problem of self-maximization by individual players in an asynchronous horizon game with durable strategies. For instance, if the players agree to share the cooperative gains proportional to the relative sizes of their non-cooperative payoffs, the imputation under cooperation that player i will receive becomes.

$$\xi^i = \frac{V^i\left(T_1^i, x_{T_1^i}^*, u_{T_1^i-}^{**}\right)}{\sum_{\theta=1}^n V^\theta\left(T_1^\theta, x_{T_1^\theta}^*, \bar{u}_{T_1^\theta-}\right)} W(1, x_1^0; \underline{u}_{1-}^*), \quad \text{for } i \in N \tag{5.1}$$

If the players agree to share the cooperative gains equally, the imputation under cooperation that player i will receive becomes

$$\xi^i = V^i\left(T_1^i, x_{T_1^i}^{**}; \underline{\bar{u}}_{T_1^i-}^{**}\right) + \frac{1}{n}\left(W(1, x_1^0; \underline{\bar{u}}_{1-}) - \sum_{\theta=1}^n V^\theta\left(T_1^\theta, x_{T_1^\theta}^{**}; \underline{\bar{u}}_{T_1^\theta-}^{**}\right)\right)\delta_1^{T_1^i},$$
$$\text{for } i \in N. \tag{5.2}$$

For a dynamically stable cooperation scheme, it is crucial to maintain individual rationality so that all players' cooperative payoffs are no less than their non-cooperative payoffs throughout the cooperation duration. The players must agree with a Payoff Distribution Procedure (PDP) such that individual rationality is maintained for all players along the cooperation state trajectory. To do this, we follow Yeung and Petrosyan (2010), Yeung and Petrosyan (2016a), Yeung and Petrosyan (2019) use $\beta_k^i(x_k^*; \underline{\bar{u}}_{k-}^*)$ to denote the payment that player i receives in stage k under the cooperative agreement along the cooperative trajectory $\{x_k^*\}_{k=1}^T$, with durable strategies executive before but still in effect being $\underline{\bar{u}}_{k-}^*$. The payment scheme involving $\beta_k^i(x_k^*; \underline{\bar{u}}_{k-}^*)$ constitutes a PDP in the sense that

$$\xi^i\left(k, x_k^*; \underline{\bar{u}}_{k-}^*\right) = \beta_k^i\left(x_k^*; \underline{\bar{u}}_{k-}^*\right)\delta_{T_1^i}^k$$
$$+ \left\{\sum_{\zeta=k+1}^{T_{z_i+1}^i - 1} \beta_\xi^i\left(x_\zeta^*; \underline{\bar{u}}_{\zeta-}^*\right)\delta_{T_1^i}^\zeta + q_{T_{z_i+1}^i}^i\left(x_{T_{z_i+1}^i}^*; \underline{\bar{u}}_{(T_{z_i}+1)-}^*\right)\delta_{T_1^i}^{T_{z_i+1}^i}\right\},$$
$$\text{for } k \in \left(T_1^i, T_1^i + 1, \cdots, T_{z_i+1}^i - 1\right); \tag{5.3}$$

with $\xi^i(T_1^i, x_{T_1^i}^*; \underline{\bar{u}}_{T_1^i-}^*)$ satisfying $\xi^i(T_1^i, x_{T_1^i}^*; \underline{\bar{u}}_{T_1^i-}^*) \geq V^i(T_1^i, x_{T_1^i}^{**}; \underline{\bar{u}}_{T_1^i-}^{**})$.

To uphold individual rationality throughout all stages $k \in \{1, 2, \cdots, T\}$, it is required that the choice of $\beta_k^i(x_k^*; \underline{\bar{u}}_{k-}^*)$, for $k \in (T_1^i, T_1^i + 1, \cdots, T_{z_i+1}^i - 1)$, must satisfy the following inequalities:

$$\xi^i\left(k, x_k^*; \underline{\bar{u}}_{k-}^*\right) \geq V^i\left(k, x_k^*; \underline{\bar{u}}_{k-}^*\right), \text{ for } k \in \left(T_1^i, T_1^i + 1, \cdots, T_{z_i+1}^i - 1\right). \tag{5.4}$$

Therefore, it is necessary to construct a set of payoff distribution $\beta_k^i(x_k^*; \underline{\bar{u}}_{k-}^*)$ that guarantees that all players' cooperative payoffs are no less than their non-cooperative payoffs along the cooperative trajectory. Using (5.3), we obtain (5.4) as:

$$\xi^i\left(k, x_k^*; \underline{\bar{u}}_{k-}^*\right) = \beta_k^i\left(x_k^*; \underline{\bar{u}}_{k-}^*\right)\delta_{T_1^i}^k$$

$$+\left\{\sum_{\zeta=k+1}^{T_{z_i+1}^i-1} \beta_\zeta^i\left(x_\zeta^*; \underline{\bar{u}}_{\zeta-}^*\right)\delta_{T_1^i}^\zeta + q_{T_{z_i+1}^i}^i\left(x_{T_{z_i+1}^i}^*; \underline{\bar{u}}_{(T_{z_i+1})-}^*\right)\delta_{T_1^i}^{T_{z_i+1}^i}\right\} > V^i\left(k, x_k^*; \underline{\bar{u}}_{k-}^*\right)$$

$$\text{for } k \in \left(T_1^i, T_1^i + 1, \cdots, T_{z_i+1}^i - 1\right). \tag{5.5}$$

Given that $V^i(k, x_k^*; \underline{\bar{u}}_{k-}^*)$, for $k \in (T_1^i, T_1^i + 1, \cdots, T_{z_i+1}^i - 1)$, can be obtained via Proposition 3.1, we can select a candidate of $\beta_k^i(x_k^*; \underline{\bar{u}}_{k-}^*)$ satisfying the following condition:

$$\beta_k^i\left(x_k^*; \underline{\bar{u}}_{k-}^*\right) \geq V^i\left(k, x_k^*; \underline{\bar{u}}_{k-}^*\right)\left(\delta_{T_1^i}^k\right)^{-1} - \left(\delta_{T_1^i}^k\right)^{-1}\left[\sum_{\zeta=k+1}^{T_{z_i+1}^i-1} \beta_\zeta^i\left(x_\zeta^*, y_\zeta^*; \underline{\bar{u}}_{\zeta-}^*\right)\delta_{T_1^i}^\zeta\right.$$

$$\left.+q_{T_{z_i+1}^i}^i\left(x_{T_{z_i+1}^i}^*; \underline{\bar{u}}_{(T_{z_i+1})-}^*\right)\delta_{T_1^i}^{T_{z_i+1}^i}\right], \quad k \in \left(T_1^i, T_1^i + 1, \cdots, T_{z_i+1}^i - 1\right) \tag{5.6}$$

The above scheme guarantees group optimality and individual rationality.

6 Cooperative Management of Cross-generational Environmental Problems

In this section, we consider an illustration of the cooperative scheme with the cross-generational environmental game presented in Sect. 3.

6.1 Group Optimal Sum of Players' Payoffs

To obtain the most efficient (group optimal) outcome, the players maximize their joint payoff in the environmental game in Sect. 3. To do this, we consider the entire game horizon $[1, T]$ in which there are z segments with a distinct set of players. For notational convenience in subsequent exposition, we adopt the notations $T_z = T$ and $T_1 = 1$. Therefore, the z game segments in the entire game horizon can be expressed as $[T_1, T_2 - 1], [T_2, T_3 - 1], \left[T_3, T_{4-1}\right], \cdots\cdots, \left[T_{z-1}, T_z - 1\right], [T_z, T_{z+1} - 1]$. We use $(\underline{u}_k^{[T_t,T_{t+1}-1]}, \bar{\underline{u}}_k^{[T_t,T_{t+1}-1]})$ to denote the set of strategies of the players in the game segment $[T_t, T_{t+1} - 1]$ in stage $k \in [T_t, T_{t+1} - 1]$, for $t \in (1, 2, \cdots, z)$.

To obtain the Pareto efficient sum of the players' payoffs, we maximize

$$\sum_{t=1}^{z} \sum_{i \in S^{[T_t,T_{t+1}-1]}} \left[\sum_{k=T_t}^{T_{t+1}-1} \left(R_k^i \bar{u}_k^i - \varphi_k^i (\bar{u}_k^i)^2 - c_k^{Ii} (u_k^{Ii})^2 + \kappa_k^j x_k^2 - c_k^{IIi} (u_k^{IIi})^2 - h_k^i x_k^1 . \right.\right.$$
$$\left. - \sum_{\theta \in S^{\left[T_t^i, T_{t+1}^i - 1\right]}} v_k^{i|k|\theta} \bar{u}_k^{\theta} - \sum_{\theta \in S^{[k-]}} \sum_{\tau=k-\omega+1}^{k-1} v_k^{i|\tau|\theta} \bar{u}_\tau^{\theta} \right) \delta^{k-1}$$
$$\left. + \sum_{j \in S^{[T_t,T_{t+1}-1]A}} \left(Q_{T_{t+1}}^{(1)j} x_{T_{t+1}}^1 + Q_{T_{t+1}}^{(2)j} x_{T_{t+1}}^2 + \varpi_{T_{t+1}}^j \right)\Big) \delta^{T_{t+1}-1} \right] \tag{6.1}$$

subject to dynamics

$$x_{k+1}^1 = x_k^1 + \sum_{\theta \in S^{[T_t,T_{t+1}-1]}} m_k^{|k|\theta} \bar{u}_k^{\theta} + \sum_{\theta \in S^{[k-]}} \sum_{\tau=k-\omega+1}^{k-1} m_k^{|\tau|\theta} \bar{u}_\tau^{\theta}$$
$$- \sum_{\vartheta \in S^{[T_t,T_{t+1}-1]}} b_k^{\theta} u_k^{I\theta} (x_k^1)^{1/2} - \lambda_k^1 x_k^1, \tag{6.2}$$

$$x_{k+1}^2 = x_k^2 + \sum_{\theta \in S^{[T_t,T_{t+1}-1]}} a_k^{\theta} u_k^{II\theta} - \lambda_k^2 x_k^2, \tag{6.3}$$

for $k \in [T_t, T_{t+1} - 1]$ and $t \in (1, 2, \cdots, z)$.

Invoking Theorem 4.1, we obtain.

Corollary 6.1 *Let $W(k, x; \bar{\underline{u}}_{k-})$ denote the maximal value of the joint payoff at stage k with state $x_k = x$ and previously executed durable strategies that are still effective $\bar{\underline{u}}_{k-}$, then the function $W(k, x; \bar{\underline{u}}_{k-})$ satisfies the following system of recursive equations:*

$$W\left(T+1, x; \bar{\underline{u}}_{(T+1)-}\right) = \sum_{i \in S^{[T_z,t]}} \left(Q_{T+1}^{(1)i} x^1 + Q_{T+1}^{(2)i} x^2 + \varpi_{T+1}^i \right) \delta^T,$$
$$\sum_{j \in S^{[T_t,T_{t+1}-1]A}} \left(Q_{T_{t+1}}^{(1)j} x_{T_{t+1}}^1 + Q_{T_{t+1}}^{(2)j} x_{T_{t+1}}^2 + \varpi_{T_{t+1}}^j \right) \tag{6.4}$$

$$
\begin{aligned}
W\left(k, x; \underline{\bar{u}}_{k-}\right) = & \max_{u_k^i,\, \bar{u}_k^i i \in S^{[T_z, T]}} \\
& \left\{ \sum_{i \in [T_z, T]} \left(R_k^i \bar{u}_k^i - \varphi_k^i \left(\bar{u}_k^i\right)^2 - c_k^{Ii}\left(u_k^{Ii}\right)^2 + \kappa_k^i x^2 - c_k^{IIi}\left(u_k^{III}\right)^2 - h_k^1 x^1 \right.\right. \\
& \left. - \sum_{\theta \in S^{[T_z, T]}} v_k^{i|k|\theta} \bar{u}_k^\theta - \sum_{\theta \in S^{[k-]}} \sum_{\tau = k-\omega+1}^{k-1} v_k^{i|\tau|\theta} \bar{u}_\tau^\theta \right) \delta^{k-1} \\
& \left. + W\left[k+1, x_{k+1}^{1[T_z, T]}, x_{k+1}^{2[T_z, T]}; \underline{\bar{u}}_k^{[T_z, T]}, \underline{\bar{u}}_{(k+1)-} \cap \underline{\bar{u}}_{k-}\right]\right\}, \\
& \text{for } k \in \left[T_z, T\right];
\end{aligned}
\tag{6.5}
$$

$$
\begin{aligned}
& W\left(T_{t+1} - 1, x; \bar{u}_{(T_{t+1}-1)-}\right) = \max_{\substack{u_{T_{t+1}-1}^i,\, \bar{u}_{T_{t+1}-1}^i \\ i \in S^{(T_t, T_{t+1}-1)}}} \\
& \left\{ \sum_{i \in S^{[T_t, T_{t+1}-1]}} \left(R_{T_{t+1}-1}^i \bar{u}_{T_{t+1}-1}^i - \varphi_{T_{t+1}-1}^i \left(\bar{u}_{T_{t+1}-1}^i\right)^2 - c_{T_{t+1}-1}^{Ii}\left(u_{T_{t+1}-1}^{Ii}\right)^2 \right.\right. \\
& + \kappa_{T_{t+1}-1}^i x^2 - c_{T_{t+1}-1}^{IIi}\left(u_{T_{t+1}-1}^{IIi}\right)^2 - h_{T_{t+1}-1}^i x^1 - \sum_{\theta \in S^{[T_t, T_{t+1}-1]}} v_{T_{t+1}-1}^{i|T_{t+1}-1|\theta} \bar{u}_{T_{t+1}-1}^\theta \\
& \left. - \sum_{\theta \in S^{[(T_{t+1}-1)-]}} \sum_{\tau = T_{t+1}-1-\omega+1}^{T_{t+1}-2} v_{T_{t+1}-1}^{i|\tau|\theta} \bar{u}_\tau^\theta \right) \delta^{T_{t+1}-2} \\
& + \sum_{j \in S^{[T_t T_{t+1}-1]c}} \left(Q_{T_{t+1}-1}^{(1)j} x_{T_{t+1}-1}^{1[T_t, T_{t+1}-1]}, Q_{T_{t+1}-1}^{(2)j} x_{T_{t+1}-1}^{2[T_t, T_{t+1}-1]}; \underline{\bar{u}}_{T_{t+1}-1}^{[T_t, T_{t+1}-1]}, \right. \\
& \left. \underline{\bar{u}}_{(T_{T+1})-} \cap \underline{\bar{u}}_{(T_{t+1}-1)-} \right) \delta^{T_t - 1} \\
& \left. + W\left[T_{t+1}, x_{T_{t+1}}^{1[T_t, T_{t+1}-1]}, x_{T_{t+1}}^{2[T_t, T_{t+1}-1]}, \underline{\bar{u}}_{T_{t+1}^{-1}}^{[T_t, T_{t+1}-1]}, \underline{\bar{u}}_{(T_{t+1})-} \cap \underline{\bar{u}}_{(T_{t+1}-1)-}\right]\right\};
\end{aligned}
\tag{6.6}
$$

$$
\begin{aligned}
W\left(k, x; \underline{\bar{u}}_{k-}\right) = & \max_{\substack{u_k^i,\, \bar{u}_k^i \\ i \in S^{[T_t, T_{t+1}-1]}}} \left\{ \sum_{i \in S^{[T_t, T_{t+1}-1]}} \left(R_k^i \bar{u}_k^i - \varphi_k^i \left(\bar{u}_k^i\right)^2 - c_k^{Ii}\left(u_k^{Ii}\right)^2 + \kappa_k^i x^2 - c_k^{IIi}\left(u_k^{IIi}\right)^2 \right.\right. \\
& \left. - h_k^i x^1 - \sum_{\theta \in S^{[T_t, T_{t+1}-1]}} v_k^{i|k|\theta} \bar{u}_k^\theta - \sum_{\theta \in S^{[k-]}} \sum_{\tau = k-\omega+1}^{k-1} v_k^{i|\tau|\theta} \bar{u}_\tau^\theta \right) \delta^{k-1} \\
& \left. + W\left[k+1, x_{k+1}^{1[T_t, T_{t+1}-1]}, x_{k+1}^{2[T_t, T_{t+1}-1]}; \underline{\bar{u}}_k^{[T_t, T_{t+1}-1]}, \underline{\bar{u}}_{(k+1)} - \cap \underline{\bar{u}}_{k-}\right]\right\} \\
& \text{for } k \in [T_t, T_t + 1, T_t + 2, \cdots, T_{t+1} - 2\} \text{ and } t \in \{1, 2, \cdots, z-1\};
\end{aligned}
\tag{6.7}
$$

where

$$x_{k+1}^{1[T_t,T_{t+1}-1]} = x_k^1 + \sum_{\theta\in S^{[T_t,T_{t+1}-1]}} m_k^{|k|\theta}\bar{u}_k^\theta + \sum_{\theta\in S^{[k-]}} \sum_{\tau=k-\omega+1}^{k-1} m_k^{|\tau|\theta}\bar{u}_\tau^\theta$$
$$- \sum_{\vartheta\in S^{[T_t,T_{t+1}-1]}} b_k^\theta u_k^{I\theta}(x_k^1)^{1/2} - \lambda_k^1 x_k^1,$$

and

$$x_{k+1}^{2[T_t,T_{t+1}-1]} = x_k^2 + \sum_{\theta\in S^{[T_t,T_{t+1}-1]}} a_k^\theta u_k^{II\theta}$$
$$-\lambda_k^2 x_k^2; \text{ for } k\in\left[T_t, T_{t+1}-1\right] \text{ and } t\in(1,2,\cdots,z).$$

Performing the indicated maximization operator in Corollary 6.1, the value function $W(k,x;\underline{\bar{u}}_{k-}^{**})$, which reflects the joint payoff of the players, be obtained as follows.

Proposition 6.1 *System* (6.4)–(6.7) *admits a solution with the joint payoff of the players being*

$$W\left(T+1,x;\underline{\bar{u}}_{(T+1)-}\right) = \sum_{i\in S^{[T_z,T]}} \left(Q_{T+1}^{(1)i}x^1 + Q_{T+1}^{(2)i}x^2 + \varpi_{T+1}^i\right)\delta^T, \tag{6.8}$$

$$W\left(k,x;\underline{\bar{u}}_{k-}\right) = \left(A_k^{(1)}x^1 + A_k^{(2)}x^2 + C_k\right)\delta^{k-1}, \text{ for } k\in[T_z,T], \text{ where.}$$

$$A_k^{(1)} = -\sum_{\theta\in S^{[T_z,T]}}\left(\frac{\left(A_{k+1}^{(1)}b_k^\theta\delta\right)^2}{4c_k^{I\theta}} + h_k^\theta\right) + \delta A_{k+1}^{(1)}\left(1+\sum_{\theta\in S^{[T_z,T]}} b_k^\theta\frac{A_{k+1}^{(1)}b_k^\theta\delta}{2c_k^{I\theta}} - \lambda_k^1\right)$$

$$A_k^{(2)} = \sum_{\theta\in S^{[T_z,T]}} \kappa_k^\theta + \delta A_{k+1}^{(2)}\left(1-\lambda_k^2\right)$$

and
C_k *is an expression which include previously executed durable strategies* $\underline{\bar{u}}_{k-}$,

$$\text{for } k\in\left[T_z,T\right], \tag{6.9}$$

$$W\left(T_{t+1}-1,x;\underline{\bar{u}}_{(T_{t+1}-1)-}\right) = \left(A_{T_{t+1}-1}^{(1)}x^1 + A_{T_{t+1}-1}^{(2)}x^2 + C_{T_{t+1}-1}\right)\delta^{k-1},$$

where

$$A^{(1)}_{T_{t+1}-1} = -\sum_{\theta\in S^{[T_t,T_{t+1}-1]}}\left(\frac{\left(A^{(1)}_{T_{t+1}}b^{\theta}_{T_{t+1}-1}\delta\right)^2}{4c^{I\theta}_{T_{t+1}-1}} + h^{\theta}_{T_{t+1}-1}\right)$$
$$+\,\delta 4^{(1)}_{T_{t+1}}\left(1+\sum_{\theta\in S^{[T_t,T_{t+1}-1]}} b^{\theta}_{T_{t+1}-1}\frac{A^{(1)}_{T_{t+1}}b^{\theta}_{T_{t+1}-1}\delta}{2c^{I\theta}_{T_{t+1}-1}} - \lambda^1_{T_{t+1}-1}\right)$$
$$+\sum_{j\in S^{[T_t,T_{t+1}-1]C}}\delta Q^{(1)j}_{T_{t+1}}\left(1+\sum_{\theta\in S^{[T_t,T_{t+1}-1]}} b^{\theta}_{T_{t+1}-1}\frac{A^{(1)}_{T_{t+1}}b^{\theta}_{T_{t+1}-1}\delta}{2c^{\theta}_{T_{t+1}-1}} - \lambda^1_{T_{t+1}-1}\right),$$
$$A^{(2)}_{T_{t+1}-1} = \sum_{\theta\in S^{[T_t,T_{t+1}-1]}} K^{\theta}_{T_{t+1}-1} + \delta A^{(2)}_{T_{t+1}}\left(1-\lambda^2_{T_{t+1}-1}\right)$$

and $C_{T_{t+1}-1}$ *is an expression which include previously executed durable strategies* $\underline{\bar{u}}_{(T_{t+1}-1)-}$,

$$\text{for } t\in(1,2,\cdots,z); \tag{6.10}$$

$W(k,x;\underline{\bar{u}}_{k-}) = (A^{(1)}_k x^1 + A^{(2)}_k x^2 + C_k)\delta^{k-1}$, for $k\in\{T_t, T_t+1,\cdots,T_{t+1}-2\}$ and $t\in(1,2,\cdots,z)$, *where*

$$A^{(1)}_k = -\sum_{\theta\in S^{[T_t,T_{t+1}-1]}}\left(\frac{\left(A^{(1)}_{k+1}b^{\theta}_k\delta\right)^2}{4c^{I\theta}_k} + h^{\theta}_k\right) + \delta 4^{(1)}_{k+1}\left(1+\sum_{\theta\in S^{[T_t,T_{t+1}-1]}} b^{\theta}_k\frac{A^{(1)}_{k+1}b^{\theta}_k\delta}{2c^{I\theta}_k} - \lambda^1_k\right),$$
$$A^{(2)}_k = \sum_{\theta\in S^{[T_t,T_{t+1}-1]}} \kappa^{\theta}_k + \delta A^{(2)}_{k+1}\left(1-\lambda^2_k\right), \tag{6.11}$$

and C_k *is an expression that includes previously executed durable strategies* $\underline{\bar{u}}_{k-}$.

Proof See Appendix IV. ■

We use $\underline{u}^*_k$ and $\underline{\bar{u}}^{Ii*}_k$ to denote the set of optimal strategies. Performing the indicated maximization operator in Corollary 6.1 and invoking Proposition 6.1, the optimal strategies can be obtained as,

for $i\in S^{[T_t,T_{t+1}-1]}$ and $k\in[T_t, T_{t+1}-1]$:

$$u^{Ii*}_k = \frac{-\delta A^{(1)}_{k+1}b^i_k}{2c^{Ii}_k}\left(x^1\right)^{1/2},\; u^{IIi*}_k = \frac{\delta A^{(2)}_{k+1}a^i_k}{2c^{IIi}_k},$$
$$\bar{u}^{i*}_k = \frac{R^i_k - \sum_{\theta\in S^{[T_t,T_{t+1}-1]}}\sum_{\eta=k}^{T_{t+1}-1}\delta^{\eta-k}v^{\theta|k|i}_{\eta} - \sum_{\rho=1}^{z-t}}{2\phi^i_k}$$
$$\frac{\sum_{\theta\in S^{\left[T_{t+\rho},T_{t+1+\rho}-1\right]}}\sum_{\eta=T_{t+\rho}}^{T_{t+\rho+1}-1}\delta^{\eta-k}v^{\theta|k|i}_{\eta}}{2\phi^i_k}$$

$$+\frac{\sum_{\eta=k}^{k+\omega-1}\delta^{\eta-k+1}A_{\eta+1}^{(1)}m_{\eta}^{|k|i}}{2\phi_k^i}, \tag{6.12}$$

where $v_{\eta}^{\theta|k|i}=0$, for $\eta>k+\omega-1$ or $\eta>T$ and.
$A_{\eta+1}^{(1)}m_{\eta}^{|k|i}=0$, for $\eta>T$.

Substituting the optimal strategies $\underline{u}_k^*$ and $\underline{\bar{u}}_k^*$ from (6.12) into (6.2) and (6.3), we obtain the Pareto optimal state dynamics as:

$$x_{k+1}^1=x_k^1+\sum_{\theta\in S^{[T_t,T_{t+1}-1]}}m_k^{|k|\theta}\bar{u}_k^{\theta*}+\sum_{\theta\in S^{[k-]}}\sum_{\tau=k-\omega+1}^{k-1}m_k^{|\tau|\theta}\bar{u}_\tau^{\theta*}-\sum_{\vartheta\in S^{[T_t,T_{t+1}-1]}}b_k^{\theta}u_k^{I\theta*}(x_k^1)^{1/2}-\lambda_k^1x_k^1$$

$$x_{k+1}^2=x_k^2+\sum_{\theta\in S^{[T_t,T_{t+1}-1]}}a_k^{\theta}u_k^{II\theta^*}-\lambda_k^2x_k^2,$$

$$\text{for } k\in\left[T_t,T_{t+1}-1\right] \text{ and } t\in(1,2,\cdots,z). \tag{6.13}$$

We use $\{x_k^*\}_{k=1}^{T+1}$ to denote the solution to the system (6.13), which depicts the Pareto optimal state trajectory.

The maximized joint payoff at stage 1 under the state of x_1^0 and previously executed durable strategies $\underline{\bar{u}}_{1-}$ is

$$W\left(1,x_1^0;\underline{\bar{u}}_{1-}\right)=\sum_{t=1}^{z}\sum_{i\in S^{[T_t,T_{t+1}-1]}}$$

$$\left[\sum_{k=T_t}^{T_{t+1}-1}\left(R_k^i\bar{u}_k^{i*}-\varphi_k^i\left(\bar{u}_k^{i*}\right)^2-c_k^{IIi}\left(u_k^{Ii^*}\right)^2+\kappa_k^jx_k^{2^*}-c_k^{IIi}\left(u_k^{IIi^*}\right)^2\right.\right.$$

$$\left.-h_k^ix_k^{1^*}-\sum_{\theta\in S^{\left[T_t^i,T_{t+1}^i-1\right]}}v_k^{i|k|\theta}\bar{u}_k^{\theta^*}-\sum_{\theta\in S^{[k-]}}\sum_{\tau=k-\omega+1}^{k-1}v_k^{i|\tau|\theta}\bar{u}_\tau^{\theta^*}\right)\delta^{k-1}$$

$$\left.+\sum_{j\in S^{[T_t,T_{t+1}-1]C}}\left(Q_{T_{t+1}}^{(1)j}x_{T_{t+1}}^{1^*}+Q_{T_{t+1}}^{(2)j}x_{T_{t+1}}^{2^*}+\varpi_{T_{t+1}}^j\right)\delta^{T_{T+1}-1}\right] \tag{6.14}$$

6.2 PoA in Cross-generational Environmental Commons

Using the result in Theorem 2.1, the sum of the players' payoffs in a non-cooperative equilibrium can be expressed as $\sum_{\theta=1}^{n}V^{\theta}(T_1^{\theta},x_{T_1^{\theta}}^{**};\underline{\bar{u}}_{T_1^{\theta}-}^{**})$.

The price of anarchy under cross-generational environmental commons can be obtained as.

$$PoA = \frac{W(1, x_1^0; \bar{u}_{1-})}{\sum_{\theta=1}^{n} V^{\theta}(T_1^{\theta}, x_{T_1^{\theta}}^{**}; \bar{u}_{T_1^{\theta}-}^{**})} \tag{6.15}$$

Given the potentially large number of players and severity of externalities in environmental analysis, self–seeking optimization by individual players has caused serious damages to the ecosystems and established a continual trend of eco-degradation. The issue becomes much more serious when the players' horizons are different and the effects of eco-degradation actions are durable.

Invoking Propositions 3.1 and 6.1, we note that the value of $A_k^{(1)}$, which represents the marginal damage of environmental degradation to all the n players under cooperation, is always more negative than $A_k^{(1)i}$, which measures the magnitude of the marginal damage of environmental degradation to player i under non-cooperation. The negative externalities of environmental degradation are internalized in the optimal cooperative solution by taking into consideration of damages to all players rather than just the damage to a single player.

The production of eco-degrading output under cooperation is always below the production of eco-degrading output under non-cooperation, that is

$$\bar{u}_k^{i*} = \frac{R_k^i - \sum_{\theta \in S^{[T_t, T_{t+1}-1]}} \sum_{\eta=k}^{T_{t+1}-1} \delta^{\eta-k} \upsilon_\eta^{\theta|k|i} - \sum_{\rho=1}^{z-t} \sum_{\theta \in S^{[T_{t+\rho} \cdot T_{t+1+\rho}-1]}} \sum_{\eta=T_{t+\rho}}^{T_{t+\rho+1}-1} \delta^{\eta-k} \upsilon_\eta^{\theta_i|k|i}}{2\phi_k^i}$$

$$+ \frac{\sum_{\eta=k}^{k+\omega-1} \delta^{\eta-k+1} A_{\eta+1}^{(1)} m_\eta^{|k|i}}{2\phi_k^i}$$

$$< \bar{u}_k^{i**} = \frac{R_k^i - \sum_{\eta=k}^{k+\omega-1} \delta^{\eta-k} \upsilon_\eta^{i|k|i} + \sum_{\eta=k}^{k+\omega-1} \delta^{\eta-k+1} A_{\eta+1}^{(1)i} m_\eta^{|k|i}}{2\phi_k^i}. \tag{6.16}$$

Note that the optimal solution takes into consideration the direct damages to the payoffs of other players $\upsilon_k^{\theta|k|i}$ within and beyond player i's game horizon.

In addition, the environmental improvement efforts under cooperation are greater than the environment improving efforts under non-cooperation, that is

$$u_k^{Ii*} = -\frac{\delta A_{k+1}^{(1)} b_k^i}{2c_k^{Ii}} x^{1/2} > u_k^{Ii**} = -\frac{\delta A_{k+1}^{(1)i} b_k^i}{2c_k^{Ii}} x^{1/2} \tag{6.17}$$

Hence, the level of environmental degradation under cooperation is lower than that under non-cooperation in every stage $k \in \{2, 3, \cdots, T+1\}$.

Invoking again Propositions 3.1 and 6.1, we note that the value of $A_k^{(2)} > A_k^{(2)i}$ and this leads to

$$u_k^{IIj*} = \frac{\delta A_{k+1}^{(2)} a_k^j}{2c_k^{IIj}} > u_k^{IIj**} = \frac{\delta A_{k+1}^{(2)j} a_k^j}{2c_k^{IIj}}, \tag{6.18}$$

which reflects that the investments in green technology are greater under cooperation than that under non-cooperation. Hence, the green technology level under cooperation is higher than that under non-cooperation in every stage $k \in \{2, 3, \cdots, T + 1\}$.

With a higher level of green technology investment, the level of green technology under cooperation will be higher than that under non-cooperation. This leads to the situation where the clean outputs produced by each player under cooperation is higher than the clean outputs produced by each of the same players under non-cooperation.

Therefore, under optimal cooperative environmental management, efficiency will be maximized, clean outputs will be higher, eco-degrading outputs lower and investments in green technology higher. The PoA under self-maximization reflected by $PoA = \frac{W(1, x_1^0; \underline{\bar{u}}_{1-})}{\sum_{\theta=1}^n V^\theta(T_1^\theta, x_{T_1^\theta}^{**}; \underline{\bar{u}}_{T_1^\theta-}^{**})}$ can be very high.

6.3 Cooperative Gain Sharing

Consider the case where the players agree to share the cooperative gains proportional to the relative sizes of their non-cooperative payoffs, the imputation under cooperation that player i will receive becomes

$$\xi^i = \frac{V^i(T_1^i, x_{T_1^i}^{**}; \underline{\bar{u}}_{T_1^i-}^{**})}{\sum_{\theta=1}^n V^\theta(T_1^\theta, x_{T_1^\theta}^{**}; \underline{\bar{u}}_{T_1^\theta-}^{**})} W(1, x_1^0; \underline{\bar{u}}_{1-}^*)$$

$$= \frac{\left(A_{T_1^i}^{(1)i} x_{T_1^i}^{1**} + A_{T_1^i}^{(2)i} x_{T_1^i}^{2**} + C_{T_1^i}^i\right)\delta^{T_1^i-1}}{\sum_{\theta=1}^n \left(A_{T_1^\theta}^{(1)\theta} x_{T_1^\theta}^{1**} + A_{T_1^\theta}^{(2)\theta} x_{T_1^\theta}^{2**} + C_{T_1^\theta}^\theta\right)\delta^{T_1^\theta-1}} \left(A_1^{(1)} x_1^1 + A_1^{(2)} x_1^2 + C_1\right), \quad \text{for } i \in N. \tag{6.19}$$

Invoking (5.5), a candidate of $\beta_k^i(x_k^*; \underline{\bar{u}}_{k-}^*)$ has to satisfy the following conditions:

$$\beta_k^i(x_k^*; \underline{\bar{u}}_{k-}^*) \geq \left(A_{T_1^i}^{(1)i} x_{T_1^i}^{1*} + A_{T_1^i}^{(2)i} x_{T_1^i}^{2*} + C_{T_1^i}^i\right)\delta^{T_1^i-1} - \left(\delta_{T_1^i}^k\right)^{-1}$$

$$\left[\sum_{\zeta=k+1}^{T_{z_i+1}^i-1} \beta_\zeta^i(x_\zeta^*, y_\zeta^*; \underline{\bar{u}}_{\zeta-}^*)\delta_{T_1^i}^\zeta \right.$$

$$\left. + q_{T_{z_i+1}^i}^i\left(x_{T_{z_i+1}^i}^*; \underline{\bar{u}}_{(T_{z_i+1})-}^*\right)\delta_{T_1^i}^{T_{z_i+1}^i}\right], k \in (T_1^i, T_1^i+1, \cdots, T_{z^i+1}^i - 1) \tag{6.20}$$

where

$$\xi^i = \frac{(A^{(1)i}_{T^i_1} x^{1**}_{T^i_1} + A^{(2)i}_{T^i_1} x^{2**}_{T^i_1} + C^i_{T^i_1})\delta^{T^i_1 - 1}}{\sum_{\theta=1}^{n} (A^{(1)\theta}_{T^\theta_1} x^{1**}_{T^\theta_1} + A^{(2)\theta}_{T^\theta_1} x^{2**}_{T^\theta_1} + C^\theta_{T^\theta_1})\delta^{T^\theta_1 - 1}} (A^{(1)}_1 x^1_1 + A^{(2)}_1 x^2_1 + C_1)$$

$$+ \left\{ \sum_{\zeta=T^i_1}^{T^i_{z_i+1}-1} \beta^i_\zeta (x^*_\zeta; \underline{u}^*_{\zeta-}) \delta^\zeta_{T^i_1} + q^i_{T^i_{z_i+1}} \left(x^*_{T^i_{z_i+1}}; \underline{u}^*_{(T_{z_i+1})-} \right) \delta^{T^i_{z_i+1}}_{T^i_1} \right\}, \tag{6.21}$$

with $\xi^i \left(T^i_1, x^*_{T^i_1}; \underline{u}^*_{T^i_1 -} \right) \geq V^i \left(T^i_1, x^{**}_{T^i_1}; \underline{u}^{**}_{T^i_1 -} \right)$.

7 Appendices

7.1 Appendix I: Proof of Theorem 2.1

Consider the last game interval of player i, that is $[T^i_{z_i}, T^i_{z_i+1} - 1]$. The payoffs of the players in stage $T^i_{z_i+1}$ are

$$V^i \left(T^i_{z_i+1}, x; \underline{\bar{u}}^{**}_{(T^i_{z_i+1})-} \right) = q^i_{T^i_{z_i+1}} \left(x_{T^i_{z_i+1}}; \underline{\bar{u}}^{i**}_{(T^i_{z_i+1})-} \right) \delta_1^{T^i_{z_i+1}},$$

$$V^j \left(T^i_{z_i+1}, x; \underline{\bar{u}}^{**}_{(T^i_{z_i+1})-} \right) = q^j_{T^i_{z_i+1}} \left(x_{T^i_{z_i+1}}; \underline{\bar{u}}^{**}_{(T^i_{z_i+1})-} \right) \delta_1^{T^i_{z_i+1}}, \quad \text{for } j \in S^{\left[T^i_{z_i}, T^i_{z_i+1}-1\right]A},$$

$$V^l \left(T^i_{z_i+1}, x; \underline{\bar{u}}^{**}_{(T^i_{z_i+1})-} \right) = V^\ell \left(T^i_{z_i+1}, x; \underline{\bar{u}}^{**}_{(T^i_{z_i+1})-} \right) =, \quad \text{for } \ell \in S^{\left[T^i_{z_i}, T^i_{z_i+1}-1\right]B}. \tag{7.1}$$

In the game interval of $[T^i_t, T^i_{t+1} - 1]$, for $t \in \{1, 2, \cdots, z_i - 1\}$, the payoffs of the players at stage T^i_{t+1} as:

$$V^i \left(T^i_{t+1}, x; \underline{\bar{u}}^{**}_{(T^i_{t+1})-} \right) = V^i \left(T^i_{t+1}, x; \underline{\bar{u}}^{**}_{(T^i_{t+1})-} \right), \text{ for } t \in \{1, 2, \cdots, z_i - 1\}$$

$$V^j \left(T^i_{t+1}, x, \underline{\bar{u}}^{**}_{(T^i_{t+1})-} \right) = q^j_{T^i_{t+1}} \left(x_{T^i_{t+1}}; \underline{\bar{u}}^{**}_{(T^i_{t+1})-} \right) \delta_1^{T^i_{t+1}}, \quad \text{for } j \in S^{[T^i_t, T^i_{t+1}-1]A},$$

$$V^\ell \left(T^\ell_{t+1}, x; \underline{\bar{u}}^{**}_{(T^i_{t+1})-} \right) = V^\ell (T^i_{t+1}, x; \underline{\bar{u}}^{**}_{(t+1)-}), \quad \text{for } \ell \in S^{[T^i_t, T^i_{t+1}-1]B}. \tag{7.2}$$

Now, we consider the maximization operations of the players in the game interval $[T_t^i, T_{t+1}^i - 1]$, for $t \in \{1, 2, \cdots, z_i\}$. In the last stage in the game interval $[T_t^i, T_{t+1}^i - 1]$, the stage $T_{t+1}^i - 1$ game problem can be expressed as

$$
\begin{aligned}
&V^i(T_{t+1}^i - 1, x; \underline{\bar{u}}^{**}_{(T_{t+1}^i-1)-}) \\
&= \max_{u^i_{T_{t+1}^i-1}, \bar{u}^i_{T_{t+1}^i-1}} \Big\{ g^i_{T_{t+1}^i-1}\Big(x, u^i_{T_{t+1}^i-1}, \bar{u}^i_{T_{t+1}^i-1}, \underline{u}^{[T_t^i, T_{t+1}^i-1](\neq i)**}_{T_{t+1}^i-1}, \underline{\bar{u}}^{[T_t^i, T_{t+1}^i-1](\neq i)**}_{T_{t+1}^i-1}; \underline{\bar{u}}^{**}_{(T_{t+1}^i-1)-}\Big)\delta_1^{T_{t+1}^i-1} \\
&+ V^i_{T_{t+1}^i}\Big[T_{t+1}^i, x^{(\neq i)**}_{T_{t+1}^i}; \bar{u}^i_{T_{t+1}^i-1}, \underline{\bar{u}}^{[T_t^i, T_{t+1}^i-1](\neq i)**}_{T_{t+1}^i-1}, \underline{\bar{u}}^{**}_{(T_{t+1}^i)-} \cap \underline{\bar{u}}^{**}_{(T_{t+1}^i-1)-}\Big]\Big\},
\end{aligned}
$$

where

$$
x^{(\neq i)**}_{T_{t+1}^i} = f_{T_{t+1}^i-1}\Big(x, u^i_{T_{t+1}^i-1}, \bar{u}^i_{T_{t+1}^i-1}, \underline{u}^{[T_t^i, T_{t+1}^i-1](\neq i)**}_{T_{t+1}^i-1}, \underline{\bar{u}}^{[T_t^i, T_{t+1}^i-1](\neq i)**}_{T_{t+1}^i-1}; \underline{\bar{u}}^{**}_{(T_{t+1}^i-1)-}\Big); \tag{7.3}
$$

$$
\begin{aligned}
&V^j(T_{z_i+1}^i - 1, x; \underline{\bar{u}}^{**}_{(T_{z_i+1}^i-1)-}) \\
&= \max_{u^j_{T_{t+1}^i-1}, \bar{u}^j_{T_{t+1}^i-1}} \Big\{ g^j_{T_{t+1}^i-1}\Big(x, u^j_{T_{t+1}^i-1}, \bar{u}^j_{T_{t+1}^i-1}, \underline{u}^{[T_t^i, T_{t+1}^i-1](\neq j)**}_{T_{t+1}^i-1}, \underline{\bar{u}}^{[T_t^i, T_{t+1}^i-1](\neq j)**}_{-T_{t+1}^i-1} \\
&; \underline{u}^{**}_{(T_{t+1}^i-1)-}\Big)\delta_1^{T_{t+1}^i-1} \\
&+ q^j_{T_{t+1}^i}\Big[x^{(\neq j)**}_{T_{t+1}^i}; \bar{u}^j_{T_{t+1}^i-1}, \underline{\bar{u}}^{[T_t^i, T_{t+1}^i-1](\neq j)**}_{T_{t+1}^i-1}, \underline{\bar{u}}^{**}_{(T_{t+1}^i)-} \cap \underline{\bar{u}}^{**}_{(T_{t+1}^i-1)-}\Big]\delta_1^{T_{t+1}^i}\Big\}, \\
&\text{for } j \in S^{[T_{z_i}^i, T_{z_i+1}^i-1]A},
\end{aligned}
$$

where

$$
x^{(\neq j)**}_{T_{t+1}^i} = f_{T_{t+1}^i-1}\Big(x, u^j_{T_{t+1}^i-1}, \bar{u}^j_{T_{t+1}^i-1}, \underline{u}^{[T_t^i, T_{t+1}^i-1](\neq j)**}_{T_{t+1}^i-1}, \underline{\bar{u}}^{[T_t^i, T_{t+1}^i-1](\neq j)**}_{T_{t+1}^i-1}; \underline{\bar{u}}^{**}_{(T_{t+1}^i-1)-}\Big); \tag{7.4}
$$

$$
\begin{aligned}
&V^\ell(T_{z_i+1}^i - 1, x; \underline{\bar{u}}^{**}_{(T_{z_i+1}^i-1)-}) \\
&= \max_{u^\ell_{T_{t+1}^i-1}, \bar{u}^\ell_{T_{t+1}^i-1}} \Big\{ g^\ell_{T_{t+1}^i-1}\Big(x, u^\ell_{T_{t+1}^i-1}, \bar{u}^\ell_{T_{t+1}^i-1}, \underline{u}^{[T_t^i, T_{t+1}^i-1](\neq \ell)**}_{T_{t+1}^i-1}, \underline{\bar{u}}^{[T_t^i, T_{t+1}^i-1](\neq \ell)**}_{T_{t+1}^i-1}; \underline{\bar{u}}^{**}_{(T_{t+1}^i-1)-}\Big)\delta_1^{T_{t+1}^i-1} \\
&+ V^\ell_{T_{t+1}^i}\Big[T_{t+1}^i, x^{(\neq \ell)**}_{T_{t+1}^i}; \bar{u}^\ell_{T_{t+1}^i-1}, \underline{\bar{u}}^{[T_t^i, T_{t+1}^i-1](\neq \ell)**}_{T_{t+1}^i-1}, \underline{\bar{u}}^{**}_{(T_{t+1}^i)-} \cap \underline{\bar{u}}^{**}_{(T_{t+1-1}^i)-}\Big]\Big\}, \\
&\text{for } \ell \in S^{[T_t^i, T_{t+1}^i-1]B},
\end{aligned}
$$

where

$$
x^{(\neq l)**}_{T_{t+1}^i} = f_{T_{t+1}^i-1}\Big(x, u^\ell_{T_{t+1}^i-1}, \bar{u}^\ell_{T_{t+1}^i-1}, \underline{u}^{[T_t^i, T_{t+1}^i-1](\neq \ell)**}_{T_{t+1}^i-1}, \underline{\bar{u}}^{[T_t^i, T_{t+1}^i-1](\neq l)**}_{T_{t+1}^i-1}; \underline{\bar{u}}^{**}_{(T_{t+1}^i-1)-}\Big). \tag{7.5}
$$

Note that in the maximization operator in stage $T_{t+1}^i - 1$, the control strategies of player i are $(u^i_{T_{z_i+1}^i-1}, \bar{u}^i_{T_{z_i+1}^i-1})$; the control strategies of player j, for

$j \in S^{[T^i_{z_i}, T^i_{z_i+1}-1]A}$, are $(u^j_{T^i_{z_i+1}-1}, \bar{u}^j_{T^i_{z_i+1}-1})$; and the control strategies of player ℓ, for $\ell \in S^{[T^i_{z_i}, T^i_{z_i+1}-1]B}$, are $(u^\ell_{T^i_{z_i+1}-1}, \bar{u}^\ell_{T^i_{z_i+1}-1})$. The state x and the previously executed strategies $\underline{\bar{u}}^{**}_{(T^i_{z_i+1}-1)-}$ appear as given. If the system of Eqs. (7.3)–(7.5) satisfies the implicit theorem, the game equilibrium strategies can be obtained as functions of $(x, \underline{\bar{u}}^{**}_{(T^i_{z_i+1}-1)-})$. Substituting these strategies into the right-hand-side of the equations in (7.3)–(7.5) yields the value functions:

$$V^i(T^i_{t+1} - 1, x; \underline{\bar{u}}^{**}_{(T^i_{t+1}-1)-})$$

$$V^j\left(T^i_{t+1} - 1, x; \underline{\bar{u}}^{**}_{(T^i_{t+1}-1)-}\right), j \in S^{\left[T^i_{z_i}, T^i_{z_i+1}-1\right]A},$$

$$V^\ell\left(T^i_{t+1} - 1, x; \underline{\bar{u}}^{**}_{(T^i_{t+1}-1)-}\right), \text{ for } \ell \in S^{[T^i_t, T^i_{t+1}-1]B}, \tag{7.6}$$

which are functions of $(x, \underline{\bar{u}}^{**}_{(T^i_{z_i+1}-1)-})$.

We move to the second last stage of the game interval $[T^i_t, T^i_{t+1} - 1]$, that is stage $T^i_{t+1} - 2$. Using (7.6), the stage $T^i_{t+1} - 2$ game problem can be expressed as:

$$V^i(T^i_{t+1} - 2, x; \underline{\bar{u}}^{**}_{(T^i_{t+1}-2)-})$$
$$= \max_{u^i_{T^i_{t+1}-2}, \bar{u}^i_{T^i_{t+1}-2}} \Bigg\{ g^i_{T^i_{t+1}-2}\Big(x, u^i_{T^i_{t+1}-2}, \bar{u}^i_{T^i_{t+1}-2}, \underline{u}^{[T^i_t, T^i_{t+1}-1](\neq i)**}_{T^i_{t+1}-2}, \underline{\bar{u}}^{[T^i_t, T^i_{t+1}-1](\neq i)**}_{T^i_{t+1}-2}$$
$$; \underline{\bar{u}}^{**}_{(T^i_{t+1}-2)-}\Big) \delta_1^{T^i_{t+1}-2}$$
$$+ V^i\Big[T^i_{t+1} - 1, x^{(\neq i)**}_{T^i_{t+1}-1}; \bar{u}^i_{T^i_{t+1}-2}, \underline{\bar{u}}^{[T^i_t, T^i_{t+1}-1](\neq i)**}_{T^i_{t+1}-2}, \underline{\bar{u}}^{**}_{(T^i_{t+1}-1)-} \cap \underline{\bar{u}}^{**}_{(T^i_{t+1}-2)-}\Big]\Bigg\}$$

where

$$x^{(\neq i)**}_{T^i_{t+1}-1} = f_{T^i_{t+1}-2}\Big(x, u^i_{T^i_{t+1}-2}, \bar{u}^i_{T^i_{t+1}-2}, \underline{u}^{[T^i_t, T^i_{t+1}-1](\neq i)**}_{T^i_{t+1}-2}, \underline{\bar{u}}^{[T^i_t, T^i_{t+1}-1](\neq i)**}_{T^i_{t+1}-2}; \underline{\bar{u}}^{**}_{(T^i_{t+1}-2)-}\Big); \tag{7.7}$$

$$V^j(T^i_{t+1} - 2, x; \underline{\bar{u}}^{**}_{(T^i_{t+1}-2)-})$$
$$= \max_{u^j_{T^i_{t+1}-2}, \bar{u}^j_{T^i_{t+1}-2}} \Bigg\{ g^j_{T^i_{t+1}-2}\Big(x, u^j_{T^i_{t+1}-2}, \bar{u}^j_{T^i_{t+1}-2}, \underline{u}^{[T^i_t, T^i_{t+1}-1](\neq j)**}_{T^i_{t+1}-2}, \underline{\bar{u}}^{[T^i_t, T^i_{t+1}-1](\neq j)**}_{T^i_{t+1}-2}; \underline{\bar{u}}^{**}_{(T^i_{t+1}-2)-}\Big) \delta_1^{T^i_{t+1}-2}$$
$$+ V^j\Big[T^i_{t+1} - 1, x^{(\neq j)**}_{T^i_{t+1}-1}; \bar{u}^j_{T^i_{t+1}-2}, \underline{\bar{u}}^{[T^i_t, T^i_{t+1}-1](\neq j)**}_{T^i_{t+1}-2}, \underline{\bar{u}}^{**}_{(T^i_{t+1}-1)-} \cap \underline{\bar{u}}^{**}_{(T^i_{t+1}-2)-}\Big]\Bigg\},$$

for $j \in S^{[T^i_t, T^i_{t+1}-1]A}$,

where

$$x_{T_{t+1}^i-1}^{(\neq j)**} = f_{T_{t+1}^i-2}\Big(x, u_{T_{t+1}^i-2}^j, \bar{u}_{T_{t+1}^i-2}^j, \underline{u}_{T_{t+1}^i-2}^{[T_t^i,T_{t+1}^i-1](\neq j)**}, \underline{\bar{u}}_{T_{t+1}^i-2}^{[T_t^i,T_{t+1}^i-1](\neq j)**}; \underline{\bar{u}}_{(T_{t+1}^i-2)-}^{**}\Big); \tag{7.8}$$

$$V^\ell\Big(T_{t+1}^i-2, x; \underline{\bar{u}}_{(T_{t+1}^i-2)-}^{**}\Big)$$
$$= \max_{u_{T_{t+1}^i-2}^\ell, \bar{u}_{T_{t+1}^i-2}^\ell} \Big\{ g_{T_{t+1}^i-2}^\ell\Big(x, u_{T_{t+1}^i-2}^\ell, \bar{u}_{T_{t+1}^i-2}^\ell, \underline{u}_{T_{t+1}^i-2}^{[T_t^i,T_{t+1}^i-1](\neq \ell)**}, \underline{\bar{u}}_{T_{t+1}^i-2}^{[T_t^i,T_{t+1}^i-1](\neq \ell)**}; \underline{\bar{u}}_{(T_{t+1}^i-2)-}^{**}\Big)\delta_1^{T_{t+1}^i-2}$$
$$+ V^\ell\Big[T_{t+1}^i-1, x_{T_{t+1}^i-1}^{(\neq \ell)**}; \bar{u}_{T_{t+1}^i-2}^\ell, \underline{\bar{u}}_{T_{t+1}^i-2}^{[T_t^i,T_{t+1}^i-1](\neq \ell)**}, \underline{\bar{u}}_{(T_{t+1}^i-1)-}^{**} \cap \underline{\bar{u}}_{(T_{t+1}^i-2)-}^{**}\Big]\Big\},$$
$$\text{for } \ell \in S^{[T_t^i,T_{t+1}^i-1]B},$$

where

$$x_{T_{t+1}^i-1}^{(\neq \ell)**} = f_{T_{t+1}^i-2}\Big(x, u_{T_{t+1}^i-2}^\ell, \bar{u}_{T_{t+1}^i-2}^\ell, \underline{u}_{T_{t+1}^i-2}^{[T_t^i,T_{t+1}^i-1](\neq \ell)**}, \underline{\bar{u}}_{T_{t+1}^i-2}^{[T_t^i,T_{t+1}^i-1](\neq \ell)**}; \bar{u}_{(T_{t+1}^i-2)-}^{**}\Big). \tag{7.9}$$

Note that in the maximization operator in stage $T_{t+1}^i - 2$, the control strategies of player i are $(u_{T_{t+1}^i-2}^i, \bar{u}_{T_{t+1}^i-2}^i)$; the control strategies of player j, for $j \in S^{[T_t^i,T_{t+1}^i-1]A}$, are $(u_{T_{t+1}^i-2}^j, \bar{u}_{T_{t+1}^i-2}^j)$; and the control strategies of player ℓ, for $\ell \in S^{[T_t^i,T_{t+1}^i-1]B}$, are $(u_{T_{t+1}^i-2}^\ell, \bar{u}_{T_{t+1}^i-2}^\ell)$. The state x and the previously executed strategies $\underline{\bar{u}}_{(T_{t+1}^i-2)-}^{**}$ appear as given. If the system of Eqs. (7.7)–(7.9) satisfies the implicit theorem, the game equilibrium strategies can be obtained as functions of $(x, \underline{\bar{u}}_{(T_{t+1}^i-2)-}^{**})$.

Substituting the game equilibrium strategies in stage $T_{t+1}^i - 2$ into the right-hand-side of the equations in (7.7)–(7.9) yields the value functions

$$V^i\Big(T_{t+1}^i - 2, x; \underline{\bar{u}}_{(T_{t+1}^i-2)-}^{**}\Big);$$

$$V^j\Big(T_{t+1}^i - 2, x; \underline{u}_{(T_{t+1}^i-2)-}^{**}\Big), \text{ for } j \in S^{[T_t^i,T_{t+1}^i-1]A};$$

$$V^\ell\Big(T_{t+1}^i - 2, x; \bar{u}_{(T_{t+1}^i-2)-}^{**}\Big), \text{ for } \ell \in S^{[T_t^i,T_{t+1}^i-1]B}; \tag{7.10}$$

which are explicit functions of $(x, \underline{\bar{u}}_{(T_{t+1}^i-2)-}^{**})$.

Following the above analysis, we move backwards sequentially from stage T_{t+1}^i-3 to stage T_t^i in the game interval $[T_t^i, T_{t+1}^i - 1]$ to obtain $V^i(k, x; \underline{\bar{u}}_{k-}^{**})$; $V^j(k, x; \underline{\bar{u}}_{k-}^{**})$, for $j \in S^{[T_t^i,T_{t+1}^i-1]A}$; and $V^\ell(k, x; \underline{\bar{u}}_{k-}^{**})$, for $\ell \in S^{[T_t^i,T_{t+1}^i-1]B}$; for $k \in \{T_{t+1}^i-3, T_{t+1}^i-4, \cdots, T_t^i\}$ and $t \in \{1, 2, \cdots, z_i\}$.

Hence Theorem 2.1 follows.

7.2 Appendix II: Proof of Proposition 3.1

We begin with the last game interval of player i, that is $[T^i_{z_i}, T^i_{z_i+1} - 1]$. Invoking Propositions 3.1 and (3.5), we obtain the payoffs of the players at $T^i_{z_i+1}$ as:

$$\left(A^{(1)i}_{T^i_{z_i+1}} x^1 + A^{(2)i}_{T^i_{z_i}+1} x^2 + C^i_{T^i_{z_i+1}}\right)\delta^{T^i_{z_i+1}-1} \text{ with } A^{(1)i}_{T^i_{z_i+1}} = Q^{(1)i}_{T^i_{z_i+1}}, A^{(2)i}_{T^i_{z_i+1}} = Q^{(2)i}_{T^i_{z_i+1}}, \quad C^i_{T^i_{z_i+1}} = \varpi^i_{T^i_{z_i+1}},$$

$$\left(A^{(1)j}_{T^i_{z_i+1}} x^1 + A^{(2)j}_{T^i_{z_i+1}} x^2 + C^j_{T^i_{z_i+1}}\right)\delta^{T^i_{z_i+1}-1} \text{ with } A^{(1)j}_{T^i_{z_i+1}} = Q^{(1)j}_{T^i_{z_i+1}}, A^{(2)j}_{T^i_{z_i+1}} = Q^{(2)j}_{T^i_{z_i+1}}, C^j_{T^i_{z_i+1}} = \varpi^j_{T^i_{z_i+1}}, \text{ for } j \in S^{\left[T^i_{z_i}, T^i_{z_i+1}-1\right]A},$$

$$\left(A^{(1)\ell}_{T^i_{z_i+1}} x^1 + A^{(2)\ell}_{T^i_{z_i+1}} x^2 + C^\ell_{T^i_{z_i+1}}\right)\delta^{T^i_{z_i+1}-1} \text{ with } A^{(1)\ell}_{T^i_{z_i+1}} = A^{(1)\ell}_{T^i_{z_i+1}}, A^{(2)\ell}_{T^i_{z_i+1}} = A^{(2)\ell}_{T^i_{z_i+1}}, \quad C^\ell_{T^i_{z_i+1}} = C^\ell_{T^i_{z_i+1}},$$

$$\text{for } \ell \in S^{\left[T^i_{z_i}, T^i_{z_i+1}-1\right]B}. \tag{7.11}$$

Performing the indicated maximization operator in stage $T^i_{z_i+1} - 1$ in (3.6)–(3.8), we obtain:

$$u^{Ii}_{T^i_{z_i+1}-1} = \frac{-\delta A^{(1)i}_{T^i_{z_{z_i}+1}} b^i_{T^i_{z_i}+1-1}}{2c^{Ii}_{T^i_{z_i+1}-1}} (x^1)^{1/2}, u^{IIi}_{T^i_{z_i+1}-1} = \frac{\delta A^{(2)i}_{T^i_{z_i+1}} a^i_{T^i_{z_i+1}-1}}{2c^{IIi}_{T^i_{z_i+1}-1}},$$

$$\bar{u}^i_{T^i_{z_i+1}-1} = \frac{R^i_{T^i_{z_i+1}-1} - v^{i|T^i_{z_i+1}-1|i}_{T^i_{z_i+1}-1} + \delta A^{(1)i}_{T^i_{z_i+1}} m^{|T^i_{z_i+1}-1|i}_{T^i_{z_i+1}-1}}{2\phi^i_{T^i_{z_i+1}-1}};$$

$$u^{Ij}_{T^i_{z_i+1}-1} = \frac{-\delta A^{(1)j}_{T^i_{z_i+1}} b^j_{T^i_{z_i+1}-1}}{2c^{Ij}_{T^i_{z_i+1}-1}} (x^1)^{1/2}, u^{IIj}_{T^i_{z_i+1}-1} = \frac{\delta A^{(2)j}_{T^i_{z_i+1}} a^j_{T^i_{z_i+1}-1}}{2c^{IIj}_{T^i_{z_i+1}-1}},$$

$$\bar{u}^j_{T^i_{z_i+1}-1} = \frac{R^j_{T^i_{z_i+1}-1} - v^{j|T^i_{z_i+1}-1|j}_{T^i_{z_i+1}-1} + \delta A^{(1)j}_{T^i_{z_i+1}} m^{\left|T^i_{z_i+1}-1\right|j}_{T^i_{z_i+1}-1}}{2\phi^j_{T^i_{T,1}-1}}, \quad j \in S^{\left[T^i_{z_i}, T^i_{z_i+1}-1\right]A};$$

$$u^{I\ell}_{T^i_{Z_i+1}-1} = \frac{-\delta A^{(1)\ell}_{T^i_{z_i+1}} b^\ell_{T^i_{z_i+1}-1}}{2c^{I\ell}_{T^i_{z_i+1}-1}} (x^1)^{1/2}, u^{II\ell}_{T^i_{z_i+1}-1} = \frac{\delta A^{(2)\ell}_{T^i_{z_i+1}} a^\ell_{T^i_{z_i+1}-1}}{2c^{II\ell}_{T^i_{z_i+1}-1}},$$

$$\bar{u}^{\ell}_{T^i_{z_i+1}-1} = \frac{R^{\ell}_{T^i_{z_i+1}-1} - \sum_{\eta=T^i_{z_i+1}-1}^{T^i_{z_i+1}-1+\omega-1} \delta^{\eta-(T^i_{z_i+1}-1)} v_{\eta}^{\ell|T_{z_i+1}-1|\ell} + \sum_{\eta=T^i_{z_i+1}-1}^{T^i_{z_i+1}-1+\omega-1} \delta^{\eta-(T^i_{z_i+1}-1)+1} A^{(1)\ell}_{\eta+1} m_{\eta}^{|T^i_{z_i+1}-1|\ell}}{2\phi^i_{T^i_{z_i+1}-1}},$$

$$\ell \in S^{\left[T^i_{z_i}, T^i_{z_i+1}-1\right]B},$$

$$\text{with } v_{\eta}^{\ell|\eta|\ell} = A^{(1)\ell}_{\eta+1} = 0 \text{ if } \eta > T^{\ell}_{EX}. \tag{7.12}$$

In particular, $\bar{u}^{\ell}_{T^i_{z_i+1}-1}$ will be vindicated later.

For expositional convenience, we use $\bar{u}^{j\sim}_{T^i_{z_i+1}-1}$ to denote $\bar{u}^{j}_{T^i_{z_i+1}-1}$ in (7.12) and $\bar{u}^{\ell\sim}_{T^i_{z_i+1}-1}$ to denote $\bar{u}^{\ell}_{T^i_{z_i+1}-1}$ in (7.12). Substituting the game equilibrium strategies of the players in (7.12) into the stage $T^i_{z_i+1} - 1$ equation for player i in (3.6), we obtain $V^i(T^i_{z_i+1} - 1, x; \underline{\bar{u}}^{**}_{(T^i_{z_i+1}-1)-})$ as:

$$\left(A^{(1)i}_{T^i_{z_i+1}-1} x^1 + A^{(2)i}_{T^i_{z_i+1}-1} x^2 + C^i_{T^i_{z_i+1}-1}\right) \delta^{T^i_{z_i+1}-2}$$

$$= \left(R^i_{T^i_{z_i+1}-1} \frac{R^i_{T^i_{z_i+1}-1} - v^{i|T^i_{z_i+1}-1|i}_{T^i_{z_i+1}-1} + \delta A^{(1)i}_{T^i_{z_i+1}} m^{|T^i_{z_i+1}-1|i}_{T^i_{z_i+1}-1}}{2\phi^i_{T^i_{z_i+1}-1}} \right)$$

$$- \frac{\left(R^i_{T^i_{z_i+1}-1} - v^{i|T^i_{z_i+1}-1|i}_{T^i_{z_i+1}-1} + \delta A^{(1)i}_{T^i_{z_i+1}} m^{|T^i_{z_i+1}-1|i}_{T^i_{z_i+1}-1}\right)^2}{4\phi^i_{T^i_{z_i+1}-1}} - \frac{\left(\delta A^{(1)i}_{T^i_{z_i+1}} b^i_{T^i_{z_i+1}-1}\right)^2}{4c^{Ii}_{T^i_{z_i+1}-1}} x^1 + \kappa^i_{T^i_{z_i+1}-1} x^2$$

$$- \frac{\left(\delta A^{(2)i}_{T^i_{z_i+1}} a^i_{T^i_{z_i+1}-1}\right)^2}{4c^{IIi}_{T^i_{z_i+1}-1}}$$

$$-h^i_{T^i_{z_i+1}-1} x^1 - v^{i\left|T^i_{z_i+1}\right|i}_{T^i_{z_i+1}-1} \frac{R^i_{T^i_{z_i+1}-1} - v^{i|T^i_{z_i+1}i|}_{T^i_{z_i+1}-1} + \delta A^{(1)i}_{T^i_{z_i+1}} m^{|T^i_{z_i+1}|i}_{T^i_{z_i+1}-1}}{2\phi^i_{T^i_{z_i+1}-1}}$$

$$- \sum_{\substack{\theta \in S^{\left[T^i_{z_i}, T^i_{z_i+1}-1\right]} \\ \theta \neq j}} v^{i|k|\theta}_{T^i_{z_i+1}-1} \bar{u}^{\theta\sim}_{T^i_{z_i+1}-1}$$

$$- \sum_{\theta \in S^{\left[\left(T^i_{z_i+1}-1\right)-\right]}} \sum_{\tau=T^i_{z_i+1}-\omega}^{T^i_{z_i+1}-2} v^{i|\tau|\theta}_{T^i_{z_i+1}-1} \bar{u}^{\theta**}_{\tau} \Bigg) \delta^{T^i_{z_i+1}-1}$$

$$+\left(A^{(1)i}_{T^i_{z_i+1}} (x^{1[T^i_{z_i}, T^i_{z_i+1}-1](\neq i)**}_{T^i_{z_i+1}}) + A^{(2)i}_{T^i_{z_i+1}} (x^{2[T^i_{z_i}, T^i_{z_i+1}-1](\neq i)**}_{T^i_{z_i+1}}) + \varpi^i_{T^i_{z_i+1}} \right) \delta^{T^i_{z_i+1}}, \tag{7.13}$$

where

$$x^{1\left[T^i_{z_i}, T^i_{z_i+1}-1\right](\neq i)**}_{T^i_{z_i+1}} = x^1 + m^{\left|T^i_{z_i+1}-1\right|i}_{T^i_{z_i+1}-1} \frac{R^i_{T^i_{z_i+1}-1} - v^{i|T^i_{z_i+1}-1|i}_{T^i_{z_i+1}-1} + \delta A^{(1)i}_{T^i_{z_i+1}-1} m^{\left|T^i_{z_i+1}-1\right|i}_{T^i_{z_i+1}-1}}{2\phi^i_{T^i_{z_i+1}-1}}$$

$$+ \sum_{\substack{\theta \in S^{\left[T^i_{z_i}, T^i_{z_i+1}-1\right]} \\ \theta \neq j}} m^{|T^i_{z_i+1}-1|\theta}_{T^i_{z_i+1}-1} \bar{u}^{\theta\sim}_{T^i_{z_i+1}-1}$$

$$+ \sum_{\theta \in S^{\left[\left(T^i_{z_i+1}-1\right)-\right]x^1}} \sum_{\tau=k-\omega+1}^{k-1} m^{|\tau|\theta}_k \bar{u}^{\theta**}_\tau + \sum_{\theta \in S^{\left[T^i_i, T^i_{i+1}-1\right]}} b^\theta_k \frac{\delta A^{(1)\theta}_{T^i_{z_i+1}-1} b^\theta_{T^i_{z_i+1}-1}}{2c^{I\theta}_{T^i_{z_i+1}-1}} x^1 - \lambda^1_k x^1,$$

$$x^{2\left[T^i_{z_i}, T^i_{z_i+1}-1\right](\neq i)**}_{T^i_{z_i+1}} = x^2 + \sum_{\theta \in S^{\left[T^i_i, T^i_{i+1}-1\right]}} a^\theta_k \frac{\delta A^{(2)\theta}_{T^i_{z_i+1}-1} a^\theta_{T^i_{z_i+1}-1}}{2c^{II\theta}_{T^i_{z_i+1}-1}} - \lambda^2_k x^2.$$

The RHS and the LHS of (7.13) are linear functions of x^1 and x^2. For (7.13) to hold, the following conditions must be satisfied:

$$A^{(1)i}_{T^i_{z_i+1}-1} = -\frac{\left(A^{(1)i}_{T^i_{z_i+1}} b^i_{T^i_{z_i+1}-1} \delta\right)^2}{4c^{Ii}_{T^i_{z_i+1}-1}} - h^i_{T^i_{z_i+1}-1}$$

$$+ \delta A^{(1)i}_{T^i_{z_i+1}} \left(1 + \sum_{\theta \in S^{\left[T^i_i, T^i_{i+1}-1\right]}} b^\theta_{T^i_{z_i+1}-1} \frac{A^{(1)\theta}_{T^i_{z_i+1}} b^\theta_{T^i_{z_i+1}-1} \delta}{2c^{I\theta}_{T^i_{z_i+1}-1}} - \lambda^1_{T^i_{z_i+1}-1} \right),$$

and

$$A^{(2)i}_{T^i_{z_i+1}-1} = \kappa^i_{T^i_{z_i+1}-1} + \delta A^{(2)i}_{T^i_{z_i+1}} \left(1 - \lambda^2_{T^i_{z_i+1}-1}\right),$$

and the corresponding $C^i_{T^i_{z_i+1}-1}$ is the expression on the RHS which does not involve x^1 and x^2 which include previously executed durable strategies

$$\underline{\bar{u}}^{**}_{(T^i_{z_i+1}-1)-}. \tag{7.14}$$

Now, we move stage $T^i_{z_i+1}-2$.

Using $V^i(T^i_{z_i+1}-1, x; \underline{\bar{u}}^{**}_{(T^i_{z_i+1}-1)-})$ as $(A^{(1)i}_{T^i_{z_i+1}-1}x^1 + A^{(2)i}_{T^i_{z_i+1}-1}x^2 + C^i_{T^i_{z_i+1}-1})\delta^{T^i_{z_i+1}-2}$ in (7.13) and performing the indicated maximization operator in stage $T^i_{z_i+1}-2$ we obtain:

$$u^{Ii}_{T^i_{z_i+1}-2} = \frac{-\delta A^{(1)i}_{T^i_{z_i+1}-1} b^i_{T^i_{z_i+1}-2}}{2c^{Ii}_{T^i_{z_i+1}-2}}\left(x^1\right)^{1/2}, u^{IIi}_{T^i_{z_i+1}-2} = \frac{\delta A^{(2)i}_{T^i_{z_i+1}-1} a^i_{T^i_{z_i+1}-2}}{2c^{IIi}_{T^i_{z_i+1}-2}},$$

$$\bar{u}^i_{T^i_{z_i+1}-2} = \frac{R^i_{T^i_{z_i+1}-2} - v^{i|T^i_{z_i+1}-2|i}_{T^i_{z_i+1}-2} + \delta A^{(1)i}_{T^i_{z_i+1}-1} m^{|T^i_{z_i+1}-2|i}_{T^i_{z_i+1}-2} + \delta^2 A^{(1)i}_{T^i_{z_i+1}} m^{|T^i_{z_i+1}-2|i}_{T^i_{z_i+1}-1}}{2\phi^i_{T^i_{z_i+1}-2}};$$

$$u^{Ij}_{T^i_{z_i+1}-2} = \frac{-\delta A^{(1)j}_{T^i_{z_i+1}-1} b^j_{T^i_{z_i+1}-2}}{2c^{Ij}_{T^i_{z_i+1}-2}}\left(x^1\right)^{1/2}, u^{II_j}_{T^i_{z_i+1}-2} = \frac{\delta A^{(2)j}_{T^i_{z_i+1}-1} a^j_{T^i_{z_i+1}-2}}{2c^{II_j}_{T^i_{z_i+1}-2}},$$

$$\bar{u}^j_{T^i_{z_i+1}-2} = \frac{R^j_{T^i_{z_i+1}-2} - v^{j\left|T^i_{z_i+1}-2\right|j}_{T^i_{z_i+1}-2} - \delta v^{j|T^i_{z_i+1}-2|j}_{T^i_{z_i+1}-1} + \delta A^{(1)j}_{T^i_{z_i+1}-1} m^{\left|T^i_{z_i+1}-2\right|j}_{T^i_{z_i+1}-2} + \delta^2 A^{(1)j}_{T^i_{z_i+1}} m^{\left|T^i_{z_i+1}-2\right|j}_{T^i_{z_i+1}-1}}{2\phi^j_{T^i_{z_i+1}-2}},$$

$$j \in S^{\left[T^i_{z_i}, T^i_{z_i+1}-1\right]A};$$

$$u^{I\ell}_{T^i_{z_i+1}-2} = \frac{-\delta A^{(1)\ell}_{T^i_{z_i+1}-1} b^\ell_{T^i_{z_i+1}-2}}{2c^{I\ell}_{T^i_{z_i+1}-2}}\left(x^1\right)^{1/2}, u^{II\ell}_{T^i_{z_i+1}-2} = \frac{\delta A^{(2)\ell}_{T^i_{z_i+1}-1} a^\ell_{T^i_{z_i+1}-2}}{2c^{II\ell}_{T^i_{z_i+1}-2}},$$

$$\bar{u}^\ell_{T^i_{z_i+1}-2} = \frac{R^\ell_{T^i_{z_i+1}-2} - \sum_{\eta=T^i_{z_i+1}-2}^{T^i_{z_i+1}-2+\omega-1} \delta^{\eta-\left(T^i_{z_i+1}-2\right)} v_\eta^{\ell|T^i_{z_i+1}-2|\ell} + \sum_{\eta=T^i_{z_i+1}-2}^{T^i_{z_i+1}-2+\omega-1} \delta^{\eta-\left(T^i_{z_i+1}-2\right)+1} A^{(1)\ell}_\eta m_\eta^{\left|T^i_{z_i+1}-2\right|\ell}}{2\phi^\ell_{T^i_{z_i+1}-2}},$$

$$\ell \in S^{\left[T^i_{z_i}, T^i_{z_i+1}-1\right]B},$$

$$\text{with } v_\eta^{\ell|\eta|\ell} = A^{(1)\ell}_{\eta+1} = 0 \text{ if } \eta > T^\ell_{EX}. \tag{7.15}$$

Substituting the game equilibrium strategies of the players in (7.15) into the stage $T^i_{z_i+1}-2$ equation of player i in (3.6), we obtain an equation system in which the RHS and the LHS of the equations are linear functions of x^1 and x^2. For the equation system to hold, the following conditions must be satisfied:

$$A^{(1)i}_{T^i_{z_i+1}-2} = -\frac{\left(A^{(1)i}_{T^i_{z_i+1}-1} b^i_{T^i_{z_i+1}-2}\delta\right)^2}{4c^{Ii}_{T^i_{z_i+1}-2}} - h^i_{T^i_{z_i+1}-2} + \delta A^{(1)i}_{T^i_{z_i+1}-1}\left(1 + \sum_{\theta \in S^{\left[T^i_{z_i}, T^i_{z_i+1}-1\right]}} b^\theta_{T^i_{z_i+1}-2} \frac{A^{(1)\theta}_{T^i_{z_i+1}-1} b^\theta_{T^i_{z_i+1}-2}\delta}{2c^{I\theta}_{T^i_{z_i+1}-2}} - \lambda^1_{T^i_{z_i+1}-2}\right),$$

and

$$A^{(2)i}_{T^i_{z_i+1}-2} = \kappa^i_{T^i_{z_i+1}-2} + \delta A^{(2)i}_{T^i_{z_i+1}-1}\left(1 - \lambda^2_{T^i_{z_i+1}-2}\right),$$

and the corresponding $C^i_{T^i_{z_i+1}-2}$ is the expression on the RHS which does not involve x^1 and x^2 which include previously executed durable strategies

$$\underline{\bar{u}}^{**}_{(T^i_{z_i+1}-2)-}. \tag{7.16}$$

Moving to stage $k \in \{T^i_{z_i+1} - 3, T^i_{z_i+1} - 4, \cdots, T^i_{z_i}\}$, we can obtain the game equilibrium strategies as:

$$u^{Ii}_k = \frac{-\delta A^{(1)i}_{k+1} b^i_k}{2c^{Ii}_k}\left(x^1\right)^{1/2}, u^{IIi}_k = \frac{\delta A^{(2)i}_{k+1} a^i_k}{2c^{IIi}_k},$$

$$\bar{u}^i_k = \frac{R^i_k - \sum_{\eta=k}^{k+\omega-1} \delta^{\eta-k} v^{i|k|i}_\eta + \sum_{\eta=k}^{k+\omega-1} \delta^{\eta-k+1} A^{(1)i}_{\eta+1} m^{|k|i}_\eta}{2\phi^i_k},$$

with $v^{i|\eta|i}_\eta = A^{(1)i}_{\eta+1} = 0$, for $\eta > T^i_{z_i+1} - 1$;

$$u^{Ij}_k = \frac{-\delta A^{(1)j}_{k+1} b^j_k}{2c^{Ij}_k}\left(x^1\right)^{1/2}, u^{IIj}_k = \frac{\delta A^{(2)j}_{k+1} a^j_k}{2c^{IIj}_k},$$

$$\bar{u}^j_k = \frac{R^j_k - \sum_{\eta=k}^{k+\omega-1} \delta^{\eta-k} v^{j|k|j}_\eta + \sum_{\eta=k}^{k+\omega-1} \delta^{\eta-k+1} A^{(1)j}_{\eta+1} m^{|k|j}_\eta}{2\phi^j_k},$$

for $j \in S^{[T^i_{z_i}, T^i_{z_i+1}-1]A}$, and with $v^{j|\eta|j}_\eta = A^{(1)j}_{\eta+1} = 0$, for $\eta > T^i_{z_i+1} - 1$;

$$u^{I\ell}_k = \frac{-\delta A^{(1)\ell}_{k+1} b^\ell_k}{2c^{I\ell}_k}\left(x^1\right)^{1/2}, u^{II\ell}_k = \frac{\delta A^{(2)\ell}_{k+1} a^\ell_k}{2c^{II\ell}_k},$$

$$\bar{u}^\ell_k = \frac{R^\ell_k - \sum_{\eta=k}^{k+\omega-1} \delta^{\eta-k} v^{\ell|k|\ell}_\eta + \sum_{\eta=k}^{k+\omega-1} \delta^{\eta-k+1} A^{(1)\ell}_{\eta+1} m^{|k|\ell}_\eta}{2\phi^\ell_k},$$

$$\text{for } \ell \in S^{\left[T^i_{z_i}, T^i_{z_i+1}-1\right]B} \text{ and with } v^{\ell|\eta|\ell}_\eta = A^{(1)\ell}_{\eta+1} = 0 \text{ if } \eta > T^\ell_{\text{Ex}}. \tag{7.17}$$

Following the above analysis, we obtain

$$V^i\left(k, x, \underline{u}^{**}_{k-}\right) = \left(A^{(1)i}_k x^1 + A^{(2)i}_k x^2 + C^i_k\right)\delta^{k-1}, \tag{7.18}$$

where

$$A_k^{(1)i} = -\frac{\left(A_{k+1}^{(1)i} b_k^i \delta\right)^2}{4c_k^{Ii}} - h_k^i + \delta 4_{k+1}^{(1)} \left(1 + \sum_{\theta \in S^{\left[T_{z_i}^T, T_{z_i+1}^i - 1\right]}} b_k^\theta \frac{A_{k+1}^{(1)\theta} b_k^\theta \delta}{2c_k^{I\theta}} - \lambda_k^1\right)$$

and

$$A_k^{(2)i} = \kappa_k^j + \delta A_{k+1}^{(2)i}(1 - \lambda_k^2), \quad \text{for } k \in \left\{T_{z_i+1}^i - 3, T_{z_i+1}^i - 4, \cdots, T_{z_i}^i\right\}. \quad (7.19)$$

Moreover, the corresponding C_k^i is an expression that includes previously executed durable strategies $\underline{\bar{u}}_{k-}^{**}$.

Now, we move the game interval $[T_t^i, T_{t+1}^i - 1]$ for $t \in \{1, 2, \cdots, z_i - 1\}$.

We first look at stage T_{t+1}^i where player $j \in S^{[T_t^i, T_{t+1}^i - 1]A}$ leaves the game. Invoking Proposition 3.1 and (3.5)–(3.8), we obtain the payoffs of the players at T_{t+1}^i as:

$$\left(A_{T_{t+1}^i}^{(1)i} x^1 + A_{T_{t+1}^i}^{(2)i} x^2 + C_{T_{t+1}^i}^i\right)\delta^{T_{t+1}^i - 1} \text{ with } A_{T_{t+1}^i}^{(1)i} = A_{T_{t+1}^i}^{(1)i}, A_{T_{t+1}^i}^{(2)i} = A_{T_{t+1}^i}^{(2)i}, C_{T_{t+1}^i}^i = C_{T_{t+1}^i}^i,$$

$$\left(A_{T_{t+1}^i}^{(1)j} x^1 + A_{T_{t+1}^i}^{(2)j} x^2 + C_{T_{t+1}^i}^j\right)\delta^{T_{t+1}^i - 1} \text{ with } A_{T_{t+1}^i}^{(1)j} = Q_{T_{t+1}^i}^{(1)j}, A_{T_{t+1}^i}^{(2)j} = Q_{T_{t+1}^i}^{(2)j}, C_{T_{t+1}^i}^{(2)j}$$
$$= \varpi_{T_{t+1}^i}^j, \text{ for } j \in S^{[T_t^i, T_{t+1}^i - 1]A},$$

$$\left(A_{T_{t+1}^i}^{(1)j} x^1 + A_{T_{t+1}^i}^{(2)j} x^2 + C_{T_{t+1}^i}^j\right)\delta^{T_{t+1}^i - 1} \text{ with } A_{T_{t+1}^i}^{(1)\ell} = A_{T_{t+1}^i}^{(1)\ell}, A_{T_{t+1}^i}^{(2)\ell}$$
$$= A_{T_{t+1}^i}^{(2)\ell}, \quad C_{T_{t+1}^i}^\ell = C_{T_{t+1}^i}^\ell,$$
$$\text{for } \ell \in S^{[T_t^i, T_{t+1}^i - 1]B}. \quad (7.20)$$

Using (7.20) and performing the indicated maximization operator in stage $T_{t+1}^i - 1$ of (3.6)–(3.8),

we obtain:

$$u_{T_{t+1}^i - 1}^{Ii} = \frac{-\delta A_{T_{t+1}^i}^{(1)i} b_{T_{t+1}^i - 1}^i}{2c_{T_{t+1}^i - 1}^{Ii}} (x^1)^{1/2}, u_{T_{t+1}^i - 1}^{IIi} = \frac{\delta A_{T_{t+1}^i}^{(2)i} a_{T_{t+1}^i - 1}^i}{2c_{T_{t+1}^i - 1}^{IIi}},$$

$$\bar{u}_{T_{t+1}^i - 1}^i = \frac{R_{T_{t+1}^i - 1}^i - \sum_{\eta = T_{t+1}^i - 1}^{T_{t+1}^i - 1 + \omega - 1} \delta^{\eta - (T_{t+1}^i - 1)} v_\eta^{i|T_{t+1}^i - 1|i} + \sum_{\eta = T_{t+1}^i - 1}^{T_{t+1}^i - 1 + \omega - 1} \delta^{\eta - (T_{t+1}^i - 1) + 1} A_{\eta+1}^{(1)i} m_\eta^{|T_{t+1}^i - 1|i}}{2\phi_{T_{t+1}^i - 1}^i},$$

with $v_\eta^{i|\eta|i} = A_{\eta+1}^{(1)i} = 0$ for $\eta > T_{EX}^i$;

$$\bar{u}_{T_{t+1}^i - 1}^{Ij} = \frac{-\delta A_{T_{t+1}^i}^{(1)j} b_{T_{t+1}^i - 1}^j}{2c_{T_{t+1}^i - 1}^{Ij}} (x^1)^{1/2}, u_{T_{t+1}^i - 1}^{IJj} = \frac{\delta A_{T_{t+1}^i}^{(2)j} a_{T_{t+1}^i - 1}^j}{2c_{T_{t+1}^i - 1}^{IJ}}, \bar{u}_{T_{t+1}^i - 1}^j$$

$$
= \frac{R^{j}_{T^i_{t+1}-1} - v^{j|T^i_{t+1}-1|j}_{T^i_{t+1}-1} + \delta A^{(1)j}_{T^i_{t+1}} m^{|T^i_{t+1}-1|j}_{T^i_{t+1}-1}}{2\phi^{j}_{T^i_{t+1}-1}},
$$

for $j \in S^{[T^i_t, T^i_{t+1}-1]A}$;

$$
u^{I\ell}_{T^i_{t+1}-1} = \frac{-\delta A^{(1)\ell}_{T^i_{t+1}} b^{\ell}_{T^i_{t+1}-1}}{2c^{I\ell}_{T^i_{t+1}-1}} \left(x^1\right)^{1/2}, u^{II\ell}_{T^i_{t+1}-1} = \frac{\delta A^{(2)\ell}_{T^i_{t+1}} a^{\ell}_{T^i_{t+1}-1}}{2c^{II\ell}_{T^i_{t+1}-1}},
$$

$$
\bar{u}^{\ell}_{T^i_{t+1}-1} = \frac{R^{\ell}_{T^i_{t+1}-1} - \sum_{\eta=T^i_{t+1}-1}^{T^i_{t+1}-1+\omega-1} \delta^{\eta-(T^i_{t+1}-1)} v^{\ell|T^i_{t+1}-1|\ell}_{\eta} + \sum_{\eta=T^i_{t+1}-1}^{T^i_{t+1}-1+\omega-1} \delta^{\eta-(T^i_{t+1}-1)+1} A^{(1)\ell}_{\eta+1} m^{|T^i_{t+1}-1|\ell}_{\eta}}{2\phi^{\ell}_{T^i_{t+1}-1}},
$$

$\ell \in S^{[T^i_t, T^i_{t+1}-1]B}$,

with $v^{\ell|\eta|\ell}_{\eta} = A^{(1)\ell}_{\eta+1} = 0$ if $\eta > T^{\ell}_{EX}$. (7.21)

Substituting the game equilibrium strategies of the players in (7.21) into the stage $T^i_{t+1} - 1$ equation of player i in (3.6) we obtain an equation system in which the RHS and the LHS of the equations are linear functions of x^1 and x^2. For the equation system to hold, the following conditions must be satisfied:

$$
A^{(1)i}_{T^i_{t+1}-1} = -\frac{\left(A^{(1)i}_{T^i_{t+1}} b^{i}_{T^i_{t+1}-1} \delta\right)^2}{4c^{Ii}_{T^i_{t+1}-1}} - h^{i}_{T^i_{t+1}-1}
$$

$$
+ \delta A^{(1)i}_{T^i_{t+1}-1} \left(1 + \sum_{\theta \in S^{[T^i_t, T^i_{t+1}-1]}} b^{\theta}_{T^i_{t+1}-1} \frac{A^{(1)\theta}_{T^i_{t+1}} b^{\theta}_{T^i_{t+1}-1} \delta}{2c^{I\theta}_{T^i_{t+1}-1}} - \lambda^{1}_{T^i_{t+1}-2}\right),
$$

and

$$
A^{(2)i}_{T^i_{t+1}-1} = \kappa^{i}_{T^i_{t+1}-1} + \delta A^{(2)i}_{T^i_{t+1}} \left(1 - \lambda^{2}_{T^i_{t+1}-1}\right),
$$

and the corresponding $C^{i}_{T^i_{t+1}-1}$ is the expression on the RHS which does not involve x^1 and x^2 which include previously executed durable strategies

$$
\underline{\bar{u}}^{**}_{(T^i_{t+1}-1)-} \tag{7.22}
$$

Following the above analysis and moving to stage $k \in \{T^i_{t+1}-2, T^i_{t+1}-4, \cdots, T^i_t\}$, we can obtain the game equilibrium strategies of the players as:

$$
u^{Ii}_{k} = \frac{-\delta A^{(1)i}_{k+1} b^{i}_{k}}{2c^{Ii}_{k}} \left(x^1\right)^{1/2}, u^{IIi}_{k} = \frac{\delta A^{(2)i}_{k+1} a^{i}_{k}}{2c^{IIi}_{k}},
$$

$$\bar{u}_k^i = \frac{R_k^i - \sum_{\eta=k}^{k+\omega-1} \delta^{\eta-k} v_\eta^{i|k|i} + \sum_{\eta=k}^{k+\omega-1} \delta^{\eta-k+1} A_{\eta+1}^{(1)i} m_\eta^{|k|i}}{2\phi_k^i},$$

with $v_\eta^{i|\eta|i} = A_{\eta+1}^{(1)i} = 0$, if $\eta > T_{z_i+1}^i - 1$;

$$u_k^{Ij} = \frac{-\delta A_{k+1}^{(1)j} b_k^j}{2c_k^{Ij}} \left(x^1\right)^{1/2}, u_k^{IIj} = \frac{\delta A_{k+1}^{(2)j} a_k^j}{2c_k^{IIJ}},$$

$$\bar{u}_k^j = \frac{R_k^j - \sum_{\eta=k}^{k+\omega-1} \delta^{\eta-k} v_\eta^{j|k|j} + \sum_{\eta=k}^{k+\omega-1} \delta^{\eta-k+1} A_{\eta+1}^{(1)j} m_\eta^{|k|j}}{2\phi_k^j},$$

for $j \in S^{[T_t^i, T_{t+1}^i - 1]A}$, with $v_\eta^{j|\eta|j} = A_{\eta+1}^{(1)j} = 0$, for $\eta > T_{EX}^j$;

$$u_k^{I\ell} = \frac{-\delta A_{k+1}^{(1)\ell} b_k^\ell}{2c_k^{I\ell}} \left(x^1\right)^{1/2}, u_k^{II\ell} = \frac{\delta A_{k+1}^{(2)\ell} a_k^\ell}{2c_k^{II\ell}}, \bar{u}_k^\ell = \frac{R_k^\ell - \sum_{\eta=k}^{k+\omega-1} \delta^{\eta-k} v_\eta^{\ell|k|\ell} + \sum_{\eta=k}^{k+\omega-1} \delta^{\eta-k+1} A_{\eta+1}^{(1)\ell} m_\eta^{|k|\ell}}{2\phi_k^\ell},$$

for $\ell \in S^{\left[T_{z_i}^i, T_{z_i+1}^i - 1\right]B}$, and with $v_\eta^{\ell|\eta|\ell} = A_{\eta+1}^{(1)\ell} = 0$ if $\eta > T_{EX}^\ell$. (7.23)

In addition, we can obtain

$$V^i\left(k, x; \underline{\bar{u}}_{k-}^{**}\right) = \left(A_k^{(1)i} x^1 + A_k^{(2)i} x^2 + C_k^i\right) \delta^{k-1}, \tag{7.24}$$

where

$$A_k^{(1)i} = -\frac{\left(A_{k+1}^{(1)i} b_k^i \delta\right)^2}{4c_k^{Ii}} - h_k^i + \delta A_{k+1}^{(1)i} \left(1 + \sum_{\theta \in S^{\left[T_{z_i}^i, T_{z_i+1}^i - 1\right]}} b_k^\theta \frac{A_{k+1}^{(1)\theta} b_k^\theta \delta}{2c_k^{I\theta}} - \lambda_k^1\right),$$

and

$$A_k^{(2)i} = \kappa_k^i + \delta A_{k+1}^{(2)i}\left(1 - \lambda_k^2\right), \quad \text{for } k \in \left\{T_{t+1}^i - 2, T_{t+1}^i - 4, \cdots, T_t^i\right\}. \tag{7.25}$$

The corresponding C_k^i is an expression that includes previously executed durable strategies $\underline{\bar{u}}_{k-}^{**}$.

Given the similarity in the game structures among players, we can obtain the game equilibrium payoffs of player $i \in N$ as:

$$V^i\left(T_{z_i+1}^i, x; \underline{\bar{u}}_{(z_i+1)-}^{**}\right) = \left(Q_{T_{z_i+1}^i}^{(1)i} x^1 + Q_{T_{z_i+1}^i}^{(2)i} x^2 + \varpi_{T_{z_i+1}^i}^i\right) \delta^{T_{z_i+1}^i - 1},$$

$$V^i(T^i_{z_i+1}, x; \underline{\bar{u}}^{**}_{(z_i+1)-}) = \left(Q^{(1)i}_{T^i_{z_i+1}} x^1 + Q^{(2)i}_{T^i_{z_i+1}} x^2 + \varpi^i_{T^i_{z_i+1}}\right)\delta^{T^i_{z_i+1}-1},$$

$$V^i(k, x; \underline{\bar{u}}^{**}_{k-}) = \left(A^{(1)i}_k x^1 + A^{(2)i}_k x^2 + C^i_k\right)\delta^{k-1}, k \in [T^i_{z_i}, T^i_{z_i+1} - 1], \tag{7.26}$$

where $A^{(1)i}_{T^i_{z_i+1}} = Q^{(1)i}_{T^i_{z_i+1}}$, $A^{(2)i}_{T^i_{z_i+1}} = Q^{(2)i}_{T^i_{z_i+1}}$ and $C^i_{T^i_{z_i+1}} = \varpi^i_{T^i_{z_i+1}}$;

$$A^{(1)i}_k = -\frac{\left(A^{(1)i}_{k+1} b^i_k \delta\right)^2}{4c^{Ii}_k} - h^i_k + \delta A^{(1)i}_{k+1}\left(1 + \sum_{\theta\in S^{\left[T^i_{z_i}, T^i_{z_i+1}-1\right]}} b^\theta_k \frac{A^{(1)\theta}_{k+1} b^\theta_k \delta}{2c^{I\theta}_k} - \lambda^1_k\right),$$

$$A^{(2)i}_k = \kappa^i_k + \delta A^{(2)i}_{k+1}(1 - \lambda^2_k),$$

and the corresponding C^i_k is an expression that includes previously executed durable strategies $\underline{\bar{u}}^{**}_{k-}$, for $k \in [T^i_{z_i}, T^i_{z_i+1} - 1]$;

$$V^i(k, x, \underline{\bar{u}}^{**}_{k-}) = \left(A^{(1)i}_k x^1 + A^{(2)i}_k x^2 + C^i_k\right)\delta^{k-1},$$
$$k \in \left[T^i_t, T^i_{t+1} - 1\right] \text{ and } t \in \{1, 2, \cdots, z_i - 1\}, \tag{7.27}$$

where

$$A^{(1)i}_k = -\frac{\left(A^{(1)i}_{k+1} b^i_k \delta\right)^2}{4c^{Ii}_k} - h^i_k + \delta A^{(1)i}_{k+1}\left(1 + \sum_{\theta\in S^{\left[T^i_t, T^i_{t+1}-1\right]}} b^\theta_k \frac{A^{(1)}_{k+1} b^\theta_k \delta}{2c^{I\theta}_k} - \lambda^1_k\right),$$

$$A^{(2)i}_k = \kappa^j_k + \delta A^{(2)i}_{k+1}(1 - \lambda^2_k),$$

and the corresponding C^i_k is an expression that includes previously executed durable strategies $\underline{\bar{u}}^{**}_{k-}$, for $k \in [T^i_t, T^i_{t+1} - 1]$.

Hence Proposition 3.1 follows.

7.3 Appendix III: Proof of Theorem 4.1

We begin with the last game interval of the entire game, that is $[T_z, T]$. The terminal joint payoff of all players at stage $T + 1$ is the sum of the payoffs of player i, for $i \in S^{[T_z,T]}$, that is

$$W(T + 1, x; \underline{\bar{u}}^i_{(T+1)-}) = \sum_{i\in S^{[T_z,T]}} q^i_{T+1}(x; \underline{\bar{u}}^i_{(T+1)-})\delta_1^{T+1} \tag{7.28}$$

Then we consider stage T in the game interval $[T_z, T]$. The stage T maximization problem can be expressed as

$$W(T, x, \underline{\bar{u}}_{T-}) = \max_{\substack{u_T^i, \bar{u}_T^i \\ i \in S^{[T_z,T]}}} \left\{ \sum_{i \in [T_z,T]} g_T^i\left(x, \underline{u}_T^{[T_z,T]}, \underline{\bar{u}}_T^{[T_z,T]}; \underline{\bar{u}}_{T-}\right)\delta_1^T \right.$$

$$\left. + \sum_{i \in S^{[T_z,T]}} q_{T+1}^i\left[f_T\left(x, \underline{u}_T^{[T_z,T]}, \underline{\bar{u}}_T^{[T_z,T]}; \underline{\bar{u}}_{T-}\right); \underline{\bar{u}}_T^{[T_z,T]}, \underline{\bar{u}}_{(T+1)-} \cap \underline{\bar{u}}_{T-}\right]\delta_1^{T+1} \right\}. \tag{7.29}$$

Note that in the maximization operator in stage T, the control strategies are $(u^i_{T^i_{z_i+1}-1}, \bar{u}^i_{T^i_{z_i+1}-1})$, for $i \in S^{[T_z,T]}$. The state x and the previously executed strategies $\underline{\bar{u}}_{T-}$ appears as given. If the Eq. (7.29) satisfies the implicit theorem, the optimal control strategies in stage T can be obtained as functions of $(x, \underline{\bar{u}}_{T-})$. Substituting the optimal control strategies into the RHS of the stage T equation in (7.29) yields the value function $W(T, x; \underline{\bar{u}}_{T-})$, which is an explicit function of $(x, \underline{\bar{u}}_{T-})$.

We move backward sequentially from stage $T-1$ to stage T_z in the interval $[T_z, T]$. Using $W(T, x; \underline{\bar{u}}_{T-})$, the maximization problem in stage $k \in \{T-1, T-2, \cdots, T_z\}$ becomes

$$W(k, x; \underline{\bar{u}}_{k-}) = \max_{\substack{u_k^i, \bar{u}_k^i \\ i \in S^{[T_z,T]}}} \left\{ \sum_{i \in [T_z,T]} g_k^i\left(x, \underline{u}_k^{[T_z,T]}, \underline{\bar{u}}_k^{[T_z,T]}; \underline{\bar{u}}_{k-}\right)\delta_1^k \right.$$

$$\left. + W\left[k+1, f_k\left(x, \underline{u}_k^{[T_z,T]}, \underline{\bar{u}}_k^{[T_z,T]}; \underline{\bar{u}}_{k-}\right); \underline{\bar{u}}_k^{[T_z,T]}, \underline{\bar{u}}_{(k+1)-} \cap \underline{\bar{u}}_{k-}\right]\right\},$$

$$\text{for } k \in \{T-1, T-2, \cdots, T_z\} \tag{7.30}$$

In the maximization operator in stage k, the control strategies are $(u_k^i, \bar{u}_k^i)$, $i \in S^{[T_z,T]}$. The state x and the previously executed strategies $\underline{\bar{u}}_{k-}$ appears as given. If the Eq. (7.30) satisfies the implicit theorem, the optimal control strategies can be obtained as functions of $(x, \underline{\bar{u}}_{k-})$. Substituting the optimal control strategies into (7.30) yields the value function $W(k, x; \underline{\bar{u}}_{k-})$, for $k \in \{T-1, T-2, \cdots, T_z\}$, which are explicit functions of $(x, \underline{\bar{u}}_{k-})$. Hence, the system of Eqs. (4.4) of Theorem 4.1 is vindicated.

Now, we move to the game interval $[T_t, T_{t+1}-1]$, for $t \in \{1, 2, \cdots, z_i - 1\}$. The set of players existing in the game interval $[T_t, T_{t+1}-1]$ is $S^{[T_t,T_{t+1}-1]}$. The set of players existing in the game interval $[T_t, T_{t+1}-1]$ and would leave the game after stage $T_{t+1}-1$ is $S^{[T_t,T_{t+1}-1]C}$. In addition, each player in $S^{[T_t,T_{t+1}-1]C}$ will receive a terminal payoff in stage T_{t+1}. The stage $T_{t+1}-1$ optimization problem in the $[T_t, T_{t+1}-1]$ interval becomes

$$W\left(T_{t+1}-1, x; \underline{\bar{u}}_{(T_{t+1}-1)-}\right)$$
$$= \max_{\substack{u^i_{T_{t+1}}-1,\, \bar{u}^i_{T+1}-1 \\ i\in S^{[T_t,T_{t+1}-1]}}} \left\{ \sum_{i\in S^{[T_t,T_{t+1}-1]}} g^i_{T_{t+1}-1}\left(x, \underline{u}^{[T_t,T_{t+1}-1]}_{T_{t+1}-1}, \underline{\bar{u}}^{[T_t,T_{t+1}-1]}_{T_{t+1}-1}, \underline{\bar{u}}_{(T_{t+1}-1)-}\right)\delta_1^{T_{t+1}-1}\right.$$
$$+ \sum_{j\in S^{[T_t,T_{t+1}-1]C}} q^j_{T_{t+1}}[f_{T_{t+1}-1}(x, \underline{u}^{[T_t,T_{t+1}-1]}_{T_{t+1}-1}, \underline{\bar{u}}^{[T_t,T_{t+1}-1]}_{T_{t+1}-1}$$
$$; \underline{\bar{u}}_{(T_{t+1}-1)-}); \underline{\bar{u}}^{[T_t,T_{t+1}-1]}_{T_{t+1}-1}, \underline{\bar{u}}_{(T_{t+1})-} \cap \underline{\bar{u}}_{(T_{t+1}-1)-}]\delta_1^{T_{t+1}}$$
$$+ W\Big[T_{t+1}, f_{T_{t+1}-1}\Big(x, \underline{u}^{[T_t,T_{t+1}-1]}_{T_{t+1}-1}, \underline{\bar{u}}^{[T_t,T_{t+1}-1]}_{T_{t+1}-1}, \underline{\bar{u}}_{(T_{t+1}-1)-}\Big);$$
$$\left.\underline{\bar{u}}^{[T_t,T_{t+1}-1]}_{T_{t+1}-1}, \underline{\bar{u}}_{(T_{t+1})-} \cap \underline{\bar{u}}_{(T_{t+1}-1)-}\Big]\right\}. \tag{7.31}$$

Note that in the maximization operator in stage $T_{t+1}-1$, the control strategies of player i are $(u^i_{T_{t+1}-1}, \bar{u}^i_{T_{t+1}-1})$, for $i \in S^{[T_t,T_{t+1}-1]}$. The state x and the previously executed strategies $\underline{\bar{u}}_{(T_{t+1}-1)-}$ appears as given. If the Eq. (7.31) satisfies the implicit theorem, the optimal strategies can be obtained as functions of $(x, \underline{\bar{u}}_{(T_{t+1}-1)-})$. Substituting the optimal strategies into (7.31) yields the value function $W(T_{t+1}-1, x; \underline{\bar{u}}_{(T_{t+1}-1)-})$, which is an explicit function of $(x, \underline{\bar{u}}_{(T_{t+1}-1)-})$. Hence Eq. (4.5) of Theorem 4.1 follows.

We move backward sequentially from stage $T_{t+1}-2$ to stage T_t in interval $[T_t, T_{t+1}-1]$. Using $W(T_{t+1}-1, x; \underline{\bar{u}}_{(T_{t+1}-1)-})$, the stage $k \in \{T_{t+1}-2, T_{T+1}-3, \cdots, T_t\}$ optimization problem can be expressed as

$$W(k, x; \underline{\bar{u}}_{k-}) = \max_{\substack{u^i_{k-1}, \bar{u}^i_k \\ i\in S^{[T_t,T_{t+1}-1]}}} \left\{ \sum_{i\in[T_t,T_{t+1}-1]} g^i_k\left(x, \underline{u}^{[T_t,T_{t+1}-1]}_k, \underline{\bar{u}}^{[T_t,T_{t+1}-1]}_k; \underline{\bar{u}}_{k-}\right)\delta_1^k\right.$$
$$\left. +W\Big[k+1, f_k\Big(x, \underline{u}^{[T_t,T_{t+1}-1]}_k, \underline{\bar{u}}^{[T_t,T_{t+1}-1]}_k; \underline{\bar{u}}_{k-}\Big); \underline{\bar{u}}^{[T_t,T_{t+1}-1]}_k, \underline{\bar{u}}_{(k+1)-} \cap \underline{\bar{u}}_{k-}\Big]\right\},$$

for $k \in \{T_{t+1}-2, T_{T+1}-3, \cdots, T_t\} \subset [T_t, T_{t+1}-1]$ and $t \in \{1, 2, \cdots, z_i-1\}$ (7.32)

In the maximization operator in stage k, the control strategies are $(u^i_k, \bar{u}^i_k)$, for $i \in S^{[T_t,T_{t+1}-1]}$. The state x and the previously executed strategies $\underline{\bar{u}}_{k-}$ appears as given. If the Eq. (7.32) satisfies the implicit theorem, the optimal control strategies can be obtained as functions of $(x, \underline{\bar{u}}_{k-})$. Substituting the optimal strategies into (7.32) yields the value function $W(k, x; \underline{\bar{u}}_{k-})$, which are explicit functions of $(x, \underline{\bar{u}}_{k-})$. Hence, Eq. (4.6) in Theorem 4.1 is vindicated. Hence Theorem 4.1 follows.

7.4   Appendix IV: Proof of Proposition 6.1

Performing the indicated maximization operator in Corollary 6.1 and invoking Proposition 6.1, the optimal strategies can be obtained as,
for $i \in S^{[T_t, T_{t+1}-1]}$ and $k \in [T_t, T_{t+1}-1]$:

$$u_k^{Ii*} = \frac{-\delta A_{k+1}^{(1)} b_k^i}{2c_k^{Ii}}(x^1)^{1/2}, \; u_k^{IIi*} = \frac{\delta A_{k+1}^{(2)} a_k^i}{2c_k^{IIi}},$$

$$\bar{u}_k^{i*} = \frac{R_k^i - \sum_{\theta=S[T_t, T_{t+1}-1]}\sum_{\eta=k}^{T_{t+1}-1}\delta^{\eta-k}v_\eta^{\theta|k|i} - \sum_{\rho=1}^{z-t}\sum_{\theta\in S}\left[T_{t+\rho}, T_{t+1+\rho}-1\right]\sum_{\eta=T_{t+\rho}}^{T_{t+\rho+1}-1}\delta^{\eta-k}v_\eta^{\theta|k|i}}{2\phi_k^i}$$

$$+ \frac{-\sum_{\eta=k}^{k+\omega-1}\delta^{\eta-k+1}A_{\eta+1}^{(1)}m_\eta^{|k|i}}{2\phi_k^i} \tag{7.33}$$

where $v_\eta^{\theta|k|i} = 0$, for $\eta > k + \omega - 1$ or $\eta > T$ and
$A_{\eta+1}^{(1)}m_\eta^{|k|i} = 0$, for $\eta > T$.
Using Propositions 6.1 and (6.4), we obtain

$$W(T+1, x; \underline{\bar{u}}_{(T+1)-}) = \left(A_{T+1}^{(1)}x^1 + A_{T+1}^{(2)}x^2 + C_{T+1}\right)\delta^T$$

$$= \sum_{i\in S^{[T_z, T]}}\left(Q_{T+1}^{(1)i}x^1 + Q_{T+1}^{(2)i}x^2 + \varpi_{T+1}^i\right)\delta^T. \tag{7.34}$$

Consider the last game interval $[T_z, T]$, Substituting the optimal strategies u_k^{Ii*}, u_k^{IIi*} and $\bar{u}_k^{i*}$, for $k \in [T_z, T]$ and $i \in S^{[T_z, T]}$ into (6.5) yields a system of equations which RHS and LHS are linear functions of x^1 and x^2. In particular, we obtain:
$W(k, x; \underline{\bar{u}}_{k-}) = (A_k^{(1)}x^1 + A_k^{(2)}x^2 + C_k)\delta^{k-1}$, for $k \in [T_z, T]$, where

$$A_k^{(1)} = -\sum_{\theta\in S^{[T_z, T]}}\left(\frac{\left(A_{k+1}^{(1)}b_k^\theta\delta\right)^2}{4c_k^{I\theta}} + h_k^\theta\right) + \delta A_{k+1}^{(1)}\left(1 + \sum_{\theta\in S^{[T_z, T]}}b_k^\theta\frac{A_{k+1}^{(1)}b_k^\theta\delta}{2c_k^{I\theta}} - \lambda_k^1\right),$$

$$A_k^{(2)} = \sum_{\theta\in S^{[T_z, T]}}\kappa_k^\theta + \delta 4_{k+1}^{(2)}(1 - \lambda_k^2) \text{ and} \tag{7.35}$$

C_k is an expression that includes previously executed durable strategies $\underline{\bar{u}}_{k-}^*$, for $k \in [T_z, T]$.

Substituting the optimal strategies u_k^{Ii*}, u_k^{IIi*} and $\bar{u}_k^{i*}$, for $k = T_{t+1} - 1$ and $i \in S^{[T_t, T_{T+1}-1]}$ into (6.6) yields an equation in which RHS and LHS are linear functions of x^1 and x^2. In particular, we obtain

$$W(T_{t+1} - 1), x; \underline{\bar{u}}_{(T_{t+1}-1)-}\Big) = \Big(A^{(1)}_{T_{t+1}-1} x^1 + A^{(2)}_{T_{t+1}-1} x^2 + C_{T_{t+1}-1}\Big)\delta^{k-1},$$

where

$$A^{(1)}_{T_{t+1}-1} = -\sum_{\theta \in S^{[T_t, T_{t+1}-1]}} \left(\frac{\left(A^{(1)}_{T_{t+1}} b^{\theta}_{T_{t+1}-1}\delta\right)^2}{4c^{I\theta}_{T_{t+1}-1}} + h^{\theta}_{T+1-1} \right)$$
$$+ \delta A^{(1)}_{T_{t+1}} \left(1 + \sum_{\theta \in S^{[T_t, T_{t+1}-1]}} b^{\theta}_{T_{t+1}-1} \frac{A^{(1)}_{T_{t+1}} b^{\theta}_{T_{t+1}-1}\delta}{2c^{I\theta}_{T_{t+1}-1}} - \lambda^1_{T_{t+1}-1} \right)$$
$$+ \sum_{i \in S^{[T_t, T_{t+1}-1]A}} \delta Q^{(1)i}_{T_{t+1}} \left(1 + \sum_{\theta \in S^{[T_t, T_{t+1}-1]}} b^{\theta}_{T_{t+1}-1} \frac{A^{(1)}_{T_{t+1}} b^{\theta}_{T_{t+1}-1}\delta}{2c^{I\theta}_{T_{t+1}-1}} - \lambda^1_{T_{t+1}-1} \right),$$
$$A^{(2)}_{T_{t+1}-1} = \sum_{\theta \in S^{[T_t, T_{t+1}-1]}} \kappa^{\theta}_{T_{t+1}-1} + \delta A^{(2)}_{T_{t+1}} \left(1 - \lambda^2_{T_{t+1}-1}\right), \tag{7.36}$$

and $C_{T_{t+1}-1}$ is an expression that includes previously executed durable strategies $\underline{\bar{u}}^*_{(T_{t+1}-1)-}$, for $t \in (1, 2, \cdots, z)$.

Consider the game interval which contains stage $k \in \{T_t, T_t + 1, \cdots, T_{t+1} - 2\}$. Substituting the optimal strategies u_k^{Ii*}, u_k^{IIi*} and $\bar{u}_k^{i*}$, for $k \in \{T_t, T_t + 1, \cdots, T_{t+1} - 2\}$ and $i \in S^{[T_t, T_{t+1}-1]}$ into (6.7) yields a system of equations which RHS and LHS are linear functions of x^1 and x^2. In particular, we obtain:
$W(k, x; \underline{\bar{u}}_{k-}) = (A_k^{(1)} x^1 + A_k^{(2)} x^2 + C_k)\delta^{k-1}$, for $k \in \{T_t, T_t + 1, \cdots, T_{t+1} - 2\}$ and $t \in (1, 2, \cdots, z)$, where

$$A_k^{(1)} = -\sum_{\theta \in S^{[T_t, T_{t+1}-1]}} \left(\frac{\left(A^{(1)}_{k+1} b^{\theta}_k \delta\right)^2}{4c^{I\theta}_k} + h^{\theta}_k \right) + \delta A^{(1)}_{k+1} \left(1 + \sum_{\theta \in S^{[T_t, T_{t+1}-1]}} b^{\theta}_k \frac{A^{(1)}_{k+1} b^{\theta}_k \delta}{2c^{I\theta}_k} - \lambda^1_k \right),$$
$$A_k^{(2)} = \sum_{\theta \in S^{[T_t, T_{t+1}-1]}} \kappa^{\theta}_k + \delta A^{(2)}_{k+1} \left(1 - \lambda^2_k\right),$$

and

C_k is an expression that includes previously executed durable strategies $\underline{\bar{u}}^*_{k-}$. Hence, Proposition 6.1 follows.

8 Chapter Notes

The Chapter presents a new class of dynamic games with asynchronous players' horizons and durable strategies. The non-cooperative game equilibrium is solved and the optimization technique for solving a cooperative optimal solution is derived. An illustration of asynchronous horizons durable strategies dynamic games in a problem which requires urgent attention—intergenerational environmental degradation—is provided. A cooperative solution with a compensatory scheme is proposed. Burton (1993) studied intertemporal preferences and intergenerational equity in resources extraction. Mourmouras (1993) examined government renewable resources policies in an overlapping generations model. Carrera and Moran (1995) presented general dynamics in overlapping generational models. Jørgensen and Yeung (1999) and Jørgensen and Yeung (2001) developed dynamic games of intergenerational and intragenerational renewable resource extraction. Jørgensen and Yeung (2005) provided a class of overlapping generations stochastic differential games. Yeung (2011) examined dynamically consistent cooperative solutions in differential games with asynchronous players' horizons. Yeung (2012) derived subgame consistent cooperative solutions in stochastic differential games with asynchronous horizons and uncertain future types of players. Sherstyuk et al. (2016) investigated intergenerational games with dynamic externalities and climate change experiments. Theorem 2.1 and Theorem 4.1 were developed by Yeung and Petrosyan (2021b).

9 Problems

(1) (Warmup exercise) Consider an asynchronous horizon dynamic durable strategies dynamic game involving public technology investments and pollution accumulation. There are 2 agents with different operating horizons. Agent 1 will start its operation in game from stage 1 till the end of the game in stage 4. Agent 2 starts having operations in the initial stage 3 till stage 6.

The agents produce an output $\bar{u}_k^{(3)j}$ that emits pollutants to add to the pollution stock over 3 stages.

The level of pollution stocks is $x_k \in X \subset R$ and the dynamics of pollution accumulation is governed by

$$x_{k+1} = x_k + \bar{u}_k^{(3)1} + \sum_{\tau=1}^{k-1} 0.5\bar{u}_\tau^{(3)1} - u_k^1(x_k)^{1/2} - 0.05x_k, \quad x_1 = 70,$$

for $k \in \{1, 2\}$;

$$x_{k+1} = x_k + \sum_{\ell\in\{1,2\}} \bar{u}_k^{(3)\ell} + \sum_{\tau=k-2}^{k-1} 0.5\bar{u}_\tau^{(3)1} + \sum_{\tau=3}^{k-1} 0.5\bar{u}_\tau^{(3)2} - \sum_{\ell\in\{1,2\}} u_k^\ell(x_k)^{1/2} - 0.05x_k,$$

for $k \in \{3, 4\}$;

$$x_{k+1} = x_k + \bar{u}_k^{(3)2} + \sum_{\tau=k-2}^{k-1} 0.05\bar{u}_\tau^{(3)1} + \sum_{\tau=k-2}^{k-1} 0.5\bar{u}_\tau^{(3)2} - u_k^2(x_k)^{1/2} - 0.05x_k,$$

for $k \in \{5, 6\}$;

where $u_k^j(x_k)^{1/2}$ is the amount of pollutants removed by u_k^j amounts of abatement effort from agent $j \in \{1, 2\}$ and $0.05x_k$ is the natural decay of the pollutants.

Moreover, $\bar{u}_k^{(3)1} = 0$ for $k < 1$ and $k > 4$ and $\bar{u}_k^{(3)2} = 0$ for $k < 3$. The cost of abatement effort is $2(u_k^j)^2$, for $j \in \{1, 2\}$. The pollution damage brought about by the pollution stock in stage k to agent 1 is $1.5x_k$ and is x_k to agent 2.

The payoff of agent 1 is

$$\sum_{k=1}^{4}\left(3\bar{u}_k^{(3)1} - \left(\bar{u}_k^{(3)1}\right)^2 - 2\left(u_k^1\right)^2 - 1.5x_k\right)0.9^{k-1} + (-x_5 + 70)0.9^4.$$

The present value (viewed at stage 1) of agent 2's payoff is

$$\sum_{k=3}^{6}\left(4\bar{u}_k^{(3)1} - \left(\bar{u}_k^{(3)1}\right)^2 - \left(u_k^1\right)^2 - x_k\right)0.9^{k-1} + (-1.5x_7 + 60)0.9^6$$

Obtain a non-cooperative game equilibrium.

(2) Consider an asynchronous horizon dynamic durable strategies dynamic game involving public technology investments and pollution accumulation. There are 4 agents with different operating horizons. Agent 1 will start its operation in game from stage 1 till the end of the game in stage 12. Agent 2 will start in the initial stage 1 till stage 6. Agent 3 will join the game in stage 7 and exit at stage 10. Agent 4 will join the game from stage 11 to the end of the game in stage 12.

The agents produce an output $\bar{u}_k^{(8)j}$ that emits pollutants to add to the pollution stock over 8 stages. Worthwhile to point out is that industrial output often emits pollutants over multiple stages. Toxic and chemical wastes from industrial production will continue to produce damaging elements that add to the pollution stock for a lengthy period.

The level of pollution stocks is $x_k \in X \subset R$ and the dynamics of pollution accumulation is governed by.

$$x_{k+1} = x_k + \sum_{\ell\in\{1,2\}} m_k^{|k|\ell}\bar{u}_k^{(8)\ell} + \sum_{l\in\{1,2\}}\sum_{\tau=1}^{k-1} m_k^{|\tau|\ell}\bar{u}_\tau^{(8)\ell}$$

$$-\sum_{l\in\{1,2\}} b_k^\ell u_k^\ell(x_k)^{1/2} - \lambda_k x_k, \quad x_1 = x_1^0,$$

for $k \in \{1, 2, \cdots 6\}$;

$$x_{k+1} = x_k + \sum_{\ell\in\{1,3\}} m_k^{|k|\ell}\bar{u}_k^{(8)\ell} + \sum_{\tau=k-7}^{k-1} m_k^{|\tau|1}\bar{u}_\tau^{(8)1} + \sum_{\tau=7}^{k-1} m_k^{|\tau|3}\bar{u}_\tau^{(8)3} + \sum_{\tau=k-7}^{6} m_k^{|\tau|2}\bar{u}_\tau^{(8)2}$$

$$-\sum_{\ell\in\{1,3\}} b_k^\ell u_k^\ell(x_k)^{1/2} - \lambda_k x_k \text{ and for } k \in \{7, 8, 9, 10\};$$

$$x_{k+1} = x_k + \sum_{\ell\in\{1,4\}} m_k^{|k|\ell}\bar{u}_k^{(8)\ell} + \sum_{\tau=k-7}^{k-1} m_k^{|\tau|1}\bar{u}_\tau^{(8)1} + \sum_{\tau=11}^{k-1} m_k^{|\tau|4}\bar{u}_\tau^{(8)4} + \sum_{\tau=k-7}^{6} m_k^{|\tau|2}\bar{u}_\tau^{(8)2}$$

$$+\sum_{\tau=7}^{10} m_k^{|\tau|3}\bar{u}_\tau^{(8)3} - \sum_{\ell\in\{1,4\}} b_k^\ell u_k^\ell(x_k)^{1/2} - \lambda_k x_k \text{ for } k \in \{11, 12\};$$

where $m_k^{|k|j}\bar{u}_k^{(8)j}$ is the amount of pollutants created in stage k by agent j's stage k polluting output $\bar{u}_k^{(8)j}$, $m_k^{|\tau|j}\bar{u}_\tau^{(8)j}$ is the amount of pollutants created in stage k by agent j's stage $\tau \in \{k-1, k-2, \cdots, k-7\}$ output, $b_k^j u_k^j(x_k)^{1/2}$ is the amount of pollutants removed by u_k^j amounts of abatement effort from agent j; $\lambda_k x_k$ is the natural decay of the pollutants.

Moreover, $\bar{u}_k^{(8)1} = 0$ for $k < 1$; $\bar{u}_k^{(8)2} = 0$ for $k < 1$ and $k > 6$; $\bar{u}_k^{(8)3} = 0$ for $k < 7$ and $k > 10$; and $\bar{u}_k^{(8)4} = 0$ for $k < 11$.

In particular, $R_k^j\bar{u}_k^{(8)j}$ is agent j's gross revenue from outputs creating pollution and $\phi_k^j(\bar{u}_k^{(8)j})^2$ is the cost of producing the output and the cost of pollution abatement are $c_k^j(u_k^j)^2$. The pollution damage brought about by the pollution stock in stage k to agent j is $h_k^j x_k$.

The payoff of agent 1 is

$$\sum_{k=1}^{12}\left(R_k^1\bar{u}_k^{(8)1} - \phi_k^1\left(\bar{u}_k^{(8)1}\right)^2 - c_k^1\left(u_k^1\right)^2 - h_k^1 x_k\right)\delta^{k-1} + (Q_{13}^1 x_{13} + \varpi_{13}^1)\delta^{12},$$

where $\left(Q_{13}^1 x_{13} + \varpi_{13}^1\right)$ is the terminal payoff of agent 1 in stage 13 and $\delta = (1+r)^{-1}$ is the discount factor and Q_{13}^1 is non-positive.

The payoff of agent 2 is

$$\sum_{k=1}^{6}\left(R_k^2\bar{u}_k^{(8)2} - \phi_k^2\left(\bar{u}_k^{(8)2}\right)^2 - c_k^2\left(u_k^2\right)^2 - h_k^2 x_k\right)\delta^{k-1} + (Q_7^2 x_7 + \varpi_7^2)\delta^6,$$

where $\left(Q_7^2 x_7 + \varpi_7^2 \right)$ is the terminal payoff of agent 2 in stage 7 and Q_7^2 is non-positive.

The present value (viewed at stage 1) of agent 3's payoff is

$$\sum_{k=7}^{10} \left(R_k^3 \bar{u}_k^{(8)3} - \phi_k^3 \left(\bar{u}_k^{(8)3} \right)^2 - c_k^3 \left(u_k^3 \right)^2 - h_k^3 x_k \right) \delta^{k-1} + (Q_{11}^3 x_{11} + \sigma_{11}^3) \delta^{10},$$

where $\left(Q_{11}^3 x_{11} + \varpi_{11}^3 \right)$ is the terminal payoff of agent 3 in stage 11 and Q_{11}^3 is non-positive.

The present value (viewed at stage 1) of agent 4's payoff is

$$\sum_{k=11}^{12} \left(R_k^4 \bar{u}_k^{(8)4} - \phi_k^4 \left(\bar{u}_k^{(8)4} \right)^2 - c_k^4 \left(u_k^4 \right)^2 - h_k^4 x_k \right) \delta^{k-1} + (Q_{13}^4 x_{13} + \varpi_{13}^4) \delta^{12},$$

where $\left(Q_{13}^4 x_{13} + \varpi_{13}^4 \right)$ is the terminal payoff of agent 4 in stage 13 and Q_{13}^4 is non-positive.

Characterize a non-cooperative game equilibrium.

(3) Now, the agents wish to enhance their payoff through cooperation.

(i) Derive the group optimal joint payoff.

(ii) Describe the agents' imputations if they agree to adopt the compensatory scheme in Sect. 5.

References

Burton P (1993) Intertemporal preferences and intergenerational equity considerations in optimal resources harvesting. J Environ Econ Manag 24:132

Carrera C, Moran M (1995) General dynamics in overlapping generational models. J Econ Dyn Control 19:813–830

Jørgensen S, Yeung DWK (1999) Inter- and intragenerational renewable resource extraction. Ann Oper Res 88:275–289

Jørgensen S, Yeung DWK (2001) Cooperative solution of a game of intergenerational renewable resource extraction. Int J Math Game Theory Algebra 11:45–64

Jørgensen S, Yeung DWK (2005) An overlapping generations stochastic differential games. Automatica 41:69–74

Mourmouras A (1993) Conservationist government policies and intergenerational equity in an overlapping generations model with renewable resources. J Public Econ 51(2):249–268

Petrosyan LA, Yeung DWK (2021) Shapley value for differential network games: theory and application. J Dynam Games 8(2):151–166. https://doi.org/10.3934/jdg.2020021

Sherstyuk K, Tarui N, Ravago MV, Saijo T (2016) Intergenerational games with dynamic externalities and climate change experiments. J Assoc Environ Resour Econ 3(2):247–281
Yeung DWK (2012) Subgame consistent cooperative solutions in stochastic differential games with asynchronous horizons and uncertain types of players. Contrib Game Theory Manage. 5:334–355
Yeung DWK (2011) Dynamically consistent cooperative solutions in differential games with asynchronous players' horizons. Ann Internat Soc Dynam Games 11:375–395
Yeung DWK, Petrosyan LA (2010) Subgame consistent solutions for cooperative stochastic dynamic games. J Optim Theory Appl 145:579–596
Yeung DWK, Petrosyan LA (2016a) A cooperative dynamic environmental game of subgame consistent clean technology development. Int Game Theo Rev 18(2):164008.01–164008.23
Yeung DWK, Petrosyan LA (2019) Cooperative dynamic games with control lags. Dynam Games Appl 9(2):550–567. https://doi.org/10.1007/s13235-018-0266-6

Chapter 9
Stochastic Durable-Strategies Dynamic Games

An essential characteristic of time—and hence decision-making over time—is that although the individual may invest in gathering past and current information, random future events are unavoidable. An empirically meaningful theory must therefore incorporate these uncertainties. This Chapter considers stochastic durable strategies dynamic games. A new class of durable strategies dynamic games of the Petrosyan and Yeung (2020a) paradigm with stochastic state dynamics and randomly furcating payoffs is formulated. Section 1 presents a dynamic optimization theorem involving durable strategies problems with stochastic elements appearing in the state dynamics and payoff structures. Section 2 gives a computational illustration of the derivation of the corresponding optimal strategies. Section 3 provides a general class of durable strategies dynamic games with stochastic state dynamics and random payoffs. The game equilibrium solution and the corresponding Hamilton–Jacobi-Bellman (HJB) equations are demonstrated. Section 8 gives an illustrative example of the computation details. Section 5 presents a general class of cooperative stochastic durable strategies dynamic game with randomly furcating payoffs. The Pareto optimal solution and conditions that guarantee individual rationality are obtained. Dynamically consistent solutions and the corresponding payment mechanism are derived. Section 6 gives an illustrative cooperative game in public capital provision. An extension to include previously executed durable strategies in the stochastic part of the state dynamics is given in Sect. 7. Mathematical appendices are relegated to Sect. 8. Chapter notes are given in Sect. 9. Problem sets are supplied in Sect. 10.

1 Stochastic Optimization with Durable Strategies

In this section, we develop an optimization technique for solving durable strategies dynamic problems with stochastic state dynamics and payoffs. We first present the basic formulation of the problem and then develop the techniques for obtaining an optimal solution.

D. W. K. Yeung and L. A. Petrosyan, *Durable-Strategies Dynamic Games*,
Theory and Decision Library C 50, https://doi.org/10.1007/978-3-030-92742-4_9

1.1 Problem Formulation

Consider a general T-stage stochastic dynamic optimization problem in which there exist non-durable strategies and durable strategies of different lag durations. The durable strategies and stochastic elements affect both the payoff and the state dynamics. We use $u_k \in U \subset R^m$ to denote the set of non-durable control strategies and $\overline{u}_k = (\overline{u}_k^{(2)}, \overline{u}_k^{(3)}, \ldots, \overline{u}_k^{(\omega)})$ to denote the set of durable control strategies, where $\overline{u}_k^{(\zeta)} \in \overline{U}^\zeta \subset R^{m_\zeta}$ for $\zeta \in \{2, 3, \ldots, \omega\}$. In particular, the strategies $\overline{u}_k^{(2)}$ are durable strategies that have effects in stages k and $k+1$. The strategies $\overline{u}_k^{(3)}$ are durable strategies that have effects within stages k, $k+1$ and $k+2$. The strategies $u_k^{(\omega)}$ are durable strategies that have effects within the duration from stage k to stage $k+\omega-1$. The state at stage k is $x_k \in X \subset R^m$ and $\overline{u}_{k-}$ is the set of durable controls which are executed before stage k but still in effect in stage k.

The evolution of the state is subjected to random shocks and the state dynamics is governed by

$$x_{k+1} = f_k(x_k, u_k, \overline{u}_k; \overline{u}_{k-}) + \vartheta_k G_k(x_k, u_k, \overline{u}_k), \quad x_1 = x_1^0, \tag{1.1}$$

where ϑ_k is a random variable with range $\{\vartheta_k^1, \vartheta_k^2, \ldots, \vartheta_k^{\mu_k}\}$ and corresponding probabilities $\{\gamma_k^1, \gamma_k^2, \ldots, \gamma_k^{\mu_k}\}$.

The single-stage payoff of the decision-maker at stage k is $g_k(\theta_k; x_k, u_k, \overline{u}_k; \overline{u}_{k-})$ which is affected by a random variable θ_k. In particular, θ_k for $k \in \{1, 2, \ldots, T+1\}$ are independent discrete random variables with range $\{\theta_k^1, \theta_k^2, \ldots, \theta_k^{\eta_k}\}$ and corresponding probabilities $\{\lambda_k^1, \lambda_k^2, \ldots, \lambda_k^{\eta_k}\}$, where η_k is a positive integer for $k \in \{1, 2, \ldots, T+1\}$. In stage 1, it is known that θ_1 equals θ_1^1 with probability $\lambda_1^1 = 1$. Moreover, the random variable θ_k is known in stage k.

The expected payoff to be maximized by the decision-maker becomes

$$E_{\theta_2,\theta_3,\ldots,\theta_{T+1};\vartheta_1,\vartheta_2,\ldots,\vartheta_T} \left\{ \sum_{k=1}^{T} g_k(\theta_k; x_k, u_k, \overline{u}_k; \overline{u}_{k-})\delta_1^k \right.$$
$$\left. + q_{T+1}(\theta_{T+1}; x_{T+1}; \overline{u}_{(T+1)-})\delta_1^{T+1} \right\}, \tag{1.2}$$

where $E_{\theta_2,\theta_3,\ldots,\theta_{T+1};\vartheta_1,\vartheta_2,\ldots,\vartheta_T}$ is the expectation operator with respect to the statistics of $\theta_2, \theta_3, \ldots, \theta_{T+1}$ and $\vartheta_1, \vartheta_2, \ldots, \vartheta_T$, the term $q_{T+1}(\theta_{T+1}; x_{T+1}; \overline{u}_{(T+1)-})$ is the terminal payoff at stage $T+1$ and δ_1^k is the discount factor from stage 1 to stage k.

The durable controls executed before the start of the operation in stage 1, that is $\overline{u}_{1-}$, are known and some or all of them can be zeros. The functions $g_k(\theta_k; x_k, u_k, \overline{u}_k; \overline{u}_{k-})$, $f_k(x_k, u_k, \overline{u}_k; \overline{u}_{k-})$, $q_{T+1}(\theta_{T+1}; x_{T+1}; \overline{u}_{(T+1)-})$ and $G_k(u_k, \overline{u}_k, x_k)$ are differentiable in $(x_k, u_k, \overline{u}_k)$.

1.2 Optimization Methodology

A solution theorem for obtaining the optimal control strategies in the stochastic dynamic optimization problem (1.1)–(1.2) can be characterized as follows.

Theorem 1.1 *Durable-strategies Dynamic Optimization*

Let $V(\theta_k^{\sigma_k}; k, x; \overline{u}_{k-})$, for $\sigma_k \in \{1, 2, , \ldots, \eta_k\}$ and $k \in \{1, 2, \ldots, T\}$, be the maximal value of the expected payoff

$$E_{\theta_{k+1},\theta_{k+2},\ldots,\theta_{T+1};\vartheta_k,\vartheta_{k+1},\ldots,\vartheta_T}\left\{\sum_{t=k}^{T} g_t(\theta_t; x_t, u_t, \overline{u}_k; \overline{u}_{t-})\delta_1^t + q_{T+1}(\theta_{T+1}; x_{T+1}; \overline{u}_{(T+1)-})\delta_1^{T+1}\right\}$$

for problem (1.1)–(1.2) starting at stage k with state $x_k = x, \theta_k = \theta_k^{\sigma_k}$ and previously executed controls $\overline{u}_{k-}$, then the function $V(\theta_k^{\sigma_k}; k, x; \overline{u}_{k-})$, for $\sigma_k \in \{1, 2, \ldots, \eta_k\}$ and $k \in \{1, 2, \ldots, T\}$, satisfies the following system of recursive equations:

$$V(\theta_{T+1}^{\sigma_{T+1}}; T+1, x; \overline{u}_{(T+1)-}) = q_{T+1}(\theta_{T+1}^{\sigma_{T+1}}; x_{T+1}; \overline{u}_{(T+1)-})\delta_1^{T+1}, \text{ for } \sigma_{T+1} \in \{1, 2, , \ldots, \eta_{T+1}\}, \tag{1.3}$$

$$\begin{aligned} V(\theta_k^{\sigma_k}; k, x; \overline{u}_{k-}) &= \max_{u_k, \overline{u}_k} E_{\theta_{k+1},\vartheta_k}\Big\{ g_k(\theta_k^{\sigma_k}; x, u_k, \overline{u}_k; \overline{u}_{k-})\delta_1^k + V[\theta_{k+1}; k+1, f_k(x, u_k, \overline{u}_k; \overline{u}_{k-}) \\ &\quad + \vartheta_k G_k(x, u_k, \overline{u}_k); \overline{u}_k, \overline{u}_{(k+1)-} \cap \overline{u}_{k-}]\Big\} \\ &= \max_{u_k, \overline{u}_k}\Big\{ \sum_{\upsilon=1}^{\eta_{k+1}} \lambda_{k+1}^{\upsilon} \sum_{\nu=1}^{\mu_k} \gamma_k^{\nu}\Big[g_k^{\sigma_k}(\theta_k; x, u_k, \overline{u}_k; \overline{u}_{k-})\delta_1^k \\ &\quad + V[\theta_{k+1}^{\upsilon}; k+1, f_k(x, u_k, \overline{u}_k; \overline{u}_{k-}) + \vartheta_k^{\nu} G_k(x, u_k, \overline{u}_k); \overline{u}_k, \overline{u}_{(k+1)-} \cap \overline{u}_{k-}]\Big]\Big\} \end{aligned}$$

for $\sigma_k \in \{1, 2, \ldots, \eta_k\}$ and $k \in \{1, 2, \ldots, T\}$. (1.4)

Proof To prove Theorem 1.1, we adopt the technique of backward induction. Consider first the last operational stage T, invoking Theorem 1.1 we have.

$$V(\theta_T^{\sigma_T}; T, x; \overline{u}_{T-})$$
$$= \max_{u_T, \overline{u}_T} \left\{ \sum_{\upsilon=1}^{\eta_{T+1}} \lambda_{T+1}^{\upsilon} \sum_{\upsilon=1}^{\mu_T} \gamma_T^{\upsilon} \Bigg[g_T(\theta_T^{\sigma_T}; x, u_T, \overline{u}_T; \overline{u}_{T-}) \delta_1^T \right.$$
$$\left. + q_{T+1}[\theta_{T+1}^{\upsilon}; f_T(x, u_T, \overline{u}_T; \overline{u}_{T-}) + \vartheta_T^{\upsilon} G_T(x, u_T, \overline{u}_T); \overline{u}_T, \overline{u}_{(T+1)-} \cap \overline{u}_{T-}] \Bigg] \right\}, \tag{1.5}$$

for $\sigma_T \in \{1, 2, \ldots, \eta_T\}$.

The maximization operator in stage T involves u_T and $\overline{u}_T$ only. The current state x and the previously executed controls $\overline{u}_{T-}$ appear in the stage T maximization problem as given parameters. If the first order conditions of the maximization problem in (1.5) satisfy the implicit function theorem, one can obtain the optimal controls u_T and $\overline{u}_T$ as functions of x and $\overline{u}_{T-}$. Substituting these optimal controls into the function on the RHS of (1.5) yields the function $V(\theta_T^{\sigma_T}; T, x; \overline{u}_{T-})$, which satisfies the optimal conditions of a maximum for given x and $\overline{u}_{T-}$.

Consider the second last operational stage $T-1$, using $V(\theta_T^{\sigma_T}; T, x; \overline{u}_{T-})$, for $\sigma_T \in \{1, 2, , \ldots, \eta_T\}$, derived from (1.5) and invoking Theorem 1.1 we have

$$V(\theta_{T-1}^{\sigma_{T-1}}; T-1, x; \overline{u}_{(T-1)-})$$
$$= \max_{u_{T-1}, \overline{u}_{T-1}} \left\{ \sum_{\upsilon=1}^{\eta_T} \lambda_T^{\upsilon} \sum_{\upsilon=1}^{\mu_{T-1}} \gamma_{T-1}^{\upsilon} \Bigg[g_{T-1}(\theta_{T-1}^{\sigma_{T-1}}; x, u_{T-1}, \overline{u}_{T-1}; \overline{u}_{(T-1)-}) \delta_1^{T-1} \right.$$
$$+ V[\theta_T^{\upsilon}; T, f_{T-1}(x, u_{T-1}, \overline{u}_{T-1}; \overline{u}_{(T-1)-})$$
$$\left. + \vartheta_{T-1}^{\upsilon} G_{T-1}(x, u_{T-1}, \overline{u}_{T-1}); \overline{u}_{T-1}, \overline{u}_{T-} \cap \overline{u}_{(T-1)-}] \Bigg] \right\}, \tag{1.6}$$

for $\sigma_{T-1} \in \{1, 2, , \ldots, \eta_{T-1}\}$.

The maximization operator in stage $T-1$ involves u_{T-1} and $\overline{u}_{T-1}$. The current state x and the previously executed controls $\overline{u}_{(T-1)-}$ appear in the stage $T-1$ maximization problem as given parameters. If the first order conditions of the maximization problem in (1.6) satisfy the implicit function theorem, one can obtain the optimal controls u_{T-1} and $\overline{u}_{T-1}$ as functions of x and previously determined controls $\overline{u}_{(T-1)-}$. Substituting these optimal controls into the function on the RHS of (1.6) yields the function $V(\theta_{T-1}^{\sigma_{T-1}}; T-1, x; \overline{u}_{(T-1)-})$.

Now consider stage $k \in \{T-2, T-3, \ldots, 2, 1\}$, invoking Theorem 1.1 we have

$$V(\theta_k^{\sigma_k}; k, x; \overline{u}_{k-})$$
$$= \max_{u_k, \overline{u}_k} \left\{ \sum_{\upsilon=1}^{\eta_{k+1}} \lambda_{k+1}^{\upsilon} \sum_{\upsilon=1}^{\mu_k} \gamma_k^{\upsilon} \Bigg[g_k(\theta_k^{\sigma_k}; x, u_k, \overline{u}_k; \overline{u}_{k-}) \delta_1^k \right.$$

$$+ V[\theta_{k+1}^{\upsilon}; k+1, f_k(x, u_k, \overline{u}_k; \overline{u}_{k-}) + \vartheta_k^{\upsilon} G_k(x, u_k, \overline{u}_k); \overline{u}_k, \overline{u}_{(k+1)-} \cap \overline{u}_{k-}] \Bigg] \Bigg\}. \tag{1.7}$$

The maximization operator involves u_k and $\overline{u}_k$. Again, the current state x and the previously executed controls $\overline{u}_{k-}$ appear in the stage k optimization problem. If the first order conditions of the maximization problem in (1.7) satisfy the implicit function theorem, one can obtain the optimal controls u_k and $\overline{u}_k$ as functions of x and $\overline{u}_{k-}$. Substituting these optimal controls into the function on the RHS of (1.7) yields the function $V(\theta_k^{\sigma_k}; k, x; \overline{u}_{k-})$. ■

Theorem 1.1 is a stochastic version of the optimization theorem for durable controls established by Petrosyan and Yeung (2020a). It yields a new optimization technique that can be used to solve stochastic durable control problems with lagged effects in both the payoffs and state dynamics of the decision-maker.

2 A Computational Illustration

Consider a monopoly in which planning horizon involves T stages. It uses knowledge-based capital $x_k \in X \subset R^+$ to produce its output. The build-up of capital stock through investment activities takes time. The accumulation process of the capital stock is governed by the dynamics

$$x_{k+1} = x_k + a_k^{|k|}\overline{u}_k^{(\omega)} + \sum_{\tau=k-\omega+1}^{k-1} a_k^{|\tau|}\overline{u}_\tau^{(\omega)} - \lambda_k x_k + \vartheta_k(G_k^1 x_k + G_k^2 \overline{u}_k^{(\omega)}) \quad , x_1 = x_1^0. \tag{2.1}$$

where $u_k^{(\omega)}$ is the investment in the knowledge-based capital in stage k which is effective within ω stages, $a_k^{|k|}\overline{u}_k^{(\omega)}$ is the addition to the capital in stage k by capital building investment $\overline{u}_k^{(\omega)}$ in stage k, $a_k^{|\tau|}\overline{u}_\tau^{(\omega)}$ is the addition to the capital in stage k by capital building investment $\overline{u}_\tau^{(\omega)}$ in stage $\tau \in \{k-1, k-2, \ldots, k-\omega+1\}$ and λ_k is the rate of obsolescence of the capital.

Moreover, G_k^1 and G_k^2 are positive parameters and ϑ_k is random variable with a range $\{\vartheta_k^1, \vartheta_k^2, \ldots, \vartheta_k^{\mu_k}\}$ and corresponding probabilities $\{\gamma_k^1, \gamma_k^2, \ldots, \gamma_k^{\mu_k}\}$. Let $\vartheta_k^{\hat{\upsilon}}$ denote the smallest (positive or negative) value of ϑ_k^{υ}, for $\upsilon \in \{1, 2, \ldots, \mu_k\}$. The value of G_k^1 satisfies the condition $\vartheta_k^{\hat{\upsilon}} G_k^1 > (\lambda_k - 1)$. The value of G_k^2 satisfies the condition $\vartheta_k^{\hat{\upsilon}} G_k^2 > -a_k^{|k|}$.

The firm uses the knowledge-based capital together with a non-durable productive input u_k to produce outputs. The non-durable input cost is $c_k u_k$ and the cost of capital investment is $\phi_k(\overline{u}_k^{(\omega)})^2$. The revenue from the output produced is $\theta_k R_k(u_k)^{1/2}(x_k)^{1/2}$, where θ_k is a random variable with range $\{\theta_k^1, \theta_k^2, \ldots, \theta_k^{\eta_k}\}$ and corresponding probabilities $\{\lambda_k^1, \lambda_k^2, \ldots, \lambda_k^{\eta_k}\}$, for $k \in \{1, 2, \ldots, T+1\}$. In stage 1, it is known that θ_1

equals θ_1^1 with probability $\lambda_1^1 = 1$. Moreover, the random variable θ_k is known in stage k.

The expected payoff of the firm is

$$E_{\theta_2,\theta_3,\ldots,\theta_{T+1};\vartheta_1,\vartheta_2,\ldots,\vartheta_T}\left\{\sum_{k=1}^{T}\left([\theta_k R_k(u_k)^{1/2}(x_k)^{1/2} - c_k u_k - \phi_k(\overline{u}_k^{(\omega)})^2]\right)\delta^{k-1}\right.$$
$$\left. +\left(\theta_{T+1}Q_{T+1}x_{T+1} + \sum_{\tau=T+1-\omega+1}^{T} v_{T+1}^{|\tau|}\overline{u}_\tau^{(\omega)} + \varpi_{T+1}\right)\delta^T\right\}, \tag{2.2}$$

where $\left(\theta_{T+1}Q_{T+1}x_{T+1} + \sum_{\tau=T+1-\omega+1}^{T} v_{T+1}^{|\tau|}\overline{u}_\tau^{(\omega)} + \varpi_{T+1}\right)$ is the salvage value of the firm in stage $T+1$ and $\delta = (1+r)^{-1}$ is the discount factor.

In particular, the salvage value of the firm in stage $T+1$ depends on the capital stock of the firm and Q_{T+1} is positive. It also depends on the previous capital investment $\sum_{\tau=T+1-\omega+1}^{T} v_{T+1}^{|\tau|}\overline{u}_\tau^{(\omega)}$. In particular, some or all $v_{T+1}^{|\tau|}$, for $\tau \in \{T, T-1, \ldots, T-\omega+2\}$, can be zero.

Invoking Theorem 1.1, we obtain.

Corollary 2.1. *Let $V(\theta_k^{\sigma_k}; k, x; \overline{u}_{k-}^{(\omega)})$ be the maximal value of the expected payoff.*

$$E_{\theta_{k+1},\theta_{k+2},\ldots,\theta_{T+1};\vartheta_k,\vartheta_{k+1},\ldots,\vartheta_T}\left\{\sum_{t=k}^{T}\left([\theta_t^{\sigma_t} R_t(u_t)^{1/2}(x_t)^{1/2} - c_t u_t - \phi_t(\overline{u}_t^{(\omega)})^2]\right)\delta^{k-1}\right.$$
$$\left. +\left(\theta_{T+1}^{\sigma_{T+1}}Q_{T+1}x_{T+1} + \sum_{\tau=T+1-\omega+1}^{T} v_{T+1}^{|\tau|}\overline{u}_\tau^{(\omega)} + \varpi_{T+1}\right)\delta^T,\right.$$

for $\sigma_{T+1} \in \{1, 2, \ldots, \eta_{T+1}\}$.

for the stochastic dynamic optimization problem (2.1)–(2.2) starting at stage k with state $x_k = x$ and previously executed controls $\overline{u}_{k-}^{(\omega)} = (\overline{u}_{k-1}^{(\omega)}, \overline{u}_{k-2}^{(\omega)}, \ldots, \overline{u}_{k-\omega+1}^{(\omega)})$, then the function $V(\theta_k^{\sigma_k}; k, x; \overline{u}_{k-}^{(\omega)})$ satisfies the following system of recursive equations:

$$V(\theta_{T+1}^{\sigma_{T+1}}; T+1, x; \bar{u}_{(T+1)-}^{(\omega)})$$
$$= \left(\theta_{T+1}^{\sigma_{T+1}}Q_{T+1}x_{T+1} + \sum_{\tau=T+1-\omega+1}^{T} v_{T+1}^{|\tau|}\bar{u}_\tau^{(\omega)} + \varpi_{T+1}\right)\delta^T, \text{ for ;} \tag{2.3}$$

$$V(\theta_k^{\sigma_k}; k, x; \overline{u}_{k-}^{(\omega)}) = \max_{u_k,\overline{u}_k^{(\omega)}}\left\{[\theta_k^{\sigma_k} R_k(u_k)^{1/2}(x)^{1/2} - c_k u_k - \phi_k(\overline{u}_k^{(\omega)})^2]\delta^{k-1}\right.$$

$$+\sum_{\upsilon=1}^{\eta_{k+1}} \lambda_{k+1}^{\upsilon} \sum_{\upsilon=1}^{\mu_k} \gamma_k^{\upsilon} \Bigg[V[\theta_{k+1}^{\upsilon}; k+1, x + a_k^{|k|} \overline{u}_k^{(\omega)} + \sum_{\tau=k-\omega+1}^{k-1} a_k^{|\tau|} \overline{u}_\tau^{(\omega)} - \lambda_k x$$

$$+ \vartheta_k^{\upsilon}(G_k^1 x + G_k^2 \overline{u}_k^{(\omega)}); \overline{u}_k^{(\omega)}, \overline{u}_{(k+1)-}^{(\omega)} \cap \overline{u}_{k-}^{(\omega)}] \Bigg]\Bigg\},$$

$$\text{for } \sigma_k \in \{1, 2, \ldots, \eta_k\} \text{ and } k \in \{1, 2, \ldots, T\}. \tag{2.4}$$

■

Performing the indicated maximization operator in Corollary 2.1, the value function $V(\theta_k^{\sigma_k} k, x; \overline{u}_{k-}^{(\omega)})$ which reflects the expected value of the firm covering stage $k \in \{1, 2, \ldots, T\}$ to stage $T + 1$ can be obtained as follows.

Proposition 2.1 *System* (2.3)–(2.4) *admits a solution with the optimal payoff of the firm being.*

$$V(\theta_k^{\sigma_k} k, x; \overline{u}_{k-}^{(\omega)}) = (A_k^{\sigma_k} x + C_k^{\sigma_k})\delta^{k-1},$$

$$\text{for } \sigma_k \in \{1, 2, \ldots, \eta_k\} \text{ and } k \in \{1, 2, \ldots, T\}; \tag{2.5}$$

where, $A_{T+1}^{\sigma_{T+1}} = \theta_{T+1}^{\sigma_{T+1}} Q_{T+1}$ and $C_{T+1}^{\sigma_{T+1}} = \sum_{\tau=T+1-\omega+1}^{T} \upsilon_{T+1}^{|\tau|} \overline{u}_\tau^{(\omega)} + \varpi_{T+1}$,, for $\sigma_{T+1} \in \{1, 2, \ldots, \eta_{T+1}\}$

$$A_k^{\sigma_k} = \frac{(\theta_k^{\sigma_k} R_k)^2}{4c_k} + \delta \sum_{z=1}^{\eta_{k+1}} \lambda_{k+1}^{z} A_{k+1}^{z} \left(1 - \lambda_k + \sum_{y=1}^{\mu_k} \gamma_k^{y} \vartheta_k^{y} G_k \right),$$

$$\textit{for } \sigma_k \in \{1, 2, \ldots, \eta_k\} \text{ and } k \in \{1, 2, \ldots, T\}, \tag{2.6}$$

and $C_k^{\sigma_k}$ is an expression that contains the previously executed controls $\overline{u}_{k-}^{(\omega)}$.

Proof See Appendix I. ■

Using Proposition 2.1 and Corollary 2.1 one can obtain the optimal strategies of the firm in stage k as (see derivation details in Appendix I):

$$u_k^{\sigma_k} = \frac{(\theta_k^{\sigma_k} R_k)^2 x}{4(c_k)^2}, \text{ for } \sigma_k \in \{1, 2, \ldots, \eta_k\} \text{ and } k \in \{1, 2, \ldots, T\}; \text{ and}$$

$$\overline{u}_k^{\sigma_k(\omega)} = \frac{\delta \sum_{z=1}^{\eta_{k+1}} \lambda_{k+1}^{z} A_{k+1}^{z} \left(a_k^{|k|} + \sum_{\upsilon=1}^{\mu_k} \gamma_k^{\upsilon} \vartheta_k^{\upsilon} G_k^2 \right) + \sum_{\tau=k+1}^{k+\omega-1} \delta^{\tau-k+1} \sum_{z=1}^{\eta_{\tau+1}} \lambda_{k+1}^{z} A_{\tau+1}^{z} a_\tau^{|k|} + \delta^{(T+1-k)} \upsilon_{T+1}^{|k|}}{2\phi_k},$$

for $k \in \{T - \omega + 2, T - \omega + 3, \ldots, T\}$ and $A_\tau = 0$ for $\tau > T + 1$;

$$\overline{u}_k^{\sigma_k(\omega)} = \frac{\delta \sum_{z=1}^{\eta_k} \lambda_{k+1}^{z} A_{k+1}^{z} \left(a_k^{|k|} + \sum_{\upsilon=1}^{\mu_k} \gamma_k^{\upsilon} \vartheta_k^{\upsilon} G_k^2 \right) + \sum_{\tau=k+1}^{k+\omega-1} \delta^{\tau-k+1} \sum_{z=1}^{\eta_{\tau+1}} \lambda_{\tau+1}^{z} A_{\tau+1}^{z} a_\tau^{|k|}}{2\phi_k}$$

for $\sigma_k \in \{1, 2, \ldots, \eta_k\}$ and $k \in \{1, 2, \ldots, T - \omega + 1\}$. (2.7)

The optimal capital investment strategies $\overline{u}_k^{\sigma_k(\omega)}$ in (2.7) are positively related to the sum of marginal benefits of investment from stage k to stage $k + \omega - 1$.

3 Stochastic Durable-Strategies Dynamic Game Formulation

In this section, we first develop a general class of non-cooperative stochastic dynamic game with durable strategies. Then, we present a theorem characterizing the equilibrium game solution.

3.1 *Game Formulation*

Consider a general class of $T-$ stage, $n-$ player nonzero-sum discrete-time non-cooperative dynamic games with durable and non-durable strategies affecting the players' payoffs and the state dynamics. We use $u_k^i \in U^i \subset R^{m^i}$ to denote the set of non-durable control strategies of player i. We use $\overline{u}_k^i = (\overline{u}_k^{(2)i}, \overline{u}_k^{(3)i}, \ldots, \overline{u}_k^{(\omega_i)i})$ to denote the set of durable strategies of player i, where $\overline{u}_k^{(\zeta)i} \in \overline{U}^{(\zeta)i} \subset R^{m_{(\zeta)i}}$ for $\zeta \in \{2, 3, \ldots, \omega_i\}$. In particular, $\overline{u}_k^{(2)i}$ are non-durable strategies that have effects in stages k and $k+1$. The strategies $\overline{u}_k^{(3)i}$ are durable strategies that have effects within stage k to stage $k + 2$. The strategies $\overline{u}_k^{(\omega)i}$ are durable strategies that have effects within stages k, $k + 1$,..., $k + \omega - 1$. The state at stage k is $x_k \in X \subset R^m$ and the state space is common for all players.

The state dynamics is characterized by a vector of stochastic difference equations:

$$x_{k+1} = f_k(x_k, \underline{u}_k, \overline{u}_k; \overline{u}_{k-}) + \vartheta_k G_k(x_k, \underline{u}_k, \overline{u}_k), \; x_1 = x_1^0,$$
$$\text{for } k \in \{1, 2, \ldots, T\}, \tag{3.1}$$

where ϑ_k is a random variable with range $\{\vartheta_k^1, \vartheta_k^2, \ldots, \vartheta_k^{\mu_k}\}$ and corresponding probabilities $\{\gamma_k^1, \gamma_k^2, \ldots, \gamma_k^{\mu_k}\}$, $\underline{u}_k = (u_k^1, u_k^2, \ldots, u_k^n)$ is the set of durable strategies of all the n players, $\overline{u}_k = (\overline{u}_k^1, \overline{u}_k^2, \ldots, \overline{u}_k^n)$ is the set of durable strategies of all the n players and $\overline{u}_{k-} = (\overline{u}_{k-}^1, \overline{u}_{k-}^2, \ldots, \overline{u}_{k-}^n)$ is the set of strategies which are executed before stage k by all players but still in effect in stage k.

The single-stage payoff of the player i at stage k is $g_k^i(\theta_k; x_k, \underline{u}_k, \overline{u}_k; \overline{u}_{k-})$ which is affected by a random variable θ_k. In particular, θ_k for $k \in \{1, 2, \ldots, T+1\}$ are independent discrete random variables with range $\{\theta_k^1, \theta_k^2, \ldots, \theta_k^{\eta_k}\}$ and corresponding probabilities $\{\lambda_k^1, \lambda_k^2, \ldots, \lambda_k^{\eta_k}\}$, where η_k is a positive integer for $k \in \{1, 2, \ldots, T+1\}$. In stage 1, it is known that θ_1 equals θ_1^1 with probability $\lambda_1^1 = 1$. Moreover, the random variable θ_k is known in stage k.

The expected payoff of player i is:

$$E_{\theta_2,\theta_3,\ldots,\theta_{T+1};\vartheta_1,\vartheta_2,\ldots,\vartheta_T}\left\{\sum_{k=1}^{T} g_k^i(\theta_k^1; x_k, \underline{u}_k, \overline{u}_k; \overline{u}_{k-})\delta_1^k + q_{T+1}^i(\theta_{T+1}; x_{T+1}; \overline{u}_{(T+1)-})\delta_1^{T+1}\right\}, \tag{3.2}$$

for $i \in \{1, 2, \ldots, n\} \equiv N$,

where $q_{T+1}^i(\theta_{T+1}; x_{T+1}; \overline{u}_{(T+1)-})$ is the terminal payoff of player i.

The controls executed before the start of the operation in stage 1, that is $\overline{u}_{1-}$, are known and some or all of them can be zeros. The $g_k^i(\theta_k; x_k, \underline{u}_k, \overline{u}_k; \overline{u}_{k-})$, $f_k^i(x_k, \underline{u}_k, \overline{u}_k; \overline{u}_{k-})$, $G_k(x_k, \underline{u}_k, \overline{u}_k)$ and $q_{T+1}^i(\theta_{T+1}; x_{T+1}; \overline{u}_{(T+1)-})$ are continuously differentiable functions.

The information set of every player includes the knowledge in:

(i) all the possible moves by himself and other players, that is u_k^i and $\overline{u}_k^i$, for $k \in \{1, 2, \ldots, T\}$ and $i \in N$;
(ii) the set of controls which are executed before stage k by all players but still in effect in stage k, that is $\overline{u}_{k-} = (\overline{u}_{k-}^1, \overline{u}_{k-}^2, \ldots, \overline{u}_{k-}^n)$, for $k \in \{1, 2, \ldots, T\}$;
(iii) the range and corresponding probabilities of the random variable ϑ_k, for $k \in \{1, 2, \ldots s, T\}$;
(iv) the range and corresponding probabilities of the random variable θ_k, for $k \in \{1, 2, \ldots, T\}$;
(v) the state dynamics $x_{k+1} = f_k(x_k, \underline{u}_k, \overline{u}_k; \overline{u}_{k-}) + \vartheta_k G_k(x_k, \underline{u}_k, \overline{u}_k)$ and the values of present and past states $(x_k, x_{k-1}, \ldots, x_1)$; and
(vi) the expected payoffs of all players, that is

$$E_{\theta_2,\theta_3,\ldots,\theta_{T+1};\vartheta_1,\vartheta_2,\ldots,\vartheta_T}\left\{\sum_{k=1}^{T} g_k^i(\theta_k^1; x_k, \underline{u}_k, \overline{u}_k; \overline{u}_{k-})\delta_1^k + q_{T+1}^i(\theta_{T+1}; x_{T+1}; \overline{u}_{(T+1)-})\delta_1^{T+1}\right\},$$

for $i \in N$.

3.2 Game Equilibria

The non-cooperative payoffs of the players in a Nash equilibrium of the dynamic game (3.1)–(3.2) can be characterized by the following theorem.

Theorem 3.1 *Let* $(\underline{u}_k^{**}, \overline{u}_k^{**})$ *be the set of Nash equilibrium strategies and* $V^i(\theta_k^{\sigma_k}; k, x; \overline{\underline{u}}_{k-}^{**})$, *for* $\sigma_k \in \{1, 2, \ldots, \eta_k\}$, *be the Nash equilibrium payoff of player* i *at stage* $k \in \{1, 2, \ldots, T\}$ *in the non-cooperative dynamic game (3.1)–(3.2), then the function* $V^i(\theta_k^{\sigma_k} k, x; \overline{\underline{u}}_{k-}^{**})$ *satisfies the following recursive equations:*

$$V^i(\theta_{T+1}^{\sigma_{T+1}}; T+1, x; \overline{\underline{u}}_{(T+1)-}^{**}) = q_{T+1}^i(\theta_{T+1}^{\sigma_{T+1}}; x; \overline{\underline{u}}_{(T+1)-}^{**})\delta_1^{T+1},$$

$$\text{for } \sigma_{T+1}^{\sigma_{T+1}} \in \{1, 2, \ldots, \eta_{T+1}\}; \tag{3.3}$$

$$V^i(\theta_k^{\sigma_k}; k, x; \overline{\underline{u}}_{k-}^{**}) = \max_{u_k^i, \overline{u}_k^i} \Bigg\{ \sum_{\upsilon=1}^{\eta_{k+1}} \lambda_{k+1}^{\upsilon} \sum_{\upsilon=1}^{\mu_k} \gamma_k^{\upsilon} \Big[g_k^i(\theta_k^{\sigma_k}; x, u_k^i, \overline{u}_k^i, \underline{u}_k^{**(\neq i)}, \overline{u}_k^{**(\neq i)}; \overline{\underline{u}}_{k-}^{**})\delta_1^k$$
$$+ V^i[\theta_{k+1}^{\upsilon}; k+1, f_k(x, u_k^i, \overline{u}_k^i, \underline{u}_k^{**(\neq i)}, \overline{u}_k^{**(\neq i)}; \overline{\underline{u}}_{k-}^{**})$$
$$+ \vartheta_k^{\upsilon} G_k(x, u_k^i, \overline{u}_k^i, \underline{u}_k^{**(\neq i)}, \overline{u}_k^{**(\neq i)}); \overline{u}_k^i, \overline{u}_k^{**(\neq i)}, \overline{\underline{u}}_{(k+1)-}^{**} \cap \overline{\underline{u}}_{k-}^{**}] \Big] \Bigg\}, \tag{3.4}$$

for $\sigma_k^{\sigma_k} \in \{1, 2, \ldots, \eta_k\}$, $k \in \{1, 2, \ldots, T\}$ and $i \in N$,
where $\underline{u}_k^{(\neq i)**} = (u_k^{1**}, u_k^{2**}, \ldots, u_k^{i-1**}, u_k^{i+1**}, \ldots, u_k^{n**})$ and

$$\overline{\underline{u}}_k^{(\neq i)**} = (\overline{u}_k^{1**}, \overline{u}_k^{2**}, \ldots, \overline{u}_k^{i-1**}, \overline{u}_k^{i+1**}, \ldots, \overline{u}_k^{n**}).$$

Proof Conditions (3.3)–(3.4) show that $V^i(\theta_k^{\sigma_k}; k, x; \overline{\underline{u}}_{k-}^{**})$ is the maximized payoff of player $i \in N$ according to Theorem 1.1, given the game equilibrium strategies of the other $n-1$ players. Hence a Nash equilibrium results. ■

System (3.3)–(3.4) can be regarded as the Hamilton–Jacobi-Bellman equations for stochastic durable-strategies dynamic games. Worth noting is that this class of games cannot be handled by the standard approach of stochastic dynamic programming.

4 An Example in Public Capital Provision

Consider an illustrative example of an oligopoly with n firms. The planning horizon of these firms involves T stages. The firms use knowledge-based public capital $x_k \in X \subset R^+$ to produce its output. Capital investment takes ω stages to complete its built up into the capital stock. The accumulation process of the capital stock is governed by the dynamics

$$x_{k+1} = x_k + \sum_{j=1}^{n} a_k^{|k|j} \overline{u}_k^{(\omega)j}$$
$$+ \sum_{j=1}^{n} \sum_{\tau=k-\omega+1}^{k-1} a_k^{|\tau|j} \overline{u}_\tau^{(\omega)j} - \lambda_k x_k + \vartheta_k \Big(G_k^1 x_k + \sum_{j=1}^{n} G_k^{2(j)} \overline{u}_k^{(\omega)j} \Big),$$

$$x_1 = x_1^0 \tag{4.1}$$

where $\overline{u}_k^{(\omega)j}$ is the investment in the knowledge-based capital by firm j in stage k, $a_k^{|k|j}\overline{u}_k^{(\omega)j}$ is the addition to the capital in stage k by firm j's capital building investment $\overline{u}_k^{(\omega)j}$ in stage k, $a_k^{|\tau|j}\overline{u}_\tau^{(\omega)j}$ is the addition to the capital in stage k by capital building investment $\overline{u}_\tau^{(\omega)j}$ in stage $\tau \in \{k-1, k-2, \ldots, k-\omega+1\}$ and λ_k is the rate of obsolescence of the capital.

Moreover, G_k^1 and $G_k^{2(j)}$ are positive parameters and ϑ_k is random variable with a range $\{\vartheta_k^1, \vartheta_k^2, \ldots, \vartheta_k^{\mu_k}\}$ and corresponding probabilities $\{\gamma_k^1, \gamma_k^2, \ldots, \gamma_k^{\mu_k}\}$. Let $\vartheta_k^{\hat{\upsilon}}$ denote the smallest (positive or negative) value of ϑ_k^{υ}, for $\upsilon \in \{1, 2, \ldots, \mu_k\}$. The value of G_k^1 satisfies the condition $\vartheta_k^{\hat{\upsilon}} G_k^1 > (\lambda_k - 1)$. The value of $G_k^{2(j)}$ satisfies the condition $\vartheta_k^{\hat{\upsilon}} G_k^{2(j)} \geq -a_k^{|k|j}$, for $j \in \{1, 2, \ldots, n\}$.

Firm $i \in N$ uses the knowledge-based capital together with a non-durable productive input u_k^i to produce outputs. The revenue from the output produced is $\theta_k^{\sigma_k} R_k^i (u_k^i)^{1/2} (x_k)^{1/2}$. The non-durable input cost is $c_k^i u_k^i$ and the cost of capital investment is $\phi_k^i (\overline{u}_k^{(\omega)i})^2$. The random variable θ_k, for $k \in \{1, 2, \ldots, T+1\}$ are independent discrete random variables with range $\{\theta_k^1, \theta_k^2, \ldots, \theta_k^{\eta_k}\}$ and corresponding probabilities $\{\lambda_k^1, \lambda_k^2, \ldots, \lambda_k^{\eta_k}\}$, where η_k is a positive integer for $k \in \{1, 2, \ldots, T+1\}$. In stage 1, it is known that θ_1 equals θ_1^1 with probability $\lambda_1^1 = 1$. Moreover, the random variable θ_k is known in stage k.

The expected payoff of the firm i is.

$$E_{\theta_2, \theta_3, \ldots, \theta_{T+1}; \vartheta_1, \vartheta_2, \ldots, \vartheta_T} \Bigg\{ \sum_{k=1}^{T} \left([\theta_k R_k^i (u_k^i)^{1/2} (x_k)^{1/2} - c_k^i u_k^i - \phi_k^i (\overline{u}_k^{(\omega)i})^2] \right) \delta^{k-1}$$
$$+ \left(\theta_{T+1} Q_{T+1}^i x_{T+1} + \sum_{\tau=T+1-\omega+1}^{T} v_{T+1}^{|\tau|i} \overline{u}_\tau^{(\omega)i} + \varpi_{T+1}^i \right) \delta^T \Bigg\}, \tag{4.2}$$

where $\left(\theta_{T+1} Q_{T+1}^i x_{T+1} + \sum_{\tau=T+1-\omega+1}^{T} v_{T+1}^{|\tau|i} \overline{u}_\tau^{(\omega)i} + \varpi_{T+1}^i \right)$ is the salvage value of firm i in stage $T+1$ and $\delta = (1+r)^{-1}$ is the discount factor.

The salvage value of the firm i in stage $T+1$ depends on the public capital stock, with Q_{T+1}^i being positive. It also depends on the previous capital investment $\sum_{\tau=T+1-\omega_3+1}^{T} v_{T+1}^{|\tau|i} \overline{u}_\tau^{(\omega)i}$. In particular, some or all $v_{T+1}^{|\tau|}$, for $\tau \in \{T, T-1, \ldots, T-\omega+2\}$, can be zero.

Invoking Theorem 3.1, we obtain.

Corollary 4.1. *Let* $(\underline{u}_k^{**}, \underline{u}_k^{(\omega)**})$ *be the set of feedback Nash equilibrium strategies and* $V^i(\theta_k^{\sigma_k}; k, x; \underline{\overline{u}}_{k-}^{(\omega)**})$, *for* $\sigma_k \in \{1, 2, \ldots, \eta_k\}$ *and* $k \in \{1, 2, \ldots, T\}$, *be the feedback Nash equilibrium payoff of player* i *in the non-cooperative game* (4.1)–(4.2), *then the function* $V^i(\theta_k^{\sigma_k}; k, x; \underline{\overline{u}}_{k-}^{(\omega)**})$ *satisfies the following recursive equations:*

$$V^i(\theta_{T+1}^{\sigma_{T+1}}; T+1, x; \overline{\underline{u}}_{(T+1)-}^{(\omega)**}) = \left(\theta_{T+1}^{\sigma_{T+1}} Q_{T+1}^i x + \sum_{\tau=T+1-\omega+1}^{T} v_{T+1}^{|\tau|i} \overline{u}_\tau^{(\omega)i} + \varpi_{T+1}^i \right) \delta^T,$$

for $\sigma_{T+1}^{\sigma_{T+1}} \in \{1, 2, \ldots, \eta_{T+1}\}$ (4.3)

$$V^i(\theta_k^{\sigma_k}; k, x; \overline{\underline{u}}_{k-}^{(\omega)**})$$
$$= \max_{u_k^i, \overline{u}_k^{(\omega)i}} \left\{ \sum_{\upsilon=1}^{\eta_{k+1}} \lambda_{k+1}^{\upsilon} \sum_{\upsilon=1}^{\mu_k} \gamma_k^{\upsilon} \left[[\theta_k^{\sigma_k} R_k^i (u_k^i)^{1/2} (x)^{1/2} - c_k^i u_k^i - \phi_k^i (\overline{u}_k^{(\omega)i})^2] \delta^{k-1} \right. \right.$$
$$+ V^i[\theta_{k+1}^{\upsilon}; k+1, x + a_k^{|k|i} \overline{u}_k^i + \sum_{j \in N, j \neq i} a_k^{|k|j} \overline{u}_k^{(\omega)j**} + \sum_{j=1}^{n} \sum_{\tau=k-\omega+1}^{k-1} a_k^{|\tau|j} \overline{u}_\tau^{(\omega)j**} - \lambda_k x$$
$$\left. \left. + \vartheta_k^{\upsilon} \left(G_k^1 x_k + G_k^{2(i)} \overline{u}_j^{(\omega)i} + \sum_{\substack{j=1 \\ j \neq i}}^{n} G_k^{2(j)} \overline{u}_k^{(\omega)j**} \right); \overline{u}_k^{(\omega)i}, \overline{\underline{u}}_k^{(\omega)(\neq i)**}, \overline{\underline{u}}_{(k+1)-}^{(\omega)**} \cap \overline{\underline{u}}_{k-}^{(\omega)**}] \right] \right\},$$

for $\sigma_k \in \{1, 2, \ldots, \eta_k\}$ and $k \in \{1, 2, \ldots, T\}$ and $i \in N$. (4.4)

■

Performing the indicated maximization operator in Corollary 4.1, the value function $V^i(\theta_k^{\sigma_k}; k, x; \overline{\underline{u}}_{k-}^{(\omega)**})$ which reflects the value of the firm covering stage $k \in \{1, 2, \ldots, T\}$ to stage $T+1$ can be obtained as follows.

Proposition 4.1. *System* (4.3)–(4.4) *admits a solution with the game equilibrium payoff of firm i being.*

$$V^i(\theta_k^{\sigma_k}; k, x; \overline{\underline{u}}_{k-}^{(\omega)**}) = (A_k^{(\sigma_k)i} x + C_k^{(\sigma_k)i}) \delta^{k-1}, \tag{4.5}$$

for $\sigma_k \in \{1, 2, \ldots, \eta_k\}$ and $k \in \{1, 2, \ldots, T\}$ and $i \in N$,
where $A_{T+1}^{(\sigma_{T+1})i} = \theta_{T+1}^{\sigma_{T+1}} Q_{T+1}^i$ and $C_{T+1}^{(\sigma_{T+1})i} = \sum_{\tau=T+1-\omega+1}^{T} v_{T+1}^{|\tau|i} \overline{u}_\tau^{(\omega)i} + \varpi_{T+1}^i$;

$$A_k^i = \frac{(\theta_k^{\sigma_k} R_k^i)^2}{4 c_k^i} + \delta \sum_{z=1}^{\eta_{k+1}} \lambda_{k+1}^z A_{k+1}^{(z)i} \left(1 - \lambda_k + \sum_{y=1}^{\mu_k} \gamma_k^y \vartheta_k^y G_k^1 \right), \tag{4.6}$$

and $C_k^{(\sigma_k)i}$ is an expression that contains the previously executed controls $\overline{\underline{u}}_{k-}^{(\omega)**}$.

Proof. See Appendix II. ■

Using Proposition 4.1 and Corollary 4.1 one can obtain the game equilibrium strategies in stage k as (see derivation details in Appendix II):
$u_k^{(\sigma_k)i**} = \frac{(\theta_k^{\sigma_k} R_k^i)^2 x}{4 (c_k^i)^2}$, for $\sigma_k \in \{1, 2, \ldots, \eta_k\}$ and $k \in \{1, 2, \ldots, T\}$; and

$$\overline{u}_k^{(\sigma_k)(\omega)i**}$$
$$= \frac{\sum_{z=1}^{\eta_{k+1}} \lambda_{k+1}^z A_{k+1}^{(z)i} \left(a_k^{|k|i} + \sum_{\upsilon=1}^{\mu_k} \gamma_k^{\upsilon} \vartheta_k^{\upsilon} G_k^{2(i)} \right) + \sum_{\tau=k+1}^{k+\omega-1} \delta^{\tau-k+1} \sum_{z=1}^{\eta_{\tau+1}} \lambda_{\tau+1}^z A_{\tau+1}^{(z)i} a_\tau^{|k|i} + \delta^{(T+1-k)} v_{T+1}^{|k|i}}{2 \phi_k^i},$$

for $k \in \{T-\omega+2, T-\omega+3, \ldots, T\}$ and $A_\tau^i = 0$ for $\tau > T+1$;

$$\bar{u}_k^{(\sigma_k)(\omega)i**} = \frac{\sum_{z=1}^{\eta_{k+1}} \lambda_{k+1}^z A_{k+1}^{(z)i}\left(a_k^{|k|i} + \sum_{\nu=1}^{\mu_k} \gamma_k^\nu \vartheta_k^\nu G_k^{2(i)}\right) + \sum_{\tau=k+1}^{k+\omega-1} \delta^{\tau-k+1} \sum_{z=1}^{\eta_{\tau+1}} \lambda_{\tau+1}^z A_{\tau+1}^{(z)i} a_\tau^{|k|i}}{2\phi_k^i}, \tag{4.7}$$

for $k \in \{1, 2, \ldots, T-\omega+1\}$.

5 Cooperative Stochastic Dynamic Games with Durable Strategies

In this section, we present a general framework of cooperative stochastic dynamic games with durable strategies and characterize the corresponding subgame consistent solutions.

5.1 Game Formulation

Consider the general class of $T-$ stage $n-$ player stochastic dynamic games with durable strategies and random elements affecting the players' payoffs and the state dynamics developed in Sect. 3. To exploit the potential gains from cooperation, the players agree to act cooperatively and distribute the payoffs among themselves according to an agreed-upon gain sharing optimality principle. Two crucial properties that a cooperative scheme must satisfy are group optimality and individual rationality. Group optimality ensures that the joint payoff of all the players under cooperation is maximized. Group optimality is a Pareto optimal condition under which yields the maximized efficiency that can be achieved. Failure to fulfil group optimality leads to the condition where the participants prefer to deviate from the agreed-upon solution plan in order to extract the unexploited gains. To achieve group optimality, the players will maximize their expected joint payoff by solving the stochastic dynamic optimization problem which maximizes

$$E_{\theta_2,\theta_3,\ldots,\theta_{T+1};\vartheta_1,\vartheta_2,\ldots,\vartheta_T}\left\{ \sum_{j=1}^{n}\sum_{k=1}^{T} g_k^j(\theta_k; x_k, \underline{u}_k, \overline{u}_k; \overline{u}_{k-})\delta_1^k + \sum_{j=1}^{n} q_{T+1}^j(\theta_{T+1}; x_{T+1}; \overline{u}_{(T+1)-})\delta_1^{T+1} \right\}, \tag{5.1}$$

subject to the dynamics

$$x_{k+1} = f_k(x_k, \underline{u}_k, \overline{u}_k; \overline{u}_{k-}) + \vartheta_k G_k(x_k, \underline{u}_k, \overline{u}_k), \quad x_1 = x_1^0, \text{ for } k \in \{1, 2, , \ldots, T\} \tag{5.2}$$

An optimal solution to the joint maximization problem of maximizing (5.1) subject to (5.2) can be characterized by the theorem below.

Theorem 5.1. *Let $W(\theta_k^{\sigma_k}; k, x; \overline{u}_{k-})$ be the maximal value of the expected joint payoff.*

$$E_{\theta_{k+1},\theta_{k+2},\ldots,\theta_{T+1};\vartheta_k,\vartheta_{k+1},\ldots,\vartheta_T}\Bigg\{\sum_{j=1}^{n}\sum_{t=k}^{T} g_t^j(\theta_t; x_t, \underline{u}_t, \overline{u}_t; \overline{u}_{t-})\delta_1^t$$
$$+\sum_{j=1}^{n} q_{T+1}^j(\theta_{T+1}; x_{T+1}; \overline{u}_{(T+1)-})\delta_1^{T+1}\Bigg\}, \tag{5.3}$$

for the expected joint payoff maximization problem (5.1)–(5.2) starting at stage k with state $x_k = x$, $\theta_k = \theta_k^{\sigma_k}$ and previously executed controls $\overline{u}_{k-}$, then the function $W(\theta_k^{\sigma_k}; k, x; \overline{u}_{k-})$ satisfies the following system of recursive equations:

$$W(\theta_{T+1}^{\sigma_{T+1}}; T+1, x; \overline{u}_{(T+1)-})$$
$$= \sum_{j=1}^{n} q_{T+1}^j(\theta_{T+1}^{\sigma_{T+1}} x; \overline{u}_{(T+1)-})\delta_1^{T+1} \tag{5.4}$$

for $\sigma_{T+1} \in \{1, 2, \upsilon, \eta_{T+1}\}$;

$$W(\theta_k^{\sigma_k}; k, x; \overline{u}_{k-})$$
$$= \max_{\underline{u}_k, \overline{u}_k} E_{\theta_{k+1},\vartheta_k}\Bigg\{\sum_{j=1}^{n} g_k^j(\theta_k^{\sigma_k}; x, \underline{u}_k, \overline{u}_k; \overline{u}_{k-})\delta_1^k$$
$$+ W[\theta_{k+1}; k+1, f_k(x, \underline{u}_k, \overline{u}_k; \overline{u}_{k-}) + \vartheta_k G_k(x, \underline{u}_k, \overline{u}_k); \overline{u}_k, \overline{u}_{(k+1)-} \cap \overline{u}_{k-}]\Bigg\},$$
$$= \max_{\underline{u}_k, \overline{u}_k}\Bigg\{\sum_{\upsilon=1}^{\eta_{k+1}} \lambda_{k+1}^{\upsilon} \sum_{\upsilon=1}^{\mu_k} \gamma_k^{\nu}\Bigg[\sum_{j=1}^{n} g_k^j(x, \underline{u}_k, \overline{u}_k; \overline{u}_{k-})\delta_1^k$$
$$+ W[\theta_{k+1}^{\upsilon}; k+1, f_k(x, \underline{u}_k, \overline{u}_k; \overline{u}_{k-}) + \vartheta_k^{\nu} G_k(x, \underline{u}_k, \overline{u}_k); \overline{u}_k, \overline{u}_{(k+1)-} \cap \overline{u}_{k-}]\Bigg]\Bigg\},$$

for $\sigma_k \in \{1, 2, \ldots, \eta_k\}$ and $k \in \{1, 2, \ldots, T\}$. (5.5)

Proof. The conditions in (5.4)–(5.5) satisfy the optimal conditions of the stochastic dynamic optimization technique with durable controls in Theorem 1.1 and hence an optimal solution to the stochastic control problem (5.1)–(5.2) results. ∎

We use $\{u_k^*, \overline{u}_k^*\}$ for $k \in \{1, 2, \ldots, T\}$ to denote the optimal cooperative control strategies derived from Theorem 5.1. Substituting these optimal controls into the state dynamics (5.2), one can obtain the dynamics of the optimal cooperative trajectory. We use X_k^* to denote the set of realizable values of x_k at stage k along the cooperative trajectory. The term $x_k^* \in X_k^*$ is used to denote elements in X_k^*.

5.2 Subgame Consistent Imputation Distribution Procedure

We let
$$\xi(\theta_k^{\sigma_k}; k, x_k^*; \underline{u}_{k-}^*) = [\xi^1(\theta_k^{\sigma_k}; k, x_k^*; \underline{u}_{k-}^*), \xi^2(\theta_k^{\sigma_k}; k, x_k^*; \underline{u}_{k-}^*), \ldots, \xi^n(\theta_k^{\sigma_k}; k, x_k^*; \underline{u}_{k-}^*)],$$
$$\text{for } \sigma_k \in \{1, 2, \ldots, \eta_k\} \text{ and } k \in \{1, 2, \ldots, T\}, \tag{5.6}$$

to denote the agreed distribution of cooperative payoffs among the players along the cooperative trajectory given the previously executed controls $\overline{u}_{k-}^*$ when $\theta_k = \theta_k^{\sigma_k}$. To satisfy subgame consistency, the players' agreed-upon optimality principle has to be effective along the cooperative state trajectory x_k^* contingent upon $\overline{u}_{k-}^*$ and $\theta_k^{\sigma_k}$. Hence, the agreed-upon imputation $\xi(\theta_k^{\sigma_k}; k, x_k^*; \overline{u}_{k-}^*)$ in (5.6) must be maintained at all stages $k \in \{1, 2, \ldots, T\}$ along the cooperative trajectory with previously executed strategies $\overline{u}_{k-}^*$ and $\theta_k = \theta_k^{\sigma_k}$.

To satisfy group optimality along the cooperative trajectory, it is required that

$$W(\theta_k^{\sigma_k}; k, x^*; \overline{u}_{k-}^*) = \sum_{j=1}^{n} \xi^j(\theta_k^{\sigma_k}; k, x_k^*; \overline{u}_{k-}^*),$$
$$\text{for } \sigma_k \in \{1, 2, \ldots, \eta_k\} \text{ and } k \in \{1, 2, \ldots, T\}. \tag{5.7}$$

The condition in (5.7) guarantees the maximal joint payoff is distributed to the players throughout the cooperation duration.

For individual rationality to be maintained throughout the game, the payoff that player i receives under cooperation along the cooperative trajectory must be greater than or equal to his non-cooperative payoff, therefore the chosen imputation must satisfy the condition:

$$\xi^i(\theta_k^{\sigma_k}; k, x_k^*; \overline{u}_{k-}^*) \geq V^i(\theta_k^{\sigma_k}; k, x_k^*; \overline{u}_{k-}^{**}), \tag{5.8}$$

for $\sigma_k \in \{1, 2, \ldots, \eta_k\}$, $k \in \{1, 2, \ldots, T\}$ and $i \in N$.

The condition in (5.8) guarantees the fulfilment of individual rationality throughout the cooperation duration so that the payoff allocated to any player under cooperation will be no less than his non-cooperative payoff. Failure to guarantee individual rationality at any stage would lead to the condition where the concerned participants would deviate from the agreed-upon solution plan and play non-cooperatively.

Crucial to the analysis is the derivation of an Imputation Distribution Procedure (IDP) leading to the realization of the agreed imputations in (5.6). To do this, we follow Yeung and Petrosyan (2010, 2016a and 2019) and use $\beta_k^i(\theta_k^{\sigma_k}; x_k^*; \underline{\overline{u}}_{k-}^*)$ to denote the payment that player i receives in stage k under the cooperative agreement along the cooperative trajectory with durable strategies executive before but still in effect being $\underline{\overline{u}}_{k-}^*$ and $\theta_k = \theta_k^{\sigma_k}$. The payment scheme involving $\beta_k^i(\theta_k^{\sigma_k}; x_k^*; \underline{\overline{u}}_{k-}^*)$ constitutes an IDP in the sense that the payoff to player i over the stages from k to $T+1$ satisfies the condition:

$$\xi^i(\theta_k^{\sigma_k}; k, x_k^*; \underline{\overline{u}}_{k-}^*) = \beta_k^i(\theta_k^{\sigma_k}; x_k^*; \underline{\overline{u}}_{k-}^*)\delta_1^k$$
$$+ E_{\theta_{k+1},\theta_{k+2},\ldots,\theta_{T+1};\vartheta_k,\vartheta_{k+1},\ldots,\vartheta_T}$$
$$\left\{ \sum_{\zeta=k+1}^{T} \beta_\zeta^i(\theta_\xi; x_\zeta^*; \underline{\overline{u}}_{\zeta-}^*)\delta_1^\zeta + q_{T+1}^i(\theta_{T+1}; x_{T+1}^*; \underline{\overline{u}}_{(T+1)-}^*)\delta_1^{T+1} \right\}, \tag{5.9}$$

for $\sigma_k \in \{1, 2, \ldots, \eta_k\}$, $k \in \{1, 2, \ldots, T\}$ and $i \in N$.

A theorem for the derivation of $\beta_k^i(\theta_k^{\sigma_k}; x_k^*; \underline{\overline{u}}_{k-}^*)$, for $k \in \{1, 2, \ldots, T\}$ and $i \in N$, that satisfies (5.9) is provided below.

Theorem 5.2. *The agreed-upon imputation* $\xi(\theta_k^{\sigma_k}; k, x_k^*; \underline{\overline{u}}_{k-}^*)$, *for* $k \in \{1, 2, \ldots, T\}$ *along the cooperative trajectory, can be realized by a payment.*

$$\beta_k^i(\theta_k^{\sigma_k}; x_k^*; \underline{\overline{u}}_{k-}^*) = (\delta_1^k)^{-1}\Big[\xi^i(\theta_k^{\sigma_k}; k, x_k^*; \underline{\overline{u}}_{k-}^*)$$
$$- E_{\theta_{k+1};\vartheta_k}\xi^i\Big(\theta_k; k+1, f_k(x_k^*, \underline{u}_k^*, \overline{u}_k^*; \underline{\overline{u}}_{k-}^*) + \vartheta_k G_k(x_k^*, \underline{u}_k^*, \overline{u}_k^*); \underline{\overline{u}}_{(k+1)-}^* \Big)\Big] \tag{5.10}$$

given to player $i \in N$ at stage $k \in \{1, 2, \ldots, T\}$ if $\theta_k^{\sigma_k}$ occurs.

Proof. Using (5.9) one can obtain.

$$\xi^i(\theta_{k+1}^{\sigma_{k+1}}; k+1, x_{k+1}^*; \underline{\overline{u}}_{k-}^*) = B_{k+1}^i(\theta_{k+1}^{\sigma_{k+1}}; x_{k+1}^*; \underline{\overline{u}}_{k-}^*)\,\delta_1^{k+1}$$
$$+ E_{\theta_{k+2},\theta_{k+3},\ldots,\theta_{T+1};\vartheta_{k+1},\vartheta_{k+2},\ldots,\vartheta_T}\left\{ \sum_{\zeta=k+2}^{T} B_\zeta^i(\theta_\xi; x_\zeta^*; \underline{\overline{u}}_{\zeta-}^*)\,\delta_1^\zeta \right.$$

$$+ q^i_{T+1}(\theta_{T+1} : x^*_{T+1}; \underline{\overline{u}}^*_{(T+1)-})\, \delta_1^{T+1} \Bigg\}, \quad \text{for } \sigma_{k+1} \in \{1, 2, \ldots, \eta_{k+1}\}. \tag{5.11}$$

Upon substituting (5.11) into (5.9) yields

$$\xi^i(\theta_k^{\sigma_k}; k, x_k^*; \underline{\overline{u}}^*_{k-}) = \beta_k^i(\theta_k^{\sigma_k}; x_k^*; \underline{\overline{u}}^*_{k-})\, \delta_1^k$$
$$+ E_{\theta_{k+1};\vartheta_k} \xi^i\left(\theta_{k+1}^{\sigma_{k+1}}; k+1,\, f_k(x_k^*, \underline{u}_k^*, \overline{u}_k^*; \underline{\overline{u}}^*_{k-}) + \vartheta_k G_k(x_k^*, \underline{u}_k^*, \overline{u}_k^*); \underline{\overline{u}}^*_{(k+1)-} \right) \Bigg],$$

which can be expressed as

$$\beta_k^i(\theta_k^{\sigma_k}; x_k^*; \underline{\overline{u}}^*_{k-}) = (\delta_1^k)^{-1} \Bigg[\xi^i(\theta_k^{\sigma_k}; k, x_k^*; \underline{\overline{u}}^*_{k-})$$
$$- E_{\theta_{k+1};\vartheta_k} \xi^i\left(\theta_{k+1}^{\sigma_{k+1}}; k+1,\, f_k(x_k^*, \underline{u}_k^*, \overline{u}_k^*; \underline{\overline{u}}^*_{k-}) + \vartheta_k G_k(x_k^*, \underline{u}_k^*, \overline{u}_k^*); \underline{\overline{u}}^*_{(k+1)-} \right) \Bigg]. \tag{5.12}$$

The payment scheme in Theorem 5.1 gives rise to the realization of the imputation guided by the agreed-upon optimality principle and constitutes a dynamically consistent payment scheme. More specifically, the payment of $\beta_k^i(\theta_k^{\sigma_k}; x_k^*; \underline{\overline{u}}^*_{k-})$ allotted to player $i \in N$ in stage $k \in \{1, 2, \ldots, T\}$, if $\theta_k^{\sigma_k}$ occurs, will establish a cooperative plan that matches with the agreed-upon imputation to every player along the cooperative path.

Finally, under cooperation, all players would use the cooperative strategies and the payoff that player i will directly receive at stage k along the cooperative trajectory with previously executed durable strategies $\underline{u}^*_{k-}$ becomes $g_k^i(\theta_k^{\sigma_k}; x_k^*, \underline{u}_k^*, \overline{u}_k^*; \underline{\overline{u}}^*_{k-})$. However, according to the agreed-upon imputation, player i will receive $\beta_k^i(x_k^*; \underline{\overline{u}}^*_{k-})$ at stage k. Therefore, a side-payment

$$\pi_k^i(\theta_k^{\sigma_k}; x_k^*; \underline{\overline{u}}^*_{k-}) = \beta_k^i(\theta_k^{\sigma_k}; x_k^*; \underline{\overline{u}}^*_{k-}) - g_k^i(\theta_k^{\sigma_k}; x_k^*, \underline{u}_k^*, \overline{u}_k^*; \underline{\overline{u}}^*_{k-}), \tag{5.13}$$

for $k \in \{1, 2, \ldots, T\}$,

has to be given to player $i \in N$, if $\theta_k^{\sigma_k}$ occurs, to yield the cooperative imputation $\xi^i(\theta_k^{\sigma_k}; k, x_k^*; \underline{\overline{u}}^*_{k-})$.

For illustration sake, we consider some examples of gain sharing optimality principles.

Case I: Consider the case of an optimality principle which divides the excess of the total cooperative payoff over the sum of individual non-cooperative payoffs equally. According to this optimality principle, the imputation to player i is

$$\xi^i(\theta_k^{\sigma_k}; k, x_k^*; \underline{\overline{u}}^*_{k-}) = V^i(\theta_k^{\sigma_k}; k, x_k^*; \underline{\overline{u}}^*_{k-})$$
$$+ \frac{1}{n}\left(W(\theta_k^{\sigma_k}; k, x_k^*; \underline{\overline{u}}^*_{k-}) - \sum_{j=1}^{n} V^j(\theta_k^{\sigma_k}; k, x_k^*; \underline{\overline{u}}^*_{k-}) \right),$$

for $i \in N$ at stage $k \in \{1, 2, \ldots, T\}$, if $\theta_k^{\sigma_k}$ occurs (5.14)

Applying Theorem 2.1, a payment

$$\begin{aligned}
&\beta_k^i(\theta_k^{\sigma_k}; x_k^*; \underline{\overline{u}}_{k-}^*) \\
&= (\delta_1^k)^{-1}\Bigg\{ V^i(\theta_k^{\sigma_k}; k, x_k^*; \underline{\overline{u}}_{k-}^*) + \frac{1}{n}\Big(W(\theta_k^{\sigma_k}; k, x_k^*; \underline{\overline{u}}_{k-}^*) - \sum_{j=1}^{n} V^j(\theta_k^{\sigma_k}; k, x_k^*; \underline{\overline{u}}_{k-}^*) \Big) \\
&- \sum_{z=1}^{\eta_{k+1}} \lambda_{k+1}^z \sum_{v=1}^{\mu_k} \gamma_k^v \Big[V^i(\theta_{k+1}^z; k+1, f_k(x_k^*, \underline{u}_k^*, \overline{u}_k^*; \underline{\overline{u}}_{k-}^*) + \vartheta_k^v G_k(x_k^*, \underline{u}_k^*, \overline{u}_k^*); \underline{\overline{u}}_{(k+1)-}^*) \\
&+ \frac{1}{n}\Big(W(\theta_{k+1}^z; k+1, f_k(x_k^*, \underline{u}_k^*, \overline{u}_k^*; \underline{\overline{u}}_{k-}^*) + \vartheta_k^v G_k(x_k^*, \underline{u}_k^*, \overline{u}_k^*); \underline{\overline{u}}_{(k+1)-}^*) \\
&- \sum_{j=1}^{n} V^j(\theta_{k+1}^z; k+1, f_k(x_k^*, \underline{u}_k^*, \overline{u}_k^*; \underline{\overline{u}}_{k-}^*) + \vartheta_k^v G_k(x_k^*, \underline{u}_k^*, \overline{u}_k^*); \underline{\overline{u}}_{(k+1)-}^*); \underline{\overline{u}}_{(k+1)-}^*) \Big) \Big] \Bigg\}
\end{aligned} \tag{5.15}$$

will be given to player $i \in N$ at stage $k \in \{1, 2, \ldots, T\}$, if $\theta_k^{\sigma_k}$ occurs.

Case II: Consider the case of an optimality principle that shares the cooperative payoff proportional to players' non-cooperative payoffs. According to this optimality principle, the imputation to player i is

$$\xi^i(\theta_k^{\sigma_k}; k, x_k^*; \underline{\overline{u}}_{k-}^*) = \frac{V^i(\theta_k^{\sigma_k}; k, x_k^*; \underline{\overline{u}}_{k-}^*)}{\sum_{j=1}^{n} V^j(\theta_k^{\sigma_k}; k, x_k^*; \underline{\overline{u}}_{k-}^*)} W(\theta_k^{\sigma_k}; k, x_k^*; \underline{\overline{u}}_{k-}^*), \tag{5.16}$$

for $i \in N$ at stage $k \in \{1, 2, \ldots, T\}$, if $\theta_k^{\sigma_k}$ occurs.

Applying Theorem 2.1, a payment

$$\begin{aligned}
&\beta_k^i(\theta_k^{\sigma_k}; x_k^*; \underline{\overline{u}}_{k-}^*) \\
&= (\delta_1^k)^{-1}\Bigg\{ \frac{V^i(\theta_k^{\sigma_k}; k, x_k^*; \underline{\overline{u}}_{k-}^*)}{\sum_{j=1}^{n} V^j(\theta_k^{\sigma_k}; k, x_k^*; \underline{\overline{u}}_{k-}^*)} W(\theta_k^{\sigma_k}; k, x_k^*; \underline{\overline{u}}_{k-}^*) \\
&- \sum_{z=1}^{\eta_{k+1}} \lambda_{k+1}^z \sum_{v=1}^{\mu_k} \gamma_k^y \Bigg[\frac{V^i(\theta_{k+1}^z; k+1, f_k(x_k^*, \underline{u}_k^*, \overline{u}_k^*; \underline{\overline{u}}_{k-}^*) + \vartheta_k^y G_k(x_k^*, \underline{u}_k^*, \overline{u}_k^*); \underline{\overline{u}}_{(k+1)-}^*); \underline{\overline{u}}_{(k+1)-}^*)}{\sum_{j=1}^{n} V^j(\theta_{k+1}^z; k+1, f_k(x_k^*, \underline{u}_k^*, \overline{u}_k^*; \underline{\overline{u}}_{k-}^*) + \vartheta_k^y G_k(x_k^*, \underline{u}_k^*, \overline{u}_k^*); \underline{\overline{u}}_{(k+1)-}^*); \underline{\overline{u}}_{(k+1)-}^*)} \\
&\times W(\theta_{k+1}^z; k+1, f_k(x_k^*, \underline{u}_k^*, \overline{u}_k^*; \underline{\overline{u}}_{k-}^*) + \vartheta_k^y G_k(x_k^*, \underline{u}_k^*, \overline{u}_k^*); \underline{\overline{u}}_{(k+1)-}^*); \underline{\overline{u}}_{(k+1)-}^*) \Bigg]
\end{aligned} \tag{5.17}$$

will be given to player $i \in N$ at stage $k \in \{1, 2, \ldots, T\}$, if $\theta_k^{\sigma_k}$ occurs.

6 An Illustration in Cooperative Public Capital Provision

Now, let us consider the stochastic durable strategies dynamic oligopoly game in Sect. 4. Given the positive externality in public capital, the firms agree to enhance their expected payoffs through cooperation. They also agree to share the total cooperative payoff proportional to their non-cooperative payoffs. To secure group optimality the participating firms seek to maximize their expected joint payoff by solving the following dynamic optimization problem:

$$\max_{\substack{u_k^i, \overline{u}_k^{(\omega)i}, i \in N \\ k \in \{1,2,\ldots,T\}}} \Bigg\{ E_{\theta_2,\theta_3,\ldots,\theta_{T+1};\vartheta_1,\vartheta_2,\ldots,\vartheta_T} \Bigg[\sum_{i=1}^{n} \Bigg(\sum_{k=1}^{T} \Big([\theta_k R_k^i (u_k^i)^{1/2} (x_k)^{1/2}$$

$$- c_k^i u_k^i - \phi_k^i (\bar{u}_k^{(\omega)i})^2] \Big) \delta^{k-1} + \Bigg(\theta_{T+1} Q_{T+1}^i x_{T+1} + \sum_{\tau=T+1-\omega+1}^{T} v_{T+1}^{|\tau|i} \bar{u}_\tau^{(\omega)i} + \varpi_{T+1}^i \Bigg) \delta^T \Bigg) \Bigg] \Bigg\} \tag{6.1}$$

subject to the stochastic public capital accumulation dynamics

$$x_{k+1} = x_k + \sum_{j=1}^{n} a_k^{|k|j} \overline{u}_k^{(\omega)j} + \sum_{j=1}^{n} \sum_{\tau=k-\omega+1}^{k-1} a_k^{|\tau|j} \overline{u}_\tau^{(\omega)j} - \lambda_k x_k$$

$$+ \vartheta_k \Bigg(G_k^1 x_k + \sum_{j=1}^{n} G_k^{2(j)} \overline{u}_k^{(\omega)j} \Bigg),$$

$$x_1 = x_1^0. \tag{6.2}$$

Invoking Theorem 5.1, a set of optimal cooperative control strategies for dynamic optimization problem (6.1)–(6.2) can be obtained by solving the following system of recursive equations in the following Corollary.

Corollary 6.1.

$$W(\theta_{T+1}^{\sigma_{T+1}}; T+1, x; \underline{\overline{u}}_{(T+1)-}^{(\omega)})$$

$$= \sum_{i=1}^{n} \Bigg(\theta_{T+1}^{\sigma_{T+1}} Q_{T+1}^i x + \sum_{\tau=T+1-\omega+1}^{T} v_{T+1}^{|\tau|i} \overline{u}_\tau^{(\omega)i} + \varpi_{T+1}^i \Bigg) \delta^T,$$

for $\sigma_{T+1} \in \{1, 2, \ldots, \eta_{T+1}\}$; (6.3)

$$W(\theta_k^{\sigma_k}; k, x; \underline{\overline{u}}_{k-}^{(\omega)}) =$$

$$\max_{\underline{u}_k, \overline{\underline{u}}_k^{(\omega)}} \Bigg\{ \sum_{z=1}^{\eta_{k+1}} \lambda_{k+1}^z \sum_{\nu=1}^{\mu_k} \gamma_k^\nu \Bigg[\sum_{i=1}^{n} \Bigg([\theta_k^{\sigma_k} R_k^i (u_k^i)^{1/2} (x)^{1/2} - c_k^i u_k^i - \phi_k^i (\overline{u}_k^{(\omega)i})^2] \Bigg) \delta^{k-1}$$

$$+ W\Bigg(\theta_{k+1}^z; k+1, x + \sum_{j=1}^{n} a_k^{|k|j} \overline{u}_k^{(\omega)j} + \sum_{j=1}^{n} \sum_{\tau=k-\omega+1}^{k-1} a_k^{|\tau|j} \overline{u}_\tau^{(\omega)j} - \lambda_k x$$

$$+ \vartheta_k^\nu \Bigg(G_k^1 x_k + \sum_{j=1}^{n} G_k^{2(j)} \overline{u}_k^{(\omega)j} \Bigg); \underline{\overline{u}}_{(k+1)-}^{(\omega)} \Bigg) \Bigg] \Bigg\},$$

for $\sigma_k \in \{1, 2, \ldots, \eta_k\}$ and $k \in \{1, 2, \ldots, T\}$, (6.4)

where $\underline{\overline{u}}_{(k+1)-}^{(\omega)} = [\overline{\underline{u}}_k^{(\omega)}, \underline{\overline{u}}_{(k+1)-}^{(\omega)} \cap \underline{\overline{u}}_{k-}^{(\omega)}]$. ■

Performing the indicated maximization in Corollary 6.1 and solving the system (6.3)–(6.4) one can obtain the maximized joint payoff under cooperation $W(\theta_k^{\sigma_k}; k, x; \overline{u}_{k-}^{(\omega)})$, which reflects the expected maximized joint payoffs of the firm covering stage k to stage $T+1$, if $\theta_k^{\sigma_k}$ occurs.

Proposition 6.1. *System* (6.3)–(6.4) *admits a solution with the expected optimal cooperative joint payoff of the firms being.*

$$W(\theta_k^{\sigma_k}; k, x; \overline{u}_{k-}^{(\omega)}) = (A_k^{\sigma_k} x + C_k^{\sigma_k})\delta^{k-1}, \tag{6.5}$$

where $A_{T+1}^{\sigma_{T+1}} = \sum_{i=1}^{n} \theta_{T+1}^{\sigma_{T+1}} Q_{T+1}^i$

and $C_{T+1}^{\sigma_{T+1}} = \sum_{i=1}^{n} \left(\sum_{\tau=T+1-\omega+1}^{T} \upsilon_{T+1}^{|\tau|i} \overline{u}_\tau^{(\omega)i} + \varpi_{T+1}^i \right)$,

for $\sigma_{T+1} \in \{1, 2, \ldots, \eta_{T+1}\}$

$$A_k^{\sigma_k} = \sum_{i=1}^{n} \frac{(\theta_k^{\sigma_k} R_k^i)^2}{4c_k^i} + \delta \sum_{z=1}^{\eta_{k+1}} \lambda_{k+1}^z A_{k+1}^z \left(1 - \lambda_k + \sum_{y=1}^{\mu_k} \gamma_k^y \vartheta_k^y G_k^1\right),$$

$$\text{for } \sigma_k \in \{1, 2, \ldots, \eta_k\} \text{ and } k \in \{1, 2, \ldots, T\}, \tag{6.6}$$

and $C_k^{\sigma_k}$ is an expression that contains the previously executed controls $\overline{u}_{k-}^{(\omega)}$.

Proof. See Appendix III. ■

Note that the values of $A_k^{\sigma_k}$, for $\sigma_k \in \{1, 2, \ldots, \eta_k\}$ and $k \in \{1, 2, \ldots, T\}$, can be obtained explicitly from (6.6). Using Proposition 6.1, one can obtain the optimal cooperative in stage k as (see derivation details in Appendix III):

$u_k^{(\sigma_k)i*} = \frac{(\theta_k^{\sigma_k} R_k^i)^2 x}{4(c_k^i)^2}$, for $\sigma_k \in \{1, 2, \ldots, \eta_k\}$ and $k \in \{1, 2, \ldots, T\}$; and

$$\bar{u}_k^{(\sigma_k)(\omega)i*} = \frac{\sum_{z=1}^{\eta_{k+1}} \lambda_{k+1}^z A_{k+1}^z \left(a_k^{|k|i} + \sum_{y=1}^{\mu_k} \gamma_k^y \vartheta_k^y G_k^{2(i)}\right) + \sum_{\tau=k+1}^{k+\omega-1} \delta^{\tau-k+1} \sum_{z=1}^{\eta_{\tau+1}} \lambda_{\tau+1}^z A_{\tau+1}^z a_\tau^{|k|i} + \delta^{(T+1-k)} \upsilon_{T+1}^{|k|i}}{2\phi_k^i}, \tag{6.7}$$

for $\sigma_k \in \{1, 2, \ldots, \eta_k\}, k \in \{T-\omega+2, T-\omega+3, \ldots, T\}$ and $A_\tau = 0$ for $\tau > T+1$;

$$\overline{u}_k^{(\sigma_k)(\omega)i*} = \frac{\sum_{z=1}^{\eta_{k+1}} \lambda_{k+1}^z A_{k+1}^z \left(a_k^{|k|i} + \sum_{y=1}^{\mu_k} \gamma_k^y \vartheta_k^y G_k^{2(i)}\right) + \sum_{\tau=k+1}^{k+\omega-1} \delta^{\tau-k+1} \sum_{z=1}^{\eta_{\tau+1}} \lambda_{\tau+1}^z A_{\tau+1}^z a_\tau^{|k|i}}{2\phi_k^i},$$

$$\text{for } \sigma_k \in \{1, 2, \ldots, \eta_k\} \text{ and } k \in \{1, 2, \ldots, T-\omega+1\} \tag{6.8}$$

We use X_k^* to denote the set of realizable values of x_k at stage k generated. The term $x_k^* \in X_k^*$ is used to denote an element in X_k^*.

Given the agreed-upon sharing of the total cooperative payoff proportional to the firms' non-cooperative payoffs, the cooperative payoff of firm i along the cooperative trajectory must fulfil:

$$\xi^i(\theta_k^{\sigma_k}; k, x_k^*; \underline{\overline{u}}_{k-}^{(\omega)*}) = \frac{V^i(\theta_k^{\sigma_k}; k, x_k^*; \underline{\overline{u}}_{k-}^{(\omega)*})}{\sum_{j=1}^n V^j(\theta_k^{\sigma_k}; k, x_k^*; \underline{\overline{u}}_{k-}^{(\omega)*})} W(\theta_k^{\sigma_k}; k, x_k^*; \underline{\overline{u}}_{k-}^{(\omega)*}), \quad (6.9)$$

for $i \in N$, $\sigma_k \in \{1, 2, \ldots, \eta_k\}$ and $k \in \{1, 2, \ldots, T\}$.

Invoking Theorem 5.2 and (5.17), to maintain the condition in (6.9), a payment

$$\beta_k^i(\theta_k^{\sigma_k}; x_k^*; \underline{\overline{u}}_{k-}^*) = \frac{\left(A_k^{(\sigma)i} x_k^* + C_k^{(\sigma_k)i}\right)}{\sum_{j=1}^n \left(A_k^{(\sigma_k)j} x_k^* + C_k^{(\sigma_k)j}\right)} (A_k^{\sigma_k} x_k^* + C_k^{\sigma_k})$$

$$- \delta \sum_{z=1}^{\eta_{k+1}} \lambda_{k+1}^z \sum_{y=1}^{\mu_k} \gamma_k^y \left[\frac{A_{k+1}^{(z)i} x_{k+1}^{(\sigma_k)y} + C_{k+1}^{(z)i}}{\sum_{j=1}^n \left(A_{k+1}^{(z)j} x_{k+1}^{(\sigma_k)y} + C_{k+1}^{(z)j}\right)} \left(A_{k+1}^z x_{k+1}^{(\sigma_k)y} + C_{k+1}^z\right) \right], \quad (6.10)$$

where $x_{k+1}^{(\sigma_k)y} = x_k + \sum_{j=1}^n a_k^{|k|j} \overline{u}_k^{(\sigma_k)(\omega)j*} + \sum_{j=1}^n \sum_{t=k-\omega+1}^{k-1} a_k^{|t|j} \overline{u}_t^{(\omega)j*} - \lambda_k x_k$

$$+\vartheta_k^y \left(G_k^1 x_k + \sum_{j=1}^n G_k^{2(j)} \overline{u}_j^{(\sigma_k)(\omega)j*} \right)$$

and $\overline{u}_t^{(\omega)j*} \in \underline{\overline{u}}_{k-}^{(\omega)*}$

will be given to player $i \in N$ at stage $k \in \{1, 2, \ldots, T\}$.

Under cooperation, all firms would use the cooperative strategies and the payoff that firm i will directly receive at stage k along the cooperative trajectory becomes.

$[\theta_k^{\sigma_k} R_k^i (u_k^{(\sigma_k)i*})^{1/2} (x_k^*)^{1/2} - c_k^i u_k^{(\sigma_k)i*} - \phi_k^i (\overline{u}_k^{(\sigma_k)(\omega)i*})^2]$.

However, according to the agreed-upon imputation, firm i will receive $\beta_k^i(\theta_k^{\sigma_k}; x_k^*; \underline{\overline{u}}_{k-}^{(\omega)*})$ at stage k. Therefore, a side-payment

$$\pi_k^i(\theta_k^{\sigma_k}; x_k^*; \underline{\overline{u}}_{k-}^{(\omega)*}) = \beta_k^i(\theta_k^{\sigma_k}; x_k^*; \underline{\overline{u}}_{k-}^*)$$

$$- \left[\theta_k^{\sigma_k} R_k^i (u_k^{(\sigma_k)i*})^{1/2} (x_k^*)^{1/2} - c_k^i u_k^{(\sigma_k)i*} - \phi_k^i (\overline{u}_k^{(\sigma_k)(\omega)i*})^2\right], \quad (6.11)$$

for $\sigma_k \in \{1, 2, \ldots, \eta_k\}$ and $k \in \{1, 2, \ldots, T\}$,

has to be given to firm $i \in N$ to yield the cooperative imputation

$$\xi^i\left(\theta_k^{\sigma_k}; k, x_k^*; \overline{u}_{k-}^{(\omega)*}\right) = \frac{V^i\left(\theta_k^{\sigma_k}; k, x_k^*; \overline{u}_{k-}^{(\omega)*}\right)}{\sum_{j=1}^n V^j\left(\theta_k^{\sigma_k}; k, x_k^*; \overline{u}_{k-}^{(\omega)*}\right)} W\left(\theta_k^{\sigma_k}; k, x_k^*; \overline{u}_{k-}^{(\omega)*}\right).$$

Finally, comparing $A_k^{(\sigma_k)i}$ in (4.6) with $A_k^{\sigma_k}$ in (6.6) shows that $A_k^{\sigma_k} > A_k^{(\sigma_k)i}$, for $i \in N$ and $k \in \{1, 2, \ldots, T\}$. Hence, the optimal cooperative investment in public capital by firm i is larger than the non-cooperative investment in public capital by firm i, that is

$$\begin{aligned}
&\overline{u}_k^{(\sigma_k)(\omega)i*} \\
&= \frac{\sum_{z=1}^{\eta_{k+1}} \lambda_{k+1}^z A_{k+1}^z \left(a_k^{|k|i} + \sum_{y=1}^{\mu_k} \gamma_k^y \vartheta_k^y G_k^{2(i)}\right) + \sum_{\tau=k+1}^{k+\omega-1} \delta^{\tau-k+1} \sum_{z=1}^{\eta_{\tau+1}} \lambda_{\tau+1}^z A_{\tau+1}^z a_\tau^{|k|i} + \delta^{(T+1-k)} v_{T+1}^{|k|i}}{2\phi_k^i} \\
&> \overline{u}_k^{(\sigma_k)(\omega)i**} \\
&= \frac{\sum_{z=1}^{\eta_{k+1}} \lambda_{k+1}^z A_{k+1}^{(z)i} \left(a_k^{|k|i} + \sum_{v=1}^{\mu_k} \gamma_k^v \vartheta_k^v G_k^{2(i)}\right) + \sum_{\tau=k+1}^{k+\omega-1} \delta^{\tau-k+1} \sum_{z=1}^{\eta_{\tau+1}} \lambda_{\tau+1}^z A_{\tau+1}^{(z)i} a_\tau^{|k|i} + \delta^{(T+1-k)} v_{T+1}^{|k|i}}{2\phi_k^i},
\end{aligned}$$

for $k \in \{T - \omega + 2, T - \omega + 3, \ldots, T\}$, $\sigma_k \in \{1, 2, \ldots, \eta_k\}$ and $A_\tau^{(z)i} = A_\tau^z = 0$ for $\tau > T + 1$; and

$$\begin{aligned}
&\overline{u}_k^{(\sigma_k)(\omega)i*} \\
&= \frac{\sum_{z=1}^{\eta_{k+1}} \lambda_{k+1}^z A_{k+1}^z \left(a_k^{|k|i} + \sum_{y=1}^{\mu_k} \gamma_k^y \vartheta_k^y G_k^{2(i)}\right) + \sum_{\tau=k+1}^{k+\omega-1} \delta^{\tau-k+1} \sum_{z=1}^{\eta_{\tau+1}} \lambda_{\tau+1}^z A_{\tau+1}^z a_\tau^{|k|i}}{2\phi_k^i} \\
&> \overline{u}_k^{(\sigma_k)(\omega)i**} \\
&= \frac{\sum_{z=1}^{\eta_{k+1}} \lambda_{k+1}^z A_{k+1}^{(z)i} \left(a_k^{|k|i} + \sum_{v=1}^{\mu_k} \gamma_k^v \vartheta_k^v G_k^{2(i)}\right) + \sum_{\tau=k+1}^{k+\omega-1} \delta^{\tau-k+1} \sum_{z=1}^{\eta_{\tau+1}} \lambda_{\tau+1}^z A_{\tau+1}^{(z)i} a_\tau^{|k|i}}{2\phi_k^i},
\end{aligned}$$

$$\text{for } k \in \{1, 2, \ldots, T - \omega + 1\} \text{ and } \sigma_k \in \{1, 2, \ldots, \eta_k\}. \tag{6.12}$$

With investments in public capital by firms under cooperation being larger than those under non-cooperation, the public capital stock under cooperation is larger than that under non-cooperation, that is $x_k^* > x_k^{**}$, for $k \in \{2, 3, \ldots, T\}$. With the internalization of the positive externalities of public capital and the individually rational payoff sharing scheme $\xi^i(\theta_k^\sigma; k, x_k^*; \overline{u}_{k-}^{(\omega)*})$ in (6.9), the firm's expected cooperative payoff is larger than its non-cooperative counterpart.

7 An Extension

In this section, we extend the stochastic state dynamics to one in which previously executed strategies $\overline{u}_{k-}$ in the state dynamics are also subject to stochastic shocks. In particular, the evolution of the state dynamics is governed by

$$x_{k+1} = f_k(x_k, \underline{u}_k, \overline{u}_k; \overline{u}_{k-}) + \vartheta_k G_k(x_k, \underline{u}_k, \overline{u}_k; \overline{u}_{k-}), \quad x_1 = x_1^0, \quad \text{for } k \in \{1, 2, \dots, T\} \tag{7.1}$$

The expected payoff of player i is:

$$E_{\theta_2, \theta_3, \dots, \theta_{T+1}; \vartheta_1, \vartheta_2, \dots, \vartheta_T} \left\{ \sum_{k=1}^{T} g_k^i(\theta_k; x_k, \underline{u}_k, \overline{u}_k; \overline{u}_{k-}) \delta_1^k + q_{T+1}^i(\theta_{T+1} x_{T+1}; \underline{u}_{(T+1)-}) \delta_1^{T+1} \right\}, \tag{7.2}$$

for $i \in \{1, 2, \dots, n\} \equiv N$,

where $q_{T+1}^i(\theta_{T+1}; x_{T+1}; \underline{u}_{(T+1)-})$ is the terminal payoff of player i.

The controls executed before the start of the operation in stage 1, that is $\overline{u}_{1-}$, are known and some or all of them can be zeros. The $g_k^i(\theta_k; x_k, \underline{u}_k, \overline{u}_k; \overline{u}_{k-})$, $f_k^i(x_k, \underline{u}_k, \overline{u}_k; \overline{u}_{k-})$, $G_k(x_k, \underline{u}_k, \overline{u}_k; \overline{u}_{k-})$ and $q_{T+1}^i(\theta_{T+1}; x_{T+1}; \underline{u}_{(T+1)-})$ are continuously differentiable functions.

7.1 Game Equilibria

The non-cooperative payoffs of the players in a Nash equilibrium of the dynamic game (7.1)–(7.2) can be characterized by the following theorem.

Theorem 7.1. *Let $(\underline{u}_k^{**}, \overline{u}_k^{**})$ be the set of Nash equilibrium strategies and $V^i(\theta_k^{\sigma_k}; k, x; \overline{u}_{k-}^{**})$ be the Nash equilibrium payoff of player i at stage k in the non-cooperative dynamic game (7.1)–(7.2) when $\theta_k^{\sigma_k}$ occurs, then the function $V^i(\theta_k^{\sigma_k}; k, x; \overline{u}_{k-}^{**})$ satisfies the following recursive equations:*

$$V^i(\theta_{T+1}^{\sigma_{T+1}}; T+1, x; \overline{\underline{u}}_{(T+1)-}^{**}) = q_{T+1}^i(\theta_{T+1}^{\sigma_{T+1}}; x; \overline{\underline{u}}_{(T+1)-}^{**}) \delta_1^{T+1}, \tag{7.3}$$

for $\sigma_{T+1} \in \{1, 2, \dots, \eta_{T+1}\}$;

$$V^i(\theta_k^{\sigma_k}; k, x; \overline{u}_{k-}^{**}) = \max_{u_k^i, \overline{u}_k^i} \left\{ \sum_{\upsilon=1}^{\eta_{k+1}} \lambda_{k+1}^{\upsilon} \sum_{\upsilon=1}^{\mu_k} \gamma_k^{\upsilon} \Big[g_k^i\Big(\theta_k^{\sigma_k}; x, u_k^i, \overline{u}_k^i, \underline{u}_k^{**(\neq i)}, \overline{u}_k^{**(\neq i)}; \overline{u}_{k-}^{**}\Big) \delta_1^k \right.$$
$$+ V^i[\theta_{k+1}^{\upsilon}; k+1, f_k(x, u_k^i, \overline{u}_k^i, \underline{u}_k^{**(\neq i)}, \overline{u}_k^{**(\neq i)}; \overline{u}_{k-}^{**})$$
$$\left. + \vartheta_k^y G_k(x, u_k^i, \overline{u}_k^i, \underline{u}_k^{**(\neq i)}, \overline{u}_k^{**(\neq i)}; \overline{u}_{k-}^{**}); \overline{u}_k^i, \overline{u}_k^{**(\neq i)}, \overline{\underline{u}}_{(k+1)-}^{**} \cap \overline{u}_{k-}^{**}] \Big] \right\}, \tag{7.4}$$

for $\sigma_k \in \{1, 2, \dots, \eta_k\}$, $k \in \{1, 2, \dots, T\}$ and $i \in N$,

where $\underline{u}_k^{(\neq i)**} = (u_k^{1**}, u_k^{2**}, \dots, u_k^{i-1**}, u_k^{i+1**}, \dots, u_k^{n**})$ and

$$\underline{\overline{u}}_k^{(\neq i)**} = (\overline{u}_k^{1**}, \overline{u}_k^{2**}, \ldots, \overline{u}_k^{i-1**}, \overline{u}_k^{i+1**}, \ldots, \overline{u}_k^{n**}).$$

Proof. Conditions (7.3)–(7.4) show that $V^i(\theta_k^{\sigma_k}; k, x; \underline{\overline{u}}_{k-}^{**})$ is the maximized payoff of player $i \in N$ according to Theorem 1.1, given the game equilibrium strategies of the other $n-1$ players. Hence a Nash equilibrium results. ■

7.2 An Example in Public Capital Provision

Consider an illustrative example of an oligopoly with n firms. The planning horizon of these firms involves T stages. The firms use knowledge-based public capital $x_k \in X \subset R^+$ to produce its output. Capital investment takes ω stages to complete its built up into the capital stock. The accumulation process of the capital stock is governed by the dynamics

$$x_{k+1} = x_k + \sum_{j=1}^{n} a_k^{|k|j} \overline{u}_k^{(\omega)j} + \sum_{j=1}^{n} \sum_{\tau=k-\omega+1}^{k-1} a_k^{|\tau|j} \overline{u}_\tau^{(\omega)j} - \lambda_k x_k$$
$$+ \vartheta_k \left(G_k^1 x_k + \sum_{j=1}^{n} G_k^{|k|j} \overline{u}_k^{(\omega)j} + \sum_{j=1}^{n} \sum_{\tau=k-\omega+1}^{k-1} G_k^{|\tau|j} \overline{u}_\tau^{(\omega)j} \right),$$
$$x_1 = x_1^0 \tag{7.5}$$

where $\overline{u}_k^{(\omega)j}$ is the investment in the knowledge-based capital by firm j in stage k, $a_k^{|k|j} \overline{u}_k^{(\omega)j}$ is the addition to the capital in stage k by firm j's capital building investment $\overline{u}_k^{(\omega)j}$ in stage k, $a_k^{|\tau|j} \overline{u}_\tau^{(\omega)j}$ is the addition to the capital in stage k by capital building investment $\overline{u}_\tau^{(\omega)j}$ in stage $\tau \in \{k-1, k-2, \ldots, k-\omega+1\}$ and λ_k is the rate of obsolescence of the capital.

Moreover, G_k^1 and $G_k^{|\tau|j}$, for $\tau \in \{k, k-1, \ldots, k-\omega+1\}$, are positive parameters and ϑ_k is a random variable with a range $\{\vartheta_k^1, \vartheta_k^2, \ldots, \vartheta_k^{\mu_k}\}$ and corresponding probabilities $\{\gamma_k^1, \gamma_k^2, \ldots, \gamma_k^{\mu_k}\}$. Let $\vartheta_k^{\hat{\nu}}$ denote the smallest (positive or negative) value of ϑ_k^{ν}, for $\nu \in \{1, 2, \ldots, \mu_k\}$. The value of G_k^1 satisfies the condition $\vartheta_k^{\hat{\nu}} G_k^1 > (\lambda_k - 1)$. The value of $G_k^{|\tau|j}$ satisfies the condition $\vartheta_k^{\hat{\nu}} G_k^{|\tau|j} > -a_k^{|\tau|j}$, for $j \in \{1, 2, \ldots, n\}$.

Firm $i \in N$ uses the knowledge-based capital together with a non-durable productive input u_k^i to produce outputs. The revenue from the output produced is $\theta_k^{\sigma_k} R_k^i (u_k^i)^{1/2} (x_k)^{1/2}$, if $\theta_k^{\sigma_k}$ occurs in stage k. The non-durable input cost is $c_k^i u_k^i$ and the cost of capital investment is $\phi_k^i (\overline{u}_k^{(\omega)i})^2$.

The expected payoff of the firm i is

$$E_{\theta_2, \theta_3, \ldots, \theta_{T+1}; \vartheta_1, \vartheta_2, \ldots, \vartheta_T} \left\{ \sum_{k=1}^{T} \left(\left[\theta_k^{\sigma_k} R_k^i (u_k^i)^{1/2} (x_k)^{1/2} - c_k^i u_k^i - \phi_k^i (\overline{u}_k^{(\omega)i})^2 \right] \right) \delta^{k-1} \right.$$

$$+\left(\theta_{T+1} Q^i_{T+1} x_{T+1} + \sum_{\tau=T+1-\omega+1}^{T} v^{|\tau|i}_{T+1} \overline{u}^{(\omega)i}_{\tau} + \varpi^i_{T+1}\right)\delta^T \Bigg\}, \tag{7.6}$$

where $\left(\theta_{T+1} Q^i_{T+1} x_{T+1} + \sum_{\tau=T+1-\omega+1}^{T} v^{|\tau|i}_{T+1} \overline{u}^{(\omega)i}_{\tau} + \varpi^i_{T+1}\right)$ is the salvage value of firm i in stage $T+1$ and $\delta = (1+r)^{-1}$ is the discount factor.

The salvage value of the firm i in stage $T+1$ depends on the public capital stock, with $\theta^{\sigma_{T+1}}_{T+1} Q^i_{T+1}$, for $\sigma_{T+1} \in \{1, 2, \ldots, \eta_{T+1}\}$, being positive. It also depends on the previous capital investment $\sum_{\tau=T+1-\omega_3+1}^{T} v^{|\tau|i}_{T+1} \overline{u}^{(\omega)i}_{\tau}$. In particular, some or all $v^{|\tau|}_{T+1}$, for $\tau \in \{T, T-1, \ldots, T-\omega+2\}$, can be zero.

Invoking Theorem 7.1, we obtain.

Corollary 7.1. *Let*

$$\left(\underline{u}^{**}_k, \underline{u}^{(\omega)**}_k\right) = \Big[\left(u^{1**}_k, u^{2**}_k, \ldots, u^{n**}_k\right),$$
$$\left(u^{(\omega)1**}_k, u^{(\omega)2**}_k, \ldots, u^{(\omega)n**}_k\right)\Big]$$

be the set of feedback Nash equilibrium strategies and $V^i(\theta^{\sigma_k}_k; k, x; \overline{\underline{u}}^{**}_{k-})$ *be the feedback Nash equilibrium payoff of player i in the non-cooperative game* (7.5)–(7.6) if $\theta^{\sigma_k}_k$ *occurs, then the function* $V^i(\theta^{\sigma_k}_k; k, x; \overline{\underline{u}}^{**}_{k-})$ *satisfies the following recursive equations:*

$$V^i\left(\theta^{\sigma_{T+1}}_{T+1}; T+1, x; \overline{\underline{u}}^{(\omega)**}_{(T+1)-}\right) = \left(\theta^{\sigma_{T+1}}_{T+1} Q^i_{T+1} x + \sum_{\tau=T+1-\omega+1}^{T} v^{|\tau|i}_{T+1} \overline{u}^{(\omega)i}_{\tau} + \varpi^i_{T+1}\right)\delta^T,$$

$$\text{for } \sigma_{T+1} \in \{1, 2, \ldots, \eta_{T+1}\}; \tag{7.7}$$

$$V^i(\theta^{\sigma_k}_k; k, x; \overline{\underline{u}}^{**}_{k-})$$
$$= \max_{u^i_k, \bar{u}^{(\omega)i}_k} \Bigg\{ \sum_{\upsilon=1}^{\eta_{k+1}} \lambda^{\upsilon}_{k+1} \sum_{\nu=1}^{\mu_k} \gamma^{\nu}_k \Big[[\theta^{\sigma_k}_k R^i_k (u^i_k)^{1/2} (x)^{1/2} - c^i_k u^i_k - \phi^i_k (\bar{u}^{(\omega)i}_k)^2] \delta^{k-1}$$
$$+ V^i[\theta^{\upsilon}_{k+1}; k+1, x + a^{|k|i}_k \bar{u}^i_k + \sum_{j\in N, j\neq i} a^{|k|j}_k \bar{u}^{(\omega)j**}_k + \sum_{j=1}^{n} \sum_{\tau=k-\omega+1}^{k-1} a^{|\tau|j}_k \bar{u}^{(\omega)j**}_{\tau} - \lambda_k x$$
$$+ \vartheta^{\nu}_k \Big(G^1_k x_k + G^{|k|i}_k \bar{u}^{(\omega)i}_j + \sum_{\substack{j=1 \\ j\neq i}}^{n} G^{|k|j}_k \bar{u}^{(\omega)j**}_k$$
$$+ \sum_{j=1}^{n} \sum_{\tau=k-\omega+1}^{k-1} G^{|\tau|j}_k \bar{u}^{(\omega)j**}_{\tau} \Big); \bar{u}^{(\omega)i}_k, \underline{\bar{u}}^{(\omega)(\neq i)**}_k, \underline{\bar{u}}^{(\omega)**}_{(k+1)-} \cap \underline{\bar{u}}^{(\omega)**}_{k-}] \Big] \Bigg\},$$

$$\text{for } \sigma_k \in \{1, 2, \cdots, \eta_k\}, k \in \{1, 2, \cdots, T\} \text{ and } i \in N. \tag{7.8}$$

■

Performing the indicated maximization operator in Corollary 7.1, the value function $V^i(\theta_k^{\sigma_k}; k, x; \overline{u}_{k-}^{**})$ which reflects the value of the firm covering stage k to stage $T+1$ can be obtained as follows.

Proposition 7.1. *System (7.7)–(7.8) admits a solution with the game equilibrium payoff of firm i being.*

$$V^i(\theta_k^{\sigma_k}; k, x; \overline{u}_{k-}^{(\omega)**}) = (A_k^{(\sigma_k)i} x + C_k^{(\sigma_k)i})\delta^{k-1}, \tag{7.9}$$

where $A_{T+1}^{(\sigma_{T+1})i} = Q_{T+1}^{(\sigma_{T+1})i}$ and $C_{T+1}^{(\sigma_{T+1})i} = \sum_{\tau=T+1-\omega+1}^{T} \upsilon_{T+1}^{|\tau|i} \overline{u}_\tau^{(\omega)i} + \varpi_{T+1}^i$,

for $\sigma_{T+1} \in \{1, 2, \ldots, \eta_{T+1}\}$ and $i \in N$;

$$A_k^{(\sigma_k)i} = \frac{(\theta_k^{\sigma_k} R_k^i)^2}{4c_k^i} + \delta \sum_{z=1}^{\eta_{k+1}} \lambda_{k+1}^z A_{k+1}^{(z)i} \left(1 - \lambda_k + \sum_{y=1}^{\mu_k} \gamma_k^y \vartheta_k^y G_k^1 \right), \tag{7.10}$$

and $C_k^{(\sigma_k)i}$ is an expression that contains the previously executed durable controls $\overline{u}_{k-}^{(\omega)**}$,
for $\sigma_k \in \{1, 2, \ldots, \eta_k\}$, $k \in \{1, 2, \ldots, T\}$ and $i \in N$.

Proof. Follow the proof of Proposition 4 in Appendix II. ■

Using Proposition 7.1 and Corollary 7.1 one can obtain the game equilibrium strategies in stage k as:
$u_k^{(\sigma_k)i**} = \frac{(\theta_k^{\sigma_k} R_k^i)^2 x}{4(c_k^i)^2}$, for $k \in \{1, 2, \ldots, T\}$; and

$$\overline{u}_k^{(\sigma_k)(\omega)i**} = \frac{1}{2\phi_k^i} \Bigg[\sum_{z=1}^{\eta_{k+1}} \lambda_{k+1}^z A_{k+1}^{(z)i} \left(a_k^{|k|i} + \sum_{y=1}^{\mu_k} \gamma_k^y \vartheta_k^y G_k^{2(i)} \right)$$

$$+ \sum_{\tau=k+1}^{k+\omega-1} \delta^{\tau-k+1} \sum_{z=1}^{\eta_{\tau+1}} \lambda_{\tau+1}^z A_{\tau+1}^{(z)i} \left(a_\tau^{|k|i} + \sum_{y=1}^{\mu_\tau} \gamma_\tau^y \vartheta_\tau^y G_k^{|\tau|} \right) + \delta^{(T+1-k)} \upsilon_{T+1}^{|k|i} \Bigg],$$

for $k \in \{T-\omega+2, T-\omega+3, \ldots, T\}$ and $A_\tau^i = 0$ for $\tau > T+1$;

$$\overline{u}_k^{(\sigma_k)(\omega)i**} = \frac{1}{2\phi_k^i} \left[\sum_{z=1}^{\eta_{k+1}} \lambda_{k+1}^z A_{k+1}^{(z)i} \left(a_k^{|k|i} + \sum_{y=1}^{\mu_k} \gamma_k^y \vartheta_k^y G_k^{2(i)} \right) \right.$$

$$+\sum_{\tau=k+1}^{k+\omega-1}\delta^{\tau-k+1}\sum_{z=1}^{\eta_{\tau+1}}\lambda_{\tau+1}^{z}A_{\tau+1}^{(z)i}\left(a_{\tau}^{|k|i}+\sum_{y=1}^{\mu_{\tau}}\gamma_{\tau}^{y}\vartheta_{\tau}^{y}G_{k}^{|\tau|}\right)\Bigg],$$

$$\text{for } k\in\{1,2,\cdots,T-\omega+1\} \qquad (7.11)$$

8 Appendices

8.1 Appendix I: Proof of Proposition 2.1

Using (2.3) and Proposition 2.1, we have

$$V(\theta_{T+1}^{\sigma_{T+1}};T+1,x;\bar{u}_{(T+1)-}^{(\omega)})=\left(\theta_{T+1}^{\sigma_{T+1}}Q_{T+1}x_{T+1}+\sum_{\tau=T+1-\omega+1}^{T}v_{T+1}^{|\tau|}\bar{u}_{\tau}^{(\omega)}+\varpi_{T+1}\right)\delta^{T}$$
$$=(A_{T+1}^{\sigma_{T+1}}x_{T+1}+C_{T+1}^{\sigma_{T+1}})\delta^{T},$$

where $A_{T+1}^{\sigma_{T+1}}=\theta_{T+1}^{\sigma_{T+1}}Q_{T+1}$ and $C_{T+1}^{\sigma_{T+1}}=\sum_{\tau=T+1-\omega+1}^{T}v_{T+1}^{|\tau|}\overline{u}_{\tau}^{(\omega)}+\varpi_{T+1}$,

$$\text{for } \sigma_{T+1}\in\{1,2,\ldots,\eta_{T+1}\}. \qquad (8.1)$$

Using Proposition 2.1 and (2.4), we can express the optimal strategies in stage T as:

$$u_{T}^{\sigma_T}=\frac{(\theta_{T}^{\sigma_T}R_T)^2x}{4(c_T)^2}$$

and

$$\overline{u}_{T}^{\sigma_T(\omega)}=\frac{\delta\sum_{\upsilon=1}^{\eta_{T+1}}\lambda_{T+1}^{\upsilon}A_{T+1}^{\upsilon}\left(a_{T}^{|T|}+\sum_{\upsilon=1}^{\mu_T}\gamma_{T}^{\upsilon}\vartheta_{T}^{\upsilon}G_{T}^{2}\right)+\delta v_{T+1}^{|T|}}{2\phi_T},$$

$$\text{for } \sigma_{T+1}\in\{1,2,\ldots,\eta_{T+1}\}. \qquad (8.2)$$

Invoking Proposition 2.1, we have $V(\theta_{T}^{\sigma_T};T,x;\overline{u}_{T-}^{(\omega)})=(A_{T}^{\sigma_T}x+C_{T}^{\sigma_T})\delta^{T-1}$. Substituting the optimal strategies in (8.2) into the stage T equation of (2.4) we obtain

$$(A_{T}^{\sigma_T}x+C_{T}^{\sigma_T})\delta^{T-1}$$
$$=\left(\frac{(\theta_{T}^{\sigma_T}R_T)^2x}{4c_T}-\frac{\left(\delta\sum_{\upsilon=1}^{\eta_{T+1}}\lambda_{T+1}^{\upsilon}A_{T+1}^{\upsilon}\left(a_{T}^{|T|}+\sum_{\upsilon=1}^{\mu_T}\gamma_{T}^{\upsilon}\vartheta_{T}^{\upsilon}G_{T}^{2}\right)+\delta v_{T+1}^{|T|}\right)^2}{4\phi_T}\right)\delta^{T-1}$$

$$+\sum_{z=1}^{\eta_{T+1}} \lambda_{T+1}^{z}\left[A_{T+1}^{z}\sum_{y=1}^{\mu_T}\gamma_T^y\left(x+a_T^{|T|}\frac{\delta\sum_{\upsilon=1}^{\eta_{T+1}}A_{T+1}^{\upsilon}\left(a_T^{|T|}+\sum_{\upsilon=1}^{\mu_T}\gamma_T^{\upsilon}\vartheta_T^{\upsilon}G_T^2\right)+\delta v_{T+1}^{|T|}}{2\phi_T}\right.\right.$$

$$+\sum_{\tau=T-\omega+1}^{T-1} a_T^{|\tau|}\bar{u}_\tau^{(\omega)}-\lambda_T x+\vartheta_T^y G_T^2\frac{\delta\sum_{\upsilon=1}^{\eta_{T+1}}A_{T+1}^{\upsilon}\left(a_T^{|T|}+\sum_{\upsilon=1}^{\mu_T}\gamma_T^{\upsilon}\vartheta_T^{\upsilon}G_T^2\right)+\delta v_{T+1}^{|T|}}{2\phi_T}$$

$$\left.+\vartheta_T^y G_T^1 x\right)+v_{T+1}^{|T|}\frac{\delta\sum_{\upsilon=1}^{\eta_{T+1}}A_{T+1}^{\upsilon}\left(a_T^{|T|}+\sum_{\upsilon=1}^{\mu_T}\gamma_T^{\upsilon}\vartheta_T^{\upsilon}G_T^2\right)+\delta v_{T+1}^{|T|}}{2\phi_T}$$

$$\left.+\sum_{\tau=T+1-\omega_3+1}^{T-1} v_{T+1}^{|\tau|}\bar{u}_\tau^{(\omega)}+\varpi_{T+1}\right]\delta^T. \tag{8.3}$$

The RHS and the LHS of (8.3) are linear functions of x. For (8.3) to hold, it is required:

$$A_T^{\sigma_T}=\frac{(\theta_T^{\sigma_T}R_T)^2}{4c_T}+\delta\sum_{z=1}^{\eta_{T+1}}\lambda_{T+1}^z A_{T+1}^z\left(1-\lambda_T+\sum_{y=1}^{\mu_T}\gamma_T^y\vartheta_T^y G_T^1\right). \tag{8.4}$$

In addition,

$$C_T=\left(-\frac{\left(\delta\sum_{\upsilon=1}^{\eta_{T+1}}\lambda_{T+1}^{\upsilon}A_{T+1}^{\upsilon}\left(a_T^{|T|}+\sum_{\upsilon=1}^{\mu_T}\gamma_T^{\upsilon}\vartheta_T^{\upsilon}G_T^2\right)+\delta v_{T+1}^{|T|}\right)^2}{4\phi_T}\right)$$

$$+\sum_{\upsilon=1}^{\eta_{T+1}}\lambda_{T+1}^{z}\left[A_{T+1}^{z}\sum_{y=1}^{\mu_T}\gamma_T^y\left(a_T^{|T|}\frac{\delta\sum_{\upsilon=1}^{\eta_{T+1}}A_{T+1}^{\upsilon}\left(a_T^{|T|}+\sum_{\upsilon=1}^{\mu_T}\gamma_T^{\upsilon}\vartheta_T^{\upsilon}G_T^2\right)+\delta v_{T+1}^{|T|}}{2\phi_T}\right.\right.$$

$$\left.+\sum_{\tau=T-\omega+1}^{T-1} a_T^{|\tau|}\bar{u}_\tau^{(\omega)}+\vartheta_T^y G_T^2\frac{\delta\sum_{\upsilon=1}^{\eta_{T+1}}A_{T+1}^{\upsilon}\left(a_T^{|T|}+\sum_{\upsilon=1}^{\mu_T}\gamma_T^{\upsilon}\vartheta_T^{\upsilon}G_T^2\right)+\delta v_{T+1}^{|T|}}{2\phi_T}\right)$$

$$+v_{T+1}^{|T|}\frac{\delta\sum_{\upsilon=1}^{\eta_{T+1}}A_{T+1}^{\upsilon}\left(a_T^{|T|}+\sum_{\upsilon=1}^{\mu_T}\gamma_T^{\upsilon}\vartheta_T^{\upsilon}G_T^2\right)+\delta v_{T+1}^{|T|}}{2\phi_T}$$

$$\left.+\sum_{\tau=T+1-\omega_3+1}^{T-1} v_{T+1}^{|\tau|}\bar{u}_\tau^{(\omega)}+\varpi_{T+1}\right]\delta, \tag{8.5}$$

which is a function of previously executed controls $\overline{u}_{T-}^{(\omega)}$.

Then we move to stage $T-1$.

Using $V(\theta_T^{\sigma_T}; T, x; \overline{u}_{T-}^{(\omega)}) = (A_T^{\sigma_T} x + C_T^{\sigma_T})\delta^{T-1}$ derived in (8.3)–(8.5) and the stage $T-1$ equation in (2.4), the optimal strategies in stage $T-1$ can be obtained as:

$$u_{T-1}^{\sigma_{T-1}} = \frac{(\theta_{T-1}^{\sigma_{T-1}} R_{T-1})^2 x}{4(c_{T-1})^2} \text{ and}$$

$$\overline{u}_{T-1}^{\sigma_{T-1}(\omega)} = \frac{\delta \sum_{\upsilon=1}^{\eta_T} \lambda_T^\upsilon A_T^\upsilon \left(a_{T-1}^{|T-1|} + \sum_{\nu=1}^{\mu_{T-1}} \gamma_{T-1}^\nu \vartheta_{T-1}^\nu G_{T-1}^2\right) + \delta^2 \left(\sum_{\upsilon=1}^{\eta_{T+1}} \lambda_{T+1}^\upsilon A_{T+1}^\upsilon a_T^{|T-1|} + \upsilon_{T+1}^{|T-1|}\right)}{2\phi_{T-1}}. \tag{8.6}$$

Invoking Proposition 2.1. and substituting the optimal strategies in (8.6) into the stage $T-1$ equation of (2.4) we obtain

$$\begin{aligned}
&(A_{T-1}^{\sigma_{T-1}} x + C_{T-1}^{\sigma_{T-1}})\delta^{T-2} = \left[\frac{(\theta_{T-1}^{\sigma_{T-1}} R_{T-1})^2 x}{4c_{T-1}} \right.\\
&\left. - \frac{\left(\delta \sum_{\upsilon=1}^{\eta_T} \lambda_T^\upsilon A_T^\upsilon \left(a_{T-1}^{|T-1|} + \sum_{\nu=1}^{\mu_{T-1}} \gamma_{T-1}^\nu \vartheta_{T-1}^\nu G_{T-1}^2\right) + \delta^2 \left(\sum_{\upsilon=1}^{\eta_{T+1}} A_{T+1}^\upsilon a_T^{|T-1|} + \upsilon_{T+1}^{|T-1|}\right)\right)^2}{4\phi_{T-1}} \right] \delta^{T-2}\\
&+ \sum_{z=1}^{\eta_T} \lambda_T^z \left[A_T^a \sum_{y=1}^{\mu_{T-1}} \gamma_{T-1}^y \right.\\
&\times \left(x + a_{T-1}^{|T-1|} \frac{\delta \sum_{\upsilon=1}^{\eta_T} \lambda_T^\upsilon A_T^\upsilon \left(a_{T-1}^{|T-1|} + \sum_{\nu=1}^{\mu_{T-1}} \gamma_{T-1}^\nu \vartheta_{T-1}^\nu G_{T-1}^2\right) + \delta^2 \left(\sum_{\upsilon=1}^{\eta_{T+1}} A_{T+1}^\upsilon a_T^{|T-1|} + \upsilon_{T+1}^{|T-1|}\right)}{2\phi_{T-1}} \right.\\
&+ \vartheta_{T-1}^y G_{T-1} \frac{\delta \sum_{\upsilon=1}^{\eta_T} \lambda_T^\upsilon A_T^\upsilon \left(a_{T-1}^{|T-1|} + \sum_{\nu=1}^{\mu_{T-1}} \gamma_{T-1}^\nu \vartheta_{T-1}^\nu G_{T-1}^2\right) + \delta^2 \left(\sum_{\upsilon=1}^{\eta_{T+1}} A_{T+1}^\upsilon a_T^{|T-1|} + \upsilon_{T+1}^{|T-1|}\right)}{2\phi_{T-1}}\\
&\left.\left. + \sum_{\tau=T-\omega}^{T-2} a_{T-1}^{|\tau|} \overline{u}_\tau^{(\omega)} - \lambda_{T-1} x + \vartheta_{T-1}^y G_{T-1} x \right) \right] \delta^{T-1}\\
&- \frac{\left(\delta \sum_{\upsilon=1}^{\eta_{T+1}} \lambda_{T+1}^\upsilon A_{T+1}^\upsilon \left(a_T^{|T|} + \sum_{\nu=1}^{\mu_T} \gamma_T^\nu \vartheta_T^\nu G_T^2\right) + \delta \upsilon_{T+1}^{|T|}\right)^2}{4\phi_T} \delta^{T-1}\\
&+ \sum_{\upsilon=1}^{\eta_{T+1}} \lambda_{T+1}^z \left[A_{T+1}^z \sum_{y=1}^{\mu_T} \gamma_T^y \left(a_T^{|T|} \frac{\delta \sum_{\upsilon=1}^{\eta_{T+1}} A_{T+1}^\upsilon \left(a_T^{|T|} + \sum_{\nu=1}^{\mu_T} \gamma_T^\nu \vartheta_T^\nu G_T^2\right) + \delta \upsilon_{T+1}^{|T|}}{2\phi_T} \right.\right.\\
&\left. + \sum_{\tau=T-\omega+1}^{T-1} a_T^{|\tau|} \overline{u}_\tau^{(\omega)} + \vartheta_T^y G_T^2 \frac{\delta \sum_{\upsilon=1}^{\eta_{T+1}} A_{T+1}^\upsilon \left(a_T^{|T|} + \sum_{\nu=1}^{\mu_T} \gamma_T^\nu \vartheta_T^\nu G_T^2\right) + \delta \upsilon_{T+1}^{|T|}}{2\phi_T} \right)\\
&+ \upsilon_{T+1}^{|T|} \frac{\delta \sum_{\upsilon=1}^{\eta_{T+1}} A_{T+1}^\upsilon \left(a_T^{|T|} + \sum_{\nu=1}^{\mu_T} \gamma_T^\nu \vartheta_T^\nu G_T^2\right) + \delta \upsilon_{T+1}^{|T|}}{2\phi_T}\\
&\left. + \sum_{\tau=T+1-\omega_3+1}^{T-1} \upsilon_{T+1}^{|\tau|} \bar{u}_\tau^{(\omega)} + \varpi_{T+1} \right] \delta^T.
\end{aligned} \tag{8.7}$$

The RHS and the LHS of (8.7) are linear functions of x. For (8.7) to hold, it is required:

$$A_{T-1}^{\sigma_{T-1}} = \frac{(\theta_{T-1}^{\sigma_{T-1}} R_{T-1})^2}{4c_{T-1}} + \delta \sum_{z=1}^{\eta_T} \lambda_T^z A_T^z \Big(1 - \lambda_{T-1} + \sum_{y=1}^{\mu_{T-1}} \gamma_{T-1}^y \vartheta_{T-1}^y G_{T-1}^1\Big). \tag{8.8}$$

In addition, C_{T-1} is the expression with terms not involving x on the right-hand-side of (8.8), which is a function of previously executed controls $\overline{u}_{(T-1)-}^{(\omega)}$.

Then we move to stage $T-2$.

Using,

$$V(\theta_T^{\sigma_T}; T, x; \overline{u}_{T-}^{(\omega)}) = (A_{T-1}^{\sigma_{T-1}} x + C_{T-1}^{\sigma_{T-1}}) \delta^{T-2}, \tag{8.9}$$

with $A_{T-1}^{\sigma_{T-1}}$ and $C_{T-1}^{\sigma_{T-1}}$ derived in (8.7)–(8.8) and the stage $T-2$ equation in (2.4), the optimal strategies in stage $T-2$ can be obtained as:

$$u_{T-2}^{\sigma_{T-2}} = \frac{(\theta_{T-2}^{\sigma_{T-2}} R_{T-2})^2 x}{4(c_{T-2})^2} \text{ and } \overline{u}_{T-2}^{\sigma_{T-2}(\omega)}$$
$$= \frac{\delta \sum_{z=1}^{\eta_{T-1}} \lambda_{T-1}^z A_{T-1}^z \Big(a_{T-2}^{|T-2|} + \sum_{v=1}^{\mu_{T-2}} \gamma_{T-2}^v \vartheta_{T-2}^v G_{T-2}^2\Big) + \delta^2 \sum_{z=1}^{\eta_T} \lambda_T^z A_T^z a_{T-1}^{|T-2|} + \delta^3 \Big(\sum_{z=1}^{\eta_{T+1}} \lambda_{T+1}^z A_{T+1}^z a_T^{|T-2|} + v_{T+1}^{|T-2|}\Big)}{2\phi_{T-2}}. \tag{8.10}$$

Substituting the optimal strategies in (8.10) into the stage $T-2$ equation of (2.4), we obtain an equation in which right-hand-side and left-hand-side are linear functions of x. For the equation to hold, it is required:

$$A_{T-2}^{\sigma_{T-2}} = \frac{(\theta_{T-2}^{\sigma_{T-2}} P_{T-2})^2}{4c_{T-2}} + \delta \sum_{z=1}^{\eta_{T-1}} \lambda_{T-1}^z A_{T-1}^z \left(1 - \lambda_{T-2} + \sum_{v=1}^{\mu_{T-2}} \gamma_{T-2}^v \vartheta_{T-2}^v G_{T-2}^1\right). \tag{8.11}$$

In addition, $C_{T-2}^{\sigma_{T-2}}$ is an expression that contains all the terms in the right-hand-side of the equation which do not involve x including the previously executed controls $\overline{u}_{(T-2)-}^{(\omega)}$.

Following the above analysis for stage $k \in \{1, 2, \ldots, T\}$ the optimal strategies in stage k can be expressed as:

$u_k^{\sigma_k} = \frac{(\theta_k^{\sigma_k} R_k)^2 x}{4(c_k)^2}$, for $\sigma_k \in \{1, 2, \ldots, \eta_k\}$ and $k \in \{1, 2, \ldots, T\}$;

$$\bar{u}_k^{\sigma_k(\omega)}$$
$$= \frac{\delta \sum_{z=1}^{\eta_{k+1}} \lambda_{k+1}^z A_{k+1}^z \Big(a_k^{|k|} + \sum_{v=1}^{\mu_k} \gamma_k^v \vartheta_k^v G_k^2\Big) + \sum_{\tau=k+1}^{k+\omega-1} \delta^{\tau-k+1} \sum_{z=1}^{\eta_{\tau+1}} \lambda_{k+1}^z A_{\tau+1}^z a_\tau^{|k|} + \delta^{(T+1-k)} v_{T+1}^{|k|}}{2\phi_k},$$

for $k \in \{T-\omega+2, T-\omega+3, \ldots, T\}$ and $A_\tau = 0$ for $\tau > T+1$;

$$\overline{u}_k^{\sigma_k(\omega)} = \frac{\delta \sum_{z=1}^{\eta_k} \lambda_{k+1}^z A_{k+1}^z \left(a_k^{|k|} + \sum_{v=1}^{\mu_k} \gamma_k^v \vartheta_k^v G_k^2\right) + \sum_{\tau=k+1}^{k+\omega-1} \delta^{\tau-k+1} \sum_{z=1}^{\eta_{\tau+1}} \lambda_{k+1}^z A_{\tau+1}^z a_\tau^{|k|}}{2\phi_k}$$

$$\text{for } \sigma_k \in \{1, 2, \ldots, \eta_k\} \text{ and } k \in \{1, 2, \ldots, T-\omega+1\}. \tag{8.12}$$

Substituting the optimal strategies in (8.12) into the stage k equation of (2.4) we obtain

$$V(\theta_k^{\sigma_k} k, x; \overline{u}_{k-}^{(\omega)}) = (A_k^{\sigma_k} x + C_k^{\sigma_k})\delta^{k-1} \quad \text{for } \sigma_k \in \{1, 2, \ldots, \eta_k\} \text{ and } k \in \{1, 2, \ldots, T\}. \tag{8.13}$$

where

$$A_k^{\sigma_k} = \frac{(\theta_k^{\sigma_k} R_k)^2}{4c_k} + \delta \sum_{z=1}^{\eta_{k+1}} \lambda_{k+1}^z A_{k+1}^z \left(1 - \lambda_k + \sum_{y=1}^{\mu_k} \gamma_k^y \vartheta_k^y G_k\right), \tag{8.14}$$

and $C_k^{\sigma_k}$ is an expression that contains the previously executed controls $\overline{u}_{k-}^{(\omega)}$. ■

8.2 Appendix II: Proof of Proposition 4.1

Using (4.3) and Proposition 4.1, we have.
$A_{T+1}^{(\sigma_{T+1})i} = \theta_{T+1}^{\sigma_{T+1}} Q_{T+1}^i$ and $C_{T+1}^{(\sigma_{T+1})i} = \sum_{\tau=T+1-\omega+1}^{T} v_{T+1}^{|\tau|i} \overline{u}_\tau^{(\omega)i} + \varpi_{T+1}^i$;

$$\text{for } \sigma_{T+1} \in \{1, 2, \ldots, \eta_{T+1}\} \text{ and } i \in N. \tag{8.15}$$

Using Proposition 4.1 and (4.4), we can express the game equilibrium strategies in stage T as:
$u_T^{(\sigma_T)i**} = \frac{(\theta_T^{\sigma_T} R_T^i)^2 x}{4(c_T^i)^2}$ and

$$\overline{u}_T^{(\sigma_T)(\omega)i**} = \frac{\delta \sum_{v=1}^{\eta_{T+1}} \lambda_{T+1}^v A_{T+1}^{(v)i} \left(a_T^{|T|i} + \sum_{v=1}^{\mu_T} \gamma_T^v \vartheta_T^v G_T^{2(i)}\right) + \delta v_{T+1}^{|T|i}}{2\phi_T^i},$$

$$\text{for } \sigma_T \in \{1, 2, \ldots, \eta_T\} \text{ and } i \in N. \tag{8.16}$$

Substituting the game equilibrium strategies in (8.16) into the stage T equations of (4.4), we can obtain

$$(A_T^{(\sigma_T)i} x + C_T^{(\sigma_T)i})\delta^{T-1}$$

$$= \left(\frac{(\theta_T^{\sigma_T} R_T^i)^2 x}{4c_T^i} - \frac{\left(\delta \sum_{v=1}^{\eta_{T+1}} \lambda_{T+1}^v A_{T+1}^{vi} \left(a_T^{|T|i} + \sum_{v=1}^{\mu_T} \gamma_T^v \vartheta_T^v G_T^{2(i)}\right) + \delta v_{T+1}^{|T|i}\right)^2}{4\phi_T^i} \right) \delta^{T-1}$$

$$+ \left[\sum_{z=1}^{\eta_{T+1}} \lambda_{T+1}^z \sum_{v=1}^{\mu_T} \gamma_T^y A_{T+1}^{(z)i} \left(x + \sum_{j=1}^{n} a_T^{|T|j} \frac{\delta \sum_{v=1}^{\eta_{T+1}} \lambda_{T+1}^v A_{T+1}^{(v)j} \left(a_T^{|T|j} + \sum_{v=1}^{\mu_T} \gamma_T^v \vartheta_T^v G_T^{2(j)}\right) + \delta v_{T+1}^{|T|j}}{2\phi_T^j} \right.\right.$$

$$+ \sum_{j=1}^{n} \sum_{\tau=T-\omega+1}^{T-1} a_T^{|\tau|j} \bar{u}_\tau^{(\omega)j**} - \lambda_T x + \vartheta_T^y G_T^1 x$$

$$\left. + \sum_{j=1}^{n} \vartheta_T^y G_T^{2(j)} \frac{\delta A_{T+1}^j \left(a_T^{|T|j} + \sum_{v=1}^{\mu_T} \gamma_T^v \vartheta_T^v G_T^{2(j)}\right) + \delta v_{T+1}^{|T|j}}{2\phi_T^j} \right)$$

$$+ v_{T+1}^{|T|i} \frac{\delta \sum_{v=1}^{\eta_{T+1}} \lambda_{T+1}^v A_{T+1}^{(v)i} \left(a_T^{|T|i} + \sum_{v=1}^{\mu_T} \gamma_T^v \vartheta_T^v G_T^{2(i)}\right) + \delta v_{T+1}^{|T|i}}{2\phi_T^i}$$

$$\left. + \sum_{\tau=T+1-\omega+1}^{T-1} v_{T+1}^{|\tau|i} \bar{u}_\tau^{(\omega)i**} + \varpi_{T+1}^i \right] \delta^T, \tag{8.17}$$

for $\sigma_T \in \{1, 2, \ldots, \eta_T\}$ and $i \in N$.

The RHS and the LHS of the equations in (8.17) are linear functions of x. For (8.17) to hold, it is required that

$$A_T^{(\sigma_T)i} = \frac{(\theta_T^{\sigma_T} R_T^i)^2}{4c_T^i} + \delta \sum_{z=1}^{\eta_{T+1}} \lambda_{T+1}^z A_{T+1}^{(z)i} \left(1 - \lambda_T + \sum_{v=1}^{\mu_T} \gamma_T^v \vartheta_T^v G_T^1\right), \; i \in N. \tag{8.18}$$

In addition,

$$C_T^{(\sigma_T)i} = \left(- \frac{\left(\delta \sum_{v=1}^{\eta_{T+1}} \lambda_{T+1}^v A_{T+1}^{vi} \left(a_T^{|T|i} + \sum_{v=1}^{\mu_T} \gamma_T^v \vartheta_T^v G_T^{2(i)}\right) + \delta v_{T+1}^{|T|i}\right)^2}{4\phi_T^i} \right)$$

$$+ \left[\sum_{z=1}^{\eta_{T+1}} \lambda_{T+1}^z \sum_{v=1}^{\mu_T} \gamma_T^v A_{T+1}^{(z)i} \left(x \right.\right.$$

$$+ \sum_{j=1}^{n} a_T^{|T|j} \frac{\delta \sum_{v=1}^{\eta_{T+1}} \lambda_{T+1}^v A_{T+1}^{(v)j} \left(a_T^{|T|j} + \sum_{v=1}^{\mu_T} \gamma_T^v \vartheta_T^v G_T^{2(j)} \right) + \delta v_{T+1}^{|T|j}}{2\phi_T^j}$$

$$+ \sum_{j=1}^{n} \sum_{\tau=T-\omega+1}^{T-1} a_T^{|\tau|j} \bar{u}_\tau^{(\omega)j**}$$

$$+\sum_{j=1}^{n} \vartheta_T^{\nu} G_T^{2(j)} \frac{\delta A_{T+1}^{j}\left(a_T^{|T|j} + \sum_{\nu=1}^{\mu_T} \gamma_T^{\nu} \vartheta_T^{\nu} G_T^{2(j)}\right) + \delta v_{T+1}^{|T|j}}{2\phi_T^{j}}\right)$$

$$+ v_{T+1}^{|T|i} \frac{\delta \sum_{\upsilon=1}^{\eta_{T+1}} \lambda_{T+1}^{\upsilon} A_{T+1}^{(\upsilon)i}\left(a_T^{|T|i} + \sum_{\nu=1}^{\mu_T} \gamma_T^{\nu} \vartheta_T^{\nu} G_T^{2(i)} \right) + \delta v_{T+1}^{|T|i}}{2\phi_T^{i}}$$

$$+ \sum_{\tau=T+1-\omega+1}^{T-1} v_{T+1}^{|\tau|i} \bar{u}_{\tau}^{(\omega)i**} + \varpi_{T+1}^{i} \Bigg]\delta, \tag{8.19}$$

which is a function of previously executed controls $\underline{\overline{u}}_{T-}^{**}$.

Then we move to stage $T-1$.

Using $(A_T^{(\sigma_T)i}x + C_T^{(\sigma_T)i})\delta^{T-1}$ in (8.17)–(8.19) and the stage $T-1$ equation in (4.4), the game equilibrium strategies in stage $T-1$ can be obtained as:

$u_{T-1}^{(\sigma_{T-1})i**} = \frac{(\theta_{T-1}^{\sigma_{T-1}} R_{T-1}^{i})^2 x}{4(c_{T-1}^{i})^2}$ and

$$\bar{u}_{T-1}^{(\sigma_{T-1})(\omega)i**} = \frac{\delta \sum_{z=1}^{\eta_T} \lambda_T^{z} A_T^{(z)i}\left(a_{T-1}^{|T-1|i} + \sum_{\upsilon=1}^{\mu_{T-1}} \gamma_{T-1}^{\upsilon} \vartheta_{T-1}^{\upsilon} G_{T-1}^{2(i)} \right) + \delta^2 (\sum_{z=1}^{\eta_{T+1}} \lambda_{T+1}^{z} A_{T+1}^{(z)i} a_T^{|T-1|i} + v_{T+1}^{|T-1|i})}{2\phi_{T-1}^{i}},$$

$$\text{for } \sigma_{T-1} \in \{1, 2, \ldots, \eta_{T-1}\} \text{ and } i \in N. \tag{8.20}$$

Invoking Proposition 4.1. and substituting the game equilibrium strategies in (8.20) into the stage $T-1$ equation of (4.4) we obtain

$$(A_{T-1}^{(\sigma_{T-1})i} x + C_{T-1}^{(\sigma_{T-1})i})\delta^{T-2} = \left(\frac{(\theta_{T-1}^{\sigma_{T-1}} R_{T-1}^{i})^2 x}{4c_{T-1}^{i}} - \phi_{T-1}^{i} (\overline{u}_{T-1}^{(\omega)i**})^2 \right)\delta^{T-2}$$

$$+\Bigg\{ \sum_{z=1}^{\eta_T} \lambda_T^{z} \sum_{y=1}^{\mu_{T-1}} \gamma_{T-1}^{y} A_T^{(z)i} \Bigg(x + \sum_{j=1}^{n} a_{T-1}^{|T-1|j} \overline{u}_{T-1}^{(\omega)j**} + \sum_{j=1}^{n} \sum_{\tau=T-\omega}^{T-2} a_{T-1}^{|\tau|j} \overline{u}_{\tau}^{(\omega)j**} - \lambda_{T-1} x$$

$$+\vartheta_{T-1}^{y} G_{T-1}^{1} x + \sum_{j=1}^{n} \vartheta_{T-1}^{y} G_{T-1}^{2(j)} \overline{u}_{T-1}^{(\omega)j**} \Bigg) - \phi_T^{i} (\overline{u}_T^{(\omega)i})^2$$

$$+\delta\Bigg[\sum_{z=1}^{\eta_{T+1}} \sum_{y=1}^{\mu_T} \gamma_T^{y} A_{T+1}^{(z)i} \Bigg(\sum_{j=1}^{n} a_T^{|T|j} \overline{u}_T^{(\omega)i**} + \sum_{j=1}^{n} \sum_{\tau=T-\omega+1}^{T-1} a_T^{|\tau|j} \overline{u}_{\tau}^{(\omega)j**} + \sum_{j=1}^{n} a_T^{|T-1|j} \overline{u}_{T-1}^{(\omega)j**}$$

$$+ \sum_{j=1}^{n} \sum_{\tau=T-\omega+1}^{T-2} a_T^{|\tau|j} \overline{u}_{\tau}^{(\omega)j**} + \sum_{j=1}^{n} \vartheta_T^{y} G_T^{2(j)} \overline{u}_T^{(\omega)j**} \Bigg) + v_{T+1}^{|T|i} \overline{u}_T^{(\omega)i**}$$

$$+v_{T+1}^{|T-1|i} \overline{u}_{T-1}^{(\omega)i**} + \sum_{\tau=T+1-\omega+1}^{T-2} v_{T+1}^{|\tau|i} \overline{u}_{\tau}^{(\omega)i**} + \delta\varpi_{T+1}^{i} \Bigg]\Bigg\}\delta^{T-1}, \text{ for } i \in N, \tag{8.21}$$

where $\overline{u}_T^{(\sigma_T)(\omega)i**} = \dfrac{\delta \sum_{\upsilon=1}^{\eta_{T+1}} \lambda_{T+1}^{\upsilon} A_{T+1}^{(\upsilon)i} \left(a_T^{|T|i} + \sum_{\nu=1}^{\mu_T} \gamma_T^{\nu} \vartheta_T^{\nu} G_T^{2(i)} \right) + \delta \upsilon_{T+1}^{|T|i}}{2\phi_T^i}$ and

$$\overline{u}_{T-1}^{(\sigma_{T-1})(\omega)i**} = \frac{\delta \sum_{z=1}^{\eta_T} \lambda_T^z A_T^{(z)i} \left(a_{T-1}^{|T-1|i} + \sum_{\nu=1}^{\mu_{T-1}} \gamma_{T-1}^{\nu} \vartheta_{T-1}^{\nu} G_{T-1}^{2(i)} \right) + \delta^2 (\sum_{z=1}^{\eta_{T+1}} \lambda_{T+1}^z A_{T+1}^{(z)i} a_T^{|T-1|i} + \upsilon_{T+1}^{|T-1|i})}{2\phi_{T-1}^i}.$$

The RHS and the LHS of the equations in (8.21) are linear functions of x. For (8.21) to hold, it is required:

$$A_{T-1}^{(\sigma_{T-1})i} = \frac{(\theta_{T-1}^{\sigma_{T-1}} R_{T-1}^i)^2}{4c_{T-1}^i} + \delta \sum_{z=1}^{\eta_T} \lambda_T^z A_T^{(z)i} \left(1 - \lambda_{T-1} + \sum_{y=1}^{\mu_{T-1}} \gamma_{T-1}^y \vartheta_{T-1}^y G_{T-1}^1 \right), \tag{8.22}$$

for $i \in N$.

In addition, $C_{T-1}^{(\sigma_{T-1})i}$ is the expression on the RHS of (8.21) which does not involve x but contains previously executed controls $\underline{\overline{u}}_{(T-1)-}^{(\omega)**}$.

Then we move to stage $T-2$.

Using $(A_{T-1}^{(\sigma_{T-1})i} x + C_{T-1}^{(\sigma_{T-1})i})\delta^{T-2}$ in (8.21)–(8.22) and the stage $T-2$ equation in (4.4), the game equilibrium strategies in stage $T-2$ can be obtained as:

$u_{T-2}^{(\sigma_{T-2})i**} = \frac{(\theta_{T-2}^{\sigma_{T-2}} R_{T-2}^i)^2 x}{4(c_{T-2}^i)^2}$ and

$$\overline{u}_{T-2}^{(\sigma_{T-2})(\omega)i**} = \frac{\delta \sum_{z=1}^{\eta_T} \lambda_{T-1}^z A_{T-1}^{(\sigma_{T-1})i} \left(a_{T-2}^{|T-1|i} + \sum_{\nu=1}^{\mu_{T-2}} \gamma_{T-2}^{\nu} \vartheta_{T-2}^{\nu} G_{T-2}^{2(i)} \right)}{2\phi_{T-2}^i}$$

$$+ \frac{\delta^2 (\sum_{z=1}^{\eta_T} \lambda_T^z A_T^{(z)i} a_{T-1}^{|T-1|i}) + \delta^3 (\sum_{z=1}^{\eta_{T=1}} \lambda_{T+1}^z A_{T+1}^{(z)i} a_T^{|T-1|i} + \upsilon_{T+1}^{|T-1|i})}{2\phi_{T-2}^i}, \tag{8.23}$$

for $\sigma_T \in \{1, 2, \ldots, \eta_T\}$ and $i \in N$.

Substituting the game equilibrium strategies in (8.23) into the stage $T-2$ equations of (4.4), we obtain a system of equations which RHS and LHS are linear functions of x. For the equation to hold, it is required:

$$A_{T-2}^{(\sigma_{T-2})i} = \frac{(\theta_{T-2}^{\sigma_{T-2}} R_{T-2}^i)^2}{4c_{T-2}^i} + \delta \sum_{z=1}^{\eta_{T-1}} A_{T-1}^{(z)i} \left(1 - \lambda_{T-2} + \sum_{y=1}^{\mu_{T-2}} \gamma_{T-2}^y \vartheta_{T-2}^y G_{T-2} \right). \tag{8.24}$$

In addition, $C_{T-2}^{(\sigma_{T-2})i}$ is an expression that contains all the terms in the RHS of the equations which do not involve x including the previously executed controls $\underline{\overline{u}}_{(T-2)-}^{(\omega)*}$.

Following the above analysis for stage $k \in \{1, 2, \ldots, T\}$ the game equilibrium strategies in stage k can be expressed as:

$u_k^{(\sigma_k)i**} = \frac{(\theta_k^{\sigma_k} R_k^i)^2 x}{4(c_k^i)^2}$, for $\sigma_k \in \{1, 2, \ldots, \eta_k\}$ and $k \in \{1, 2, \ldots, T\}$; and

$\bar{u}_k^{(\sigma_k)(\omega)i**}$

$$= \frac{\sum_{z=1}^{\eta_{k+1}} \lambda_{k+1}^z A_{k+1}^{(z)i}\left(a_k^{|k|i} + \sum_{v=1}^{\mu_k} \gamma_k^v \vartheta_k^v G_k^{2(i)}\right) + \sum_{\tau=k+1}^{k+\omega-1} \delta^{\tau-k+1} \sum_{z=1}^{\eta_{\tau+1}} \lambda_{\tau+1}^z A_{\tau+1}^{(z)i} a_\tau^{|k|i} + \delta^{(T+1-k)} v_{T+1}^{|k|i}}{2\phi_k^i},$$

for $k \in \{T - \omega + 2, T - \omega + 3, \ldots, T\}$ and $A_\tau^i = 0$ for $\tau > T + 1$;

$\bar{u}_k^{(\sigma_k)(\omega)i**}$

$$= \frac{\sum_{z=1}^{\eta_{k+1}} \lambda_{k+1}^z A_{k+1}^{(z)i}\left(a_k^{|k|i} + \sum_{v=1}^{\mu_k} \gamma_k^v \vartheta_k^v G_k^{2(i)}\right) + \sum_{\tau=k+1}^{k+\omega-1} \delta^{\tau-k+1} \sum_{z=1}^{\eta_{\tau+1}} \lambda_{\tau+1}^z A_{\tau+1}^{(z)i} a_\tau^{|k|i}}{2\phi_k^i},$$

$$\text{for } k \in \{1, 2, \ldots, T - \omega + 1\}. \tag{8.25}$$

Substituting the game equilibrium strategies in (8.25) into the stage k equation of (4.4), we can obtain the game equilibrium payoff of firm i as

$$(A_k^{(\sigma_k)i} x + C_k^{(\sigma_k)i})\delta^{k-1}, \quad \text{for } \sigma_k \in \{1, 2, \ldots, \eta_k\}, k \in \{1, 2, \ldots, T\} \text{ and } i \in N, \tag{8.26}$$

where

$$A_k^i = \frac{(\theta_k^{\sigma_k} R_k^i)^2}{4c_k^i} + \delta \sum_{z=1}^{\eta_{k+1}} \lambda_{k+1}^z A_{k+1}^{(z)i}\left(1 - \lambda_k + \sum_{y=1}^{\mu_k} \gamma_k^y \vartheta_k^y G_k^1\right), \tag{8.27}$$

and $C_k^{(\sigma_k)i}$ is an expression that contains the previously executed controls $\underline{\bar{u}}_{k-}^{(\omega)**}$. ∎

8.3 Appendix III: Proof of Proposition 6.1

Using (6.3) and Proposition 6.1, we have

$$A_{T+1}^{\sigma_{T+1}} = \sum_{i=1}^n \theta_{T+1}^{\sigma_{T+1}} Q_{T+1}^i \text{ and } C_{T+1}^{\sigma_{T+1}} = \sum_{i=1}^n \left(\sum_{\tau=T+1-\omega+1}^{T} v_{T+1}^{|\tau|i} \bar{u}_\tau^{(\omega)i} + \varpi_{T+1}^i\right),$$

$$\text{for } \sigma_{T+1} \in \{1, 2, \ldots, \eta_{T+1}\}. \tag{8.28}$$

Using Proposition 6.1 and (6.4), we can express the optimal cooperative strategies in stage T as:

$$u_T^{(\sigma_T)i} = \frac{(\theta_T^{\sigma_T} R_T^i)^2 x}{4(c_T^i)^2} \text{ and}$$

$$\overline{u}_T^{(\sigma_T)(\omega)i} = \frac{\delta \sum_{\upsilon=1}^{\eta_{T+1}} \lambda_{T+1}^{\upsilon} A_{T+1}^{\upsilon}\left(a_T^{|T|i} + \sum_{\upsilon=1}^{\mu_T} \gamma_T^{\upsilon} \vartheta_T^{\upsilon} G_T^{2(i)}\right) + \delta v_{T+1}^{|T|i}}{2\phi_T^i},$$

$$\text{for } \sigma_T \in \{1, 2, \ldots, \eta_T\} \text{ and } i \in N \tag{8.29}$$

Substituting the optimal cooperative strategies in (8.29) into the stage T equations of (6.4), we can obtain

$$(A_T^{\sigma_T} x + C_T^{\sigma_T})\delta^{k-1}$$
$$= \sum_{i=1}^{n} \left(\frac{(\theta_T^{\sigma_T} R_T^i)^2 x}{4c_T^i} - \frac{\left(\delta \sum_{\upsilon=1}^{\eta_{T+1}} \lambda_{T+1}^{\upsilon} A_{T+1}^{\upsilon}\left(a_T^{|T|i} + \sum_{\upsilon=1}^{\mu_T} \gamma_T^{\upsilon} \vartheta_T^{\upsilon} G_T^{2(i)}\right) + \delta v_{T+1}^{|T|i}\right)^2}{4\phi_T^i} \right) \delta^{T-1}$$
$$+ \left[\sum_{z=1}^{\eta_{T+1}} \lambda_{T+1}^{z} \sum_{y=1}^{\mu_T} \gamma_T^{y} A_{T+1}^{z} \left(x + \sum_{j=1}^{n} a_T^{|T|j} \frac{\delta \sum_{\upsilon=1}^{\eta_{T+1}} A_{T+1}\left(a_T^{|T|i} + \sum_{\upsilon=1}^{\mu_T} \gamma_T^{\upsilon} \vartheta_T^{\upsilon} G_T^{2(i)}\right) + \delta v_{T+1}^{|T|i}}{2\phi_T^i} \right.\right.$$
$$+ \sum_{j=1}^{n} \sum_{\tau=T-\omega+1}^{T-1} a_T^{|\tau|j} \overline{u}_\tau^{(\omega)j} - \lambda_T x + \vartheta_T^{y} G_T^{1} x$$
$$\left. + \vartheta_T^{y} \sum_{j=1}^{n} G_T^{2(j)} \frac{\delta \sum_{\upsilon=1}^{\eta_{T+1}} \lambda_{T+1}^{\upsilon} A_{T+1}^{\upsilon}\left(a_T^{|T|j} + \sum_{\upsilon=1}^{\mu_T} \gamma_T^{\upsilon} \vartheta_T^{\upsilon} G_T^{2(j)}\right) + \delta v_{T+1}^{|T|j}}{2\phi_T^j} \right)$$
$$+ \sum_{i=1}^{n} \left(v_{T+1}^{|T|i} \frac{\delta \sum_{\upsilon=1}^{\eta_{T+1}} A_{T+1}^{\upsilon}\left(a_T^{|T|i} + \sum_{\upsilon=1}^{\mu_T} \gamma_T^{\upsilon} \vartheta_T^{\upsilon} G_T^{2(i)}\right) + \delta v_{T+1}^{|T|i}}{2\phi_T^i} \right.$$
$$\left.\left. + \sum_{\tau=T+1-\omega+1}^{T-1} v_{T+1}^{|\tau|i} \overline{u}_\tau^{(\omega)i} + \varpi_{T+1}^i \right)\right] \delta^T. \tag{8.30}$$

The RHS and the LHS of the equations in (8.30) are linear functions of x. For (8.30) to hold, it is required:

$$A_T^{\sigma_T} = \sum_{i=1}^{n} \frac{(\theta_T^{\sigma_T} R_T^i)^2}{4c_T^i} + \delta \sum_{z=1}^{\eta_{T+1}} \lambda_{T+1}^{z} A_{T+1}^{z} \left(1 - \lambda_T + \sum_{y=1}^{\mu_T} \gamma_T^{y} \vartheta_T^{y} G_T^{1} \right). \tag{8.31}$$

In addition,

$$C_T^{\sigma_T} = -\sum_{i=1}^{n} \frac{\left(\delta \sum_{\upsilon=1}^{\eta_{T+1}} \lambda_{T+1}^{\upsilon} A_{T+1}^{\upsilon}\left(a_T^{|T|i} + \sum_{\upsilon=1}^{\mu_T} \gamma_T^{\upsilon} \vartheta_T^{\upsilon} G_T^{2(i)}\right) + \delta v_{T+1}^{|T|i}\right)^2}{4\phi_T^i}$$

$$+ \left[\sum_{z=1}^{\eta_{T+1}} \lambda_{T+1}^{z} \sum_{y=1}^{\mu_T} \gamma_T^{y} A_{T+1}^{z} \right.$$

$$\left(\sum_{j=1}^{n} a_T^{|T|j} \frac{\delta \sum_{\upsilon=1}^{\eta_{T+1}} A_{T+1}\left(a_T^{|T|i} + \sum_{\upsilon=1}^{\mu_T} \gamma_T^{\upsilon} \vartheta_T^{\upsilon} G_T^{2(i)}\right) + \delta v_{T+1}^{|T|i}}{2\phi_T^i}\right.$$

$$+ \sum_{j=1}^{n} \sum_{\tau=T-\omega+1}^{T-1} a_T^{|\tau|j} \bar{u}_\tau^{(\omega)j}$$

$$\left. + \vartheta_T^y \sum_{j=1}^{n} G_T^{2(j)} \frac{\delta \sum_{\upsilon=1}^{\eta_{T+1}} \lambda_{T+1}^{\upsilon} A_{T+1}^{\upsilon}\left(a_T^{|T|j} + \sum_{\upsilon=1}^{\mu_T} \gamma_T^{\upsilon} \vartheta_T^{\upsilon} G_T^{2(j)}\right) + \delta v_{T+1}^{|T|j}}{2\phi_T^j}\right)$$

$$+ \sum_{i=1}^{n} \left(v_{T+1}^{|T|i} \frac{\delta \sum_{\upsilon=1}^{\eta_{T+1}} A_{T+1}^{\upsilon}\left(a_T^{|T|i} + \sum_{\upsilon=1}^{\mu_T} \gamma_T^{\upsilon} \vartheta_T^{\upsilon} G_T^{2(i)}\right) + \delta v_{T+1}^{|T|i}}{2\phi_T^i} \right.$$

$$\left.\left. + \sum_{\tau=T+1-\omega+1}^{T-1} v_{T+1}^{|\tau|i} \bar{u}_\tau^{(\omega)i} + \varpi_{T+1}^i \right)\right] \delta, \tag{8.32}$$

which is a function of previously executed controls $\underline{\bar{u}}_{T-}^{(\omega)}$.

Then we move to stage $T-1$.

Using $(A_T^{\sigma_T} x + C_T^{\sigma_T})\delta^{T-1}$ in (8.30)–(8.32) and the stage $T-1$ equation in (6.4), the optimal cooperative strategies in stage $T-1$ can be obtained as:

$u_{T-1}^{(\sigma_{T-1})i} = \frac{(\theta_{T-1}^{\bar{\sigma}-1} R_{T-1}^i)^2 x}{4(c_{T-1}^i)^2}$ and

$$\bar{u}_{T-1}^{(\sigma_{T-1})(\omega)i}$$

$$= \frac{\delta \sum_{z=1}^{\eta_T} \lambda_T^z A_T^z\left(a_{T-1}^{|T-1|i} + \sum_{y=1}^{\mu_{T-1}} \gamma_{T-1}^y \vartheta_{T-1}^y G_{T-1}^{2(i)}\right) + \delta^2\left(\sum_{z=1}^{\eta_{T+1}} \lambda_{T+1}^z A_{T+1}^z a_T^{|T-1|i} + v_{T+1}^{|T-1|i}\right)}{2\phi_{T-1}^i},$$

$$\text{for } \sigma_{T-1} \in \{1, 2, \ldots, \eta_{T-1}\} \text{ and } i \in N. \tag{8.33}$$

Invoking Proposition 6.1 and substituting the optimal cooperative strategies in (8.33) into the stage $T-1$ equation of (6.4) we obtain

$$(A_{T-1}^{\sigma_{T-1}} x + C_{T-1}^{\sigma_{T-1}})\delta^{T-2} = \sum_{i=1}^{n} \left(\frac{(\theta_{T-1}^{\sigma_{T-1}} R_{T-1}^i)^2 x}{4c_{T-1}^i} - \phi_{T-1}^i (\bar{u}_{T-1}^{(\omega)i})^2 \right)$$

$$+ \left[\sum_{z=1}^{\eta_T} \lambda_T^z \sum_{y=1}^{\mu_{T-1}} \gamma_{T-1}^y A_T^z \left(x + \sum_{j=1}^{n} a_{T-1}^{|T-1|j} \bar{u}_{T-1}^{(\omega)j} \right.\right.$$

$$\left. + \sum_{j=1}^{n} \sum_{\tau=T-\omega}^{T-2} a_{T-1}^{|\tau|j} \bar{u}_\tau^{(\omega)j} - \lambda_{T-1} x + \vartheta_{T-1}^y G_{T-1}^1 x + \vartheta_{T-1}^y \sum_{j=1}^{n} G_{T-1}^{2(j)} \bar{u}_{T-1}^{(\omega)j} \right)$$

$$
- \delta \sum_{i=1}^{n} \phi_T^i (\bar{u}_{T-1}^{(\omega)i})^2
$$

$$
+ \delta^2 \sum_{z=1}^{\eta_{T+1}} \lambda_{T+1}^z A_{T+1}^z \left(\sum_{j=1}^{n} a_T^{|T|j} \overline{u}_T^{(\omega)j} + \sum_{j=1}^{n} a_T^{|T-1|j} \overline{u}_{T-1}^{(\omega)j} + \sum_{j=1}^{n} \sum_{\tau=T-\omega+1}^{T-2} a_T^{|\tau|j} \overline{u}_\tau^{(\omega)j} \right.
$$

$$
\left. + \sum_{y=1}^{\mu_T} \gamma_T^y \vartheta_T^y \sum_{j=1}^{n} G_T^{2(j)} \overline{u}_T^{(\omega)j} \right)
$$

$$
\left. + \delta^2 \sum_{i=1}^{n} \left(v_{T+1}^{|T|i} \overline{u}_T^{(\omega)i} + v_{T+1}^{|T-1|i} \overline{u}_\tau^{(\omega)i} + \sum_{\tau=T+1-\omega+1}^{T-2} v_{T+1}^{|\tau|i} \overline{u}_\tau^{(\omega)i} + \varpi_{T+1}^i \right) \right],
$$

$$
\text{for } \sigma_{T-1} \in \{1, 2, \ldots, \eta_{T-1}\} \text{ and } i \in N, \tag{8.34}
$$

where $\overline{u}_T^{(\sigma_T)(\omega)i} = \frac{\delta \sum_{z=1}^{\eta_T} \lambda_{T+1}^z A_{T+1}^z \left(a_T^{|T|i} + \sum_{y=1}^{\mu_T} \gamma_T^y \vartheta_T^y G_T^{2(i)} \right) + \delta v_{T+1}^{|T|i}}{2\phi_T^i}$ and

$$
\bar{u}_{T-1}^{(\sigma_{T-1})(\omega)i}
$$

$$
= \frac{\delta \sum_{z=1}^{\eta_T} \lambda_T^z A_T^z \left(a_{T-1}^{|T-1|i} + \sum_{y=1}^{\mu_{T-1}} \gamma_{T-1}^y \vartheta_{T-1}^y G_{T-1}^{2(i)} \right) + \delta^2 \left(\sum_{z=1}^{\eta_{T+1}} \lambda_{T+1}^z A_{T+1}^z a_T^{|T-1|i} + v_{T+1}^{|T-1|i} \right)}{2\phi_{T-1}^i}.
$$

The RHS and the LHS of the equations in (8.34) are linear functions of x. For (8.34) to hold, it is required:

$$
A_{T-1}^{\sigma_{T-1}} = \sum_{i=1}^{n} \frac{(\theta_{T-1}^{\sigma_{T-1}} R_{T-1}^i)^2}{4c_{T-1}^i} + \delta \sum_{z=1}^{\eta_T} \lambda_T^z A_T^z \left(1 - \lambda_{T-1} + \sum_{y=1}^{\mu_{T-1}} \gamma_{T-1}^y \vartheta_{T-1}^y G_{T-1}^1 \right),
$$

$$
\sigma_{T-1} \in \{1, 2, \ldots, \eta_{T-1}\}. \tag{8.35}
$$

In addition, $C_{T-1}^{\sigma_{T-1}}$ is the expression on the RHS of (8.34) which does not involve x but contains previously executed controls $\underline{u}_{(T-1)-}^{(\omega)}$.

Then we move to stage $T-2$.

Using $(A_{T-1}^{\sigma_{T-1}} x + C_{T-1}^{\sigma_{T-1}}) \delta^{T-2}$ in (8.34)–(8.35) and the stage $T-2$ equation in (6.4), the optimal cooperative strategies in stage $T-2$ can be obtained as:

$$
u_{T-2}^{(\sigma_{T-2})i} = \frac{(\theta_{T-2}^{\sigma_{T-2}} R_{T-2}^i)^2 x}{4(c_{T-2}^i)^2}
$$

and

$$
\bar{u}_{T-2}^{(\sigma_{T-2})(\omega)i}
$$

$$
= \frac{\delta \sum_{z=1}^{\eta_{T-1}} \lambda_{T-1}^z A_{T-1}^z \left(a_{T-2}^{|T-1|i} + \sum_{y=1}^{\mu_{T-2}} \gamma_{T-2}^y \vartheta_{T-2}^y G_{T-2}^{2(i)} \right) + \delta^2 \left(\sum_{z=1}^{\eta_T} \lambda_T^z A_T^z a_{T-1}^{|T-1|i} \right) + \delta^3 \left(\sum_{z=1}^{\eta_{T+1}} \lambda_{T+1}^z A_{T+1}^z a_T^{|T-1|i} + v_{T+1}^{|T-1|i} \right)}{2\phi_{T-2}^i},
$$

$$\text{for } \sigma_{T-1} \in \{1, 2, \ldots, \eta_{T-1}\} \text{ and } i \in N. \tag{8.36}$$

Substituting the optimal cooperative strategies in (8.36) into the stage $T - 2$ equations of (4.4), we obtain a system of equations in which RHS and LHS are linear functions of x. For the equation to hold, it is required:

$$A_{T-2}^{\sigma_{T-2}} = \sum_{i=1}^{n} \frac{(\theta_{T-2}^{\sigma_{T-2}} R_{T-2}^{i})^2}{4c_{T-2}^{i}} + \delta \sum_{z=1}^{\eta_{T-1}} \lambda_{T-1}^{z} A_{T-1}^{z} \left(1 - \lambda_{T-2} + \sum_{y=1}^{\mu_{T-2}} \gamma_{T-2}^{y} \vartheta_{T-2}^{y} G_{T-2}^{1} \right). \tag{8.37}$$

In addition, $C_{T-2}^{\sigma_{T-2}}$ is an expression that contains all the terms in the RHS of the equations which do not involve x including the previously executed controls $\underline{\overline{u}}_{(T-2)-}^{(\omega)}$.

Following the above analysis for stage $k \in \{1, 2, \ldots, T\}$ the optimal cooperative strategies in stage k can be expressed as:

$u_k^{(\sigma_k)i} = \frac{(\theta_k^{\sigma_k} R_k^i)^2 x}{4(c_k^i)^2}$, for $\sigma_k \in \{1, 2, \ldots, \eta_k\}$ and $k \in \{1, 2, \ldots, T\}$; and

$$\overline{u}_k^{(\sigma_k)(\omega)i} = \frac{\delta \sum_{z=1}^{\eta_{k+1}} \lambda_{k+1}^{z} A_{k+1}^{z} \left(a_k^{|k|i} + \sum_{y=1}^{\mu_k} \gamma_k^y \vartheta_k^y G_k^{2(i)} \right) + \sum_{\tau=k+1}^{k+\omega-1} \delta^{\tau-k+1} \sum_{z=1}^{\eta_{\tau+1}} \lambda_{\tau+1}^{z} A_{\tau+1}^{z} a_{\tau}^{|k|i} + \delta^{(T+1-k)} v_{T+1}^{|k|i}}{2\phi_k^i},$$

for $\sigma_k \in \{1, 2, \ldots, \eta_k\}$, $k \in \{T - \omega + 2, T - \omega + 3, \ldots, T\}$ and $A_{\tau}^{\sigma_\tau} = 0$ for $\tau > T + 1$;

$$\overline{u}_k^{(\sigma_k)(\omega)i} = \frac{\delta \sum_{z=1}^{\eta_{k+1}} \lambda_{k+1}^{z} A_{k+1}^{z} \left(a_k^{|k|i} + \sum_{y=1}^{\mu_k} \gamma_k^y \vartheta_k^y G_k^{2(i)} \right) + \sum_{\tau=k+1}^{k+\omega-1} \delta^{\tau-k+1} \sum_{z=1}^{\eta_{\tau+1}} \lambda_{\tau+1}^{z} A_{\tau+1}^{z} a_{\tau}^{|k|i}}{2\phi_k^i},$$

$$\text{for } \sigma_k \in \{1, 2, \ldots, \eta_k\} \text{ and } k \in \{1, 2, \ldots, T - \omega + 1\}. \tag{8.38}$$

Substituting the optimal cooperative strategies in (8.38) into the stage k equation of (4.4) we can obtain the optimal cooperative joint payoff as

$$(A_k^{\sigma_k} x + C_k^{\sigma_k})\delta^{k-1}, \quad \text{for } i \in N. \tag{8.39}$$

where

$$A_k^{\sigma_k} = \sum_{i=1}^{n} \frac{(\theta_k^{\sigma_k} R_k^i)^2}{4c_k^i} + \delta \sum_{z=1}^{\eta_{k+1}} \lambda_{k+1}^{z} A_{k+1}^{z} \left(1 - \lambda_k + \sum_{y=1}^{\mu_k} \gamma_k^y \vartheta_k^y G_k^1 \right), \tag{8.40}$$

and $C_k^{\sigma_k}$ is an expression that contains the previously executed controls $\underline{\overline{u}}_{k-}^{(\omega)}$. ■

9 Chapter Notes

Durable controls that have influences lasting over certain periods of time and uncertain elements in the evolution of the future state conditions are frequently observed in real-life situations. Yeung (2001 and 2003) introduced the class of randomly furcating stochastic differential games which allows the future payoff structures of the game to furcate (branch-out) randomly in addition to the game's stochastic dynamics. The presence of random elements in future payoff structures and stock dynamics are prevalent in many practical game situations like regional economic cooperation, corporate joint ventures and transboundary environmental management. Yeung and Petrosyan (2013) presented a general class of randomly furcating stochastic dynamic games. In the presence of these elements, modification of the dynamic optimization techniques has to be made to accommodate these phenomena. Based on the work of Petrosyan and Yeung (2020a), this Chapter presents a new class of randomly furcating stochastic dynamic games with durable controls of different lag durations affecting both the players' payoffs and the state dynamics. Worth noting is that this class of games cannot be handled by the standard approach of dynamic programming. This Chapter also presents a generic class of cooperative stochastic dynamic games with durable controls in which the lagged effects are allowed to affect both the player's payoffs and the state dynamics. A subgame consistent solution for cooperative stochastic dynamic games with durable controls is derived. It expands the Yeung and Petrosyan (2019) game to a comprehensive theory of durable control games with the new features of (i) multiple durable controls with lags of different durations, (ii) durable controls affecting both the players' payoffs and state dynamics and (iii) stochastic dynamics and randomly furcating payoffs. The analysis widens the application of dynamic games to many problems where durable controls and stochastic dynamics appear.

10 Problems

1. (Warmup exercise) Consider a 5-stage 2-player stochastic dynamic game with durable strategies.

The state dynamics is governed by the dynamics

$$x_{k+1} = x_k + \sum_{\ell=1}^{2} \overline{u}_k^{(2)\ell} + \sum_{\ell=1}^{2} \sum_{\tau=k-\omega+1}^{k-1} 0.5\overline{u}_\tau^{(2)\ell} + 0.2\vartheta_k x_k, \quad x_1 = 80,$$

where $u_k^{(2)\ell}$ is player ℓ's durable strategy which lasts for 2 stages, ϑ_k is a random variable with a range $\{2, 3, 5\}$ and corresponding probabilities $\{0.4, 0.2, 0.4\}$.

The cost of $u_k^{(2)\ell}$ is $2(u_k^{(2)\ell})^2$ for player $\ell \in \{1, 2\}$.

The payoff of player 1 is

$$\sum_{k=1}^{5}\left(\left[5x_k - 2(\overline{u}_k^{(2)1})^2\right]\right)0.9^{k-1} + (6x_6 + 20)0.9^5.$$

The payoff of player 2 is

$$\sum_{k=1}^{5}\left(\left[4x_k - 2(\overline{u}_k^{(2)1})^2\right]\right)0.9^{k-1} + (5x_6 + 25)0.9^5.$$

Derive (i) the game equilibrium strategies of the game and (ii) the expected payoff of firms in stage 1.

2. Consider an oligopoly with n firms. The planning horizon of these firms involves T stages. The firms use knowledge-based public capital $x_k \in X \subset R^+$ to produce its output. Capital investment takes time to be built up into the capital stock. The accumulation process of the public capital stock is governed by the dynamics

$$x_{k+1} = x_k + \sum_{\ell=1}^{n} a_k^{|k|}\overline{u}_k^{(\omega)\ell} + \sum_{\ell=1}^{n}\sum_{\tau=k-\omega+1}^{k-1} a_k^{|\tau|}\overline{u}_\tau^{(\omega)\ell} - \lambda_k x_k + \vartheta_k G_k x_k,$$

$$x_1 = x_1^0$$

where $u_k^{(\omega)\ell}$ is firm ℓ's investment in the knowledge-based capital in stage k which is effective within ω stages, $a_k^{|k|}\overline{u}_k^{(\omega)\ell}$ is the addition to the capital in stage k by capital building investment $u_k^{(\omega)\ell}$ in stage k, $a_k^{|\tau|}\overline{u}_\tau^{(\omega)\ell}$ is the addition to the capital in stage k by capital building investment $\overline{u}_\tau^{(\omega)\ell}$ in stage $\tau \in \{k-1, k-2, \ldots, k-\omega+1\}$ and λ_k is the rate of obsolescence of the capital.

Moreover, G_k is a positive parameter and ϑ_k is a random variable with a range $\{\vartheta_k^1, \vartheta_k^2, \ldots, \vartheta_k^{\mu_k}\}$ and corresponding probabilities $\{\gamma_k^1, \gamma_k^2, \ldots, \gamma_k^{\mu_k}\}$. The most negative value of ϑ_k^ν, for $\nu \in \{1, 2, \ldots, \mu_k\}$, is larger than $(\lambda_k - 1)/G_k$.

Firm i's revenue of the firm is related to the capital stock and yields a net revenue $R_k^i x_k$ and the cost of capital investment is $\phi_k^i(\overline{u}_k^{(\omega)})^2$. The payoff of the firm i is

$$\sum_{k=1}^{T}\left(\left[R_k^i x_k - \vartheta_k^i(\overline{u}_k^{(\omega)i})^2\right]\right)\delta^{k-1} + (Q_{T+1}^i x_{T+1} + \varpi_{T+1}^i)\delta^T,$$

where $\left(Q_{T+1}^i x_{T+1} + \varpi_{T+1}^i\right)$ is the salvage value of firm i in stage $T+1$.

The above game is a durable-strategies and stochastic version of the game in Yeung and Petrosyan (2013) and a discrete-time durable-strategies version of the games in Fershtman and Nitzan (1991) and Wirl (1996).

Derive (i) the game equilibrium investment strategies of the game and (ii) the expected payoff of firms in stage k.

3. Consider an oligopoly with n firms in Problem 2. Derive (i) the optimal cooperative investment strategies and (ii) the expected optimal cooperative joint payoff.
4. Consider Problem 2 above with the state dynamics being

$$x_{k+1} = x_k + \sum_{j=1}^{n} a_k^{|k|j} \overline{u}_k^{(\omega)j} + \sum_{j=1}^{n} \sum_{\tau=k-\omega+1}^{k-1} a_k^{|\tau|j} \overline{u}_\tau^{(\omega)j} - \lambda_k x_k$$
$$+ \vartheta_k \left(G_k^1 x_k + \sum_{j=1}^{n} G_k^{|k|j} \overline{u}_k^{(\omega)j} + \sum_{j=1}^{n} \sum_{\tau=k-\omega+1}^{k-1} G_k^{|\tau|j} \overline{u}_\tau^{(\omega)j} \right).$$

Derive (i) the equilibrium investment strategies of the game and (ii) the expected payoff of firms in stage k.

References

Fershtman C, Nitzan S (1991) Dynamic voluntary provision of public goods. Eur Econ Rev 35:1057–1067

Petrosyan LA, Yeung DWK (2020) Cooperative dynamic games with durable controls: theory and application. Dyn Games Appl 10:872–896. https://doi.org/10.1007/s13235-019-00336-w

Wirl F (1996) Dynamic voluntary provision of public goods: extension to nonlinear strategies. Eur J Polit Econ 12:555–560

Yeung DWK (2001) Infinite horizon stochastic differential games with branching payoffs. J Optim Theor Appl 111(2):445–460

Yeung DWK, Petrosyan LA (2010) Subgame consistent solutions for cooperative stochastic dynamic games. J Optim Theory Appl 145:579–596

Yeung DWK, Petrosyan LA (2013) Subgame-consistent cooperative solutions in randomly furcating stochastic dynamic games. Math Comput Model 57(3–4):976–991

Yeung DWK, Petrosyan LA (2019) Cooperative dynamic games with control lags. Dyn Games Appl 9(2):550–567. https://doi.org/10.1007/s13235-018-0266-6

Yeung DWK, Petrosyan LA (2016a) A cooperative dynamic environmental game of subgame consistent clean technology development. Int Game Theor Rev 18(2):164008.01–164008.23

Yeung DWK (2003) Randomly furcating stochastic differential games. In: Petrosyan L, Yeung DWK (eds) ICM Millennium Lectures on Games. Springer, Berlin, pp 107–126. ISBN: 3-540-00615-X

Index

D. W. K. Yeung and L. A. Petrosyan, *Durable-Strategies Dynamic Games*, Theory and Decision Library C 50, https://doi.org/10.1007/978-3-030-92742-4

Printed in the United States
by Baker & Taylor Publisher Services